International Association of Fire Chiefs

Vehicle Extrication

Levels I & II: Principles and Practice

David A. Sweet, BS

Fire Inspector/Investigator
Broward Sheriffs Office Fire Rescue Department
Broward County, Florida

Division Chief (Retired)
North Lauderdale Fire Rescue
North Lauderdale, Florida

JONES & BARTLETT
LEARNING

Jones & Bartlett Learning
World Headquarters
5 Wall Street
Burlington, MA 01803
978-443-5000
info@jblearning.com
www.jblearning.com

Jones & Bartlett Learning Canada
6339 Ormindale Way
Mississauga, Ontario L5V 1J2
Canada

Jones & Bartlett Learning International
The Exchange
Express Park
Bristol Road
Bridgewater TA6 4RR

National Fire Protection Association
1 Batterymarch Park
Quincy, MA 02169-7471
www.NFPA.org

International Association of Fire Chiefs
4025 Fair Ridge Drive
Fairfax, VA 22033
www.IAFC.org

Jones & Bartlett Learning books and products are available through most bookstores and online booksellers. To contact Jones & Bartlett Learning directly, call 800-832-0034, fax 978-443-8000, or visit our website, www.jblearning.com.

Substantial discounts on bulk quantities of Jones & Bartlett Learning publications are available to corporations, professional associations, and other qualified organizations. For details and specific discount information, contact the special sales department at Jones & Bartlett Learning via the above contact information or send an email to specialsales@jblearning.com.

Production Credits
Chairman, Board of Directors: Clayton Jones
Chief Executive Officer: Ty Field
President: James Homer
SVP, Chief Operating Officer: Don Jones, Jr.
VP, Manufacturing and Inventory Control: Therese Connell
Executive Publisher: Kimberly Brophy
VP, Sales, Public Safety Group—Matthew Maniscalco
Executive Acquisitions Editor—Fire: William Larkin
Editor: Amanda J. Green
Production Manager: Jenny L. Corriveau
Marketing Manager: Brian Rooney
Cover Design: Kristin E. Parker
Rights and Permissions Manager: Katherine Crighton
Photo Research Supervisor: Anna Genoese
Photo Research Assistant: Lian Bruno
Cover Image: Courtesy of David Sweet
Composition: Publishers' Design and Production Services, Inc.
Printing and Binding: Courier Companies
Cover Printing: Courier Companies

The procedures and protocols in this book are based on the most current recommendations of responsible sources. The International Association of Fire Chiefs (IAFC), National Fire Protection Association (NFPA), and the publisher, however, make no guarantee as to, and assume no responsibility for, the correctness, sufficiency, or completeness of such information or recommendations. Other or additional safety measures may be required under particular circumstances.

Notice: The individuals described in "You Are the Technical Rescuer" throughout this text are fictitious.

Additional illustration and photographic credits appear on page 397, which constitutes a continuation of the copyright page.

To order this product, use ISBN: 978-1-4496-4882-4

Library of Congress Cataloging-in-Publication Data
Sweet, David.
  Vehicle extrication levels I & II : principles and practice / National Fire Protection Association, International Association of Fire Chiefs, David Sweet.
       p. cm.
   ISBN-13: 978-0-7637-5802-8 (pbk.)
   ISBN-10: 0-7637-5802-7 (pbk.)
  1. Traffic accidents. 2. Crash injuries. 3. Transport of sick and wounded. 4. Rescue work. I. National Fire Protection Association. II. International Association of Fire Chiefs. III. Title.
   RC88.9.T7S94 2012
   617.1'028—dc22
        2011004059
6048

Printed in the United States of America
15 14 13 12 11   10 9 8 7 6 5 4 3 2 1

Dedication

This text is dedicated to Chief Brian Nolte, a man who exemplified true leadership through his compassion for others and his undying willingness to serve; I miss you brother . . . "There is no greater love than one that lays his life down for another."

Brief Contents

Contents

Skill Drills

Resource Preview

www.Fire.jbpub.com

Special Privileges Access Code

www.Fire.jbpub.com provides a wealth of resources that are free and available to all, including:

- Crossword Puzzles
- Flashcards
- Glossary
- Ready for Review

Your personal access code gives you special user privileges and provides free admission to additional interactive educational resources such as:

- Appendix D: *Ropes and Rigging*
  - Ropes and rigging is a critical technical rescuer skill. Get up-to-speed on the skills required of all technical rescues with this special comprehensive appendix covering all the JPRs from Section 5.5 Ropes/Rigging as required from NFPA 1006, *Standard for Technical Rescuer Professional Qualifications*, 2008 Edition.
- Chapter Pretests
  - Discover your strengths and weaknesses.
- Interactive Skill Drills
  - Practice your skills in the safety of a virtual environment.
- Skill Evaluation Sheets
  - Keep track of your psychomotor skill progress.

Vehicle Extrication Levels I & II: Principles and Practice

Jones & Bartlett Learning, the National Fire Protection Association®, and the International Association of Fire Chiefs have joined forces to raise the bar for fire service once again with the release of *Vehicle Extrication Levels I & II: Principles and Practice.*

Chapter Resources

Vehicle Extrication Levels I & II: Principles and Practice thoroughly supports instructors and prepares future extrication specialists for the job. This text meets and exceeds the requirements as outlined in Chapters 4, 5, and 10 of NFPA 1006, *Technical Rescuer Professional Qualifications,* 2008 Edition. The text also addresses the requirements of Chapters 4, 8, and 12 of NFPA 1670, *Operations and Training for Technical Search and Rescue Incidents,* 2009 Edition.

Vehicle Extrication Levels I & II: Principles and Practice serves as the core of a highly effective teaching and learning system. Its features reinforce and expand on essential information. These features include:

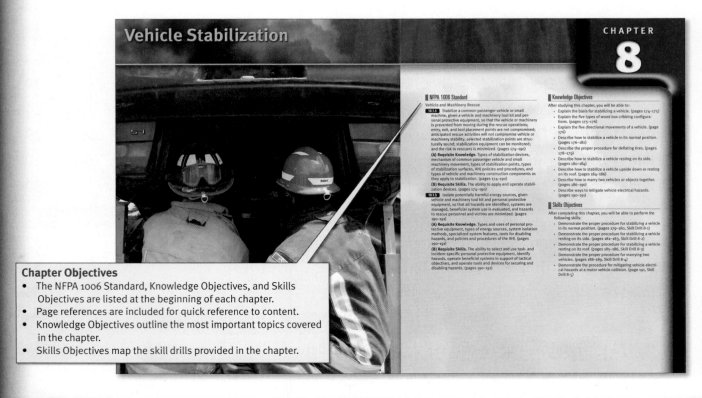

Chapter Objectives

- The NFPA 1006 Standard, Knowledge Objectives, and Skills Objectives are listed at the beginning of each chapter.
- Page references are included for quick reference to content.
- Knowledge Objectives outline the most important topics covered in the chapter.
- Skills Objectives map the skill drills provided in the chapter.

You Are the Technical Rescuer

ou are on the scene of an extrication incident with a car on its side and one victim trapped. The officer asks you and your partner to set up a triangle configuration utilizing a tension buttress system with a chain package. You and your partner have attempted this setup only one time on a training drill and do not feel confident performing this technique.

1. Do you advise the officer that you are not familiar with this technique and suggest another, simpler technique that is equally effective?

2. Which section of this vehicle will you crib first?

You Are the Technical Rescuer

Each chapter opens with a case study intended to stimulate classroom discussion, capture students' attention, and provide an overview for the chapter. An additional case study is provided in the end-of-chapter Wrap-Up material.

Introduction

"Few things are impossible to diligence and skill. Great works are performed not by strength, but perseverance."
—Samuel Johnson

Chapter 7, *Site Operations*, Chapter 8, *Vehicle Stabilization*, and Chapter 9, *Victim Access and Management*, outline a successive three-phase process that the technical rescuer should follow at every extrication incident Figure 8-1. This chapter will discuss the second step of this process, vehicle stabilization.

If not controlled, unstable vehicles are serious threats to rescuers and to those injured in a motor vehicle collision (MVC). The shape, size, and resting positions of vehicles after a collision can create many challenges for rescuers. Proper vehicle stabilization provides a solid foundation from which to work, ensuring safety for the emergency personnel as well as the victims Figure 8-2.

There are numerous methods for cribbing and stabilizing vehicles, such as box cribbing, struts, step chocks, wedges, shims, ratchet lever jacks, stabilizer jacks, rope, chain, cable, winches, ratchet straps, and tow trucks. This chapter will dis-

Victim
Access and
Management
Access points, using
hydraulics to gain
access, providing initial
medical care, packaging
and removal

Vehicle Stabilization
Cribbing, vehicle positioning, stabilization of vehicle in its
normal position, on its side, resting on its roof,
or on another object

Site Operations
Safety, responding to the scene, scene size-up, scene safety zones,
hazards, inner/outer survey

Figure 8-1 Vehicle extrication is a technical process that requires structured successive steps to produce favorable results. This chapter will discuss the second step of this process, vehicle stabilization.

Voices of Experience

A sedan carrying an older couple left a divided highway one afternoon. The area next to the highway was a deep depression bordered by rock outcroppings about 20 to 30 feet (6 to 9 m) high. The car left the roadway at a high rate of speed—about 60 miles per hour (mph) (97 kilometers per hour [kph])—and nosed into the rock. The rear of the car came to rest on the upslope of the depression, causing the car to turn into a U-shape, with the dashboard firmly entrapping both occupants.

Because the ground was relatively soft, the front of the car was actually embedded in the ground, with the bumper resting on the ground. Stabilization looked like it would be a breeze. After all, the front of the car was rock stable, and the back of the car was on the ground as well. We placed elevated step chocks under the sides and felt pretty confident we had a stable work platform.

Due to the nature of the entrapment, we had to perform a dash-lift technique. With the front end on the ground, we had to do a modified raise, cutting the structural beam in the driver's side front quarter panel. We first addressed popping open the front doors … and found our first problem. We'd placed our cribbing on the sides of the car, effectively blocking the door from opening. The entire rescue came to a grinding halt while our stabilization was reevaluated, removed, replanned, and replaced. The outcome was 50-50: one dead, one critical but alive. We'll never know if the time lost in restabilizing the vehicle was a factor in our patient's death. By the time we were able to put a medic into the car, he was already dead.

When you stabilize a vehicle for a rescue, keep your strategic and tactical goals in mind. Don't let your stabilization block access or disentanglement efforts. Time is life.

W. Buck Heath
Overland Park Fire Department
Overland Park, Kansas

> **"**We'd placed our cribbing on the sides of the car, effectively blocking the door from opening. **"**

Voices of Experience

In the Voices of Experience essays, veteran extrication specialists share their accounts of memorable incidents while offering advice and encouragement. These essays highlight what it is truly like to be a technical rescuer.

Skill Drills

Skill Drills provide written step-by-step explanations and visual summaries of important skills and procedures. This clear, concise format enhances student comprehension of sometimes complex procedures. In addition, each Skill Drill identifies the corresponding NFPA job performance requirement.

Chapter 8   Vehicle Stabilization   183

Skill Drill 8-2

NFPA 1006, (10.1.4)

Stabilizing a Vehicle Resting On Its Side

1. Don PPE. Enter the secure work area safely. Assess the scene for hazards and complete the inner and outer scene surveys. Lay out a tarp at the edge of the secure work area for staging tools and equipment, if indicated. Position an officer at the front or rear of the vehicle. The officer should position a free hand on the vehicle to look and feel for movement or shifting of the vehicle. Build up cribbing under the hood and rear section of the vehicle using step chocks, wood cribbing, and wedges.

2. Place a tensioned buttress strut at a solid section of the undercarriage at the front of the vehicle. Adjust the strut height to maintain an angle of not less than 45 degrees to the vehicle and lock it into place.

3. Move to the opposite (hood) side of the vehicle. Measure and then mark a purchase point location in the hood. Create a purchase point in the hood by using the spike end of a Halligan bar and by rotating the tool 180 degrees and prying or pulling down on the bar to create a lip on the top of the purchase point.

4. Place the tip of the strut into the purchase point, adjust the strut height, and lock it into place. Attach the hooks of the ratchet strap to the base of each strut. Double-check the placement of the struts before ratcheting. Tighten the ratchet strap, locking the struts into place. Reseat all cribbing to be sure the vehicle is stabilized.

16. Once these steps have been completed, reseat all cribbing by striking each section firmly with the butt end of a four-by-four or rubber mallet. (**Step 4**)

Initial crib placement will focus on the most unstable area, which, in this particular scenario will be the front roof side of the vehicle. The objective here is to set up an A-frame configuration using a tension buttress system, thus building up cribbing

under the hood and rear sections of the vehicle and leaving the roof area unobstructed and open to work on.

As an additional safety factor, another set of struts can be applied in the same manner to the rear section of the vehicle for extra stability, but generally one set of struts in the front section of the vehicle with the cribbing configurations at the rear is sufficient to accomplish the task. Also, if the grade level of

---

122   Vehicle Extrication Levels I & II: Principles and Practice

Near Miss REPORT

Report Number: 08-533
Report Date: 10/27/2008

Synopsis: The tool had overpressurized due to an incorrect connection on the return hose couplings.

Event Description: During a routine training meeting, we removed our extrication spreader from our rescue truck and placed the tool on the ground. Our gas-powered power unit was started and left at idle. Our mission was to open and close the tool to check the operation.

My fire fighter was in full PPE, with the shield down on his helmet, when starting the tool. When the tool opened to approximately 4 inches (102 mm), the spreader ruptured the length of the housing. The fire fighter received a blast of mineral oil–based hydraulic fluid in the face, knocking his helmet off and hitting him in the eyes and cheek.

When the tool was analyzed, the report stated that the tool had overpressurized due to an incorrect connection on the return hose couplings. The tool had been run just the night before in the same manner. The dealer has since replaced all the couplings with bleeder-style couplings. The report also stated that the relief valve had opened but could not keep up to the pressure, and that is why the tool ruptured.

The fire fighter received emergency care with a follow-up doctor and eye doctor appointment. No long-lasting effects were reported. Had this been an actual incident, the pump would have been at full throttle and would have been much worse.

Lessons Learned:

- The basic lesson learned was to have full PPE and also goggles when operating high-pressure tools.
- A second lesson, which we felt we had in place, is to have a repair/maintenance plan that covers the entire tool with your dealer. We now know of more "areas" that need to be serviced, which are now serviced annually by our request.
- A third lesson is to check the couplings each and every time before using the tools. Checking them after we remove the tools from storage will verify that the couplings are not knocked loose when preparing to use the tools.

Electric Tools

Electric-powered tools utilize a standard household generator to operate. Electrical generators may be fixed and are primarily used to power scene light power tools and equipment. Generators [capacity less than 1000 watts (1 kilowatt [k watts (75 kW), and larger. Depending on size output as 120 or 240 volts. Maintenance tasks include checking for evidence of leaks or dama those powered by an attached engine, check eng ensure proper operation. Familiarize yourself wi recommendations provided by the manufacture ment. One obvious rule and general disadvanta portable generators is that they cannot be ope structure because of the lethal carbon monoxi and build-up. Carbon monoxide gas is known as because unsuspecting victims normally pass o they realize they are in danger and need to esca

Figure 6-42 NT h... ...ure rescue-lift air bag.

You Are the Technical Rescuer

...ou are on an extrication training exercise where you have been instructed to perform a roof removal procedure. In the process of removing the roof, you attempt to cut through the A-post and encounter some difficulty with the hydraulic cutters; they cannot cut all the way through the post. You make several attempts to no avail.

1. Does the reinforced post contain High Strength Steel (HSS) or metal alloy? Check the make, model, and year of the vehicle.
2. What is the rated cutting force of your hydraulic cutter?
3. Does the hydraulic cutter need servicing? Are the blades separating as you attempt the cut?

Introduction

"Man's brain may be compared to an electric battery . . . A group of electric batteries will provide more energy than a single battery." —Napoleon Hill

This chapter explores the application of energy in relation to a motor vehicle collision and the anatomical parts that make up a vehicle system. Understanding these two components is integral in the extrication process.

Energy

Merriam-Webster defines energy as "a fundamental entity of nature that is transferred between parts of a system in the production of physical change within the system and usually regarded as the capacity for doing work." Energy is all around us; it is a constant force. In fact, the Law of Conservation of Energy states that energy can neither be created nor destroyed; it can only change from one form to another. For example, the forward movement of a vehicle (which is kinetic energy) can be changed into heat energy from the friction caused by the application of the brakes. Energy comes in different forms: electrical, mechanical, heat (thermal), light (radiant), chemical, gravitational, and nuclear. When dealing with the science of a motor vehicle collision, mechanical energy is the driving force behind the dynamics of what actually occurs.

Mechanical energy can be broken down into two types of energy, kinetic and potential. With the combination of work (transfer of energy), a mechanical energy system is produced.

Kinetic energy is the energy of motion, which is based on vehicle mass (weight) and the speed of travel (velocity). Kinetic energy is expressed as:

$$\text{Kinetic energy} = \frac{\text{mass}}{2} \times \text{velocity}^2$$

$$\text{or,} \quad \text{KE} = \frac{m}{2} \times v^2$$

Potential energy is stored energy, or the energy of position. To better illustrate this definition, take two bricks (brick 1 and brick 2), each of equal size and weight. Now hold brick 1 one inch (25 millimeters [mm]) over the top of a glass table, and elevate brick 2 three feet (1 meter [m]) over the top of the glass table Figure 5-1 . Which brick, if dropped, has the most energy force, or potential, to cause more damage to the glass table? Brick 2, based on the height, speed of travel, and gravitational force, will have the highest level of stored, or potential, energy. When the brick is actually released, the potential energy is transferred to kinetic energy because the brick is in motion. Once the brick strikes the glass table, a force is applied to stop, displace, or alter the brick's kinetic path of travel. This force is known as work.

Work, in its most basic definition, is a mechanism for the transfer of energy. Work is said to be applied to an object when energy is transferred to the object and the object is displaced. Work can be applied in the form of a positive (with the direction of travel) or negative (against the path of travel) force. To better illustrate this, let's look at two vehicles that are the same type, make, weight, and model. Vehicle 1 is traveling in a forward motion at 20 miles per hour (mph) (32 kilometers per hour, or kph), and vehicle 2 is coming up from behind vehicle 1 at a speed of 40 mph (64 kph). Vehicle 2 eventually crashes into the rear of vehicle 1. Through the application of work, vehicle 2 has transferred its forward kinetic energy into vehicle 1, causing a positive work displacement of vehicle 1's path of travel. Vehicle 2 has also experienced a negative work force because its path of travel at 40 mph (64 kph) has been displaced by crashing into

Near-Miss Reports

Utilizing incident data, National Fire Fighter Near-Miss Reporting System cases are discussed to highlight important points about safety and the lessons learned from real-life incidents.

Comprehensive Measurements

Both U.S. Imperial units and Metric units are used throughout the text.

Chapter 3 Mechanical Energy and Vehicle Anatomy 37

Understanding the series of events that occur with these three collisions will help you make the connections between the amount of damage to the exterior of the vehicle and potential injury to the passenger. For example, in a high-speed collision that results in massive damage to the vehicle, you should suspect serious injuries to the passengers, even if the injuries are not readily apparent. A number of potential physical problems may develop as a result of trauma or injuries. Your initial general impression of the patient and the evaluation of the MOI can help direct lifesaving care and provide critical information to the appropriate medical facility. Therefore, if you see a contusion on the patient's forehead and the windshield is broken and pushed out, you should care for this patient as if he or she has suffered an injury to the brain and communicate this concern to the medical providers.

Rescue Tips

When you are assessing trauma victims at a motor vehicle collision, the MOI is a crucial element to consider for the potential and type of injuries that can be sustained by the occupants. Be alert to the extent of damage to the interior and exterior of the vehicles involved in crashes. Use this observation to paint a picture of the scene in written and verbal communication, especially when consulting with a trauma or medical facility.

Front Impact Collisions

A front impact collision occurs when the vehicle strikes an object head-on, whether that object is stationary or in motion. With this initial impact, there can be two other events that can occur: The vehicle can travel under the object (which is known as an under-ride collision) or the vehicle can travel on top of the object (which is known as an over-ride collision). Understanding the MOI after a frontal collision first involves evaluation of the vehicle's restraint systems, which include seat belts (standard three-point harness and a pretensioning system) and air bags. You should determine whether the occupants were restrained by a full and properly applied standard three-point restraint harness or pretensioning system. In addition, you should determine whether the air bag deployment impacted the occupant, which could cause crushing injuries or burn injuries to the face, arms, and upper torso area due to the high temperatures that the air bag generates on deployment. Other MOIs to look for consist of bent or deformed steering wheels and broken or penetrated windshields (imbedded blood, hair, or teeth fragments are all positive confirmation of an impact).

Rear-End Collisions

Rear-end collisions are known to cause whiplash-type injuries, particularly when the passenger's head and/or neck is not restrained by an appropriately placed headrest Figure 3-6 ▶. On impact, the passenger's body and torso move forward by the transfer of kinetic energy. As the body is propelled forward, the head and neck are left behind because the head is relatively heavy, and they appear to be whipped back relative to the torso. As the vehicle comes to rest, the unrestrained passenger moves

Figure 3-6 Rear-end impacts often cause whiplash-type injuries, particularly when the head and/or neck is not restrained by a headrest.

forward, striking the dashboard. In this type of collision, the cervical spine and surrounding area may be injured. Due to the anatomical position of the spine, the cervical portion of the spine is less tolerant of damage when it is bent back. Headrests decrease extension of the head and neck during a collision and, therefore, help reduce injury. Other parts of the spine and the pelvis may also be at risk for injury. In addition, the patient may sustain an acceleration-type injury to the brain—that is, the third collision of the brain within the skull. Passengers in the backseat wearing only a lap belt might have a higher incidence of injuries to the thoracic and lumbar spine.

Lateral (Side-Impact) Collisions

Because of the limited protection to the occupants, lateral or side impacts (commonly called T-bone collisions) are a very common cause of fatalities associated with motor vehicle crashes. When a vehicle is struck from the side, the impact results in the passenger sustaining a lateral whiplash injury Figure 3-7 ▶.

Figure 3-7 In a lateral collision, where the vehicle is struck from the side, the impact results in the passenger sustaining a lateral whiplash injury where the movement is to the side, and the passenger's shoulders and head whip toward the intruding vehicle.

---

38 Vehicle Extrication Levels I & II: Principles and Practice

The movement is to the side, and the passenger's shoulders and head whip toward the intruding vehicle. This action may thrust the shoulder, thorax, upper extremities, and, most importantly, skull against the doorpost or the window. Due to the anatomical position of the spine, the cervical spine has little tolerance for lateral bending.

Rollovers

Certain vehicles, such as large trucks and some sport utility vehicles (SUVs), are more prone to rollover crashes because of their high center of gravity. Injury patterns that are commonly associated with rollover crashes differ, depending on whether the passenger was restrained or unrestrained. The most unpredictable types of injuries are caused by rollover crashes in which an unrestrained passenger may have sustained multiple strikes within the interior of the vehicle as it rolled one or more times. The most common life-threatening event in a rollover is ejection or partial ejection of the passenger from the vehicle Figure 3-8 ▼. Passengers who have been ejected may have struck the interior of the vehicle many times before ejection. The passenger may also have struck several objects, such as trees, a guardrail, or the vehicle's exterior, before landing. Passengers who have been partially ejected may have struck both the interior and exterior of the vehicle and may have been sandwiched between the exterior of the vehicle and the environment as the vehicle rolled. Ejection and partial ejection are significant MOIs; in these cases, you should prepare to care for life-threatening injuries.

Rotational Collisions

Rotational collisions (spins) are conceptually similar to rollovers. The rotation of the vehicle as it spins provides opportunities for the vehicle to experience secondary impacts. For example, as a vehicle spins and strikes a pole on the driver's side, the driver experiences not only the rotational impact and motion but also a secondary lateral impact.

Figure 3-8 Passengers who have been ejected or partially ejected may have struck the interior of the car many times before ejection.

The Vehicle System

Before a rescuer can properly apply any extrication procedures to a vehicle, he or she must understand the inner and outer components that make up a vehicle system. Just as a surgeon thoroughly understands the inner workings of the human body well before making that first incision, the technical rescuer should know the components or basic parts that make up various kinds of vehicles well before starting to extricate. To better illustrate this statement, place yourself on the scene of an extrication incident where the officer in charge tells you to perform a dash-lift technique to gain access to the patient. During this process, you are told to make a relief cut through the upper rail section between the strut tower and firewall. If you do not have a thorough understanding of vehicle anatomy, you will not have any idea where this relief cut needs to be made. This may make you a burden on scene and a hindrance to the operation. Do not try to improvise if you do not understand the technique! Step aside and pass the tool to a more experienced person.

Vehicle Classifications

Vehicles can be classified in several different ways. The Department of Transportation (DOT) classifies vehicles based on whether the vehicle transports passengers or commodities, with a nonpassenger vehicle being further classified by the number of axles and unit attachments it has. A passenger vehicle is defined by the DOT as all sedans, coupes, and station wagons manufactured primarily for the purpose of carrying passengers, including those passenger cars pulling recreational or other light trailers.

The Department of Energy (DOE) classifies vehicles by size utilizing a cubic feet system (passenger and cargo volume) and gross weight system. A type of passenger vehicle that is termed a sedan, for example, is classified or known as a sedan based on the cubic feet of space it has for passengers or cargo. Sedan types, according to the DOE, range in size of less than 85 cubic feet (2 cubic meters) to up to 130 cubic feet (4 cubic meters) depending on whether the sedan is minicompact, subcompact, compact, midsize, or large. Table 3-1 ▶ describes the different classifications of vehicles based on size and weight.

Vehicle Identification Numbers

A vehicle identification number, or VIN, is a unique identification system composed of a 17-character sequence containing both numbers and letters with the exclusion of the letters I, O, and Q to avoid confusion with the numbers 1 and 0. In 1981, the United States enacted the VIN identification system under the Code of Federal Regulations (CFR), Title 49, Chapter V, Part 565, *Vehicle Identification Requirements*, which mandated that every passenger vehicle, SUV, truck, or trailer manufactured be identified and tracked utilizing the VIN identification system. A VIN is affixed to every type of vehicle manufactured in the United States and many other countries. The VIN is normally etched on a plate and attached or embossed on the driver's side dashboard, labeled on the driver's side vehicle door, or affixed to the inside of the glove compartment. The VIN is

---

Rescue Tips
Rescue Tips reinforce safety-related concerns.

Hot Terms
Hot Terms are easily identifiable within the chapter and define key terms that the student must know. A comprehensive glossary of Hot Terms also appears in the chapter Wrap-Up or in the Glossary at the end of the text.

Technical Rescuer in Action
This feature promotes critical thinking through the use of case studies and provides instructors with discussion points for the classroom presentation.

Technical Rescuer in Action

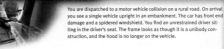

You are dispatched to a motor vehicle collision on a rural road. On arrival you see a single vehicle upright in an embankment. The car has front end damage and a spidered windshield. You find an unrestrained driver sitting in the driver's seat. The frame looks as though it is a unibody construction, and the hood is no longer on the vehicle.

1. What type of collision was this?
   A. Rear-end collision
   B. Rollover crash
   C. Frontal collision
   D. Both B and C

2. There are two frame systems that are most common in today's vehicles; they are the unibody construction and the_____ construction.
   A. ladder-type
   B. aluminum
   C. body-over-frame
   D. synthetic wrapped frame

3. You know the unibody frame:
   A. has the ability to absorb or redirect energy during a collision.
   B. has a formal frame structure.
   C. consists of two large beams tied together by cross member beams.
   D. is sometimes referred to as a ladder frame.

4. Kinetic energy is:
   A. the energy of motion.
   B. the force times the speed.
   C. the body in motion remaining in motion.
   D. the energy that can neither be created nor destroyed.

5. Window spidering occurs with which type of glass?
   A. Polycarbonate
   B. Laminated safety glass
   C. Tempered safety glass
   D. Plate glass

6. The hood is off the vehicle. What is the main structural component that assists in holding the hood in place?
   A. B-post
   B. Console area
   C. Dash area
   D. Upper rail

7. Roof posts, also known as roof pillars, are designed to add vertical support to the roof of the vehicle. The posts are generally labeled with:
   A. a color-coding system of red-green-blue.
   B. a basic numbering system of 1-2-3.
   C. an alpha system of A-B-C.
   D. There is no system of identification.

8. The swing bar located on a vehicle door is designed to:
   A. assist the door in opening and closing.
   B. retain the door in place.
   C. keep the door locked.
   D. be used in two door models.

9. Boron is alloyed with steel during processing for its unique:
   A. welding properties.
   B. hardening properties.
   C. stress resistance.
   D. ability to remain flexible.

10. Crumple zones are found in what type of vehicle frame?
    A. Unibody construction
    B. Ladder frame structure
    C. Monocoque-type frame
    D. Aluminum frame

Wrap-Up
End-of-chapter activities reinforce important concepts and improve students' comprehension. Additional instructor support and answers for all questions are available in the Instructor's ToolKit CDs.

Wrap-Up

Ready for Review
- Three concepts of energy are typically associated with injury: potential energy, kinetic energy, and work.
- Motor vehicle collisions are classified traditionally by the area of initial impact: front impact (head-on), lateral impact (T-bone, side impact), rear-end, rotational (spins), and rollovers.
- In every crash three collisions occur:
  - The collision of the vehicle against an object
  - The collision of the passenger against the interior of the vehicle
  - The collision of the passenger's internal organs against the solid structures of the body
- Before a rescuer can properly apply any extrication procedures to a vehicle, he or she must understand the inner and outer components that make up a vehicle system.
- The Department of Transportation (DOT) classifies vehicles based on whether the vehicle transports passengers or commodities, with a nonpassenger vehicle being further classified by the number of axles and unit attachments it has.
- The Department of Energy (DOE) classifies vehicles by size utilizing a cubic foot system (passenger and cargo volume) and gross weight system.
- Most vehicles on the road today are conventional-type vehicles. These types of vehicles utilize internal combustion engines for power. Other types of vehicles include hybrid electric vehicles, hydrogen fuel cell vehicles, and electric-powered vehicles.
- Electrical power in conventional-type vehicles with internal combustion engines is supplied by a basic 12-volt lead acid battery system. In hybrids, fuel cell vehicles, and electric vehicles, a different, advanced electrical design is used.
- The development of strong, crash-resistant vehicles requires engineers in the steel industry to develop stronger and lighter steel to meet demands.
- There are two frame systems that are most common in today's vehicles: body-over-frame construction and unitized or unibody construction. These frames can be composed of steel (most common), aluminum, or carbon fiber/composite. Another type of frame system that is less common today is space frame, which can consist of aluminum construction or tubular steel construction.
- Several key components make up the body portion of the vehicle.
- The technician can encounter several types of glass in a vehicle, including laminated safety glass, tempered safety glass, enhanced protective glass (EPG), polycarbonate, and ballistic glass.

Hot Terms
Advanced High Strength Steel (AHSS) Steel with a *minimum* tensile strength of 73 ksi to 116 ksi (500 MPa to 800 MPa) or greater.
Alloyed steel Steel composed of a mixture of various metals and elements.
A-post A vertical support member located closest to the front windshield of a vehicle.
Automatic seat belt system A seat belt system that uses a shoulder harness that automatically slides on a steel or aluminum track system on the door window frame. When the door is closed, the shoulder harness automatically slides into place. The lap section of the harness has to be manually engaged.
Ballistic glass Glass that utilizes multiple layers of tempered glass, laminate material, and polycarbonate thermoplastics, all sandwiched together to the desired thickness. The weight and thickness of the glass will increase depending on each increased level of protection, which can be as high as 3 or more inches (76 or more mm).
Body-over-frame construction Vehicle design where the body of the vehicle is placed onto a frame skeleton and the frame acts as the foundation for the vehicle. The design consists of two large beams tied together by cross member beams.
B-post A vertical support member located between the front and rear doors of a vehicle.
Bumper system A feature located at the front and rear of a vehicle that helps a vehicle withstand the impact of a collision.
Chassis The frame, braking, steering, and suspension system of a vehicle.
Conventional-type vehicle A vehicle that utilizes an internal combustion engine (ICE) for power.
C-post A vertical support member located behind the rear doors of a vehicle.
Crumple zones Engineered collapsible zones that are incorporated into the frame of a vehicle to absorb energy during a collision.
Dash bar A steel beam or bar that runs partway or the entire width of the dash.
Dash brackets Two brackets that are bolted or welded into the floorboard of the vehicle that are designed to lock the dash in place in order to minimize any movement resulting from an impact.
Dicing A term used to describe the small pieces of glass that are produced when tempered glass is broken.
Door hinge A mechanism that provides the opening and closing movements for a door. Door hinges commonly

Hot Terms
Hot Terms provide key terms and definitions from the chapter.

Ready for Review
The Ready for Review section highlights critical information from the chapter in a bulleted format to help students prepare for exams.

Instructor Resources

A complete teaching and learning system developed by educators with an intimate knowledge of the obstacles that instructors face each day supports this text. These resources provide practical, hands-on, time-saving tools such as PowerPoint presentations, customizable lesson plans, test banks, skill sheets, and image/table banks to better support instructors and students.

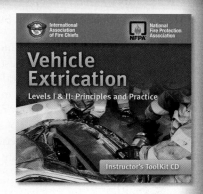

Instructor's ToolKit CD-ROM
ISBN: 978-0-7637-9844-4

Preparing for class is easy with the resources on this CD-ROM. The CD-ROM includes the following resources:

- **Adaptable PowerPoint Presentations:** Provides instructors with a powerful way to create presentations that are educational and engaging to their students. These slides can be modified and edited to meet instructors' specific needs.
- **Detailed Lesson Plans:** The lesson plans are keyed to the PowerPoint presentations with sample lectures, lesson quizzes, and teaching strategies. Complete, ready-to-use lesson plans include all of the topics covered in the text. The lesson plans can be modified and customized to fit any course.
- **Electronic Test Bank:** Contains multiple-choice questions and allows instructors to create tailor-made classroom tests and quizzes quickly and easily by selecting, editing, organizing, and printing a test along with an answer key, including page references to the text. All multiple choice questions include NFPA 1006, Job Performance Requirement (JPR) references.
- **Image and Table Bank:** Offers a selection of the most important images and tables found in the text. Instructors can use these graphics to incorporate more images into the PowerPoint presentations, make handouts, or enlarge a specific image for further discussion.
- **Skill Sheets:** Provides you with a resource to track students' skills and conduct skill proficiency exams. Each sheet is customizable.

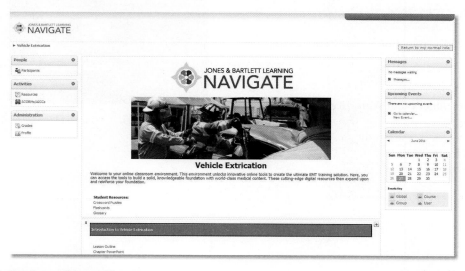

Navigate Course Manager
ISBN: 978-1-4496-4831-2

Combining our robust teaching and learning materials with an intuitive and customizable learning platform, Navigate Course Manager enables instructors to create an online course quickly and easily. The system allows instructors to readily complete the following tasks:

- Customize preloaded content or easily import new content
- Provide online testing
- Offer discussion forums, real-time chat, group projects, and assignments
- Organize course curricula and schedules
- Track student progress, generate reports, and manage training and compliance activities

To learn more about Navigate Course Manager, contact your sales specialist at 1-800-832-0034.

Student Resources

JBTest Prep: Vehicle Extrication Success
ISBN: 978-0-7637-9845-1

JBTest Prep: Vehicle Extrication Success is a dynamic program designed to prepare students to sit for certification examinations by including the same type of questions they will likely see on the actual examination.

It provides a series of self-study modules, organized by chapter and level, offering practice examinations and simulated certification examinations using multiple-choice questions. All questions are page referenced to *Vehicle Extrication Levels I & II: Principles and Practice* for remediation to help students hone their knowledge of the subject matter.

Students can begin the task of studying for certification examinations by concentrating on those subject areas where they need the most help. Upon completion, students will feel confident and prepared to complete the final step in the certification process–passing the examination.

Also Available:

Exam Prep: Technical Rescue–Vehicle/Machinery and Water/Ice
ISBN: 978-0-7637-2851-9

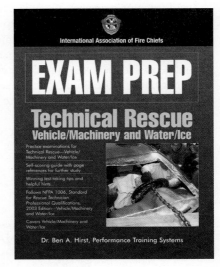

Exam Prep: Technical Rescue–Vehicle/Machinery and Water/Ice is designed to thoroughly prepare you for a Technical Rescue certification, promotion, or training examination by including the same type of multiple-choice questions you are likely to encounter on the actual exam.

To help improve examination scores, this preparation guide follows Performance Training Systems, Inc.'s Systematic Approach to Examination Preparation. *Exam Prep: Technical Rescue–Vehicle/Machinery and Water/Ice* is written by fire personnel explicitly for fire personnel, and all content has been verified with the latest reference materials and by a technical review committee.

This manual includes coverage of the NFPA 1006, *Standard for Technical Rescuer Professional Qualifications*.

Benefits of the Systematic Approach to Examination Preparation include:

- Emphasizing areas of weakness
- Providing immediate feedback
- Learning material through context and association

Exam Prep: Technical Rescue–Vehicle/Machinery and Water/Ice includes:

- Practice Rescue Specialist examinations
- Self-scoring guide with page references for further study
- Winning test-taking tips and helpful hints
- FREE 150-question online practice examination!

Acknowledgments

Contributors

Thank you to the following people for their chapter contributions:

Jim Dobson (Chapter 11, *Commercial Vehicles*)
Broward Sheriff's Office Fire Rescue Department
Broward County, Florida

Andrew S. Gurwood (Appendix D, *Ropes and Rigging*, available on the text website)
Bucks County Public Safety Training Center
Doylestown, Pennsylvania

Davis E. Hill (Chapter 12, *Agricultural Extrication*)
Pennsylvania State University
University Park, Pennsylvania

Mike Nugent (Chapter 11, *Commercial Vehicles*, and Appendix D, *Ropes and Rigging*, available on the text website)
Broward Sheriff's Office Fire Rescue Department
Broward County, Florida

Eric J. Rickenbach (Chapter 12, *Agricultural Extrication*)
Vehicle & Machinery Rescue Instructor
Rehrersburg, Pennsylvania

Rev. Paul J. Schweinler (Chapter 13, *Terminating the Incident*)
LMHC, NCC, DAPA
Bd. Certified Expert in Traumatic Stress
Clinical Director - Broward Region X CISM
Florida Crisis Consortium – Bd. Mbr., Consultant, Trainer, Team Leader
Coral Springs, Florida

Thank you to the following people for submitting a *Voice of Experience:*

J. T. Cantrell
Pulaski County Office of Emergency Management
Little Rock, Arkansas

Glenn Clapp
High Point Fire Department
High Point, North Carolina

Mike Daley
Monroe Township Fire District #3
Monroe Township, New Jersey

Keith Davis
Collingswood Fire Department
Collingswood, New Jersey

Paul Hasenmeier
Huron Fire Department
Huron, Ohio

W. Buck Heath
Overland Park Fire Department
Overland Park, Kansas

Rob Hitt
Glassy Mountain Fire Department
Landrum, South Carolina

Matt Johnston
Big Knob Fire Company
Rochester, Pennsylvania

Gerard S. London
Broward Sheriff's Office Department of Fire Rescue
Broward County, Florida

Mike Nugent
Broward Sheriff's Office Fire Rescue Department
Broward County, Florida

Mike Stanley
Aurora Fire Department
Aurora, Colorado

Brian Staska
Riverland Community College
Austin, Minnesota

Reviewers

Thank you to the following people for completing reviews:

Bryan Altman
Ashburn Fire & Emergency Services
Ashburn, Georgia

Raul A. Angulo
Seattle Fire Department, Ladder Company 6
Seattle, Washington

David W. Boyd
Anniston Fire Department
Anniston, Alabama

Sean M. Canto
Harrods Creek Fire Protection District
Prospect, Kentucky

J.T. Cantrell
Pulaski County Office of Emergency Management
Little Rock, Arkansas

Ian Cassidy
Northwest Fire District Training Division
Tucson, Arizona

Glenn Carmon Clapp
High Point Fire Department
High Point, North Carolina

Darin Clark
Hastings Fire Department
Hastings, Nebraska

Michael Daley
Monroe Township Fire District #3
Monroe Township, New Jersey

David Dalrymple
Roadway Rescue LLC/Transportation Emergency Rescue
   Committee-US (TERC-US)
Clinton, New Jersey

John Dean
Phoenix Fire Department
Phoenix, Arizona

Donna Kaye Dingler
Dekalb Technical College
Georgia Public Safety Training Center
Fulton County Fire Rescue
Atlanta, Georgia

Mark J. Doty
West Virginia State Fire Academy
Weston, West Virginia

Mark A. Elias
ARC of NW Indiana
Munster, Indiana

Mark J. Enright
Western Dakota Technical Institute
Rapid City, South Dakota

Jerry D. Eubank
Glenwood, Alabama

Bruce Evans
National Fire Academy
Henderson, Nevada

RC Fellows
Tolles Career and Technical Center
Plain City, Ohio

J. Michael Freeman
RESA 7
Clarksburg, West Virginia

Joey D. Fowler
Moultrie Fire Department
Moultrie, Georgia

Mark B. Fowler
Flagler County Fire Rescue
Palatka, Florida

Fred Halazon
Cunningham Fire Protection District
Denver, Colorado

Michael L. Hancock
Walls Fire Department
Hernando, Mississippi

Paul Hasenmeier
Huron Fire Department
Huron, Ohio

Brad Havrilla
Palm Beach County Fire Rescue
West Palm Beach, Florida

Kurt Heindrichs
Prince William County Department of Fire and Recue
Prince William, Virginia

W. Buckley Heath
Overland Park Fire Department Training Center
Overland Park, Kansas

Rob Hitt
Glassy Mountain Fire Department
Taylors, South Carolina

Jack Holliday, Jr.
Marion Township Fire Department
Marion, Ohio

Jason Hoover
Martinsburg Fire Department
Martinsburg, West Virginia

Walter D. Idol
The University of Tennessee, Knoxville
Knoxville, Tennessee

Gary Klaus
Warrensville Heights Fire Department
Warrensville Heights, Ohio

Daniel G. Klein
Cologne Fire & Rescue Department
Cologne, Minnesota

John Lankford
Auburn Fire Division
Auburn, Alabama

Donald L. Longerbeam
West Virginia Public Service Training – RESA 8
Martinsburg, West Virginia

Raymond Lussier
Auburn Fire Department
Auburn, Maine

Bill McCombs
Trans-Care Rescue Ltd.
Langham, Saskatchewan
Canada

Jerry L. McGhee
RESA V/Marmet Community Fire Department, Inc.
Marmet, West Virginia

Stephen McKenna
New Hampshire Department of Safety, Division of Fire
Standards and Training
Concord, New Hampshire

Matthew McLean
City of Dover, New Hampshire Fire and Rescue
Rochester, New Hampshire

Michael J. McNamara
Burlington Fire Department
Burlington, Ontario
Canada

Chris Michaelson
Walnut Township Fire & EMS
Crawfordsville, Indiana

Dan Neenan
National Education Center for Agricultural Safety
Peosta, Iowa

David Newberry
Dekalb Technical College
Covington, Georgia

Daryl Newport
Tennessee Fire and Codes Academy
Hohenwald, Tennessee

Gerald H. Phipps II
Wyoming Fire Academy
Riverton, Wyoming

Chad Ponder
Texas Engineering Extension Service/Emergency Services
Training Institute
College Station, Texas

Michael Powell
Columbus Ohio Division of Fire Rescue 16
The Ohio Fire Academy
Columbus Ohio

Mark A. Rivero
Las Vegas Fire & Rescue
Las Vegas, Nevada

Adam Roberts
Rockdale County Fire & Rescue
Conyers, Georgia

Earl Rudolph
Occupational Health and Safety Services
Papillion, Nebraska

Keith S. Schultz
General Motors Company
Warren, Michigan

Stephen Scionti
Tucson Fire Department
Public Safety Academy
Tucson, Arizona

Rodney Slaughter
California Office of State Fire Marshal
Chico, California

Jeffrey Smith
Wyomissing Fire Department
Wyomissing, Pennsylvania

Joshua J. Smith
Mitchell Community College
Statesville, North Carolina

John Smoot
Teays Valley Fire Department
Charleston, West Virginia

Michael G. Stanley
Aurora Fire Department
Elizabeth, Colorado

Brian Staska
Riverland Community College
Austin, Minnesota

Mike Stone
City of Lawrence Fire Department
Lawrence, Indiana

Alan Tresemer
Painted Rocks Fire & Rescue Company
Darby, Montana

H. Jeffrey Turner
Mohave Community College
Bullhead City, Arizona

William J. Vandevort
Monterey Fire Department
Monterey, California

W. Douglas Whittaker
Onondaga Community College Public Safety
Training Center
Syracuse, New York

Keith Wilson
Lancaster County Fire Service
Lancaster County, South Carolina

Gray Young
Louisiana State University Fire and Emergency
Training Institute
Minden, Louisiana

Photographic Contributors

We would like thank Glen E. Ellman, the photographer for this project. Glen is a commercial photographer and fire fighter based in Fort Worth, Texas. His expertise and professionalism are unmatched!

We would also like to thank the Broward Sheriff's Office Fire Rescue Department and Davie Fire Rescue for participating in the photo shoots. We would also like to thank the following people who contributed their time and expertise during the photo shoots: Mark Ouellette, Mike Nugent, Bud McMahon, Steve Stillwell, Dean Shepherd, Rod DelVecchio, José Rodriquez, Joseph Dorsette, Jaime Blandon, Roger Gonzalez, and Miguel A. Ferrer.

Author Note

First and foremost I want to give thanks to my Lord and savior; You are the rock for which I stand. To my wife Devon, thank you for your love, support, and understanding, through all those late nights of typing and long days at the junkyard. I also want to give thanks to Mom and Dad (Karen and Ray), and my children, Austin, Cameron, and Grace. I want to thank all my friends, especially those who helped with the text. All of you, and I mean all of you, would come to help at the drop of a hat; I will never forget your generosity and friendship: Steve Stillwell, Mike Nugent, Mike and Ralph at M & L Auto Salvage, Brett at Westway Towing, Bill Larkin and Amanda Green at Jones & Bartlett Learning, Ruben Parker and Jerry Graziose at the Broward County School Board, Bud McMahon, Dean Shepherd, and the rest of the TNT family, Jaime Blandon, Rev. Paul J Schweinler, Joey Dorsette, Jim Dobson, Marc Ouellette, Mike Korte, Roger Gonzalez, Mike Jachles, Gregg Pagliarulo, Rob Ruel, Joe Fortnash, Mike Tumminello, Bob Ricciardi, John Cavallo, Chief Rodney Turpel, Chief Neil de Jesus, the Broward Sheriff's Office Fire Rescue Department, and Davie Fire Rescue. Thank you also to Milwaukee Tools, AJAX Tool Works, Inc., Mechanix Wear, Inc., Rescue 42, Inc., and Hi-Lift Jack Company. Thank you all.

Foreword

Vehicle extrication is one of the most common rescue operations we perform as fire rescue personnel. As technology changes, so must our tactics and strategies. As the saying goes "there is more than one way to skin a cat"; likewise, there is more than one way to extricate a patient.

Fortunately there are seasoned rescuers like Dave Sweet who are willing to graciously share their knowledge and experience with others. Dave has captured the essence of vehicle extrication practices and procedures in a very clear, concise, and particular manner.

This text brings us back to the basics for bread and butter operations and also provides us with practical insights into the advanced techniques necessary for technical rescues.

When applied correctly, the proven techniques within this text will provide for a safe and effective means for fire rescue personnel to execute rapid and safe vehicle extrications that are sure to save many lives of those we serve.

Neal de Jesus
Fire Chief, Broward Sheriff's Office
Department of Fire Rescue & Emergency Services
Broward County, Florida

Introduction to Vehicle Extrication

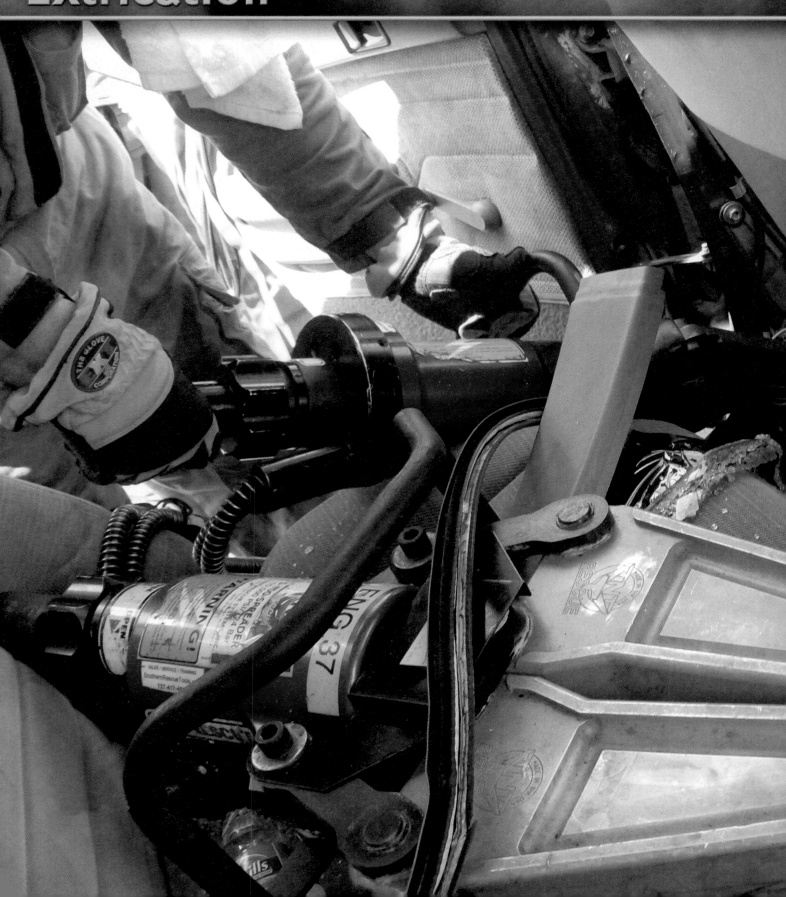

NFPA 1006 Standard

Chapter 4, Technical Rescuer

4.1 **General Requirements.**

4.1.1 Because technical rescue is inherently dangerous and technical rescuers are frequently required to perform rigorous activities in adverse conditions, regional and national safety standards shall be included in agency policies and procedures. (page 4)

4.1.2 Technical rescuers shall complete all activities in the safest possible manner and shall follow national, federal, state, provincial, and local safety standards as they apply to the technical rescuer. (page 4)

4.2 **Entrance Requirements.** Before beginning training activities or engaging in rescue operations, technical rescuers shall comply with the following requirements:

(1) Age requirement established by the AHJ (page 4)

(2) Medical requirements established by the AHJ (page 4)

(3) Minimum physical fitness as required by the AHJ (page 4)

(4) Emergency medical care performance capabilities for entry-level personnel developed and validated by the AHJ (page 4)

(5) Minimum educational requirements established by the AHJ (page 4)

(6) Minimum requirements for hazardous material incident and contact control training for entry-level personnel, validated by the AHJ (page 4)

4.3 **Minimum Requirements.** Qualification is specific to a specialty area. For qualification, a rescuer shall perform all of the job performance requirements in Chapter 5 and all job performance requirements listed in at least one level of a specialty area (Chapters 6 through 14). Technical rescuers will be identified by their specialty area and level of qualification (i.e., Rope Rescuer—Level I, Confined Space Rescuer—Level II, etc.). (page 3)

4.3.1 **Level I.** This level shall apply to individuals who identify hazards, use equipment, and apply limited techniques specified in this standard to perform technical rescue operations. (page 3)

4.3.2 **Level II.** This level shall apply to individuals who identify hazards, use equipment, and apply advanced techniques specified in this standard to perform technical rescue operations. (page 3)

Chapter 10, Vehicle and Machinery Rescue

10.1 **Level I General Requirements.** Level I rescue skills are applicable to vehicle or machinery events involving common passenger vehicles, simple small machinery, and environments where rescuer intervention does not constitute a high level of risk based upon the environment or other factors. The job performance requirements defined in 10.1.1 through 10.1.10 shall be met prior to Level I qualification in vehicle and machinery rescue. (page 3)

10.2 **Level II General Requirements.** Level II skills apply to those incidents where commercial or heavy vehicles are involved, complex extrication processes have to be applied, or multiple uncommon concurrent hazards are present, or that involve heavy machinery or more than digital entrapment of a victim. The job performance requirements defined in Section 10.1 and 10.2.1 through 10.2.5 shall be met prior to Level II qualification in vehicle and machinery rescue. (page 3)

Knowledge Objectives

After studying this chapter, you will be able to:

- Define the term extrication as it applies to vehicle and machinery rescue. (page 2)
- Define the term disentanglement as it applies to vehicle and machinery rescue. (page 2)
- List the major technical rescue specialties. (page 3)
- Describe the standards that affect the rescue community. (page 3)
- Describe Level I and Level II Technical Rescuer general job performance requirements for performing technical rescue operations as outlined in NFPA 1006. (page 3)
- Identify the three competency levels for safely and effectively conducting operations at technical rescue incidents as outlined in NFPA 1670. (pages 4–5)
- Identify the training requirements for each of the three competency levels as outlined in NFPA 1670. (pages 4–5)

Skills Objectives

There are no skills objectives for this chapter.

ou respond to a multi-vehicle accident with victims trapped. As you approach the scene, your heart is pounding from the rush of adrenaline that's building inside you. You see a twisted ton of metal that resembles a vehicle. You hear screams coming from the wreckage as you step off the rig. Can you handle this call with confidence knowing that the training you received will produce the most successful outcome? This is a common extrication scenario that many technical rescuers encounter daily.

1. Do you have a plan of action?
2. What's your process going to be?
3. What resources do you have?

Introduction

"Excellence is an art won by training and habituation."
—Aristotle

According to the National Highway Traffic Safety Administration (NHTSA), in the United States, an average of 6 million motor vehicle collisions have occurred each year since 1998 **Figure 1-1 ▼**.

With these staggering figures, vehicle extrication training should be a priority for fire and emergency medical service (EMS) organizations across the globe. With the modern-day advancement in new vehicle technology, basic skills in vehicle extrication are no longer sufficient to successfully manage a motor vehicle collision involving the extrication of trapped victims. The old standard of just "poppin' doors and rippin' roofs"

Figure 1-1 The number of motor vehicle collisions is on the rise.

without the proper training and knowledge of this advancement in vehicle technology can potentially cause further injury to the patient or injury to yourself or your crew. In today's world of litigations, fire and EMS organizations need to adhere to the goal of providing the best level of care and service that is possible while keeping all emergency personnel safe. In doing so, many organizations turn to agencies such as the National Fire Protection Association (NFPA) or the Occupational Safety and Health Administration (OSHA) for standards and regulations to follow.

To maintain an optimal level of proficiency, vehicle extrication training should be scheduled on a continual basis, qualifying each member to the appropriate skill level. Qualifying training means to utilize and adopt accredited measurable standards for providing the best level of service and safety to personnel. Vehicle extrication requires continuous training to become proficient; how you train directly correlates with how you will perform on the street.

According to the NFPA, a **technical rescuer** is a person who is trained to perform or direct a technical rescue. A **technical rescue** is the application of knowledge, skills, and equipment to safely resolve unique and/or complex rescue situations.

Extrication is defined as the process of removing a trapped victim from a vehicle or machinery. **Disentanglement** is the spreading, cutting, or removal of a vehicle and/or machinery away from trapped or injured victims. It is vital for the technical rescuer to fully understand that vehicle extrication is a step-by-step technical process requiring stabilization of the scene, stabilization of the vehicle, and stabilization of the patient **Figure 1-2 ▶**. This text is intended to be used as a training resource that explains this process in detail. It outlines a no-nonsense street-level perspective on the fundamentals of vehicle extrication utilizing structured procedures and adhering to NFPA standards for handling a vehicle extrication incident.

Figure 1-2 Vehicle extrication is a process that requires stabilization of the scene, stabilization of the vehicle, and stabilization of the patient.

Levels of Training

<u>Standards</u> are typically developed to provide guidance on the performance of processes, products, individuals, or organizations. Compliance with standards is considered voluntary, unless they are formally adopted by an organization or government agency. Once a standard is adopted, it takes on the force of law. Where not officially adopted, standards can be viewed as generally accepted practice. A common example of such a standard is NFPA 1001, *Standard for Fire Fighter Professional Qualifications*. Many states have adopted this standard as a requirement for some or all fire fighters within their state.

The following standards are addressed in this text:

- NFPA 1001, *Standard for Fire Fighter Professional Qualifications*, Sections 6.4.1 and 6.4.2.
- NFPA 1006, *Standard for Technical Rescuer Professional Qualifications*, Chapter 4, Chapter 5, and Chapter 10.
- NFPA 1670, *Standard on Operations and Training for Technical Search and Rescue Incidents*, Chapter 4; Chapter 8, Sections 8.1–8.34; and Chapter 12, Sections 12.1–12.34.

Rescue Tips

All organizations that conduct vehicle and machinery extrication operations should be trained, at a minimum, to an awareness level in accordance with NFPA 1670.

NFPA 1006, *Standard for Technical Rescuer Professional Qualifications*, 2008 Edition

<u>NFPA 1006</u>, *Standard for Technical Rescuer Professional Qualifications*, 2008 Edition, establishes the minimum job performance requirements (JPRs) necessary for fire service and other emergency response personnel who perform technical rescue operations, including:

- Rope rescue
- Confined space rescue
- Trench rescue
- Structural collapse
- Vehicle and machinery rescue
- Surface water rescue
- Swiftwater rescue
- Dive rescue
- Ice rescue
- Surf rescue
- Wilderness rescue
- Mine and tunnel rescue
- Cave rescue

NFPA 1006 contains a core set of requirements common to all forms of rescue, specifically, the knowledge and skills used in site operations, victim management, maintenance of equipment, and basic rope and rigging techniques. Beyond this core set, each discipline listed previously has its own set of knowledge and skills unique to that discipline. Technical rescuers are identified by their specialty area.

Technical rescuers are also identified by their level of qualification. NFPA 1006 defines two skill levels for qualifying a technical rescuer. A <u>Level I Technical Rescuer</u> is an individual who identifies hazards, uses equipment, and applies limited techniques specified in NFPA 1006 to perform technical rescue operations. Level I rescue skills for vehicle and machinery rescue apply to those incidents that involve *common passenger vehicles*, *simple small machinery*, and environments where rescuer intervention does not constitute a high level of risk based upon the environment or other factors. <u>Common passenger vehicles</u> are defined as light- or medium-duty passenger and commercial vehicles commonly encountered in the jurisdiction and presenting no unusual construction, occupancy, or operational characteristics to rescuers during an extrication event. A <u>commercial vehicle</u> is a type of vehicle that may be used for transporting passengers or goods. <u>Small machinery</u> is considered to be equipment or machinery that can be disassembled simply or that is constructed of lightweight materials, presenting simple hazards, which the rescuers can control.

A <u>Level II Technical Rescuer</u> is an individual who identifies hazards, uses equipment, and applies advanced techniques specified in NFPA 1006 to perform technical rescue operations. Level II rescue skills for vehicle and machinery rescues apply to those incidents where *commercial* or *heavy vehicles* are involved, complex extrication processes are applied, multiple uncommon concurrent hazards are present, or heavy machinery or more than digital entrapment of a victim is involved. <u>Heavy vehicles</u> are defined as heavy-duty highway, off-road, construction, or mass transit vehicles constructed of materials presenting resistance to common extrication procedures, tactics, and resources and posing multiple concurrent hazards to rescuers from occupancy, cargo, size, construction, weight, or position.

NFPA 1006 states in Section 4.1.1 that "because technical rescue is inherently dangerous and technical rescuers are frequently required to perform rigorous activities in adverse conditions, regional and national safety standards shall be included in agency policies and procedures." Also included in the standard is Section 4.1.2, which states that "technical rescuers shall complete all activities in the safest possible manner and shall follow national, federal, state, provincial, and local safety standards as they apply to the technical rescuer." In addition to this, it is the responsibility of the <u>authority having jurisdiction (AHJ)</u> to establish entrance requirements for the technical rescuer before beginning any training activities or engaging in rescue operations. NFPA 1006 outlines the following entrance requirements as outlined in Section 4.2 of NFPA 1006:

- Age requirement established by the AHJ
- Medical requirements established by the AHJ
- Minimum physical fitness as required by the AHJ
- Emergency medical care performance capabilities for entry-level personnel developed and validated by the AHJ
- Minimum educational requirements established by the AHJ
- Minimum requirements for hazardous material incident and contact control training for entry-level personnel, validated by the AHJ

This text discusses the core set of requirements common to all forms of rescue as well as the job performance and general requirements for the Level I and Level II Technical Rescuer for vehicle and machinery rescue.

NFPA 1670, *Standard on Operations and Training for Technical Search and Rescue Incidents*, 2009 Edition

<u>NFPA 1670</u>, *Standard on Operations and Training for Technical Search and Rescue Incidents*, 2009 Edition, was established to identify and qualify levels of functional capabilities for safely and effectively conducting operations at technical rescue incidents, including vehicle and machinery extrication incidents. NFPA 1670 states that the technical rescuer shall meet the requirements specified in Chapter 4, *Competencies for Awareness Level Personnel*, and/or Chapter 5, *Core Competencies for Operations Level Responders*, of NFPA 472, *Standard for Competence of Responders to Hazardous Materials/Weapons of Mass Destruction Incidents*.

Like NFPA 1006, the NFPA 1670 standard includes a core set of requirements common to all forms of rescue, including requirements related to hazard identification and risk assessment, written procedures, safety protocols, training level of personnel, availability of resources, and equipment. Specific specialty requirements are included for the following:

- Rope rescue
- Structural collapse search and rescue
- Confined space search and rescue
- Vehicle search and rescue
- Water search and rescue
- Wilderness search and rescue
- Trench and excavation search and rescue
- Machinery search and rescue
- Cave search and rescue
- Mine and tunnel search and rescue
- Helicopter search and rescue

Each of the specialties in NFPA 1670 includes three response levels—awareness level, operations level, and technician level.

Awareness Level

The <u>awareness level</u> represents the minimum capability of organizations that provide response to technical search and rescue incidents. The awareness level, as it pertains to vehicle and machinery extrication, is a basic competency level that stresses recognition of the basic components in vehicle extrication rather than the actual application of them.

NFPA 1670 Awareness-Level Tasks for Vehicles and Machinery

Implement procedures for the following:
- Recognizing the need for vehicle and machinery search and rescue
- Identifying the resources necessary to conduct operations
- Initiating the emergency response system for vehicle and machinery search and rescue incidents
- Initiating site control and scene management
- Recognizing general hazards associated with vehicle and machinery search and rescue incidents
- Initiating traffic control

Source: Reprinted with permission from NFPA 1670-2009: *Operations and Training for Technical Search and Rescue Incidents*, Copyright © 2008, National Fire Protection Association, Quincy, MA. This reprinted material is not the complete and official position of the NFPA on the referenced subject, which is represented only by the standard in its entirety.

Rescue Tips

NFPA 1670 requires that the minimum training for an organization shall be at the awareness level for *vehicle* search and rescue emergencies and that the organization's training at this level shall meet the requirements of Section 8.2, Awareness Level, of NFPA 1670. Organizations operating at the awareness level for *machinery* emergencies shall meet the requirements of Section 12.2, Awareness Level, of NFPA 1670. Organizations operating at the operations level for vehicle emergencies shall meet the requirements of Sections 8.2, Awareness Level, and 8.3, Operations Level, of NFPA 1670. Organizations operating at the operations level for machinery emergencies shall meet the requirements of Sections 12.2, Awareness Level, and 12.3, Operations Level, of NFPA 1670. Organizations operating at the technician level shall meet the requirements of Sections 8.2, Awareness Level; 8.3, Operations Level; and 8.4, Technician Level, of NFPA 1670.

Operations Level

The operations level represents the capability of organizations to respond to technical search and rescue incidents and to recognize and identify hazards, use equipment, and apply and implement *limited* techniques, specified in NFPA 1670 to support and participate in technical search and rescue incidents involving persons injured or entrapped in a vehicle or a small machine. The operations level is an intermediate competency skills level. This level, as it pertains to vehicle and machinery extrication, covers all of the necessary components to operate safely and efficiently on the most common passenger vehicle and

NFPA 1670 Operations-Level Tasks for Vehicles and Machinery

Develop and implement procedures for the following:

- Sizing up existing and potential conditions at vehicle and machinery search and rescue incidents
- Identifying probable victim locations and survivability
- Making the search and rescue area safe, including the stabilization and hazard control of all vehicles involved
- Making the search and rescue area safe, including the stabilization and isolation (e.g., lockout/tagout) of all machinery involved
- Identifying, containing, and stopping fuel release
- Identifying and controlling the hazards presented by the release of fluids or gases associated with the machinery, which include, but are not limited to, fuel, cutting or lubricating oil, and cooling water
- Protecting a victim during extrication or disentanglement
- Packaging a victim prior to extrication or disentanglement
- Accessing victims trapped in a common passenger vehicle or machinery
- Performing extrication and disentanglement operations involving packaging, treating, and removing victims trapped in common passenger vehicles through the use of hand and power tools
- Performing extrication and disentanglement operations involving packaging, treating, and removing victims trapped in machinery where the entrapment is limited to digits or where the machine can be disassembled simply or is constructed of lightweight materials that can be cut, spread, or lifted and has only simple hazards that are readily controlled
- Mitigating and managing general and specific hazards (i.e., fires and explosions) associated with vehicle and machinery search and rescue incidents
- Procuring and utilizing the resources necessary to conduct vehicle and machinery search and rescue operations
- Maintaining control of traffic at the scene of vehicle search and rescue incidents

Source: Reprinted with permission from NFPA 1670-2009: *Operations and Training for Technical Search and Rescue Incidents*, Copyright © 2008, National Fire Protection Association, Quincy, MA. This reprinted material is not the complete and official position of the NFPA on the referenced subject, which is represented only by the standard in its entirety.

machinery accident scenarios. This would be comparable to a Level I Technical Rescuer as outlined in NFPA 1006.

Technician Level

The technician level represents the capability of organizations to respond to technical search and rescue incidents, identify hazards, use equipment, and apply *advanced* techniques to coordinate, perform, and supervise technical search and rescue incidents. The technician level, as it pertains to vehicle and machinery extrication, outlines how to manage incidents involving larger commercial vehicles, such as semi-trucks, school buses, commercial buses, and trains, as well as how to work with large tow units that use winching, mechanical advantage, and other pulley systems. This would be comparable to a Level II Technical Rescuer as outlined in NFPA 1006.

NFPA 1670 Technician-Level Tasks for Vehicles and Machinery

Develop and implement procedures for the following:

- Evaluating existing and potential conditions at vehicle and machinery search and rescue incidents
- Performing extrication and disentanglement operations involving packaging, treating, and removing victims injured or trapped in large, heavy vehicles
- Performing extrication and disentanglement operations from large machines
- Stabilizing in advance of unusual vehicle search and rescue situations
- Stabilizing in advance machine elements at machinery search and rescue incidents
- Using all specialized search and rescue equipment immediately available and in use by the organization

Source: Reprinted with permission from NFPA 1670-2009: *Operations and Training for Technical Search and Rescue Incidents*, Copyright © 2008, National Fire Protection Association, Quincy, MA. This reprinted material is not the complete and official position of the NFPA on the referenced subject, which is represented only by the standard in its entirety.

These three competency levels are guidelines that give benchmarks to all emergency services on what functional capability they currently meet. If your agency does not meet any of the three levels, then the AHJ will need to reevaluate the response protocols, train personnel to meet these guidelines, or call for qualified mutual aid to handle any incident that is beyond your department's operational level. A plan that maximizes efficiency and uses the most appropriate resources should be in place and documented. NFPA 1670 warns departments that meeting the requirements of NFPA 1670 is not sufficient for an organization to consider itself at the advanced level. This would require consistent standardized training, self evaluation through skills testing, and frequency of operation in vehicle extrication.

At a minimum, all emergency services should adopt and write standard operating procedures (SOPs) for vehicle extrication, establishing guidelines that minimize risk and threats and that provide consistency and safety for all personnel to follow.

Voices of Experience

We responded to a motor vehicle collision with little dispatch information given other than the location. Upon arrival, a semi tanker truck was rolled over into a field and a small heavily damaged passenger car was sitting in the front yard of a residence 50 yards (46 meters) away. The truck driver was ambulatory and stated his diesel tanks were leaking. The tanks had leaked about 75 gallons (300 quarts) of fuel into the field, which would require a private clean-up company. We called in a heavy wrecker company to upright the semi tractor and trailer. It was confirmed that the tank on the trailer was loaded with can coating but was not leaking. The can coating was pumped to another tanker, while fire suppression crews stood by.

While our ambulance and rescue truck went to the passenger car, an incoming engine handled the hazardous materials incident at the semi-truck. We found a single passenger in the car. The car had sustained heavy front end and driver's side damage. The driver was semiconscious and was pinned by 12 inches (305 millimeters) of door intrusion and a crumpled dash. We knew that this extrication would require stabilization, popping the door, rolling the dash, and removing the roof. In addition, due to the prolonged extrication and patient injuries, an additional crew would be needed to set up a landing zone for a medical helicopter to transport the patient to definitive care.

At this incident, multiple crews were assigned varying tasks, including EMS, rescue, stabilization, fire suppression, establishing a landing zone, and spill containment. Your trainings for crashes involving multiple vehicles, vehicles into buildings or waterways, and large vehicles such as school buses, cement mixers, and semis should incorporate probable assignments, realism, and challenge, but ultimately prepare your department for advanced- and technician-level extrication scenes.

Paul Hasenmeier
Huron Fire Department
Huron, Ohio

> **"At this incident, multiple crews were assigned varying tasks, including EMS, rescue, stabilization, fire suppression, establishing a landing zone, and spill containment."**

Vehicle Extrication Resources

In addition to this text, there are a number of great resources and opportunities for acquiring vehicle extrication information and training outlines. The Internet is probably the most economical and convenient way for acquiring such information. Search engines can query photos of extrication incidents from all across the globe. With a little time and patience, finding and piecing together an accident scene as it evolved can pay off in dividends for the technical rescuer. This is a great opportunity to study a vehicle extrication incident and determine what went right and what went wrong. There are numerous training articles, outlines, and links on vehicle extrication that can be downloaded and utilized at your convenience; all it takes is time and a few strokes on the keyboard **Figure 1-3 ▸** .

One of the best ways to get involved in training and to learn new skills is to organize extrication training or extrication competitions with area departments. These activities will enable you to become familiar with area tools, equipment, and practices. The positive results will also emphasize the importance of training and motivate others to train.

Extrication competitions are an excellent way to gain practical hands-on skills through real-life scenario-based evolutions. Participating in a team event immediately elevates skills, circulates through an agency, encouraging each member to improve by wanting to provide a better, more efficient level of service to the community. This desire for improvement occurs because of the sharing of ideas and the visualization of other agency's performance and skills at a competition. Organizing a local extrication competition with surrounding departments is not difficult; it takes cooperation between agencies to share

Figure 1-3   Great resources for finding training scenarios, outlines, and articles can be found on an Internet search engine.

resources, tow agencies/auto salvage yards for junk vehicles, and a desire to learn. Competition forms, judging sheets, and tool requirements are all free to download on the Internet through various competition organizations or through other departments or agencies that have hosted events in the past.

See it!

Teach it!

Learn it!

Master it!

Figure 1-4   Maximum learning is best acquired when the four processes in the cycle of learning are followed: See it! Learn it! Master it! Teach it!

Wrap-Up

■ Ready for Review

- Extrication is the process of removing a trapped victim from a vehicle or machinery. It is a step-by-step technical *process* that requires continuous training to become proficient.
- Major technical rescue specialties include rope rescue, water search and rescue, mine and tunnel rescue, and wilderness rescue.
- Training should be conducted by a qualified instructor following NFPA 1006 and NFPA 1670. These standards establish the job performance skill levels and the functional capability levels for conducting operations at a technical rescue incident.
- NFPA 1006 outlines Level I and Level II rescue skills.
  - Level I: Involve *common passenger vehicles*, *simple small machinery*, and environments where rescuer intervention does not constitute a high level of risk based upon the environment or other factors.
  - Level II: Involve incidents where *commercial* or *heavy vehicles* are involved, complex extrication processes are applied, multiple uncommon concurrent hazards are present, or heavy machinery or more than digital entrapment of a victim is involved.
- NFPA 1670 outlines three operational levels—awareness, operations, and technician.
  - Awareness level: A basic competency level that stresses recognition of the basic components in vehicle extrication.
  - Operations level: Capability of organizations to respond to technical search and rescue incidents and to recognize and identify hazards, use equipment, and apply and implement *limited* techniques.
  - Technician level: Capability of organizations to respond to technical search and rescue incidents, to identify hazards, use equipment, and apply *advanced* techniques to coordinate, perform, and supervise technical search and rescue incidents.
- Each agency should, at a minimum, be trained to the awareness level.

■ Hot Terms

<u>Authority having jurisdiction (AHJ)</u> An organization, office, or individual responsible for enforcing the requirements of a code or standard, or for approving equipment, materials, an installation, or a procedure.

<u>Awareness level</u> The level that represents the minimum capability of organizations that provide response to technical search and rescue incidents.

<u>Commercial vehicle</u> A type of vehicle that may be used for transporting passengers or goods.

<u>Common passenger vehicle</u> A light- or medium-duty passenger and commercial vehicle commonly encountered in the jurisdiction and presenting no unusual construction, occupancy, or operational characteristics to rescuers during an extrication event.

<u>Disentanglement</u> The spreading, cutting, or removal of a vehicle and/or machinery away from trapped or injured victims.

<u>Extrication</u> The process of removing a trapped victim from a vehicle or machinery.

<u>Heavy vehicle</u> A heavy-duty highway, off-road, construction, or mass transit vehicle constructed of materials presenting resistance to common extrication procedures, tactics, and resources and posing multiple concurrent hazards to rescuers from occupancy, cargo, size, construction, weight, or position.

<u>Level I Technical Rescuer</u> The level that applies to individuals who identify hazards, use equipment, and apply limited techniques specified in NFPA 1006 to perform technical rescue operations. Level I rescue skills for vehicle and machinery rescue apply to those incidents that involve *common passenger vehicles*, *simple small machinery*, and environments where rescuer intervention does not constitute a high level of risk based upon the environment or other factors.

<u>Level II Technical Rescuer</u> The level that applies to individuals who identify hazards, use equipment, and apply advanced techniques specified in NFPA 1006 to perform technical rescue operations. Level II rescue skills for vehicle and machinery rescues apply to those incidents where *commercial* or *heavy vehicles* are involved, complex extrication processes are applied, multiple uncommon concurrent hazards are present, or heavy machinery or more than digital entrapment of a victim is involved.

<u>NFPA 1006</u> The standard that establishes the minimum requirements/qualifications necessary for fire service and other emergency response personnel that perform technical rescue. This standard establishes two skill

levels (Level I and II Technical Rescuer) that describe job performance requirements.

NFPA 1670 The standard that identifies and qualifies levels of functional capabilities for safely and effectively conducting operations at technical rescue incidents. This standard outlines three skill levels—awareness, operations, and technician.

Operations level The level that represents the capability of organizations to respond to technical search and rescue incidents and to identify hazards, use equipment, and apply limited techniques specified in this standard to support and participate in technical search and rescue incidents.

Small machinery Equipment or machinery that can be disassembled simply or that is constructed of light-weight materials, presenting simple hazards, which the rescuer(s) can control.

Standards Documents developed to provide guidance on the preformance of processes, products, individuals, or organizations. Compliance is voluntary, unless formally adopted by an organization or government agency.

Technical rescue The application of special knowledge, skills, and equipment to safely resolve unique and/or complex rescue situations.

Technical rescuer A person who is trained to perform or direct a technical rescue.

Technician level The level that represents the capability of organizations to respond to technical search and rescue incidents, to identify hazards, use equipment, and apply advanced techniques specified in this standard necessary to coordinate, perform, and supervise technical search and rescue incidents.

Technical Rescuer *in Action*

As a technical rescuer, the only way to become proficient in this specialized field is through continuous training and dedication.

1. According to the National Highway Traffic Safety Administration, an average of _____ vehicle accidents occur each year in the United States.
   A. 3 million
   B. 4 million
   C. 6 million
   D. 12 million

2. NFPA 1670 outlines three levels of functional capabilities for the technical rescuer; they are awareness level, operations level, and:
   A. special master level.
   B. master level.
   C. superior level.
   D. technician level.

3. The operations level is the minimum level at which all agencies must be able to operate at a technical rescue incident.
   A. True
   B. False

4. The learning process that combines the benefits of different learning styles to produce maximum understanding and comprehension of a skill requires the technical rescuer to See it! Learn it!
   A. Master it! Teach it!
   B. Show it! Copy it!
   C. Teach it! Understand it!
   D. Comprehend it! Forget it!

5. NFPA 1006 contains a core set of requirements common to _____ rescue.
   A. vehicle and machinery
   B. trench
   C. dive
   D. all forms of

6. NFPA 1006 Level I rescue skills for vehicle and machinery rescue apply to those incidents that involve:
   A. common passenger vehicles and small machinery.
   B. heavy machinery.
   C. interventions that require a high level of risk.
   D. commercial vehicles.

7. NFPA 1006 Level II rescue skills for vehicle and machinery rescue apply to those incidents that involve:
   A. commercial or heavy vehicles.
   B. small machinery.
   C. interventions that require a low level of risk.
   D. agricultural equipment.

8. The NFPA defines *extrication* as the _____ of removing a trapped victim from a vehicle or machinery.
   A. technique
   B. process
   C. intervention
   D. standard of care

9. According the NFPA, a technical rescuer is a person who is trained to perform or direct a:
   A. situation.
   B. technical extrication.
   C. technical rescue.
   D. scene.

10. An organization, office, or individual responsible for enforcing the requirements of a code or standard, or for approving equipment, materials, an installation, or a procedure, is the definition for the:
    A. code of ethics.
    B. technical rescue procedure.
    C. standard operation procedures.
    D. authority having jurisdiction.

Rescue Incident Management

NFPA 1006 Standard

There are no objectives for this chapter.

Knowledge Objectives

After studying this chapter, you will be able to:

- Describe the characteristics and organization of the incident command system and functions of positions within this system. (pages 14–21)
- Describe the function of, and explain the basis for, an incident action plan. (page 14)
- Describe the purpose, components, and benefits of performing a needs assessment. (pages 21–22)
- Describe the components of an operational risk–benefit analysis. (page 22)
- List various methods of personnel and equipment accountability, and describe the importance of having accountability systems. (pages 23–25)
- Describe the administrative and operational aspects of an organization health and safety program. (page 26)
- List and describe the components of incident response planning. (page 26)
- Describe the purpose and benefits of standard operating procedures. (pages 26–27)

Skill Objectives

After studying this chapter, you will be able to perform the following skills:

- Demonstrate an understanding of the incident command system and its functional areas as they relate to rescue. (pages 14–21)
- Perform a needs assessment for your response area. (pages 21–22)
- Perform a hazard and risk analysis for a given incident. (pages 21–22)

During the middle of the night, you are jarred out of a sound sleep by the alert tones. Your engine is called for a reported multi-vehicle collision with unknown injuries on a highway. You arrive on scene and discover chaos; there are victims who have been thrown from their vehicles, fluids in the roadway, and electrical lines down.

1. Will this incident require the Incident Management System to properly manage the incident?
2. What resources will be needed to manage this incident?

Introduction

"If your actions inspire others to dream more, learn more, do more and become more, you are a leader."
—John Quincy Adams

During extrication incidents, at times, it is easy for the basic fundamentals of extrication to be tossed aside and replaced by emotionally driven actions such as, *"Just rip the door off and yank the patient out!"* At an extrication incident, a rescuer's *fight-or-flight* response mechanism of the central nervous system will always be active, with emotions running high. This is not an excuse for poor technique or lack of a structured plan. As professionals, we need to be skill driven and follow a plan of action. An **incident action plan (IAP)** is an oral (informal) or written (formal) plan containing general objectives reflecting the overall strategy for managing an incident **Figure 2-1 ▶**. IAPs are discussed in more detail in Chapter 7, *Site Operations*.

Rescue Tips

Vehicle extrication should never be emotion driven; it should be skill driven.

The technical rescuer must possess the skills of managing an emergency incident, whether large or small. This chapter is an overview of the incident command system (ICS) and the components that make up the incident command structure. The chapter also discusses planning and its importance in ensuring that everyone involved in the incident is following the same overall plan. The ICS used by an organization must meet the requirements of the National Incident Management System (NIMS) model. This model allows for a standardized approach to incident management; standard command and management structures; and emphasis on preparedness, mutual aid, and resource management. This chapter will address these subjects and show how all of these subjects are interrelated. This chapter is not, however, intended to replace ICS training such as that offered by the U.S. Department of Homeland Security's Federal

Figure 2-1 Have a structured plan of action for all personnel to follow.

Emergency Management Agency (FEMA). It is expected that at least basic ICS training will have already occurred prior to reading this chapter.

In sports, the winning individual or team is usually the one that does the most preparation. The athlete or team fully understands its challenge or opponent before stepping onto the ball field or into the ring. The great football coach Vince Lombardi based his winning philosophy on focusing and drilling his team on perfecting the basics and building a strong foundation. He believed that if his team mastered the basics, then success would always follow. This philosophy is no different for the technical rescuer; the more preparation, training, and study that is done before the call, the more success you will have.

Incident Command System

It is not enough to be proficient in various rescue skills; you must also be able to operate as a member of a team. For a team to function properly, there must be a system in place where roles and responsibilities are identified and assigned. To this purpose, we use an <u>incident command system (ICS)</u>. As part of this hierarchy of responsibilities and duties, a technical rescuer must also be proficient with certain concepts such as standard operating procedures (SOPs) or standard operating guidelines (SOGs), risk–benefit analysis, hazard analysis, personnel accountability systems, and equipment inventory, capability, and tracking systems. It is also expected that those involved in the management of the team will also be knowledgeable in subjects such as needs identification and response planning. Although most emergency situations are handled by local response organizations, during a major incident, help may be needed from other local jurisdictions, state responders, and/or the federal government. The NIMS management plan allows responders from different jurisdictions and disciplines to work together more efficiently.

Jurisdictional Authority

An effective ICS clearly defines the agency that will be in charge of each incident. Although determination of the jurisdiction in charge is rarely a problem for a small incident, larger-scale incidents may cross geographic and/or statutory boundaries or other disciplines may need to be involved such as local, county, or state law enforcement. When situations arise where there are overlapping responsibilities, the ICS may employ a <u>unified command</u>. Unified command is often used in multijurisdictional or multiagency incident management. It allows agencies with different legal, geographic, and functional responsibilities to coordinate, plan, and interact effectively. In a unified command structure, multiple agency representatives make command decisions, instead of a single incident commander making all decisions on scene. A unified command allows for representatives from each agency participating in the response to share command authority at large, complex emergencies; this power-sharing arrangement helps ensure cooperation, avoids confusion, and guarantees agreement on goals and objectives.

All-Risk and All-Hazard System

By design, an ICS can and should be used during training and at all types of incidents, including nonemergency events, such as public gatherings Figure 2-2 ▶. Regular use of the system builds familiarity with standard procedures and terminology.

Unity of Command

In a properly run ICS, each person working at an incident has only one direct supervisor. All orders and assignments come directly from that supervisor, and all reports are made to the same supervisor. That supervisor then reports to his or her supervisor and so on up the chain of command.

Span of Control

The ICS allows for a manageable span of control of people and resources. In most situations, one person can effectively supervise only three to seven people (with five being the optimal number). In the ICS setting, the <u>incident commander (IC)</u> communicates with and receives information from a maximum of five people, rather than assuming responsibility for the

Figure 2-2 An ICS can be used for both emergency and nonemergency events.

assignment of all personnel at the scene. Individual managers of personnel and resources within the ICS are also working within a manageable span of control. The actual span of control will depend on the complexity of the incident and the nature of the work being performed.

Modular Organization

An ICS is designed to be modular. Not all components of an ICS need to be utilized at every incident—only what is appropriate given the incident's nature and size. Additional components can be added or eliminated as needed as the incident unfolds. Some components are used on almost every incident, while others apply to only the largest and most complex situations.

Rescue Tips

Not all components of an ICS need to be utilized at every incident—only what is appropriate given the incident's nature and size.

Common Terminology

The ICS promotes the use of common terminology both within an organization and among all the agencies involved in emergency incidents. This shared language eliminates confusion about what is intended when different things are called by the same name in different jurisdictions, countries, areas, or departments. For example, a phrase that is commonly used to identify hydraulic tools is "the jaws of life." This phrase was originally established by a hydraulic tool company to describe its hydraulic spreader. The phrase is now inaccurately used by rescue personnel to describe all hydraulic tools. To avoid confusion, a hydraulic spreader should be referred to as a hydraulic spreader, a hydraulic cutter should be referred to as a hydraulic cutter, and a hydraulic ram should be referred to as a hydraulic ram. Using common terminology increases the level of understanding among the various response agencies working at an incident site.

Integrated Communications

The ICS must support communication up and down the chain of command at every level. Messages must move efficiently throughout the system. This consideration is especially important because it ensures that control objectives established by the command staff are effectively implemented by task-level resources. Integrated communications are also necessary so that outcomes produced by these task-level units are reported back up the chain of command, allowing progress toward incident goals to be measured as the incident unfolds.

Consolidated Incident Action Plans

An ICS ensures that everyone involved in the incident is following the same overall plan. The IAP may be developed by the IC alone on smaller incidents or in collaboration with all agencies involved in larger incidents.

Designated Incident Facilities

Development of a standard terminology for commonly needed operational facilities improves operations because everyone knows what occurs at each facility. Examples of such standard terms include the following:

- Base
- Command post
- Staging area

For incidents that involve only a few vehicles, the command post at an extrication incident may be established outside of the hazard zones but in close proximity to maintain control and a visual of the incident. A larger incident may require the command post to be established within a structure with an established base, where logistical and administrative functions, such as traffic control management, are coordinated. A staging area is vital for a large incident and may be used to stage incoming apparatus. The staging area should be established close to the incident in a location such as a parking lot.

Resource Management

A standard system of assigning and tracking the resources involved in the incident is of critical importance in ensuring an efficient and safe operation. At small-scale incidents, units and personnel usually respond directly to the scene and receive their assignments there. At large-scale incidents, units are often dispatched to a staging area, rather than going directly to the incident scene. Some units are assigned upon arrival, whereas others may be held in reserve, ready to be assigned if needed. A mass-casualty incident (MCI) involves more than one victim and places great demand on equipment or personnel, stretching resources to their limit Figure 2-3 ▶ . It is imperative that proper resource management is established in the beginning stages of the incident to ensure any chance of a successful outcome. Delaying the management of resources will cause confusion, freelancing, and possible gridlock of incoming units.

ICS Organization

The ICS structure identifies a full range of duties, responsibilities, and functions that are performed at emergency incidents. Its hierarchy is best illustrated by the standard organizational chart shown in Figure 2-4 ▶ , which clearly defines the positions within the ICS and the chain of command.

The IC is the individual with overall responsibility for the management of all incident operations, including rotating personnel to limit stress. This command position is always filled and is initially established by the first unit on the scene. Ultimately, command is likely to be transferred to the senior arriving officer, unless organization policy or circumstances dictate that someone else would be more appropriate in this position. During the transfer of command at a vehicle extrication incident, a brief current situation status report is given to the new IC and includes the following elements:

- **Tactical priorities:** involve actions that need to be enacted or are presently being performed. For example, rescuers may need to mitigate a spill or leak that is preventing operational groups from entering areas to perform extrication.
- **Action plans:** describe what the overall operation encompasses and establish objectives, including what, where, and how resources are allocated. For example, an action plan may demonstrate that five extrication groups under the operations section have been established and are currently operating at several geographic locations on the highway.
- **Hazardous or potentially hazardous conditions:** areas that could potentially jeopardize the safety of personnel on the scene. An example is a petroleum tanker truck leaking fuel.

Figure 2-3 This vehicle incident will likely stretch resources to their limit.

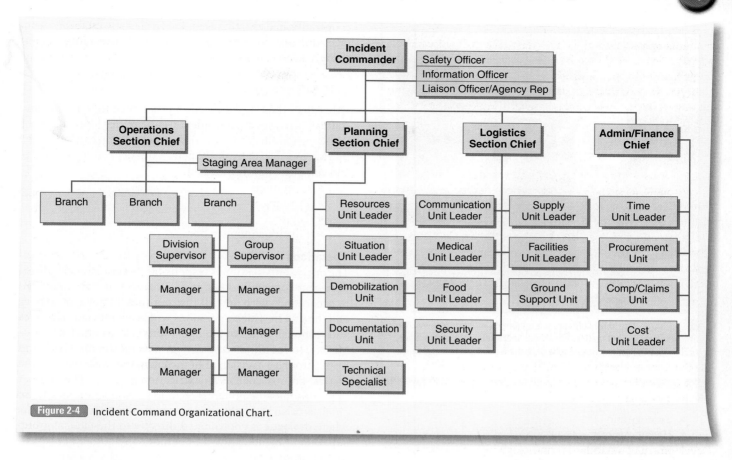

Figure 2-4 Incident Command Organizational Chart.

- **Accomplishments:** objectives that have been completed at the time of transfer. For example, 15–20 patients have been transported from the scene to several area hospitals.
- **Assessment of effectiveness of operations:** may consist of a report that, for example, relays that three extrications have been completed and 15 patients have been transported to an area hospital within 20 minutes of the first arriving apparatus.
- **Current status of resources:** outlines what resources are allocated, where the resources are allocated, and how the resources are allocated. This provides information about the units that are in staging ready to deploy and units that may be out of service.

The IC does not need to be well versed in technical rescue; however, he or she should be thoroughly familiar with the ICS management process. The IC should be stationed at a command post that is nearby (relative to the size and scope of the incident) but is set apart from the immediate incident scene. The command post does not need to be directly on scene but should be protected with restricted access so the command staff can function without needless distractions or interruptions.

The IC also communicates directly with command staff:

- Public information officer
- Safety officer
- Liaison officer

Public Information Officer

The **public information officer (PIO)** interfaces with the media and provides a single point of contact for information related to the incident, thereby allowing the IC to focus on managing the incident. The PIO prepares for IC approval any press releases to be issued. Also, prior to any press briefings, the PIO may provide the IC with background information, suggest questions that may be asked, and assist with selection and coordination of photographers.

Safety Officer

The **safety officer (SO)** is responsible for enforcing general safety rules and developing measures for ensuring personnel safety. This role includes the identification, evaluation, and correction of hazards and unsafe practices. The SO can bypass the chain of command when necessary to correct unsafe acts immediately and has the authority to stop or suspend unsafe operations, as is clearly stated in national standards such as NFPA 1500, *Standard on Fire Department Occupational Safety and Health Program*, NFPA 1521, *Standard for Fire Department Safety Officer*, and NFPA 1561, *Standard on Emergency Services Incident Management System*. In this event, the SO must immediately report to the IC any action taken that may affect operations **Figure 2-5 ▶**.

The SO, at a minimum, must be knowledgeable in the following areas:

Figure 2-5 The safety officer, at a minimum, must be knowledgeable in strategy and tactics, hazardous materials, rescue practices, building construction and collapse potential, and departmental safety rules and regulations.

- Strategy and tactics
- Hazardous materials
- Rescue practices
- Departmental safety rules and regulations

More specific responsibilities for extrication incidents include:

- Monitoring environmental hazards
- Ensuring personnel working in hazard zones are wearing appropriate PPE
- Conducting an ongoing evaluation of the psychological, physical, and mental state of rescuers
- Ensuring that personnel working near the edge of an elevation are securely tied off or the area is closely monitored
- Developing measures for ensuring personnel safety
- Ensuring the stability of vehicles is maintained and monitored throughout the incident

The safety officer's role and responsibilities at an extrication incident are no different than they are at any emergency incident. Remember that the main objective is to oversee that the operation is conducted in the safest manner, whether it is by ensuring that personnel performing their assignment are wearing their protective gear inside a hazard zone or by calling a freeze on an operation because cribbing support on a vehicle shifted when a tool was applied. This is not a seated position; the safety officer has to remain very active, moving around the vehicles and the scene.

Liaison Officer

The liaison officer (LO) is the IC's point of contact for outside agencies and coordinates information and resources between cooperating and assisting agencies. The LO also establishes contacts with agencies that may be capable of providing support or are available to do so.

ICS Sections

Other than those tasks specifically assigned to the command staff, all activities at an incident can be relegated to one of the following four major ICS sections. Each section can be headed by a section chief or by the IC personally, depending on the size and complexity of the incident.

- Operations
- Planning
- Logistics
- Finance/administration

Operations

The operations section is responsible for development, direction, and coordination of all tactical operations conducted in accordance with an IAP that outlines strategic objectives and the way in which operations will be conducted. The roles and responsibilities of the operations section chief at a large MCI are to coordinate and disseminate information back and forth through a division or branch manager who oversees the tactical groups, such as an extrication group(s). The number of groups must be no larger than five. An operations section chief can also be established on smaller extrication incidents where dialogue is coordinated directly to the tactical group supervisor. Establishing an operations section chief's position at an incident is a great way to effectively manage the on-scene tactical work, keeping the responsibilities of the IC to a manageable level. Remember, the great thing about the ICS is that it is flexible; it can expand or contract throughout the incident depending on what is required or how the incident evolves.

Under the operations chief, there are functional areas and positions that need to be established at larger incidents to allow for a manageable span of control and to organize resources based on incident needs. The first of these areas is staging. All resources in the staging area are available and should be ready for assignment. This area should not be used for locating out-of-service resources or for performing logistics functions. Staging areas may be relocated as necessary. After a staging area has been designated and named, a staging area manager will be assigned. The staging area manager will report to the operations section chief (or to the IC if an operations section chief has not been designated).

Divisions, groups, and branches are established when the number of resources exceeds the manageable span of control of the IC and the operations section chief. Divisions are established to divide an incident into physical or geographic areas of operation. For example, if a vehicle incident affected two different streets, then Division 1 might be Street A and Division 2 might be Street B. Groups are established to divide the incident into functional areas of operation (e.g., medical groups, search and rescue groups).

Branches may serve several purposes and may either be functional or geographic in nature. In general, they are established when the number of divisions or groups exceeds the recommended span of control for the operations section chief or for geographic reasons. Branches are identified by functional name or Roman numerals and are managed by a branch director. They may have deputy positions as required.

Planning

The planning section is responsible for the collection, evaluation, dissemination, and use of information and intelligence critical to the incident (unless the IC places this function elsewhere). One of the most important functions of the planning section is to look beyond the current and next operational period and anticipate potential problems, events, and logistical needs to execute the upcoming IAP. Other responsibilities include the following:

- Developing and updating the IAP
- Examining the current situation
- Reviewing available information
- Predicting the probable cause of events
- Preparing recommendations for strategies and tactics
- Maintaining resource status
- Maintaining and displaying situation status
- Providing documentation services

For a smaller MCI with extrication, a planning section may not have to be created, and the responsibilities can be carried out by the IC. Remember, the IAP can be a formalized written plan or it can be an informal verbal plan depending on the incident size, complexity, and length of the operational period.

Technical specialists initially report to the planning section and work within that section or are reassigned to another part of the organization. These advisors have the special skills required at the incident. These skills can be in any discipline required—for example, aviation, environment, hazardous materials, and engineering.

Logistics

The logistics section is responsible for all support requirements needed to facilitate effective and efficient incident management, including providing supplies, services, facilities, and materials during the incident. Key responsibilities include the following duties:

- Communications
- Medical support to incident personnel
- Food for incident personnel
- Supplies
- Facilities
- Ground support

At a large MCI involving multiple extrication groups and required transports, the logistics section will work very closely with all general staff in supplying resources to accomplish and/or satisfy the incident objectives. Are additional hydraulic tools or specialty tools needed? Should heavy equipment or operators be allocated, such as large tow units? How many transport-capable units will be required? Is there a temporary rehabilitation facility that can be established for personnel? Are the communications sufficient to handle the size or complexity of the incident?

Even at a smaller incident, a logistics section can help manage resources, such as by maintaining area coverage while jurisdiction units are operating at the incident.

The service branch within the logistics section may include the following units:

- **Communications Unit:** Develops plans for use of incident communications equipment and facilities, and installs, tests, distributes, and maintains communication equipment. This unit also supervises the incident communications center.
- **Medical Unit:** Develops the medical plan, obtains medical aid and transportation for injured and ill incident personnel, and prepares reports and records.
- **Food Unit:** Supplies the food needs for the incident.

The support branch within the logistics section may include the following units:

- **Supply Unit:** Obtains, stores, and maintains inventory of supplies needed for an incident and services supplies and equipment as necessary. This unit also orders personnel, equipment, and supplies.
- **Facilities Unit:** Is responsible for the layout and use of incident facilities and provides sleeping and sanitation facilities for incident personnel. The facilities unit is also responsible for managing base and camp operations.
- **Ground Support Unit:** Transports personnel, supplies, food, and equipment and implements the traffic plan for the incident. This unit also fuels, services, maintains, and repairs vehicles and other ground support equipment.

Finance/Administration

The finance/administration section is responsible for the accounting and financial aspects of an incident, as well as any legal issues that may arise. While not staffed at most incidents, this position accounts for all activities and ensures that enough money is made available to keep operations running. The following units are contained within the finance/administration section:

- **Time Unit:** Is responsible for equipment and personnel time recording.
- **Procurement Unit:** Administers all financial matters pertaining to vendor contracts, leases, and fiscal agreements.
- **Compensation/Claims Unit:** Handles financial concerns resulting from property damage, injuries, or fatalities at the incident.
- **Cost Unit:** Tracks costs, analyzes cost data, creates cost estimates, and recommends cost-saving measures.

A finance section is another management asset that can be utilized regardless of the size of an incident. Even at an MCI on a large interstate highway, the finance section can immediately procure monies to acquire needed equipment that may not be readily available or that may take a lot of time to establish. An incident on a highway during a hot summer day may require additional personnel to be called in to replace or cover for those requiring rehabilitation. This will be an unforeseen expenditure that an IC will not have time to negotiate during an incident.

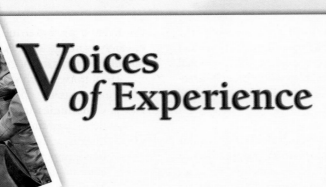

Voices of Experience

Several years ago, I was off duty and returning home one night from eating dinner at a restaurant in a neighboring jurisdiction. The weather was clear and the traffic was light. As I rounded a curve in the four-lane street, I saw what appeared to be a vehicle accident in the roadway. I pulled up at a safe distance from the scene, performed a size-up, and surveyed the scene for hazards. I quickly realized that I was first on scene at a three-vehicle collision. The occupants of two of the vehicles had exited their cars and were standing in the street. The occupants of the remaining vehicle remained inside their car.

As I approached the vehicle with the occupants still inside, I was met by the driver of one of the other vehicles, who was clearly upset and was shouting, "We have to get them out of there!" While trying to calm him down, I looked inside the vehicle to find the driver and her infant child. The situation then became more complex when I discovered that the driver did not speak English. The infant was secured in a car seat; however, the car seat had become dismounted and was displaced over the rear seat and into the cargo area of the hatchback car. Through further investigation and the use of my very rusty Spanish, I was able to determine that they were both entrapped. The driver had no injuries, and I was able to ascertain visually that the infant was breathing normally and appeared uninjured. There were no other hazards present, and it was confirmed that emergency responders had been requested.

> **"I explained that the occupants were in no danger and we would wait for emergency responders to arrive before getting them out."**

During the time these actions were occurring, the upset driver of the other vehicle became more and more distraught and continued to shout that we had to get the occupants out of the vehicle. I explained that the occupants were in no danger and we would wait for emergency responders to arrive before getting them out. He continued to hinder my actions at the scene and I finally told him, in a somewhat less than cordial way, to sit on the curb and be quiet so that I could do what needed to be done. The fire department, EMS, and law enforcement personnel arrived and an ICS was established. A fire department battalion chief assumed command, and a unified command was established when personnel from other agencies arrived. A safety officer was appointed and a proper span of control was maintained. I gave a briefing to the IC upon his arrival detailing the situation at hand, and the occupants of the vehicle were quickly extricated and checked for injuries by EMS.

This incident illustrates the need for establishing a proper ICS and the use of sound incident management processes. As witnessed here, we do not always know what we may encounter in terms of a vehicle extrication incident, even when off duty. As emergency responders, we are often called upon to make order out of a chaotic scene and often have to force ourselves to refrain from the "just rip the door off and yank the patient out" mentality mentioned in this chapter.

Glenn Clapp
High Point Fire Department
High Point, North Carolina

◼ Additional ICS Terminology

Single Resources and Crews

A <u>single resource</u> is an individual vehicle and its assigned personnel. A <u>crew</u> is a group of personnel working without apparatus and led by a leader or boss.

Task Forces and Strike Teams

Task forces and strike teams are groups of single resources assigned to work together for a specific purpose or for a specific period of time under a single leader. A <u>task force</u> is a group of up to five single resources of any type. A <u>strike team</u> is a group of five units of the same type working on a common task or function.

◼ Needs Assessment

Why should organizations perform a needs assessment? Often there is a perceived need within an organization that it should have a capability to handle every possible call that could arise. Sometimes this perception occurs because of specific incidents to which the organization has responded. At other times the "need" is based on the availability of grant money to improve response to a specific problem, even though the facts are insufficient to support the need. While the sentiment of being "all things to all people" is laudable when it comes to rescue capabilities, the reality of the response need can be radically different than what is perceived. The provision of technical rescue services is a rather involved endeavor and should not be taken lightly. Organizations that plan to develop response assets must concentrate on addressing those emergencies that are a priority in their response district.

A number of considerations must be taken into account before an organization commits to providing these services. These considerations include actual need, cost, personnel requirements, and political climate. Additionally, four separate components must be addressed when performing a needs assessment:

- Hazard analysis
- Organizational analysis
- Risk–benefit analysis
- Level of response analysis

◼ Hazard Analysis

<u>Hazard analysis</u> is the identification of situations or conditions that may injure people or damage property or the environment. In other words, it determines whether there is a hazard to protect against. There is a host of possible situations where specialized rescue services could be needed, and vehicle and machinery extrication is one such situation.

Some incidents might be more likely to occur than others. The organization needs to determine not only the possibility, but more importantly the *probability* of various types of incidents occurring within its jurisdiction. In such cases, the authority having jurisdiction must provide equipment and training to responders.

Surveys should be conducted regularly to identify the types of incidents that are likely to occur. A good place to begin a hazard analysis is to look at all of the major roadways within the jurisdiction and determine the type of traffic flow; is the roadway mainly used for residential traffic with an occasional petroleum tanker passing through, or is it mainly used for industrial traffic where all types of hazardous transport cargo travel along with residential flows? What are some ancillary access routes and containment plans that can be set in place? Again, when conducting a hazard analysis, it is always a good idea to secure your roadways first and then continue on to the jurisdictional area.

◼ Organizational Analysis

The next step in the needs assessment process, termed <u>organizational analysis</u>, seeks to determine whether an organization could establish and maintain the capability and whether the organization could comply with appropriate rules, regulations, laws, and standards.

The first step is to look at the personnel requirements. Some types and levels of technical rescue are very labor intensive. How many responders are required? What is the availability of the personnel? Can you get the people when you need them? Do the members have the time to develop, train for, and maintain the capability?

Training—both initial and ongoing—is a major concern with technical rescue capabilities. What is the existing versus required level of training? Is the training available? What will it cost to attain and maintain the required level?

Providing technical rescue services can require a significant amount of resources Figure 2-6 ▼ . You need to determine

Figure 2-6   Organizational analysis examines the available resources and determines which other resources would be required for different rescue incidents.

which resources are available in-house and what is available from outside sources, such as mutual aid, local industry, or private vendors. Does your organization have the financial resources necessary to support the proposed capabilities? How will you acquire the external resources? Concerns in this area also include equipment, tools and consumable supplies, personal protective equipment (PPE), vehicles, and maintenance. A list of internal and external resources should be made, updated yearly, and kept available at all times.

Risk–Benefit Analysis

Risk–benefit analysis entails an assessment of the risk to the rescuers compared to the benefits that might result from the rescue. All rescue work is a relative risk, but some operations have higher risks than others **Figure 2-7 ▾**. Environmental, physical, social, and cultural factors are taken into consideration during this analysis. Two important questions should be considered at this phase of the analysis when weighing risk versus benefit:

1. What is the probable (and possible) danger to rescuers?
2. Are the victims likely to be salvageable? Or, phrased differently, is this a rescue or a recovery operation?

On an extrication incident, regardless of the size, a risk–benefit analysis must be quickly and consistently conducted every time to determine a safe operation. An example would be an underride incident involving a common passenger vehicle colliding with a petroleum tanker, trapping the occupant in the passenger vehicle and causing a tank break with an active leak. Would this incident be a cause for rushing in and pulling the victim out without assessing the risk to personnel and determining the appropriate action plan to stabilize the scene before entering the hazard zone? Absolutely not! The scene would have to be properly assessed for the risks versus the benefits, and then an action plan would be established and put in place.

For the hazards identified in your response area, if there is a high level of risk associated with a low probability of a favorable outcome, you should consider not performing that type of rescue. The hazard analysis and risk–benefit analysis should be reviewed and updated regularly, and all changes should be documented.

Level of Response Analysis

After the determination is made that there is a need for a specific technical rescue capability, the next step in the process is to decide which level of service will be provided. Based on an evaluation of the information obtained from the hazard, organizational, and risk–benefit analyses, a three-level system is used to determine the level of response:

- Awareness (or basic)
- Operations (or medium)
- Technician (or heavy)

These terms are consistent with the levels of capability outlined in NFPA 1670. Because this is an organizational standard, it can be argued that every organization should have a capability at one of these levels. In some cases there may be no possibility of certain types of incidents occurring (such as ice rescue in the tropics). For example, if there is only a low risk of having a certain type of incident, your organization should be trained to at least the awareness level. Organizations with a response area that mainly consists of residential communities may only respond to extrication incidents involving common passenger vehicles, with a rare entrapment involving a commercial vehicle. The organization should be trained to an operations or technician level to properly address this type of incident. A mutual aid agreement may be established with another jurisdiction that can provide a technician-level response to any incident that is beyond the organization's capabilities. The organization provides personnel who are expected to perform at a higher level with the necessary training or the continuing education to operate at that level. Performance levels and training are evaluated and documented annually to determine whether personnel are prepared for rescue, and during difficult situations. Training documentation is available to those who are authorized to view it.

Figure 2-7 The high risks associated with some incidents require the scene to be stabilized and the hazards mitigated prior to the performance of the rescue.

Rescue Tips

Operations at vehicle extrication incidents are no different from operations at other emergency situations in that a consistent approach in dealing with an incident will produce a more favorable outcome. With the use of the ICS, a step-by-step approach to preparing for, assessing, and responding to vehicle extrication incidents in a safe, effective, and efficient manner is crucial. An example of a step-by-step approach may be as follows:

1. Preparation
2. Response
3. Arrival and size-up
4. Stabilization
5. Access
6. Disentanglement
7. Removal
8. Transport
9. Incident termination
10. Post-incident debriefing and analysis

Resource Management

Personnel Accountability

The accountability system is one of the most important tools an IC can utilize to ensure safety at an incident and should be implemented at all emergencies. Implementing accountability on a vehicle extrication incident can be as basic as ensuring that personnel operating hydraulic tools or conducting other strenuous tasks be rotated out to rehabilitation on a continual basis to prevent heat-related stress/emergencies. Training on the implementation of a personnel accountability system should meet the requirements of NFPA 1561. An accountability system tracks the personnel on the scene, including the following information:

- Responders' identities
- Their assignments
- Their locations

This system ensures that only qualified rescuers who have been given specific assignments are operating within the area where the rescue is taking place. By using an accountability system and working within ICS, an IC can track the resources at the scene, hand out the proper assignments, and ensure that every person at the scene operates safely. This system also allows the IC to rotate personnel as needed to prevent stress and fatigue.

A number of accountability systems exist, including lists, boards, tags, badges, T cards, bar-code systems, and radio frequency identification (RFID) tags Figure 2-8. Bar-code systems and RFIDs are used with electronic systems, which can give real-time information on personnel such as qualifications and assignment. This information can even be transmitted to multiple computers so more than one individual or location can have real-time access to this information (such as the safety officer, IC, or even the emergency operations center).

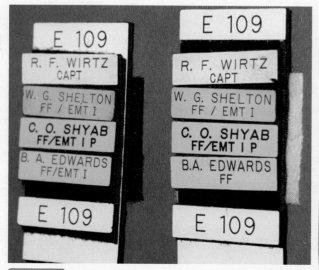

Figure 2-8 Accountability systems such as the accountability tag and the passport system are available to keep track of personnel at a rescue.

Rescue Tips

RFID tags use radio waves to transmit personnel data. With some models, it is not necessary to be within line of sight of the receiver for the equipment to work.

Equipment Inventory and Tracking Systems

In addition to documentation on the purchase, maintenance, and use of equipment, many technical rescue organizations maintain certain records during an incident Figure 2-9. The combination of these records provides a comprehensive resource and accountability system that can be manually or electronically tracked. Manual systems may include lists, sign-out sheets, and T cards. Computerized or electronic systems may consist of bar-code systems and RFID tags. In some cases, both types of systems may be used simultaneously. Fully integrated systems include all information on all equipment and may use bar codes or RFID tags to identify both the equipment and the personnel signing out that equipment.

Agency	ST	Kind	Type	I.D. No.
Order/ Request No.	Date/time Check In			
Home Base				
Departure point				
Leader Name				
Crew I.D. No./Name (For Strike Teams)				
No. Personnel	Manifest ☐Yes ☐No	Weight		
Method of Travel ☐Own ☐Bus ☐Air				
Other				
Destination Point	ETA			
Transportation Needs ☐Own ☐Bus ☐Air				
Other				
Ordered Date/Time	Confirmed Date/Time			
Remarks				

Agency	TF	Kind	Type	I.D. No./ Name
Incident Location	Time			
Status ☐Assigned ☐O/S REST ☐O/S PERS. ☐Available ☐O/S MECH ☐ ETR				
Note				
Incident location	Time			
Status ☐Assigned ☐O/S REST ☐O/S PERS. ☐Available ☐O/S MECH ☐ ETR				
Note				
Incident Location	Time			
Status ☐Assigned ☐O/S REST ☐O/S PERS. ☐Available ☐O/S MECH ☐ ETR				
Note				
Incident Location	Time			
Status ☐Assigned ☐O/S REST ☐O/S PERS. ☐Available ☐O/S MECH ☐ ETR				
Note				

Figure 2-9 Resource tracking systems are used to keep accurate records during an incident.

Near Miss REPORT

Report Number: 07-747
Report Date: 02/26/2007

Synopsis: An Operations Chief or Safety Officer has not been assigned.

Event Description: I responded to a motor vehicle collision with entrapment in an engine with one other fire fighter. There was no officer on board. We were fourth or fifth due. The first unit on scene was an EMT in a POV [personally owned vehicle] who advised via radio that power lines were down and that the scene was unsafe. Upon arrival, I was told to park my engine at the end of the emergency vehicle line and send my crew to assist with extrication.

I observed a car that was crashed into a sheared-off utility pole. High-voltage lines (> 10,000 volts) were hanging low over the road and across the road. Lines were less than 6 feet (2 meters) from the road at the low point. There was a transformer on the ground and it was leaking. The house lights in surrounding residences were all on. There were bystanders in several residential driveways. Utility pole wires were on the ground next to the vehicle. Extrication had just finished and the patient was being carried to the ambulance.

The IC was in a POV located behind a high-body rescue truck, and he did not have visual contact with the scene. The utility company had not yet arrived. The police were on scene and had made contact with the patient, but had pulled back from the downed power lines. Fire and EMS were milling around the scene, accident vehicle, and under the hanging wires. There were multiple foot tracks through the transformer oil. There was no assigned Operations Chief or Safety Officer. The first due officer turned over command to an arriving chief, but there was no face-to-face meeting for the exchange of command. The IC was trying to determine who was handling operations via radio. The officer charged with operations refused to take the position as it was out of control and he was fearful of scene safety.

I did not approach the scene to closer than 100 feet (30 meters); however, my crewman left our engine as soon as I set the brake and rushed into the scene. When I saw the scene, and chaotic situation, I inquired as to who was the Operations and/or Safety Officer and was told that there was no one in those positions. When I tried to direct personnel to remain clear until the power was confirmed off, I was ignored. I then went to the IC and told him to go to view the scene as it was out of control and very dangerous.

I requested from the IC to be released from the scene and left as the power company was arriving. (I later learned that the power company confirmed that the lines were in fact energized during the entire incident.) Fortunately there were no fire fighter, EMS, or bystander injuries. There was uncertainty about the nature of the transformer oil; it was suspected to be hazardous material. I was never notified of this. It was not until several days later that I learned of the hazardous materials situation and the fact that my crewman as well as our engine could have been contaminated.

Lessons Learned:

- There was no Command and Control on the scene.
- The exchange of command was not accomplished effectively.
- The IC did not take control of the scene until after the patient was treated and extricated.
- IC was reluctant to leave the radio in his vehicle even though he was not getting good information from the scene.
- Personnel were not aware of the danger of energized high-voltage lines.
- Personnel were not acting as a cohesive unit, but were instead acting as individuals.
- There was no accountability for personnel or their actions on the scene. Crew integrity was not maintained.
- There was no communication regarding the hazardous materials situation.
- At the debriefing, most personnel on the scene felt that they had acted appropriately, that there was no danger, and that the outcome is all that was important.
- The department has a severe problem with scene and fire fighter discipline. It is staffed with inexperienced officers who are oblivious to situational awareness and fire fighter safety.

(Continues)

Near Miss REPORT (Continued)

Lessons Learned Continued:

- The department does not train for accountability or fire fighter safety.
- All five hazardous attitudes were present on the scene. This was evident at the debriefing that was held at the next training meeting.

   1) **Impulsive Behavior:** The first EMT arrived and declared the scene unsafe. He then entered the scene with a fire fighter and decided to conduct the rescue prior to arrival of the power company.

   2) **Anti-authority:** My crewman left the engine even though he was told to remain with me. Per department policy, if an officer is not on an apparatus, the driver is in charge. Others at the scene would not listen to the first arriving officer; nor did they listen to the officer initially tasked with Operations. They were fixated on the extrication and doing their own thing.

   3) **Invulnerability:** Personnel believed they would not be hurt by the high voltage. Some expressed their opinion that they were not exposed for too long. They did not understand, nor do they want to understand, the dangers of high voltage.

   4) **Macho:** Some fire fighters said during debriefing that they knew that it was dangerous but had to do something because the bystanders expected them to initiate the rescue. When asked at debriefing how the fire fighters knew that the lines were not energized or that the scene was safe, they said that it looked OK and that others went in and didn't get hurt.

   5) **Resignation:** Two different officers gave up. The first arriving officer had the situation get out of hand and then assisted in the extrication, even though he knew that the lines could be energized. The officer tasked with Operations declined because he felt that the situation was out of his control and people were not listening to him. This also shows the anti-authority attitude of the department.

A large-scale MCI requiring the operation of multiple extrication groups can quickly deplete equipment and resources through equipment use, failure, and/or misplacement. Equipment such as hydraulic tools will fail at some point, hand tools such as an axe or Halligan bar may go missing or end up on another apparatus, and backboards and immobilization devices may be left at a trauma facility. A tracking system such as an RFID system can quickly inventory available assets before and after an incident occurs, better preparing the organization's readiness to respond.

Long-Term Operations

Fire fighters are accustomed to incidents that terminate quickly, often within an hour of onset. Many short-term incidents can be effectively directed and controlled using a basic ICS system consisting of an IC, safety officer, and operations section chief, with the various group functional assignments being established under the operations section as needed.

By comparison, technical rescue incidents can extend well beyond these relatively short time periods and be much more complex. The IC needs to consider both immediate and anticipated need for all phases of such an incident. Remember that whichever functions and responsibilities the IC does not assign to others in the ICS system stay with the IC. The IC could become overwhelmed easily in a technical rescue incident if he or she attempts to handle all of these issues for the duration of the incident.

Use of an expanded ICS system will assist the IC with planning and functioning in the long-term incident environment. This should include staffing of the planning, logistics, and finance/administration sections.

A rescue team can typically work for up to 24 hours without downtime (depending on the type of rescue), but then must have at least 24 hours for rest and equipment maintenance. If the rescue effort is expected to go beyond 24 hours, plans must be made for long-term operations. This determination can depend on any of the following factors:

- The extent of rescue required
- The training level of the rescue team
- The available resources
- The rescuers' physical condition
- The rescuers' psychological condition
- Needs supported by other ICS functions

Safety

To all organizations and ICs, the health and safety of their personnel is of the highest importance. Members who are injured not only reduce the available personnel by their loss but also can divert other personnel and resources when they must come to the injured individuals' aid. And, of course, no commander wants to be the one to explain to the member's family what happened, and most likely why it should not have happened.

NFPA 1500, *Standard on Fire Department Occupational Safety and Health Program*, is a critical document that relates to health and safety programs within the firefighting organization. Specifically, all personnel with functions at technical search and rescue incidents must meet the relevant requirements of the Special Operations, PPE, and Emergency Operations sections of this document. Contained within this standard are provisions that can be invaluable not only to developing a health and safety program but also in the development of response plans, SOPs/SOGs, and risk assessments. NFPA 1500, in concert with NFPA 1250, 1521, and 1561, provide the tools necessary to run a safe and effective organization.

Incident Response Planning

Sound, timely planning provides the foundation for effective incident management. **Response planning (preincident planning)**, is the process of compiling, documenting, and dispersing information that will assist the organization should an incident occur at a particular location. While there are frequently variables in any incident type, this document provides the framework of the organization's response. The major components of a response plan include the identification of the potential problem, resource identification and allocation, and, in some cases, suggested or mandatory procedures. Additional administrative elements include plan authority and approval, plus determining who gets a copy of the plan and when the plan is to be implemented and revised (including those with responsibilities outlined in the plan). A system should be in place for changes or revisions made to the plan.

Problem Identification

To write any plan, you must first identify the problem. This is accomplished by conducting the needs assessment discussed earlier in this chapter. This assessment, along with the organization's identified operational capability, forms the basis of what the plan will try to accomplish. Whether a single plan or multiple plans are used will depend on the complexity of your organization's involvement in a given rescue incident type.

Resource Identification and Allocation

Response planning will be worthless if there are no resources to implement it. The necessary resources (personnel, equipment, and materials) must be identified and must be assured of being made available in a timely manner. Frequently, this goal is accomplished through the use of mutual aid agreements, although some resources may require other arrangements. For example, agreements may be made with private vendors such as contractors for heavy equipment, lumber yards, and other private resources (e.g., industrial rescue team). When at all possible, these arrangements should be made before the incident occurs and should be put in writing to try to avoid any confusion.

Once you have identified these resources, collect the necessary information pertaining to them. This includes procedures for getting the needed resource and contact information such as names and phone numbers. Review this information at least annually to ensure it is still accurate. The preincident plan should also include any necessary forms or agreements that might be necessary during the incident.

Operational Procedures

The preincident plan should identify not only what resources are needed, but how they will be used (deployed). Should any resource not be available, the plan should identify alternative resources. It should identify possible scenarios where needed resources may not be available or may be delayed because of response difficulties or responder capabilities. In this event, alternate plans should address the needs of the incident.

Once the preincident plan has been developed, it is important that it be tested to ensure that it will work and accomplish the desired goals. Should any deficiencies be found, revise and retest the plan. Periodic testing is also important because of changing personnel or conditions.

Standard Operating Procedures

A **standard operating procedure (SOP)** is an organizational directive that establishes a standard course of action; in other words, it provides written guidelines that explain what is expected and required of emergency services personnel while performing their job **Figure 2-10 ▶**. SOPs are not intended to tell you how to do the job (technical knowledge and skills) but rather are designed to describe related considerations (the rules for doing the job) such as safety, evacuation, communication and notification procedures, command structures, and reporting requirements. Technical knowledge and skills, such as how to rappel or operate a specific tool, are obtained through training and technical protocols and are not normally the subject of SOPs.

FIRE AND RESCUE DEPARTMENT STANDARD OPERATING PROCEDURE		
SUBJECT: RESOURCE DEPLOYMENT	**S.O.P. 05.04.01**	
	PAGE 1 OF **4**	
CATEGORY: Suppression	**SUBCATEGORY:** Transfer and Fill-In Procedures	
APPROVED BY: CHIEF, FIRE AND RESCUE DEPARTMENT	**EFFECTIVE DATE:**	
FORMS REQUIRED: None **NOTE:** Current forms are located on the department's Intranet.		

PURPOSE:

To ensure that apparatus and personnel resources are deployed in a consistent and effective manner.

I. PREFACE

Operations staff shall be responsible for ensuring that personnel and apparatus are strategically positioned to provide efficient and effective service within response times consistent with department objectives.

II. GUIDELINES FOR APPARATUS DEPLOYMENT

A. The Assistant Chief of Operations ultimately shall be responsible for managing Operations' resources and informing the Operations' deputy chiefs and the Staff Duty Officer of conditions that will affect operations.

B. Battalion chiefs shall be responsible for managing resources within their respective battalions and for informing the Operations' deputy chief, or his or her designee, of conditions that may affect countywide Fire and Rescue Department service.

C. The uniformed fire officer (UFO) at the Department of Public Safety Communications (DPSC) shall be consulted any time units need to be placed out of service or relocated, i.e., units going to the Radio Shop or units being moved for training.

D. The UFO, in coordination with the field battalion chiefs, shall assume responsibility for relocating units for coverage shortages created by working incidents, vehicles out of service, or other short-term or temporary situations.

E. When units are out of service for an extended period of time due to prearranged training exercises (OARs, EMSCEP, etc.), staff shortages, or long-term mechanical problems, the Operations deputy chief shall consult with the UFO at DPSC to coordinate unit relocations.

F. The Apparatus Section's vehicle coordinator (or duty apparatus officer after normal business hours) shall be informed any time reserve apparatus is placed in or out of service.

Figure 2-10   A sample SOP regarding apparatus resource deployment.

Wrap-Up

■ Ready for Review

- Vehicle extrication is a technical process that requires structured successive steps to produce favorable results.
- The technical rescuer needs to develop a plan of action or incident action plan for mitigating an incident; this gives direction for personnel to follow.
- The incident command system (ICS) is a management structure that provides a standard approach and structure to managing operations. The use of ICS ensures that operations are coordinated, safe, and effective, especially when multiple agencies are working together.
- There are a number of considerations that should be taken into account before committing your organization to providing rescue services, including actual need, cost, personnel requirements, and political climate. The components of a needs assessment are hazard, organizational, risk–benefit, and level of response analyses.
- Sound, timely planning provides the foundation for effective incident management. A clear, concise incident action plan (IAP) is essential to guide the incident commander (IC) and subordinate staff and help the continuing collective planning activities of incident management teams.
- Resource management is a critical tool for managing any incident and includes not only personnel but equipment and other assets.
- Rescue incidents many times extend into long-term operations (greater than a few hours), and consideration must be given to planning for such a possibility.
- The health and safety of members are of the highest importance, and a health and safety program, developed in concert with NFPA 1500, 1250, 1521, and 1561, will provide your organization with the tools necessary to run a safe and effective organization.
- Response planning is a process of compiling information that will assist the organization should an incident occur. The major components of a response plan include the identification of the problem, resource identification and allocation, and, in some cases, suggested or mandatory procedures.
- Standard operating procedures (SOPs) are designed to describe related considerations (the rules for doing the job) such as safety, evacuation, communication and notification procedures, command structures, and reporting requirements.

■ Hot Terms

<u>Branches</u> A segment within the ICS that may be functional or geographic in nature. Branches are established when the number of divisions or groups exceeds the recommended span of control for the operations section chief or for geographic reasons.

<u>Crew</u> A group of personnel working without apparatus and led by a leader or boss.

<u>Divisions</u> A segment within the ICS established to divide an incident into physical or geographic areas of operation.

<u>Finance/administration section</u> ICS function responsible for the accounting and financial aspects of an incident, as well as any legal issues that may arise.

<u>Group</u> A segment within the ICS established to divide an incident into functional areas of operation.

<u>Hazard analysis</u> The process of identifying situations or conditions that have the potential to cause injury to people, damage to property, or damage to the environment.

<u>Incident action plan (IAP)</u> An oral or written plan containing general objectives reflecting the overall strategy for managing an incident. It may include the identification of operational resources and assignments. It may also include attachments that provide direction and important information for management of the incident during one or more operational periods.

<u>Incident commander (IC)</u> The individual with overall responsibility for the management of all incident operations.

<u>Incident command system (ICS)</u> A management structure that provides a standard approach and structure to managing operations, ensuring that operations are coordinated, safe, and effective, especially when multiple agencies are working together.

<u>Liaison officer (LO)</u> The incident commander's point of contact for outside agencies. This officer coordinates information and resources among cooperating and assisting agencies and establishes contacts with agencies that may be capable or available to provide support.

<u>Logistics section</u> ICS function responsible for all support requirements needed to facilitate effective and efficient incident management, including providing supplies, services, facilities, and materials during the incident.

Mass-casualty incident (MCI) An emergency situation that involves more than one victim that places great demand on equipment or personnel, stretching the system to its limit or beyond.

Operations section ICS position responsible for development, direction, and coordination of all tactical operations conducted in accordance with an IAP.

Organizational analysis A process to determine if it is possible for an organization to establish and maintain a given capability.

Planning section ICS function responsible for the collection, evaluation, dissemination, and use of information and intelligence critical to the incident.

Public information officer (PIO) ICS position that interfaces with the media and provides a single point of contact for information related to an incident.

Response planning (preincident planning) The process of compiling, documenting, and dispersing information that will assist the organization should an incident occur at a particular location.

Risk–benefit analysis An assessment of the risk to the rescuers versus the benefits that can be derived from their intended actions.

Safety officer (SO) ICS position responsible for enforcing general safety rules and developing measures for ensuring personnel safety.

Single resource An individual vehicle and its assigned personnel.

Staging area manager ICS position responsible for ensuring that all resources in the staging area are available and ready for assignment.

Standard operating procedure (SOP) An organizational directive that establishes a standard course of action (also referred to as a standard operating guideline [SOG]).

Strike team A group of five units of the same type working on a common task or function.

Task force A group of up to five single resources of any type.

Technical specialists Advisors who have the special skills required at a rescue incident.

Unified command An incident management tool and process that allows agencies with different legal, geographic, and functional responsibilities to coordinate, plan, and interact effectively to manage emergencies or events. In a unified command structure, multiple agency representatives make command decisions instead of just a single IC.

Technical Rescuer *in Action*

Your unit is the first on scene at a multiple vehicle accident. It is apparent that there are several victims. Fluid is noticeably leaking from one of the vehicles and the odor of gasoline is in the air. Several agencies have already been dispatched to the scene, including law enforcement, fire rescue departments, and transport units from different jurisdictions. A media van is also on scene already, and there is a crowd gathering.

1. The first action of the rescuer should be to:
A. begin crowd control.
B. initiate patient triage.
C. conduct size-up and establish command.
D. identify a staging area.

2. In order to identify the strategies of the incident, the incident commander should develop a formal or informal:
A. tactical priority system.
B. incident action plan.
C. strategic overview policy.
D. task assignment method.

3. The optimal number of individuals within a span of control is:
A. three.
B. four.
C. five.
D. six.

4. When multiple agencies have responsibility for the incident, it may be advisable to establish a command system that is:
A. cooperative.
B. shared.
C. collaborative.
D. unified.

5. One member of the command staff is the:
   A. media officer.
   B. medical officer.
   C. logistics officer.
   D. safety officer.

6. The command post should be away from the immediate scene.
   A. True
   B. False

7. In larger incidents, the incident commander may choose to establish a section in charge of:
   A. purchasing.
   B. planning.
   C. performance.
   D. probabilities.

8. The medical unit is responsible for:
   A. triage of patients.
   B. care of responders.
   C. identifying available hospitals.
   D. treating the critical patients.

9. A group of up to five single resources of any type is also known as a:
   A. task force.
   B. strike team.
   C. strike force.
   D. tactical force.

10. One of the things the accountability system is meant to recognize is a responder's:
    A. liability.
    B. experience.
    C. certification.
    D. assignment.

Mechanical Energy and Vehicle Anatomy

NFPA 1006 Standard

There are no objectives for this chapter.

Knowledge Objectives

After studying this chapter, you will be able to:

- Discuss the three concepts of energy that are typically associated with injury. (pages 34–35)
- Describe the three collisions that occur during every motor vehicle collision. (pages 35–37)
- Explain the motor vehicle collision area of impact classifications. (pages 37–38)
- Discuss the various methods used to power vehicles. (pages 39–40)
- Explain the various types of steel designations. (pages 40–41)
- Describe the three forms of frame systems. (pages 41–42)
- Describe the inner and outer components that make up a vehicle system. (pages 42–49)
- Discuss the various types of vehicle glass. (pages 49–50)

Skills Objectives

There are no skills objectives for this chapter.

ou are on an extraction training exercise where you have been instructed to perform a roof removal procedure. In the process of removing the roof, you attempt to cut through the A-post and encounter some difficulty with the hydraulic cutters; they cannot cut all the way through the post. You make several attempts to no avail.

1. Does the reinforced post contain High Strength Steel (HSS) or metal alloy? Check the make, model, and year of the vehicle.
2. What is the rated cutting force of your hydraulic cutter?
3. Does the hydraulic cutter need servicing? Are the blades separating as you attempt the cut?

Introduction

"Man's brain may be compared to an electric battery . . . . A group of electric batteries will provide more energy than a single battery." —Napoleon Hill

This chapter explores the application of energy in relation to a motor vehicle collision and the anatomical parts that make up a vehicle system. Understanding these two components is integral in the extrication process.

Energy

Merriam-Webster defines **energy** as "a fundamental entity of nature that is transferred between parts of a system in the production of physical change within the system and usually regarded as the capacity for doing work." Energy is all around us; it is a constant force. In fact, the **Law of Conservation of Energy** states that energy can neither be created nor destroyed; it can only change from one form to another. For example, the forward movement of a vehicle (which is kinetic energy) can be changed into heat energy from the friction caused by the application of the brakes. Energy comes in different forms: electrical, mechanical, heat (thermal), light (radiant), chemical, gravitational, and nuclear. When dealing with the science of a motor vehicle collision, mechanical energy is the driving force behind the dynamics of what actually occurs.

Mechanical energy can be broken down into two types of energy, kinetic and potential. With the combination of work (transfer of energy), a mechanical energy system is produced.

Kinetic energy is the energy of motion, which is based on vehicle mass (weight) and the speed of travel (velocity). Kinetic energy is expressed as:

$$\text{Kinetic energy} = \frac{\text{mass}}{2} \times \text{velocity}^2$$

$$\text{or,} \quad KE = \frac{m}{2} \times v^2$$

Potential energy is stored energy, or the energy of position. To better illustrate this definition, take two bricks (brick 1 and brick 2), each of equal size and weight. Now hold brick 1 one inch (25 millimeters [mm]) over the top of a glass table, and elevate brick 2 three feet (1 meter [m]) over the top of the glass table **Figure 3-1 ▶**. Which brick, if dropped, has the most energy force, or the *potential*, to cause more damage to the glass table? Brick 2, based on the height, speed of travel, and gravitational force, will have the highest level of stored, or potential, energy. When the brick is actually released, the potential energy is transferred to kinetic energy because the brick is in motion. Once the brick strikes the glass table, a force is applied to stop, displace, or alter the brick's kinetic path of travel. This force is known as work.

Work, in its most basic definition, is a mechanism for the transfer of energy. Work is said to be applied to an object when energy is transferred to the object and the object is displaced. Work can be applied in the form of a positive (with the direction of travel) or negative (against the path of travel) force. To better illustrate this, let's look at two vehicles that are the same type, make, weight, and model. Vehicle 1 is traveling in a forward motion at 20 miles per hour (mph) (32 kilometers per hour, or kph), and vehicle 2 is coming up from behind vehicle 1 at a speed of 40 mph (64 kph). Vehicle 2 eventually crashes into the rear of vehicle 1. Through the application of work, vehicle 2 has transferred its forward kinetic energy into vehicle 1, causing a *positive* work displacement of vehicle 1's path of travel. Vehicle 2 has also experienced a *negative* work force because its path of travel at 40 mph (64 kph) has been displaced by crashing into

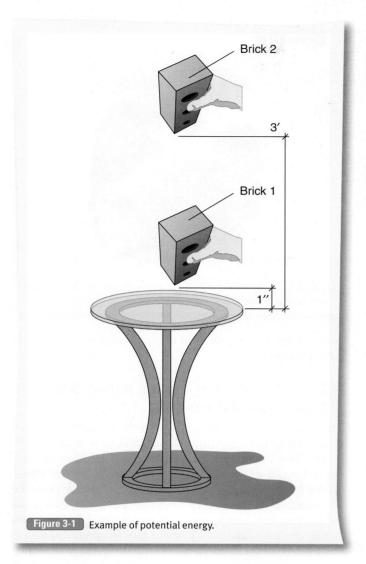

Figure 3-1   Example of potential energy.

Figure 3-2   The kinetic energy of a speeding car is transferred by the work force of stopping the car.

The science of a motor vehicle collision through the application of energy can be directly applied to the understanding of how an occupant sustains injuries.

Traumatic injury occurs when the body's tissues are exposed to energy levels beyond their tolerance. The **mechanism of injury (MOI)** is the way in which traumatic injuries occur; it describes the forces (or energy transmission) acting on the body that cause injury. The same release of energy that occurs during a vehicle collision, on and from the vehicle through the mechanical system of potential energy, kinetic energy, and work, is also applied directly to the human body. The exception to this is that the human body will experience both a positive and negative work force when the internal organs are bounced back and forth against the inside of the human body. The types of injuries that the human body can potentially sustain relate not only to the release of energy on it but also the type of collision that occurs during the vehicle crash.

Vehicular Collision Classifications

Motor vehicle collisions are classified traditionally by the area of initial impact. They consist of the following: front impact (head-on), including under-rides and over-rides; lateral impact (T-bone, side impact); rear-end; rollovers; and rotational (spins), which can also consist of secondary collisions with multiple impacts. The sequence of events that occur during a motor vehicle collision typically consists of three separate collisions. Understanding the sequence of events during each one of these three collisions will help you be alert for certain types of injury patterns.

For example, let's examine the sequence of events involving the three separate collisions that occur during a front impact vehicle collision.

Event 1. Vehicle Impact with Object

The first event is when the vehicle impacts an object, whether the object is stationary or in motion. Depending on the speed of travel, weight of the vehicle, braking distance, and whether the object struck was stationary or in motion, damage to the vehicle is perhaps the most dramatic part of the

the rear of vehicle 1, which slowed vehicle 2 down or stopped it altogether.

This work or transfer of energy can be extreme or lessened based on a number of factors. These factors consist of the stored potential energy of the vehicles, which is based on the speed of travel, weight of the vehicles, whether the brakes are applied (which would transfer and displace some energy by means of releasing frictional heat when the brakes are applied), and whether the vehicle design is of a unibody type construction (which is designed to collapse and absorb and/or displace energy as a crash occurs) **Figure 3-2 ▶**.

Newton's First **Law of Motion** states that objects at rest tend to stay at rest and objects in motion tend to stay in motion unless acted on by an outside force. This outside force at the moment of impact is considered the applied work force, which causes displacement. To better illustrate this, if a vehicle in motion crashes into a stationary wall (an outside force containing potential energy and a negative work force, which occurs at the moment of impact), the forward energy from the vehicle (kinetic energy) cannot be destroyed by the wall, but it will change form and be transferred by work being diverted and distributed throughout the vehicle and its occupants.

Figure 3-3 The first collision in a frontal impact is that of the vehicle against another object, whether the object is stationary or in motion (in this case, a utility pole). The appearance of the vehicle can provide you with critical information about the severity of the crash and can often determine the mechanism of injury (MOI) sustained by the occupants.

Figure 3-4 The second collision in a frontal impact is the occupant striking or impacting the interior of the car.

collision. The amount of damage can provide information about the severity of the collision and the resulting injuries of the occupants **Figure 3-3 ▲**. The greater the damage to the car, the greater the release of energy that was involved and, therefore, the greater the potential to cause injury to the patient. By assessing the vehicle that has crashed, you can often determine the MOI, which may allow you to predict what injuries may have happened to the passengers at the time of impact according to forces that acted on their bodies. A great amount of force is required to crush and deform a vehicle, cause intrusion into the passenger compartment, tear seats from their mountings, and collapse steering wheels. Such damage suggests the presence of high-energy trauma.

Event 2. Occupant Impact with Vehicle

The second event during the collision is the occupant striking or impacting the interior of the vehicle. Just as the kinetic and potential energy of the vehicle's mass and velocity is converted into the work of bringing the vehicle to a stop, the kinetic and potential energy of the passenger's mass and velocity is converted into the work of stopping his or her body. This work can be extreme or lessened depending on whether the occupants were restained by a three-point seat belt harness or air bag. Remember Newton's First Law of Motion, which states that an object in motion will remain in motion until acted upon by an outside force. If the vehicle was traveling at 40 mph (64 kph) at impact, then the occupant's body will still travel at 40 mph (64 kph) until his or her impact on the interior of the vehicle stops this forward motion. With the amount of damage to the exterior of the car, the injuries that result are often dramatic and can often be immediately apparent during your scene size-up and primary assessment of the occupants. Common passenger injuries that can occur from this type of impact include lower extremity fractures (knees into the dashboard) **Figure 3-4 ▶**, flail chest (rib cage into the steering wheel), and head trauma (head into the windshield). These types of injuries can occur more frequently and be more severe if the passenger is not restrained. However, even when the passenger is restrained with a properly

adjusted seat belt, injuries can occur, especially in lateral and rollover impacts.

Event 3. Occupant Organs Impact Solid Structures of the Body

The final event during the collision is the occupant's internal organs striking or impacting the solid internal structures of the body. These organs can impact back and forth several times depending on when the body's motion finally comes to rest. The injuries that occur during the third collision may not be as obvious as external injuries, but they are often the most life threatening. For example, as the passenger's head hits the windshield, the brain continues to move forward until it comes to rest by striking the inside of the skull **Figure 3-5 ▼**. This results in a compression injury (or bruising) to the front of the brain and stretching (or tearing) of the back of the brain. Similarly, in the thoracic cage, the heart may slam into the sternum, which may tear the aorta, the largest artery in the body, and cause fatal hemorrhaging.

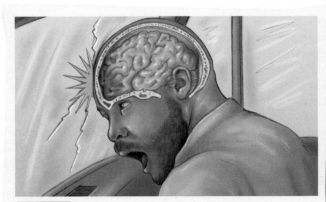

Figure 3-5 The third collision in a frontal impact is the occupant's internal organs striking or impacting the solid internal structures of the body. In this illustration, the brain continues its forward motion and strikes the inside of the skull, resulting in a compression injury to the anterior portion of the brain and stretching of the posterior portion.

Understanding the series of events that occur with these three collisions will help you make the connections between the amount of damage to the exterior of the vehicle and potential injury to the passenger. For example, in a high-speed collision that results in massive damage to the vehicle, you should suspect serious injuries to the passengers, even if the injuries are not readily apparent. A number of potential physical problems may develop as a result of trauma or injuries. Your initial general impression of the patient and the evaluation of the MOI can help direct lifesaving care and provide critical information to the appropriate medical facility. Therefore, if you see a contusion on the patient's forehead and the windshield is broken and pushed out, you should care for this patient as if he or she has suffered an injury to the brain and communicate this concern to the medical providers.

Figure 3-6   Rear-end impacts often cause whiplash-type injuries, particularly when the head and/or neck is not restrained by a headrest.

Rescue Tips

When you are assessing trauma victims at a motor vehicle collision, the MOI is a crucial element to consider for the potential and type of injuries that can be sustained by the occupants. Be alert to the extent of damage to the interior and exterior of the vehicles involved in crashes. Use this observation to paint a picture of the scene in written and verbal communication, especially when consulting with a trauma or medical facility.

■ Front Impact Collisions

A front impact collision occurs when the vehicle strikes an object head-on, whether that object is stationary or in motion. With this initial impact, there can be two other events that can occur: The vehicle can travel under the object (which is known as an under-ride collision) or the vehicle can travel on top of the object (which is known as an over-ride collision). Understanding the MOI after a frontal collision first involves evaluation of the vehicle's restraint systems, which include seat belts (standard three-point harness and a pretensioning system) and air bags. You should determine whether the occupants were restrained by a full and properly applied standard three-point restraint harness or pretensioning system. In addition, you should determine whether the air bag deployment impacted the occupant, which could cause crushing injuries or burn injuries to the face, arms, and upper torso area due to the high temperatures that the air bag generates on deployment. Other MOIs to look for consist of bent or deformed steering wheels and broken or penetrated windshields (imbedded blood, hair, or teeth fragments are all positive confirmation of an impact).

■ Rear-End Collisions

Rear-end collisions are known to cause whiplash-type injuries, particularly when the passenger's head and/or neck is not restrained by an appropriately placed headrest **Figure 3-6 ▶**. On impact, the passenger's body and torso move forward by the transfer of kinetic energy. As the body is propelled forward, the head and neck are left behind because the head is relatively heavy, and they appear to be whipped back relative to the torso. As the vehicle comes to rest, the unrestrained passenger moves

forward, striking the dashboard. In this type of collision, the cervical spine and surrounding area may be injured. Due to the anatomical position of the spine, the cervical portion of the spine is less tolerant of damage when it is bent back. Headrests decrease extension of the head and neck during a collision and, therefore, help reduce injury. Other parts of the spine and the pelvis may also be at risk for injury. In addition, the patient may sustain an acceleration-type injury to the brain—that is, the third collision of the brain within the skull. Passengers in the backseat wearing only a lap belt might have a higher incidence of injuries to the thoracic and lumbar spine.

■ Lateral (Side-Impact) Collisions

Because of the limited protection to the occupants, lateral or side impacts (commonly called T-bone collisions) are a very common cause of fatalities associated with motor vehicle crashes. When a vehicle is struck from the side, the impact results in the passenger sustaining a lateral whiplash injury **Figure 3-7 ▼**.

Figure 3-7   In a lateral collision, where the vehicle is struck from the side, the impact results in the passenger sustaining a lateral whiplash injury where the movement is to the side, and the passenger's shoulders and head whip toward the intruding vehicle.

The movement is to the side, and the passenger's shoulders and head whip toward the intruding vehicle. This action may thrust the shoulder, thorax, upper extremities, and, most importantly, skull against the doorpost or the window. Due to the anatomical position of the spine, the cervical spine has little tolerance for lateral bending.

Rollovers

Certain vehicles, such as large trucks and some sport utility vehicles (SUVs), are more prone to rollover crashes because of their high center of gravity. Injury patterns that are commonly associated with rollover crashes differ, depending on whether the passenger was restrained or unrestrained. The most unpredictable types of injuries are caused by rollover crashes in which an unrestrained passenger may have sustained multiple strikes within the interior of the vehicle as it rolled one or more times. The most common life-threatening event in a rollover is ejection or partial ejection of the passenger from the vehicle Figure 3-8 ▾ . Passengers who have been ejected may have struck the interior of the vehicle many times before ejection. The passenger may also have struck several objects, such as trees, a guardrail, or the vehicle's exterior, before landing. Passengers who have been partially ejected may have struck both the interior and exterior of the vehicle and may have been sandwiched between the exterior of the vehicle and the environment as the vehicle rolled. Ejection and partial ejection are significant MOIs; in these cases, you should prepare to care for life-threatening injuries.

Rotational Collisions

Rotational collisions (spins) are conceptually similar to rollovers. The rotation of the vehicle as it spins provides opportunities for the vehicle to experience secondary impacts. For example, as a vehicle spins and strikes a pole on the driver's side, the driver experiences not only the rotational impact and motion but also a secondary lateral impact.

Figure 3-8 Passengers who have been ejected or partially ejected may have struck the interior of the car many times before ejection.

The Vehicle System

Before a rescuer can properly apply any extrication procedures to a vehicle, he or she must understand the inner and outer components that make up a vehicle system. Just as a surgeon thoroughly understands the inner workings of the human body well before making that first incision, the technical rescuer should know the components or basic parts that make up various kinds of vehicles well before starting to extricate. To better illustrate this statement, place yourself on the scene of an extrication incident where the officer in charge tells you to perform a dash-lift technique to gain access to the patient. During this process, you are told to make a relief cut through the upper rail section between the strut tower and firewall. If you do not have a thorough understanding of vehicle anatomy, you will not have any idea where this relief cut needs to be made. This may make you a burden on scene and a hindrance to the operation. Do not try to improvise if you do not understand the technique! Step aside and pass the tool to a more experienced person.

Vehicle Classifications

Vehicles can be classified in several different ways. The Department of Transportation (DOT) classifies vehicles based on whether the vehicle transports passengers or commodities, with a nonpassenger vehicle being further classified by the number of axles and unit attachments it has. A **passenger vehicle** is defined by the DOT as all sedans, coupes, and station wagons manufactured primarily for the purpose of carrying passengers, including those passenger cars pulling recreational or other light trailers.

The Department of Energy (DOE) classifies vehicles by size utilizing a cubic feet system (passenger and cargo volume) and gross weight system. A type of passenger vehicle that is termed a sedan, for example, is classified or known as a sedan based on the cubic feet of space it has for passengers or cargo. Sedan types, according to the DOE, range in size of less than 85 cubic feet (2 cubic meters) to up to 130 cubic feet (4 cubic meters) depending on whether the sedan is minicompact, subcompact, compact, midsize, or large. Table 3-1 ▸ describes the different classifications of vehicles based on size and weight.

Vehicle Identification Numbers

A vehicle identification number, or VIN, is a unique identification system composed of a 17-character sequence containing both numbers and letters with the exclusion of the letters I, O, and Q to avoid confusion with the numbers 1 and 0. In 1981, the United States enacted the VIN identification system under the Code of Federal Regulations (CFR), Title 49, Chapter V, Part 565, *Vehicle Identification Requirements*, which mandated that every passenger vehicle, SUV, truck, or trailer manufactured be identified and tracked utilizing the VIN identification system. A VIN is affixed to every type of vehicle manufactured in the United States and many other countries. The VIN is normally etched on a plate and attached or embossed to the driver's side dashboard, labeled on the driver's side vehicle door, or affixed to the inside of the glove compartment. The VIN is

Table 3-1 Vehicle Classifications

CARS

Class	Passenger and Cargo Volume (Cu. Ft.)	
Two-Seaters	Any (cars designed to seat only two adults)	
Sedans		
Minicompact	< 85 (< 2.40 cubic meters)	
Subcompact	85–99 (2.40–2.80 cubic meters)	
Compact	100–109 (2.83–3.08 cubic meters)	
Midsize	110–119 (3.11–3.36 cubic meters)	
Large	≥ 120 (≥ 3.39 cubic meters)	
Station Wagons		
Small	< 130 (< 3.68 cubic meters)	
Midsize	130–159 (3.68–4.50 cubic meters)	
Large	≥ 160 (≥ 4.53 cubic meters)	

TRUCKS

Class	Gross Vehicle Weight Rating (GVWR)	
Pick-up Trucks	**Through Model Year 2007**	**Beginning Model Year 2008**
Small	< 4500 pounds (2041 kilograms)	< 6000 pounds (2722 kilograms)
Standard	4500–8500 pounds (2041–3856 kilograms)	6000–8500 pounds (2722–3856 kilograms)
Vans		
Passenger	< 8500 pounds (< 3856 kilograms)	
Cargo	< 8500 pounds (< 3856 kilograms)	
Minivans	< 8500 pounds (< 3856 kilograms)	
Sport Utility Vehicles (SUVs)	< 8500 pounds (< 3856 kilograms)	
Special Purpose Vehicles	< 8500 pounds (< 3856 kilograms)	

Source: Modified from U.S. Environmental Protection Agency, Vehicle Size Classes Used in the Fuel Economy Guide [http://www.fueleconomy.gov/feg/info.shtml]. Accessed February 11, 2011. Courtesy of U.S. Department of Energy.

also listed in vehicle documents such as insurance documents and registration. Each alphanumeric designation of a VIN has a specific meaning, such as the type and make of the vehicle, country of origin where the vehicle was manufactured, and year of manufacturing.

Vehicle Propulsion Systems

Conventional Vehicles

The overwhelming majority of vehicles on the road today are conventional-type vehicles; these types of vehicles utilize internal combustion engines for power. An internal combustion engine (ICE) can be designed to burn a multitude of petroleum-based fuels and alternative fuels, with gasoline and diesel fuel being the most commonly used. Fuel tanks for conventional-type vehicles are more commonly constructed of steel or aluminum, but high-density plastic fuel tanks are also used because they are lighter than the steel or aluminum designs. In addition, there has been a rise in the use of compressed gas systems that utilize high-pressure storage tanks composed of steel, aluminum, or a combination of both with a carbon fiber or fiberglass wrapping. These tanks can

range in pressures from 3500 pounds per square inch (psi) to 10,000 psi (24,115 to 68,900 kilopascals [kPa]), depending on the type of gas being stored. Pounds per square inch (psi) is a unit of measure to describe pressure; it is the amount of force exerted on an area equaling 1 square inch. Alternative powered vehicles and fuel systems will be discussed in depth in Chapter 4, *Advanced Vehicle Technology: Alternative Powered Vehicles*.

Hybrid Vehicles

A hybrid vehicle (HV) is defined as a vehicle that combines two or more power sources for propulsion. This combination of power sources generally consists of generated electricity through a high-voltage electrical system and a petroleum-based fuel or alternative fuel system through the process of an internal combustion engine. This is known as a hybrid electric vehicle (HEV).

Rescue Tips

The basic components of a hybrid electric vehicle system consist of an electric motor, generator, internal combustion engine, and battery pack.

■ Hydrogen Fuel Cell Vehicles

A hydrogen fuel cell is an electrochemical device that utilizes a catalyst-facilitated chemical reaction of hydrogen and oxygen to create electricity that is then used to power an electric motor. A fuel cell vehicle by definition is a hybrid vehicle system by design where two separate sources of power are utilized or combined as a propulsion mechanism for the vehicle. The first fuel cell was developed in 1839.

> ### Rescue Tips
>
> Hydrogen fuel cell vehicles have the potential to be two to three times more efficient than conventional vehicles, emitting little to no greenhouse gas emissions.

■ Electric-Powered Vehicles

All electric-powered vehicles utilize an electric motor for propulsion and are powered by batteries in a rechargeable battery pack. Also known as battery electric vehicles (BEVs), many can travel an average distance of 100 to 200 miles (161 to 322 kilometers) between recharging. Recharging can be accomplished through dedicated charging stations or a general plug-in house current (120/240 volts).

▌ Electricity

Electrical power in standard conventional-type, internal combustion engine vehicles utilizes a basic 12-volt lead acid battery system for starting and powering various electrical components within the vehicle. Hybrids, fuel cells, and electric vehicles also use a 12-volt battery lead acid battery system for starting purposes but additionally use an advanced electrical design to power the vehicle. These types of vehicles will be discussed in detail in Chapter 4, *Advanced Vehicle Technology: Alternative Powered Vehicles*. There are some larger conventional-type vehicles that utilize two 12-volt batteries for starting in cold weather climates, for heavy towing assignments, or for operating or assisting in powering additional electrical components; these batteries are normally wired in a parallel-type system and still only operate as a 12-volt system, not 24 volts. Larger commercial trucks as well as military vehicles use a true 24-volt power electrical system; these batteries are wired in series with the system being designed to operate at 24 volts.

The 12-volt lead acid battery system consists of six cells in an electrolyte solution of sulfuric acid and water. This electrolyte solution causes a chemical reaction with the lead plates, producing electrons that then flow through conductors, thus generating the power for starting the vehicle. Each cell stores roughly 2.1 volts or more for approximately 12.5 volts. When a 12-volt lead acid battery is overcharged and without the proper venting, a by-product can be generated consisting of a highly explosive mixture of hydrogen and oxygen gas.

Other main sources of power within the vehicle consist of the alternator and voltage regulator. The alternator is basically a belt-driven generator that produces current used to operate various electrical components within the vehicle. A voltage regulator then regulates the flow of electricity coming from the alternator, which keeps the voltage to a safe 12.5–14.5 volts.

▌ Steel

Today, vehicle manufacturers are on an endless quest to make a vehicle more fuel efficient, lighter, stronger, and more crash resistant. The development of strong, crash-resistant vehicles requires engineers and the steel industry to develop stronger and lighter steels to meet these demands. They are utilizing many of the same concepts that the race car industry utilizes, such as incorporating a "safety cage" into the design so the vehicle has reinforced sides, roof, floor panels, rocker panels, and seat structures that are designed to protect occupants from rollovers and various types of impacts. These are just a few of the safety measures that the technical rescuer must be aware of.

Conventional vehicle designs utilizing low- and medium-strength steels are becoming obsolete; unfortunately, these are the vehicles that most technical rescuers have been or are presently still training on; the junkyards are loaded with them. Understanding new techniques and the capabilities and limitations of rescue tools in relation to modern steel is imperative for successful technical rescue operations.

Steel is measured by tensile strength and yield strength, which are depicted in thousands of pounds per square inch (ksi) or megapascals (MPa). The **tensile strength** of steel measures the amount of force that is required to tear a section of steel apart. The **yield strength** is the amount of force or stress that a section of steel can withstand before permanent deformation occurs.

Alloyed steels are composed of a mixture of various metals and elements. They are classified by both their strength range in MPa and by their metallurgical type designation. The metallurgical type designations can vary throughout the world. Some examples include Mild/Low/Medium Strength Steel, High Strength Steel (HSS), High Strength Low Alloy Steel (HSLA) or Micro-alloyed Steel, Ultra High Strength Steel (UHSS), and the newer Advanced High Strength Steel (AHSS), including Dual Phase Steel (DP), Transformation Induced Plasticity (TRIP), Complex Phase Steel (CP), and Martensitic Steel (MS). With these many variations, it can be quite confusing.

The World Steel Association classifies a **High Strength Steel (HSS)** as any steel with a tensile strength between 39 ksi and 102 ksi (270 MPa and 700 MPa). The average hydraulic cutter yields a range of 60 ksi and 80 ksi (414 MPa and 552 MPa) of cutting force. **Ultra High Strength Steel (UHSS)** is classified as any steel with a tensile strength of 102 ksi (700 MPa) or greater. **Advanced High Strength Steel (AHSS)** is classified as any steel with a *minimum* tensile strength of 73 ksi to 116 ksi (500 MPa to 800 MPa). An example of an AHSS-type steel is boron-alloyed steel. Boron is alloyed with steel during processing for its unique hardening properties. It is now widely used in the auto industry to reinforce specific areas of a vehicle frame that are most subject to impacts from a collision or rollover. This boron reinforcement may come in the form of rods or tailored blanks that are welded onto existing framing. Tailored blanks are sections of steel manufactured in different gauges of thickness, sizes, strengths, or coatings to fit perfectly in a specified location; they are welded to the existing frame structure for reinforcement purposes. The boron-alloyed steel used in modern vehicles has a yield point of about 196 ksi to 203 ksi (1350 MPa to 1400 MPa) or greater depending on the processing. Volvo, Audi, Mercedes, Subaru, Land Rover, BMW, Porsche, and Ford are just a few of the auto manufacturers utilizing boron-alloyed steel in their vehicles for reinforcement and improved safety standards. With the auto industry rapidly replacing conventional vehicle designs with advanced technology, so must the world of technical rescue replace conventional extrication practices to keep up with the pace of the future.

Frame Systems

There are two frame systems that are most common in today's vehicles, the body-over-frame construction and the unitized or unibody construction. These frames can be composed of steel (most common), aluminum, or carbon fiber/composite. Another type of frame system that is less common today is the space frame, which can consist of aluminum construction or tubular steel.

Body-Over-Frame Construction

The **body-over-frame construction** design is just as it states, the body of the vehicle is placed onto a frame skeleton and the frame acts as the foundation for the vehicle. This basic design consists of two large beams tied together by cross member beams. This type of frame is sometimes referred to as a **ladder frame** because of its similarity **Figure 3-9**. Most of the heavier vehicles, such as a larger pick-up truck or SUV, utilize this type of frame construction. The potential for the body-over-frame design to be split in half with a severe collision is low, but be aware that the force distribution from the impact will be greater on the occupants.

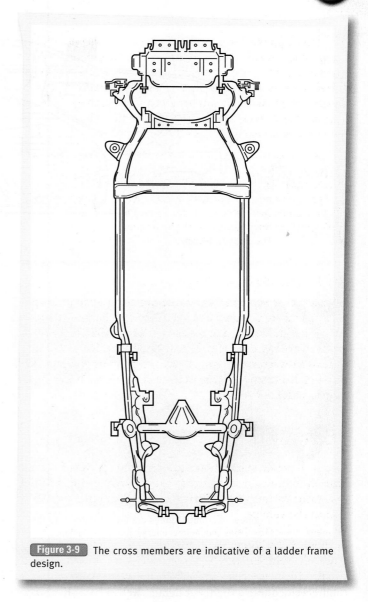

Figure 3-9 The cross members are indicative of a ladder frame design.

Unibody Construction

The **unibody construction** or unitized structure is basically one piece. There is no formal frame structure; the body parts supply the structural integrity of the vehicle. The vehicle body is merged with the **chassis**, which consists of the braking, steering, and suspension system **Figure 3-10**.

The main difference of the unibody construction as compared to the body-over-frame design is its ability to absorb or redirect energy during a collision. Unibody construction incorporates **crumple zones** into the front and sometimes the rear of the vehicle to redirect energy away from the passenger compartment **Figure 3-11**. When an impact occurs, these crumple zones collapse, absorbing and diverting the force or energy of the collision and preventing intrusion into the cab of the vehicle. To fully understand this, you must understand the Law of Conservation of Energy and particularly the mechanical

energy system as it relates to the vehicle crash. Understanding the science dealing with the laws and theories related to energy will help clarify what actually occurs during a vehicle crash. A negative aspect of the unibody construction is the potential, because there is no frame structure, for the vehicle to be split in half with a severe collision. This can occur from another vehicle or from a stationary object, such as a tree or pole.

Space Frame Construction

A true **space frame**–constructed vehicle was predominately designed for the auto racing industry because of its lighter weight and rigid structure **Figure 3-12**. The high costs in production kept most vehicle manufacturers from mass producing any sustainable models. The Saturn Corporation is one manufacturer that used the space frame design. The body design of the space frame is made up of multiple lengths and angles of

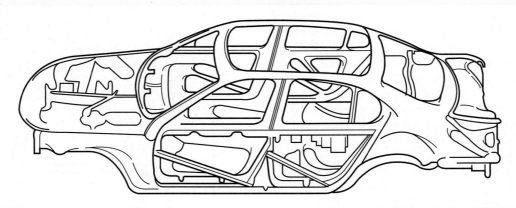

Figure 3-10 The unibody design.

Figure 3-11 Engineered crumple zones are incorporated into the unibody frame to absorb and redirect energy during a collision.

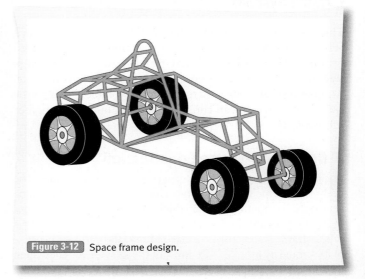

Figure 3-12 Space frame design.

tubing welded into a rigid, but lighter, web or truss-like structure; the vehicle's outer panels are attached independently to the frame after its completion. The major difference of a space frame–constructed vehicle compared to the unibody and body-over-frame design is that a space frame can be driven in its skeleton form, void of any body panels.

Vehicle Anatomy and Structural Components

Several key components make up the body portion of the vehicle. At the front and rear of the vehicle is the **bumper system**. The bumper system helps the vehicle to withstand the impact of a collision. Federal transportation regulation 49 CFR 581, *Bumper Standard*, mandates performance standards for bumpers on passenger cars. One of these standards states that the vehicle bumper must be able to absorb a front or rear impact collision at a speed of 2.5 mph (4 kph) without sustaining damage to the vehicle body. The federal bumper standard is applied only to passenger vehicles; it does not apply to SUVs, minivans, or pick-up trucks. Several types of bumper system designs can be

found on passenger vehicles. One bumper system utilizes a gas strut telescoping-type design where two cylinders reside inside the bumper. These cylinders act as shock absorbers when an impact occurs. One cylinder holds nitrogen gas and the other hydraulic oil or mineral oil. As an impact occurs, the nitrogen gas in the first cylinder compresses and is forced into the second cylinder, which displaces the hydraulic or mineral oil through small valves.

Another, more common type of bumper system is composed of polypropylene foam or plastic material that is constructed in a cratelike design. In this type of bumper system, the plastic or foam compresses and absorbs force during an impact.

Located on the front top section of the vehicle is the **upper rail**. The upper rail consists of two beams located on both sides of the vehicle that hold the hood section in place and attach the front wheel strut system to the chassis **Figure 3-13 ▶**. The upper rail runs from the front bumper area to the firewall or dash area of the vehicle. Located within the beams of the upper rail may be the crumple zones.

An important area of the upper rail section is located between the **strut tower**, which is a structural component of

Figure 3-13 The upper rail is the main support structure for the front upper section of the vehicle.

Figure 3-15 Dash brackets are steel brackets designed to hold the dash in place.

the suspension system, and the dash area. This is a critical relief cut area that is associated with the dash-lift technique, which will be discussed in detail throughout Chapter 9, *Victim Access and Management*.

Moving up to the dash area are several support structures that make up the passenger compartment. Some manufacturers reinforce the dash area by installing a steel beam or bar that runs the entire length of the dash or width of the car; this is known as a **dash bar** **Figure 3-14 ▾**. Attached to the dash bar are two steel brackets sometimes called **dash brackets**. These dash brackets are located in the center console area where the radio, air conditioning control unit, and other various components are located. The brackets are bolted or welded into the floorboard of the vehicle and are designed to lock the dash in place to minimize any movement resulting from an impact **Figure 3-15 ▸**.

This causes a significant problem when the dash area needs to be displaced to gain patient access; these brackets may need to be cut in order to create enough space to lift or push the dash section away from the patient. The proper technique for gaining

access and cutting through these brackets will be discussed in Chapter 9, *Victim Access and Management*.

Rocker Panel

Running along the outermost sections of the floorboard area on both the driver and passenger sides is a channel where the doors rest; this is known as the **rocker panel**. The rocker panel is a hollow section of metal. Various items such as wiring, fuel lines, and, in hybrid vehicles, high-voltage lines can run underrneath or in close proximity to it **Figure 3-16 ▾**. There is very little structural support in this section, so be aware that it will tear or collapse very easily under the force of a hydraulic tool or the impact of a collision.

Doors

There are several key components of a door that the technical rescuer needs to be aware of when gaining access or performing a door removal. The **door hinges** allow the doors to swing

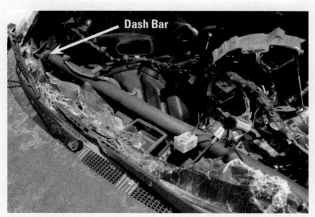

Figure 3-14 Dash bars are added to give structural support and integrity to the passenger compartment.

Figure 3-16 The rocker panel is a hollow section of metal running along the outermost sections of the floorboard area on both the driver and passenger sides. Various items, such as wiring and, in hybrid vehicles, high-voltage lines can run underneath or in close proximity to it.

A.

B.

Figure 3-17   Common door hinge designs. **A.** A leaf system. **B.** Full body system.

Figure 3-18   Door hinges sometimes come equipped with spring attachments that can fire off violently when forced.

hinges have is a spring attachment that can come off violently without warning under extreme force, such as when cutting or spreading with hydraulic tools Figure 3-18 ▲. Full personal protective equipment (PPE) should be worn at all times.

In some vehicles, a swing bar can be located between the top and bottom hinge. The **swing bar** is designed to assist the door in opening and closing Figure 3-19 ▼. The swing bar can be composed of hardened steel or other alloyed-type metals. At times, this bar can be very difficult to cut through, but it can be easily detached with very little spreading pressure from a hydraulic tool.

open or closed. The **hinges**, latching mechanisms, and locks come in various sizes as far as thickness and gauge, and also in various types. The more commonly designed hinges can be layered in a leaf system or full body with hardened steel or HSS Figure 3-17 ▲. The leaf system has two separate pieces that make up one hinge. One piece is attached to the vehicle door, and the other is attached to the body of the vehicle. The piece attached to the door slips into the center of the other piece that is attached to the vehicle body. It has a pin in the center of it that holds the two pieces together, or holds the door closed. When you look at the two pieces attached together it looks as if there are two sections of metal on the top (top leaf) and two sections of metal on the bottom (bottom leaf) of the one hinge.

The full body hinge is made up of two solid pieces of HSS and has a large pin in the center that holds them together, or holds the door closed. An example may be found on the rear doors of large vans. This hinge design is important to recognize because if a hydraulic tool is not rated to cut through the hardened steel or HSS, it will fail, sometimes breaking or shattering the blade of the tool. Another important feature that some

Figure 3-19   Swing bars are designed to assist the door in opening and closing.

There are multiple types of latching mechanisms that the technical rescuer can encounter on a door. Two common types are the Nader bolt and the U-bolt **Figure 3-20 ▼**. The Nader bolt, which is named after consumer rights advocate Ralph Nader, is composed of heavy-gauge HSS, and is round in shape with a cap at the end of it. This cap is designed to hold the door in place. This is one of the most difficult types of latching mecha-

Figure 3-21   Door impact beams are structured to absorb the impact energy of another vehicle or object and to lessen the intrusion into the passenger compartment.

nism to cut through or release from the latch mechanism of the door. The u-bolt is generally smaller-gauge steel, which makes it easier to cut through and/or release from the latch mechanism of the door.

Located inside the door is an **impact beam** that runs the entire length of the door. The impact beam can be located in both the front and rear doors. It is structured to absorb the impact energy of another vehicle or object and to lessen the intrusion into the passenger compartment. This beam can come in several different designs, including round, flat, hardened steel, and boron or titanium microalloy **Figure 3-21 ▲**.

The probability that an impact beam would need to be cut is low; because it is very difficult to cut through, doing so should be avoided if possible. Still, there always is the possibility that a front impact collision can cause an impact beam (particularly the round tubular type) to breach its outer wall and enter into the rear door or rear panel, locking the front door in place. Check with your hydraulic tool company to see if your tools are rated to cut this type of steel.

The outer skin or panel of the door is usually composed of light-gauge steel, fiberglass, or polycarbonate/plastic-type material. Saturn was a company that offered polycarbonate/plastic-type panels. These panels can be easily removed with hand tools when needed.

◼ Roof Posts

Roof posts, also known as roof pillars, are designed to add vertical support to the roof structure of the vehicle. The posts are generally labeled with an alphanumeric type description (A-B-C) **Figure 3-22 ▶**.

A-Post

The posts closest to the front windshield are known as the A-posts. The A-post can consist of several layers of steel or aluminum with some manufacturers reinforcing them with HSS, thicker-gauge metals, bars, rods, micro-alloy steel sections, or talored blanks. In addition, some manufacturers will utilize a polyurethane or similar type of foam that is either injected or placed in the roof posts and hollow cavities of the vehicle body

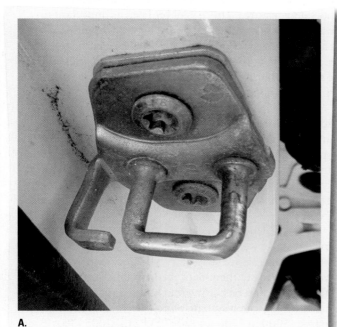

A.

B.

Figure 3-20   Two types of latching mechanisms that are commonly found in a door design. **A.** The U-bolt. **B.** Nader bolt.

Voices of Experience

**As a state fire training coordinator, I know how important it is to understand a vehicle's construction and its safety features. This has never been as important to understand as it is now. There was a day when we could approach a vehicle with our hydraulic spreaders and cutters and feel very confident that the tools would do just about anything we wanted them to do. With today's vehicle construction standards in place, the problem comes with the fire service still thinking we can do just about anything to the vehicles we are cutting apart.

Since the auto industry upgraded some of the construction standards to make the vehicles we drive safer, fire departments across the country have seen increased difficulty in extricating patients. In many cases, the auto industry is unaware that the safety features intended for the drivers and occupants exist in locations that fire fighters feel very comfortable in cutting. If we continue to treat vehicles the same, there could be disastrous results for us or the patients we are extricating.

> **"Always strip or pull the inside trim; it will tell you where you can and cannot cut."**

Now, no matter what type of vehicle we encounter, it is very important for our safety and the safety of our patients that we strip or pull the inside trim of the A, B, or C posts before we attempt to cut them with the hydraulic cutters. If we don't, there is a possibility we could be cutting into a stored gas inflator, which is part of the vehicle's air bag safety system. Stored gas inflators contain a nonflammable gas and can range from 3000 to 4500 psi (20670 kPa to 31005 kPa). If cut into, the inflators could explode and shrapnel could injure us or our patients. Be aware that these types of inflators are being used in more and more locations in the vehicle, not just the roof posts.

If you pull the inside roof trim and it comes out of the vehicle, it most likely does not have a side curtain air bag. If you pull the inside roof trim and you find it is attached to the vehicle and will not come out, it most likely has a side curtain air bag. This is because the makers of the vehicle tether the trim into place so when the air bags deploy, they do not throw the trim throughout the vehicle. Always strip or pull the inside trim; it will tell you where you can and cannot cut.

Brian Staska
Riverland Community College
Austin, Minnesota

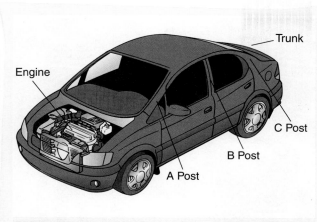

Figure 3-22 Roof posts are generally labeled with an alphanumeric type description (A-B-C).

to give added strength, support, and energy-absorbing qualities when exposed to a collision or rollover type of impact. The structural foam is designed to withstand compression and limit the vehicle body frame from folding on collision. The technical rescuer may experience difficulties in cutting through a post filled with structural foam because the product can impede the cutting action of some low-pressure hydraulic tools by resisting the compression force exerted from the tool's blades.

The technical rescuer must also be prepared for the possibility of air bag cylinders installed in any section of the A-post. All posts need to be exposed and examined for supplemental restraint system (SRS) components by removing interior liners prior to cutting. SRS components will be discussed further in Chapter 5, *Supplemental Restraint Systems*.

Rescue Tips

All posts need to be exposed and examined for SRS components by removing interior liners prior to cutting.

B-Post

Most side-impact collisions occur at the B-post area. In four-door vehicles, the **B-posts** are located between the front and rear doors of a vehicle. Understanding that this area is frequently impacted in a collision has made more manufacturers reinforce sections of the B-post from the rocker panel to the roof rail utilizing HSS, such as boron or other types of AHSS. These reinforced sections can come in the form of plates, bars, rods, tailored blanks, or multiple layered sheets of steel.

The technical rescuer can encounter several different types of seat belt mechanisms within the B-post. The most common seat belt mechanisms are the standard seat belt harness, the pretensioner seat belt system, and the automatic seat belt system.

As a collision occurs, the forward movement of the vehicle and of the passenger are independent of each other. Without the seat belt, the force that impacted the vehicle, either from another car or from striking a pole, will stop the forward movement of

that vehicle, but the passenger will keep traveling forward at the current rate of speed prior to the impact. The seat belt is designed to merge the passenger and vehicle together, combining the forward energy of the vehicle and passenger to one using the vehicle to distribute most of the force.

The **standard seat belt harness** includes a shoulder and lap belt, known as a three-point harness system. This system helps distribute the energy of a collision over larger areas of the body such as the chest, pelvis, and shoulders. The three-point belt mechanism uses a retractor gear that locks in place when activated.

The **pretensioner seat belt system** is designed to pull back and tighten when activated by a collision. The more common pretensioner seat belt system uses a pyrotechnic propulsion device to engage a retractor gear, which pulls back on the belt. The pretensioning seat belt system is normally tied in with the SRS air bag system, utilizing the same crash impact sensors to activate that SRS air bags use when a collision occurs.

The **automatic seat belt system** uses a shoulder harness that automatically slides on a steel or aluminum track system on the door window frame. When the door is closed, the shoulder harness automatically slides into place. The lap section of the harness must be manually engaged.

All seat belts have an anchoring device that is most commonly attached in two areas, the top of the B-post and either the floorboard or the bottom of the B-post. Be aware that these areas around the anchors are reinforced with heavier-gauge steel, so try to avoid cutting into these areas if possible **Figure 3-23 ▾** .

Some manufacturers install a height-adjusting anchor at the upper section of the B-post. The anchor is attached to a slide track that adjusts the seat belt to the height of the occupant. This slide track is normally heavy-gauge steel or HSS, which can cover several inches in the upper section of the B-post; this will cause cutting problems with low-pressure hydraulic tools or reciprocating saws **Figure 3-24 ▸** .

Figure 3-23 Remember when cutting through a post that the areas around the seat belt anchoring devices are reinforced with heavier-gauge steel to add support.

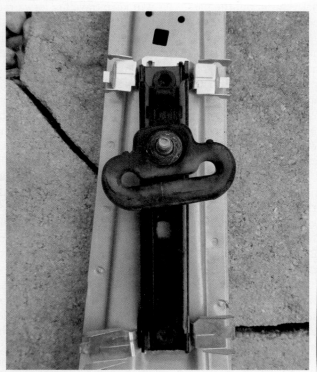

Figure 3-24   Heavy-gauge steel slide tracks for a seat belt anchoring device can extend down a post, causing cutting problems for low-pressure hydraulic tools or reciprocating saws.

Figure 3-25   Wide posts are generally two separate sections of metal that are molded together.

C-Post

The C-post in most standard two- and four-door vehicles is the rear post, with the exception of larger vehicles, such as an SUV or wagon-type vehicle, which may have numerous roof posts. In these larger vehicles, the midposts between the B-post and the rear post can contain the same components and materials as described for the previous roof posts.

Be aware that every vehicle manufacturer and vehicle design is different. Most will have the same layout, with the A-B-C post, but one manufacturer may place a steel rod in the B-post for safety, and another manufacturer may put one in the A-post or add an air bag cylinder. It is very difficult to keep up with all of the safety upgrades and changes from one vehicle to the next. You need to be cognizant of the potential to encounter one or more of these safety features anytime you cut into a vehicle. Oftentimes these added safety features are reported by technical rescuers who discover the safety features while engaged in an extrication incident; they may have cut into a post, had problems with the post, and eventually found that they were dealing with some type of HSS, boron rod, or other AHSS.

Rear posts can be wide or narrow, depending on the make and model of the vehicle. The wide posts are generally two separate sections of metal that are welded together. This forms a hollow pocket that manufacturers may use to insert structural foam, speakers, wires, or various other items Figure 3-25 ▸ . Air bag cylinders can be present in almost any area of the vehicle, so always remember to expose the post or area prior to any cutting or spreading.

Piston Struts

Hydraulic or gas-filled piston struts are becoming commonplace in most passenger vehicles, SUVs, and small trucks on the road today. Piston struts are used to assist in the lifting and support of vehicle components, such as hatchbacks, hoods, trunks, hard "tonneau" truck bed covers, tailgates, toolboxes, vehicle seatbacks, and more. Hydraulic or gas-filled piston struts are designed with a basic theory of compression: Nitrogen gas, or a petroleum-based or mineral-type hydraulic fluid, is placed in a steel or aluminum tubular cylinder. A steel rod is inserted in the cylinder with seals and compression fittings to support the movable rod from being fully ejected. The cylinder has a smaller chamber inside itself with small holes designed to slowly bleed out the fluid or gas into this chamber when the piston is compressed. Pressure is naturally built up inside the cylinder through the compression of the hydraulic fluid or nitrogen gas when the steel rod is pushed into the cylinder. This can be noticed when a hatchback is placed in the closed position. The pressure is now against the rod inside the cylinder; once the hatchback is released from the latching mechanism, the fluid or gas fills the chamber again and pushes against the rod, forcing it open, either rapidly or gradually, based on the design of the piston strut. The hatchback in the open position is now supported by the extended piston rod from the pressure release of the previously compressed fluid or gas.

The location of piston struts will vary with each manufacturer; they can be in clear view or hidden within the contour of a roof post or upper rail section of the engine compartment hood. These are areas that are normally cut into with hydraulic cutters when performing certain techniques; these areas must be exposed prior to cutting and the piston strut dealt with accordingly. Some types of struts utilize a nylon ball joint to connect the base of the cylinder to one of the sections (normally the nonmovable section) of the vehicle. The ball joints can be easily separated from their housing with a pry tool such as a

Halligan bar with very little leveraging pressure applied under the rod closest to the area of attachment. It is advised that the piston strut be removed in this fashion or cut with a hydraulic cutter at the area of attachment; avoid cutting into the pressure-filled cylinder. Accidentally cutting the cylinder section of the piston strut with a hydraulic cutter will cause a rapid release of pressure or hydraulic fluid, which can possibly cause the piston rod or a section of the piston rod to fire out of the housing unit. Proper removal of these devices will be discussed in Chapter 9, *Victim Access and Management*.

Another area of concern when dealing with piston struts is vehicle fires. Fire that impinges on a piston strut will cause rapid pressure build-up and expansion of the cylinder walls, causing the cylinder wall to breach and possibly separate or fire the piston rod out of the cylinder housing unit. It has been documented that some piston rods have penetrated through vehicle bodies and impaled walls or rescuers. Extreme caution and proper procedures must be observed when dealing with vehicle fires.

Vehicle Glass

There are several types of glass that the technical rescuer can encounter in a vehicle, including laminated safety glass, tempered safety glass, enhanced protective glass, polycarbonate, and ballistic-type glass.

Laminated Safety Glass

In 1929, the first <u>laminated safety glass</u> was produced. It was created by heating a layer of clear plastic film between two layers of plate glass. This process held the two pieces of glass together and prevented big shards of glass from flying in on the occupant. It also stopped occupants from being ejected out of the vehicle Figure 3-26 ▶ . <u>Window spidering</u> is an effect caused when an object breaks the laminated safety glass and causes spiraling rings at the area of impact resembling a spider's web.

Tempered Safety Glass

<u>Tempered safety glass</u> goes through a process where the glass is heated and then quickly cooled; this process gives the glass its strength and resistance to impacts. When tempered glass is fractured, it is designed to break into small pieces with no long shards. This is known as <u>dicing</u> Figure 3-27 ▶ .

Enhanced Protective Glass

<u>Enhanced protective glass (EPG)</u> is a modern glass that uses both the laminating and tempering process. EPG is used primarily on side and rear windows because of its security and sound-proofing qualities.

Polycarbonate

<u>Polycarbonate</u> is a clear plastic material that is very strong and can endure impacts without breaking. The big advantage of the polycarbonate material is that it is very pliable on impact, with the flexibility to conform better to the impact rather than shat-

Figure 3-26   When broken, laminated safety glass prevents big shards of glass from flying at occupants and also stops occupants from being ejected out of the vehicle.

tering as glass would. Polycarbonates can also be up to 50 percent lighter than regular glass, providing better fuel economy as well as safety features (e.g., lowering a vehicle's center of gravity and lessening the chance of vehicle rollover).

Lexan® and Makrolon® are two of the brand names used to label polycarbonate material. Until recently, polycarbonate had limited use in vehicle window design because of several problems with the material itself; it was not scratch resistant, nor could it overcome the noise vibrations from wind and basic driving that regular glass provided. In addition, polycarbonate materials used for vehicle windows could not pass all of the strict safety standards established by the National Highway Traffic Safety Administration (NHTSA).

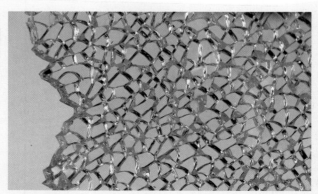

Figure 3-27   Dicing is a term used to describe the small pieces of glass that are produced from breaking tempered glass.

Newer technological advances include a system that puts a thin layer of glass over the polycarbonate material, thus reducing the scratch potential and giving the material rigidity to reduce noise and vibration problems. In 2005, this material was approved for use in all vehicle window applications as long as all of the testing requirements pass. This means that more and more manufacturers are going to be utilizing this newer material in their vehicles to promote better fuel efficiency and safety.

This material poses some difficult access problems for the technical rescuer. It cannot be broken by an impact, using a forcible entry tool will only cause the tool to bounce off. You will shatter the outer glass coating, but the flexibility of the plastic can potentially cause the tool to rebound out of your hands causing possible injury to yourself or someone else. Cutting into the material with a tool such as a reciprocating saw can, at times, cause the cut to reseal itself because of the heat generated from the blade.

The better technique to handle polycarbonate material is to treat it as a part of the vehicle body and leave it in place, removing the entire section as one, whether it is an entire roof or door structure. If there happens to be a purchase point section caused by a vehicle's crash deformity, you may be able to place the tips of a hydraulic spreader into the opening and release the section containing the polycarbonate material from its casing. A purchase point is an access area where you can insert and better position a tool for operation. Utilizing the hydraulic spreader correctly to create an opening eliminates the need for some of the traditional techniques that are taught using hand tools. The process of creating a purchase point is described in Chapter 9, *Victim Access and Management*.

Beware that any polycarbonate material that has a bend or some type of deformity caused by an impact can be potentially loaded and can release from its casing suddenly on its own or from the force of a tool. It is designed to conform back to its original shape. Chapter 9, *Victim Access and Management*, discusses glass removal techniques in detail.

■ Ballistic Glass

Ballistic glass for vehicles can be composed of several different types of materials and can vary in thickness depending on the level of protection needed. The U.S. Department of Justice, National Institute of Justice (NIJ), is one of several entities that establish the minimum performance requirements and methods of testing for levels of protection against ballistic-type weaponry. The NIJ Standard 0108.01, *Ballistic Resistant Protective Materials*, establishes five armor classifications based on the munitions size and type. Ballistic or bullet-resistant types of glass utilize multiple layers of tempered glass, laminate material, and polycarbonate thermoplastics, all sandwiched together to the desired thickness. The weight and thickness of the glass will increase depending on each increased level of protection, which can be as high as 3 or more inches (76 or more mm). Any attempts to remove or cut into this type of material are not advised; this type of glass should be handled just as the polycarbonate material, by treating it as a part of the vehicle body and leaving it in place. When removing an entire section or part of the vehicle with ballistic-rated or bullet-resistant glass, the area should be viewed as one unit, whether it is an entire roof or door structure.

Wrap-Up

■ Ready for Review

- Three concepts of energy are typically associated with injury: potential energy, kinetic energy, and work.
- Motor vehicle collisions are classified traditionally by the area of initial impact: front impact (head-on), lateral impact (T-bone, side impact), rear-end, rotational (spins), and rollovers.
- In every crash three collisions occur:
  - The collision of the vehicle against an object
  - The collision of the passenger against the interior of the vehicle
  - The collision of the passenger's internal organs against the solid structures of the body
- Before a rescuer can properly apply any extrication procedures to a vehicle, he or she must understand the inner and outer components that make up a vehicle system.
- The Department of Transportation (DOT) classifies vehicles based on whether the vehicle transports passengers or commodities, with a nonpassenger vehicle being further classified by the number of axles and unit attachments it has.
- The Department of Energy (DOE) classifies vehicles by size utilizing a cubic foot system (passenger and cargo volume) and gross weight system.
- Most vehicles on the road today are conventional-type vehicles; these types of vehicles utilize internal combustion engines (ICEs) for power. Other types of vehicles include hybrid electric vehicles, hydrogen fuel cell vehicles, and electric-powered vehicles.
- Electrical power in conventional-type vehicles with internal combustion engines is supplied by a basic 12-volt lead acid battery system. In hybrids, fuel cell vehicles, and electric vehicles, a different, advanced electrical design is used.
- The development of strong, crash-resistant vehicles requires engineers and the steel industry to develop stronger and lighter steels to meet demands.
- There are two frame systems that are most common in today's vehicles: body-over-frame construction and unitized or unibody construction. These frames can be composed of steel (most common), aluminum, or carbon fiber/composite. Another type of frame system that is less common today is the space frame, which can consist of aluminum construction or tubular steel construction.
- Several key components make up the body portion of the vehicle.
- The technical rescuer can encounter several types of glass in a vehicle, including laminated safety glass, tempered safety glass, enhanced protective glass (EPG), polycarbonate, and ballistic glass.

■ Hot Terms

Advanced High Strength Steel (AHSS) Steel with a *minimum* tensile strength of 73 ksi to 116 ksi (500 MPa to 800 MPa) or greater.

Alloyed steel Steel composed of a mixture of various metals and elements.

A-post A vertical support member located closest to the front windshield of a vehicle.

Automatic seat belt system A seat belt system that uses a shoulder harness that automatically slides on a steel or aluminum track system on the door window frame. When the door is closed, the shoulder harness automatically slides into place. The lap section of the harness has to be manually engaged.

Ballistic glass Glass that utilizes multiple layers of tempered glass, laminate material, and polycarbonate thermoplastics, all sandwiched together to the desired thickness. The weight and thickness of the glass will increase depending on each increased level of protection, which can be as high as 3 or more inches (76 or more mm).

Body-over-frame construction Vehicle design where the body of the vehicle is placed onto a frame skeleton and the frame acts as the foundation for the vehicle. The design consists of two large beams tied together by cross member beams.

B-post A vertical support member located between the front and rear doors of a vehicle.

Bumper system A feature located at the front and rear of a vehicle that helps a vehicle withstand the impact of a collision.

Chassis The frame, braking, steering, and suspension system of a vehicle.

Conventional-type vehicle A vehicle that utilizes an internal combustion engine (ICE) for power.

C-post A vertical support member located behind the rear doors of a vehicle.

Crumple zones Engineered collapsible zones that are incorporated into the frame of a vehicle to absorb energy during a collision.

Dash bar A steel beam or bar that runs partway or the entire width of the dash.

Dash brackets Two brackets that are bolted or welded into the floorboard of the vehicle that are designed to lock the dash in place in order to minimize any movement resulting from an impact.

Dicing A term used to describe the small pieces of glass that are produced when tempered glass is broken.

Door hinge A mechanism that provides the opening and closing movements for a door. Door hinges commonly range from 8- to 15-gauge metal and can be layered in a leaf system or full body.

Energy A fundamental entity of nature that is transferred between parts of a system in the production of physical change within the system and is usually regarded as the capacity for doing work.

Enhanced protective glass (EPG) A modern glass that uses both the laminating and tempering process.

High Strength Steel (HSS) Steel with a tensile strength between 39 ksi and 102 ksi (270 MPa and 700 MPa).

Hinge A mechanism that allows movable objects such as doors to join and swing open or closed.

Impact beam A steel section located within a door frame designed to absorb the impact energy of another vehicle or object and lessen the intrusion into the passenger compartment.

Internal combustion engine An engine designed to burn a multitude of petroleum-based fuels and alternative fuels, with gasoline and diesel fuel being the most commonly used.

Kinetic energy The energy of motion, which is based on vehicle mass (weight) and the speed of travel (velocity).

Ladder frame Body-over-frame construction that is referred to as a ladder frame because the cross members and beams resemble a ladder.

Laminated safety glass Glass that contains a layer of clear plastic film between two layers of glass.

Law of Conservation of Energy A law of physics stating that energy can neither be created nor destroyed; it can only change from one form to another.

Law of Motion A law of physics describing momentum, acceleration, and action/reaction.

Mechanism of injury (MOI) The way in which traumatic injuries occur; it describes the forces (or energy transmission) acting on the body that cause injury.

Nader bolt Named after consumer rights advocate Ralph Nader, a bolt composed of heavy-gauge metal that is round in shape with a cap at the end of it. It is a section of the latching mechanism.

Passenger vehicle All sedans, coupes, and station wagons manufactured primarily for the purpose of carrying passengers, including those passenger cars pulling recreational or other light trailers.

Polycarbonate A clear plastic material that is very strong and can endure impacts without breaking.

Potential energy Stored energy or the energy of position.

Pounds per square inch (psi) A unit of measure used to describe pressure; it is the amount of force that is exerted on an area equaling 1 square inch.

Pretensioner seat belt system A seat belt system designed to pull back and tighten when activated by a collision. The most common pretensioner seat belt system uses

a pyrotechnic propulsion device to engage a gear that pulls back on the belt.

Rocker panel A hollow section of metal running along the outer sections of the floorboard on the driver and passenger sides.

Roof posts Posts designed to add vertical support to the roof structure of the vehicle. These are generally labeled with an alphanumeric type description (A-B-C), starting with the first post closest to the front windshield, which is known as the A-post. Also referred to as roof pillars.

Space frame A frame made up of multiple lengths and angles of tubing welded into a rigid, but light, web or truss-like structure; the vehicle's outer panels are attached independently to the frame after its completion.

Standard seat belt harness A seat belt system that helps distribute the energy of a collision over larger areas of the body such as the chest, pelvis, and shoulders. The three-point belt mechanism uses a retractor gear that locks in place when activated. Also known as a three-point harness system.

Strut tower A structural component of the suspension system that normally has both a coil spring and shock absorber. Its main function is to resist compression.

Swing bar A hardened section of steel that is designed to assist the door in opening and closing. It can be located between the top and bottom hinge.

Tempered safety glass A type of glass that has been heated and then quickly cooled; this process gives the glass its strength and resistance to impact.

Tensile strength A measurement of the amount of force required to tear a section of steel apart.

U-bolt A latching mechanism made of a light-gauge steel (as compared to the Nader bolt). It is generally easy to cut through and/or release from the latch mechanism of the door.

Ultra High Strength Steel (UHSS) Steel with a tensile strength of 102 ksi (700 MPa) or greater.

Unibody construction A vehicle design with no formal frame structure; the body and frame are one piece, which is considered to be the structural integrity of the vehicle. The vehicle body is merged with the chassis. Also known as a unitized structure.

Upper rail Two side beams located in the front of the vehicle that hold the hood in place and attach the front wheel strut system to the chassis.

Window spidering An effect caused when an object breaks laminated glass and causes spiraling rings at the area of impact, resembling a spider's web.

Work A mechanism for the transfer of energy.

Yield strength The amount of force or stress that a section of steel can withstand before permanent deformation occurs.

Technical Rescuer *in Action*

You are dispatched to a motor vehicle collision on a rural road. On arrival you see a single vehicle upright in an embankment. The car has front end damage and a spidered windshield. You find an unrestrained driver sitting in the driver's seat. The frame looks as though it is a unibody construction, and the hood is no longer on the vehicle.

1. What type of collision was this?
A. Rear-end collision
B. Rollover crash
C. Frontal collision
D. Both B and C

2. There are two frame systems that are most common in today's vehicles; they are the unibody construction and the_____ construction.
A. ladder-type
B. aluminum
C. body-over-frame
D. synthetic wrapped frame

3. You know the unibody frame:
A. has the ability to absorb or redirect energy during a collision.
B. has a formal frame structure.
C. consists of two large beams tied together by cross member beams.
D. is sometimes referred to as a ladder frame.

4. Kinetic energy is:
A. the energy of motion.
B. the force times the speed.
C. the body in motion remaining in motion.
D. the energy that can neither be created nor destroyed.

5. Window spidering occurs with which type of glass?
A. Polycarbonate
B. Laminated safety glass
C. Tempered safety glass
D. Plate glass

6. The hood is off the vehicle. What is the main structural component that assists in holding the hood in place?
A. B-post
B. Console area
C. Dash area
D. Upper rail

7. Roof posts, also known as roof pillars, are designed to add vertical support to the roof of the vehicle. The posts are generally labeled with:
A. a color-coding system of red-green-blue.
B. a basic numbering system of 1-2-3.
C. an alpha system of A-B-C.
D. There is no system of identification.

8. The swing bar located on a vehicle door is designed to:
A. assist the door in opening and closing.
B. retain the door in place.
C. keep the door locked.
D. be used in two door models.

9. Boron is alloyed with steel during processing for its unique:
A. welding properties.
B. hardening properties.
C. stress resistance.
D. ability to remain flexible.

10. Crumple zones are found in what type of vehicle frame?
A. Unibody construction
B. Ladder frame structure
C. Monocoque-type frame
D. Aluminum frame

CHAPTER 4

NFPA 1006 Standard

There are no objectives for this chapter.

Knowledge Objectives

After studying this chapter, you will be able to:

- Explain what a manufacturer's emergency response guide is and how it is utilized by the technical rescuer. (page 56)
- Explain what a pressure release device (PRD) is. (page 57)
- Describe standard safety procedures that may be applied to all vehicles. (pages 57–58)
- Define what an alternative powered vehicle is. (pages 58–59)
- Explain the various alternative fuels as defined by The Energy Policy Act of 1992. (pages 58–67)
- Explain and identify what a vehicle identification badge is. (pages 58–59)
- Describe flexible fuel vehicles (FFVs) and emergency procedures. (page 59)
- Explain the properties of natural gas, its use as an alternative fuel, storage procedures, and emergency procedures. (pages 60–63)
- Explain the properties of liquid petroleum gas (LPG), its use as an alternative fuel, storage procedures, and emergency procedures. (pages 63–65)
- Explain the properties of biodiesel, its use as an alternative fuel, and emergency procedures. (page 65)
- Explain the properties of hydrogen, its use as an alternative fuel, storage procedures, and emergency procedures. (pages 65–67)
- Explain what a hydrogen fuel cell vehicle is and the components that make up a hydrogen fuel cell. (pages 67–70)
- Explain the emergency procedures for a hydrogen fuel cell vehicle. (pages 70–72)
- Explain what a hybrid electric vehicle (HEV) is and the components that make up a hybrid system. (pages 72–75)
- Explain the emergency procedures for a hybrid electric vehicle. (pages 75–76)
- Explain what an electric vehicle (EV) is and the components that make up an EV system. (pages 77–79)
- Explain the emergency procedures for an electric vehicle. (page 79)

Skills Objectives

After studying this chapter, you will be able to perform the following skill:

- Apply standard safety procedures to a vehicle. (pages 57–58, Skill Drill 4-1)

ou and your crew respond to a vehicle accident involving a hybrid electric vehicle that has been severely damaged.

1. Does the vehicle contain hybrid identification labeling or badging on the outside body to confirm that it is a hybrid electric vehicle?
2. Is there exposed wiring?
3. Is the vehicle still running?
4. Does the vehicle contain a smart key?

Introduction

"I do not think much of a man who is not wiser today than he was yesterday." —Abraham Lincoln

The world's reliance on petroleum-based fuels has always dominated the auto industry, but the future holds many promises for the development and use of alternative fuels. Many of these alternative fuels are in use today, and it is the responsibility of rescue personnel to familiarize themselves with these fuel types, vehicle systems, and the various emergency procedures that are utilized to manage them.

When responding to incidents involving new technology such as new vehicle design and engineering changes, responders must realize that some safety equipment designed to protect the driver and passengers actually creates additional hazards and extrication issues for the rescuers. Examples of these hazards include multiple batteries; high-voltage power cables; advanced air bag protection systems; various types of alloyed metals, including advanced high-strength steels; a reinforced passenger compartment "safety cage"; and advanced energy management systems for collisions—just to name a few.

Rescue Tips

It is important that rescuers be trained to recognize all of the hazards associated with alternative fuel vehicles and how to mitigate or neutralize these hazards before the proper rescue techniques can be applied.

The term "new technology" may be a poor choice of words; given the exponential rate at which technology grows, something labeled "new technology" may very well be obsolete the minute such words are put down on paper. It is almost impossible to keep current with all of the latest advances in the development of vehicle technology. This chapter will discuss several types of alternative fuels, vehicle propulsion systems, and other prominent advancements in vehicle technology in order to expand the rescuer's knowledge base.

The most common issue in dealing with advanced vehicle technology in vehicle extrication is trying to demystify misinformation that may be circulating around the emergency services community. Misinformation can cause emergency personnel to overreact or not react at all. Knowledge is empowering, and the best way to start acquiring the right information is by visiting a car dealership and asking about the latest hybrid vehicle, alternative fueled vehicle, or fuel cell vehicle that they have. Most dealerships will go out of their way to accommodate emergency personnel; some will provide and send a vehicle and a technician out to departments to present the information directly. Another great resource for learning about alternative fuel vehicles and various propulsion systems is the manufacturer **emergency response guides** that most vehicle manufacturers offer for emergency personnel; these guides are packed with information about the vehicle and can be accessed free of charge by download or by contacting the manufacturer. With multiple variations in the different types of fuels utilized today as well as the varying propulsion systems, it is highly recommended that technical rescuers take advantage of these learning opportunities. It is also a good idea to spend time conducting research utilizing the Internet.

Using the same emergency procedures on every incident involving one of these advanced vehicle systems is not practical and can be very dangerous; each vehicle system will require some of the steps in its emergency procedures that are unique to the particular type of fuel or propulsion system used. There are some similar steps in emergency procedures that are universal and can be applied to each vehicle type, but remember that the best practice model is to preplan by studying and training so you are better prepared to recognize differences in vehicle types, thus enabling you to adjust and apply your tactics appropriately.

Some of the standard safety procedures that can be applied to every vehicle, regardless of the type of propulsion system, are defined in the following Skill Drill. While these basic safety guidelines can be followed for all vehicle types, be aware that various vehicles will have additional or unique emergency procedures that are specific to that particular type of vehicle. It must be pointed out that some of these steps are not necessarily completed in succession; some of them can be completed simultaneously, depending on the number of personnel on scene who are available to be assigned to the step.

Follow the steps in **Skill Drill 4-1 ▾** to apply some of the standard safety procedures to a vehicle, regardless of the type of propulsion system the vehicle utilizes:

1. Don the appropriate personal protective equipment (PPE), including self-contained breathing apparatus (SCBA) if called for, and clear the scene of all hazards and bystanders.

2. Approach the scene from upwind and uphill, and conduct the inner and outer scene surveys with atmospheric monitoring, if available, if alternative fuel is suspected.

3. Set the hazard control zones (hot, warm, and cold), appropriate to the type of hazard recognized/confirmed. Utilize the **Department of Transportation's (DOT)** *Emergency Response Guidebook (ERG)* to determine the initial zone diameter.

4. Look for any visible vapor clouds and listen for a loud hissing noise, which may indicate product release through a leak or through a **pressure release device (PRD)**.

5. Stage two charged 1¾-inch (44-millimeter [mm]) hose lines. The first line is for protection of personnel, and the second line, if needed, is to control the dispersion of escaping vapors. One charged 1¾-inch (44-mm) hose line is the minimum standard of protection for all vehicle extrication incidents, regardless of vehicle type.

6. Turn off the ignition switch of the vehicle. This simple action turns off the engine and the electric motor(s), preventing the high-voltage electrical current from flowing into the cables and thus shutting down the fuel supply. After you turn off the ignition switch, remove the key so the car cannot be accidentally restarted. **Smart keys** must be moved out of operational range, a minimum of 15 feet (5 meters [m]) from the vehicle. (**Step 1**)

7. Stabilize the vehicle from movement with cribbing. Avoid placing any cribbing under high-voltage wiring, alternative fuel supply lines or cylinders, and high-voltage battery packs.

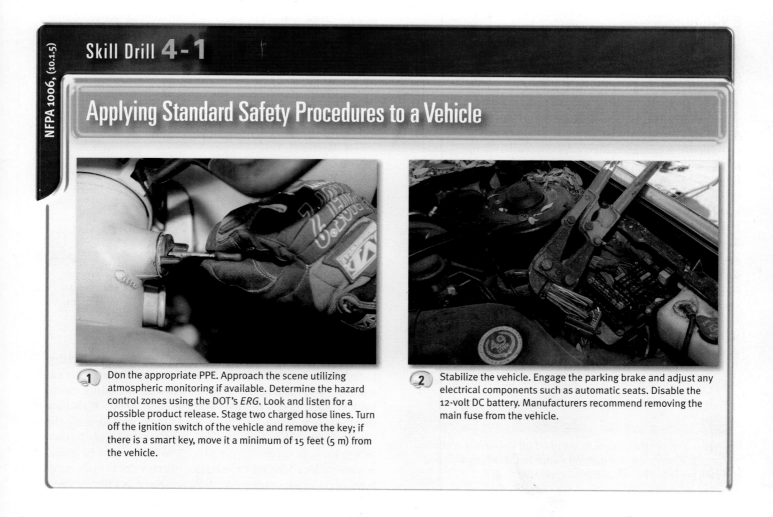

NFPA 1006, (10.1.5)

Skill Drill 4-1

Applying Standard Safety Procedures to a Vehicle

1 Don the appropriate PPE. Approach the scene utilizing atmospheric monitoring if available. Determine the hazard control zones using the DOT's *ERG*. Look and listen for a possible product release. Stage two charged hose lines. Turn off the ignition switch of the vehicle and remove the key; if there is a smart key, move it a minimum of 15 feet (5 m) from the vehicle.

2 Stabilize the vehicle. Engage the parking brake and adjust any electrical components such as automatic seats. Disable the 12-volt DC battery. Manufacturers recommend removing the main fuse from the vehicle.

8. Engage the parking brake. Some parking brakes are electrically controlled, so this procedure should be done before disabling the 12-volt battery.

9. Attempt any necessary component adjustments before disabling the power to the vehicle such as activating the electric parking brake, adjusting power seats, and releasing hatchbacks.

10. Disable the 12-volt direct current (DC) battery by disconnecting the negative and then the positive battery cables.

11. Manufacturers recommend removal of the main fuse from the vehicle to ensure that the electrical system is disabled. This will be a decision for the authority having jurisdiction (AHJ), as location of fuses can vary greatly among models. **(Step 2)**

12. Some manufacturer's recommend using service disconnects as indicated in the manufacturer's emergency response guides. This will be up to the AHJ, as locations of service disconnects can vary greatly among models.

> **Rescue Tips**
>
> Never stand in front of or to the rear of a hybrid electric vehicle before determining that the power has been shut off.

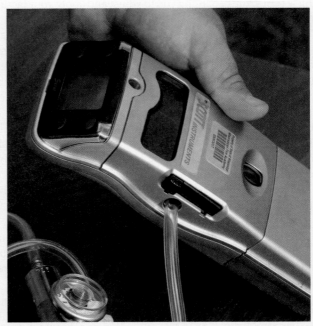

Figure 4-1 Multi-gas meters provide information about hazardous atmospheres.

Safety

The practice of atmospheric monitoring at vehicle accidents should be considered when dealing with alternative vehicles and fuel cell vehicles, especially with a heavier-than-air fuel such as liquid petroleum gas (LPG). Unfortunately, not every agency carries the proper equipment to conduct atmospheric monitoring. A multi-gas meter will detect the presence of hazards in the air **Figure 4-1 ▶**. Note, some variables may produce inaccurate readings or no reading at all, such as high wind, high humidity, or lack of proper calibration; even so, it still should be considered as a safety measure for ensuring a safe working environment. Additional sensors that detect the presence of hydrogen can be added to some models as an optional feature. Some may say that atmospheric monitoring is unnecessary because high-pressure fuel systems under compression, such as hydrogen, will rapidly empty their contents and disperse upward into the atmosphere before emergency personnel arrive on scene. Nonetheless, atmospheric monitoring is an added safety practice; if you have the equipment on the apparatus, then it should be utilized. Atmospheric monitoring for combustible gases should be reviewed and decided upon by the agency's AHJ. If utilized, it should be incorporated into any emergency procedures for dealing with alterative fueled or fuel cell vehicles.

> **Rescue Tips**
>
> Never assume that a vehicle is turned off because it is silent or still. Always turn the ignition switch to the off position and then move the key away from the vehicle.

Alternative Fuels

<u>Alternative powered vehicles</u> are vehicles that use fuels other than petroleum or a combination of petroleum and another fuel for power. The Energy Policy Act of 1992 outlines a list of fuels that can be classified as an "alternative fuel" for vehicles:

- Hydrogen
- Electricity
- Biodiesel
- LPG (Propane)
- Methanol
- Ethanol
- Natural gas

> **Rescue Tips**
>
> Rescue personnel are responsible for situational awareness and for notifying command of the type of vehicle or aftermarket products that may be in use, such as nitrous oxide or any other add-ons.

A multitude of alternative fuel variations are available today. This text outlines a few of the more prevalent alternative fuels and the alternative fueled vehicles, including transit vehicles such as school buses and commercial buses, that are in production. Most vehicle manufacturers will identify the vehicle type or fuel type through a labeling process known as a <u>vehicle identification badge</u>. The badges can be found at various locations on the exterior of the vehicle body, with some of the more common symbols presented as a blue triangle

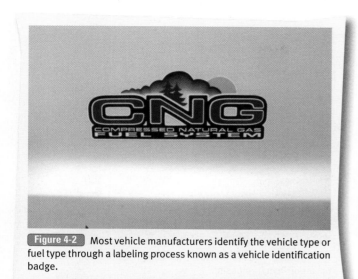

Figure 4-2 Most vehicle manufacturers identify the vehicle type or fuel type through a labeling process known as a vehicle identification badge.

with the letters "CNG" for compressed natural gas, a green leaf with the letters "FCV" for fuel cell vehicle, or simply the word "Hybrid" on the sides or rear of the vehicle Figure 4-2 . Be aware that badge labeling is not standardized and can vary in design from one manufacturer to the other, or may not be used at all. For some vehicles, there are NFPA requirements for the use of identification labeling. This will be discussed later in the chapter. It is recommended that the technical rescuer scan the outside of the vehicle to identify any vehicle identification badges prior to working on the vehicle.

Rescue Tips

Vehicle identification badges are not standardized and can vary in design from one manufacturer to the other, or may not be used at all.

▇ Ethanol and Methanol

Ethanol is a fuel comprised of an alcohol base that is normally processed from crops such as corn, sugar, trees, or grasses. Ethanol is also known as a grain alcohol and is denatured to prevent human consumption. This fuel can be blended in several different percentages with other fuels such as gasoline.

E10, better known as gasohol, is a blend of 10 percent ethanol and 90 percent gasoline. This blend is classified by the Environmental Protection Agency as "substantially similar" to gasoline and is not considered an alternative fuel. All auto manufacturers approve the use of blends of 10 percent or less in their gasoline vehicles.

Another ethanol option is E85 flex fuel. E85 contains 85 percent ethanol and 15 percent gasoline. E85 is classified as an alternative fuel and is used to fuel E85-capable flexible fuel vehicles (FFVs). FFVs are capable of running on gasoline alone or on the E85 blend of up to 85 percent ethanol and 15 percent gasoline.

Methanol, like ethanol, is an alcohol-based fuel. Methanol is known as a wood alcohol because it is processed from natural wood sources such as trees and yard clippings. Methanol can also be utilized as a flex fuel in a ratio of 85 percent methanol and 15 percent gasoline, better known as M85. Since the early 1990s, the use of methanol has dramatically declined in the United States; it is, however, widely used outside of the United States. Methane is the most common gas used to extract or separate the hydrogen gas from within it through a steam reforming or electrolysis process. This process will be discussed in the hydrogen gas section.

More vehicle manufacturers are offering FFVs, and most label the vehicle with a flex fuel badge on the side or rear of the vehicle Figure 4-3 ; as of 2008, most manufacturers have also started using yellow gas caps to indicate this distinction, but remember that this is not standardized among manufacturers.

Ethanol and Methanol Emergency Procedures

Because both ethanol and methanol are alcohol-based fuels, when their content is greater than 10 percent of the fuel mixture, they require an alcohol-resistant foam as the effective method of fire extinguishment. In the case of a breached fuel tank, vapor suppression and the use of diking procedures may be necessary to contain run-off. Diking is the placement of materials to form a barrier that will keep a liquid hazardous material from entering an area or that will hold a liquid hazardous material in a given area. Both fuels are also miscible in water and will separate from the gasoline blend when water is applied. With the exception of a vehicle fire, the emergency procedures will be the same as a standard conventional vehicle.

▇ Natural Gas

Natural gas is a fossil fuel primarily composed of methane that can be utilized as compressed natural gas or liquefied natural gas. Although known as one of the cleanest burning alternative fuels, natural gas vehicles are not produced commercially in large numbers but are steadily growing each year.

Figure 4-3 A flex fuel identification badge.

When natural gas is processed and cooled to a temperature below −259° Fahrenheit (F) (−161° Celsius [C]), it turns into a cryogenic liquid and can be utilized as liquefied natural gas in vehicles that have been modified to run on this fuel. Liquefied natural gas (LNG) is a colorless, odorless, nontoxic gas that floats on water and is lighter than air when released as a vapor. LNG has an expansion vapor ratio that is 600 to 1 and a flammability ratio of 5 to 15 percent. When LNG flows out of a tank due to a leak, it will form a liquid pool and then boil off into gas form. Because of its cryogenic state and large expansion ratio, as LNG changes to a gas and releases into the atmosphere, it condenses the dry ambient air surrounding it, causing a visible vapor cloud to appear. This vapor cloud is not necessarily comprised of natural gas, as the actual gas can travel well ahead or behind the cloud depending upon the general wind currents.

Avoid using water on an LNG leak or fire because the warm water will cause the liquid gas to violently react with an instant boil-off, causing a sudden expansion and vaporization of the liquid, which will intensify a fire or cause an explosion. High-expansion foam or dry chemicals are best utilized for this type of incident in place of water. A hose stream should only be directed toward a vapor cloud to disperse the product; be cognizant of the PRD location and avoid directing the stream toward this area, which could freeze the device and render it inoperable. PRDs rapidly release product through a small metal tube attachment when detecting excessive amounts of heat at a preset temperature.

The fuel tank needed to store LNG to keep it in its cryogenic liquid state must be double-walled and well insulated to prevent a boil-off. This makes the tanks very bulky, limiting space in a vehicle; these types of tanks are commonly placed in trunks. LNG is stored at low pressures in tanks up to a maximum pressure of 230 pounds per square inch (psi) (1585 kilopascals [kPa]) and is regulated back down to an operating pressure up to 120 psi (827 kPa). Remember that LNG is stored as a cryogen, which is a liquid that boils at temperatures below −256° F (−160° C). It is in a constant state of boil-off and is consistently changing into a gas. As a result, there will always be an increase in the tank vessel pressure, which is regulated by a PRD. As the liquid is drawn from the tank to the vehicle's internal combustion engine, it travels through the fuel lines and is heated by a heating mechanism to change it to its useable gaseous state for the internal combustion engine. NFPA 52, *Vehicular Gaseous Fuel Systems Code*, covers all of the requirements for fuel storage and use of LNG-fueled vehicle systems.

Compressed natural gas (CNG) is much more practical and is utilized primarily as a fuel type for many fleet vehicles found on the roadways today. CNG storage tanks are comprised of steel, aluminum, or carbon fiber/composite Figure 4-4 ▶. According to the DOT, these tanks must go through extensive crash and drop tests to ensure durability. In passenger vehicles, these fuel tanks are normally located behind the rear passenger seat or in the trunk area or under the vehicle in some fleet vehicles.

To achieve the desired pressures for storage, natural gas is compressed to pressures ranging from 3000 to 3600 psi (20,684

Figure 4-4   CNG storage tanks are comprised of steel, aluminum, or carbon fiber/composite.

to 24,821 kPa) and may have to be stored in several onboard tanks to achieve the same mile range as gasoline Figure 4-5 ▼. These high storage pressures are regulated down at the engine to workable pressures. Stainless steel high-pressure lines run under the vehicle from the tanks to the engine compartment Figure 4-6 ▶. Several safety features are built into both a CNG fuel system and an LNG fuel system; these features will vary depending on the manufacturer. One such safety design occurs when the vehicle's ignition is turned off; a sensing unit or valve will turn off the fuel at the tanks, stopping any flow of fuel from escaping. This sensing unit or valve will also engage and shut off the tank when any leak is detected. Depending on the manufacturer, each tank will have its own manual shut-off valve that can be accessed from under the vehicle or through the trunk area, if the tank is placed in the trunk Figure 4-7 ▶. This will cut off all

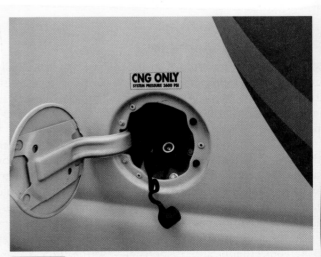

Figure 4-5   To achieve the desired pressures for storage, natural gas is compressed to pressures ranging from 3000 to 3600 psi (20,684 to 24,821 kPa) and may have to be stored in several tanks to achieve the same mile range as gasoline.

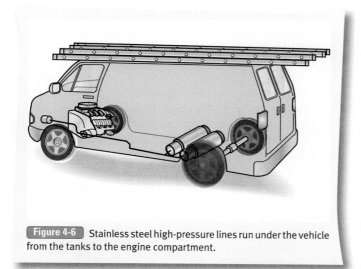

Figure 4-6 Stainless steel high-pressure lines run under the vehicle from the tanks to the engine compartment.

fuel from the tanks to the engine. Another safety feature is the PRD. The PRD is designed to rapidly release all of the gas when exposed to high temperatures, such as exposure to fire, which causes an overpressurization of the cylinder and actuates the PRD. PRDs are also required to be vented to the outside of the vehicle with the relief valve discharge points directed upward or downward within 45 degrees of vertical. Emergency personnel should be aware of these discharge points. Remember, natural gas is lighter than air and will dissipate when released into the atmosphere. As another safety measure for CNG-type fuels, a chemical odorant called ethyl-mercaptan is added, primarily to detect any potential leaks.

According to NFPA 52, *Vehicular Gaseous Fuel Systems Code*, vehicles that utilize both CNG and LNG as fuels are required to be clearly marked with an identification label adhered to the right lower rear section of the vehicle. This identification label should have a diamond shape with the letters CNG or LNG in white or silver reflective lettering against a blue or dark background Figure 4-8 ▶ .

Natural Gas Emergency Procedures

It is highly recommended that the technical rescuer review some of the various manufacturer emergency response guides for dealing with vehicles that utilize LNG or CNG as a fuel. Also the DOT's *ERG*, particularly guide 115, is a great reference for initial planning Figure 4-9 ▶ . Some of the guides recommend steps for mitigating any potential problems or leaks before attempting to extricate any patients. This will be determined by the incident commander; quick hazard and risk analyses will have to be determined prior to taking action. The hazard analysis identifies situations or conditions that may injure people or personnel, or may damage property or the environment. The risk analysis assesses the risk to the rescuers compared to the benefits that might come from the rescue. Hazard and risk analyses are continual processes that are reevaluated throughout the entire incident.

General emergency procedures when dealing with CNG or LNG incidents include the following. Note, some of these steps are not necessarily completed in succession; some of them

A.

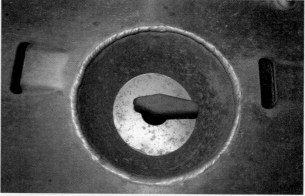

B.

Figure 4-7 Depending on the manufacturer, each tank will have its own manual shut-off valve. **A.** The manual shut-off valve can be accessed from under the vehicle or through the trunk area, if the tank is placed in the trunk. **B.** The manual shut-off valve.

can be completed simultaneously, depending on the number of personnel on scene who are available to be assigned to complete each step.

- Don the appropriate PPE, including SCBA, and clear the scene of all bystanders and hazards.
- Establish hazard control zones (hot, warm, and cold). Utilize the DOT's *ERG* book as a reference guide to determine initial zone sizing.
- Conduct the inner and outer surveys. If possible, approach the vehicle from upwind and uphill and from

Figure 4-8 The natural gas identification label should have a diamond shape with the letters CNG or LNG in white or silver reflective lettering against a blue or dark background.

the sides because the CNG and LNG tanks are commonly stored in the trunk area, behind the rear seat, or under the vehicle. Look for a visible vapor cloud and listen for a hissing noise, which may indicate product release through a leak or through a PRD.

■ Utilize a combustible gas meter, if carried on an apparatus, to detect possible leaks and concentrations of combustible gases or vapors in the surrounding atmosphere. This is a continuous process until the leak has been contained and stopped.

■ Look for a vehicle identification badge; in this situation, it may be a blue or dark-colored diamond shape with the letters "CNG" or "LNG" in white or silver.

■ Two 1¾-inch (44-mm) charged hose lines should be deployed to protect personnel and to disperse any significant release of LNG vapor or CNG product to keep it below the flammable range. Fires involving vehicles utilizing CNG as the fuel source should not be extinguished until the leak can be isolated and eliminated or the fuel tank containing the product can be shut down.

■ When the scene is safe, ensure that the vehicle's ignition is turned off, the keys are out of the ignition, and the vehicle is placed in park.

■ Stabilize the vehicle from movement with cribbing. Avoid placing any cribbing under high-voltage wiring, alternative fuel supply lines or cylinders, and high-voltage battery packs.

■ Attempt, if needed, any necessary component adjustments before disabling the power to the vehicle, such as activating the electric parking brake, adjusting power seats, and releasing hatchbacks.

■ Disconnect the 12-volt DC battery starting with the negative line first. (Location of the 12-volt DC battery will vary on each model of vehicle.)

■ Manually turn off the gas at the tanks utilizing the shut-off valves.

■ Manufacturers recommend removal of the main fuse from the vehicle to ensure that the electrical system is disabled. This will be an AHJ decision, as location of fuses can vary greatly among models.

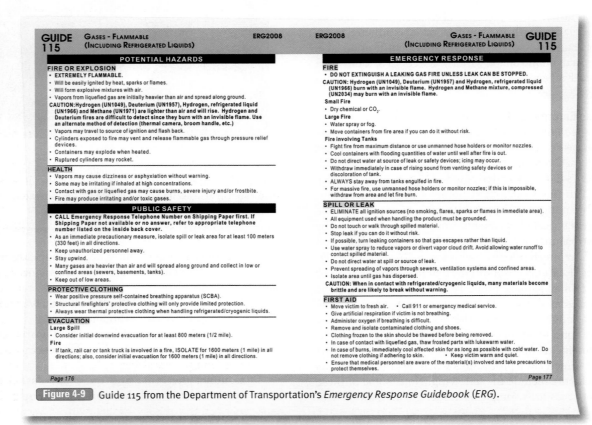

Figure 4-9 Guide 115 from the Department of Transportation's *Emergency Response Guidebook* (ERG).

For CNG to be combustible, it must fall within its flammability range. Flammability range refers to the amount of a gas that must be present in the surrounding air for combustion to occur. For CNG, this amount is 5 to 15 percent. Thus, if an air mixture contains 4 percent CNG, it will not support combustion; likewise, if an air mixture contains 16 percent CNG, it will also not support combustion. The same procedures are to be followed for PRDs that are releasing product with visible flame. Do not extinguish the flame; instead cool the tank or eliminate the flame impingement, and the PRD will reset and self-extinguish once the tank and product are cooled. The high pressure of CNG tanks, 3000 to 3600 psi (20,684 to 24,821 kPa), should cause the contents to release quickly, and the fire at the point of product release should self-extinguish. Other exposures such as the vehicle itself, additional vehicles, or surrounding structures can be protected and/or extinguished with hose streams if needed. Until the leak can be isolated and eliminated, it is safer to let the product burn itself out.

When attempting an extrication technique on a CNG- or LNG-fueled vehicle utilizing power or hand tools, it is best practice to ensure that the emergency procedures discussed in Skill Drill 4-1 are followed, which includes wearing full protective gear, including SCBA, to ensure safety. Also, examine the vehicle carefully for fuel line locations before attempting any cutting. The techniques that will require extra caution will be any of the dash displacement techniques where relief cuts are made or where tearing will occur when the dash is released or pushed forward. In addition, techniques that involve cutting and dropping the floorboard area under the brake and gas pedal should not be attempted. Such maneuvers could expose high-voltage wires or gas lines running under the rocker panel channel area, which should be avoided regardless of the type of fuel system the vehicle utilizes, or whether the power supply has been secured. Alternative techniques that are safer should be considered.

■ Liquid Petroleum Gas

Liquid petroleum gas (LPG), also known as propane, is a fossil fuel produced from the processing of natural gas and is also produced as part of the refining process of crude oil. Propane is the third most widely utilized fuel source behind gasoline and diesel; it is commonly used with forklifts and other similar work units.

Propane is heavier than air (1.5 times as dense) and will sink and pool at floor level when released into the atmosphere. Propane fuel tanks that are designed for passenger vehicles are built according to the specifications and standards set by the American Society of Mechanical Engineers (ASME) *Boiler and Pressure Vessel Code.*

In compliance with the *Boiler and Pressure Vessel Code* of the ASME, propane vehicle tanks are constructed from carbon steel. These tanks are designed to be 20 times more puncture-resistant than a standard gasoline tank. ASME tanks used for passenger vehicles vary in size but cannot exceed 200 gallons (757 liters) according to NFPA 58, *Liquefied Petroleum Gas Code.* The fill capacity of an ASME tank uses water in gallons as a unit of measurement; this is expressed in the letter markings on the tank normally stenciled in WC (water capacity). Because of the high expansion rate of propane (270 parts gas to 1 part liquid), when a tank is filled there is a mandatory 20-percent reduction in product to account for this expansion space. To ensure that the proper expansion space is provided, all tanks are designed with an overfill prevention device that limits a tank to 80- to 85-percent capacity, leaving a 20-percent vapor space just above the fuel line. With the 20-percent reduction, a tank that has a rated WC of 100 gallons can be filled to hold 80 gallons (303 liters) of propane. Propane weighs approximately 4.24 pounds per gallon (1.9 kilograms [kg] per 3.8 liters).

As used with CNG fuels, a chemical odorant called ethylmercaptan is added to propane as a safety measure to detect any potential leaks. Propane tanks are also equipped with a PRD to release any overpressure due to high temperature exposure such as flame impingement. Pressure relief valves are also required to be vented to the outside of the vehicle with the relief valve discharge points directed upward or downward within 45 degrees of vertical. Emergency personnel should be aware of these discharge points.

Propane is normally a vapor gas at temperatures above its boiling point (−44° F or −42° C). When stored under pressure, it is compressed into a liquid state and will remain in a liquid state. When a leak occurs, propane will rapidly convert back to a gaseous state, expanding 270 times its original liquid volume. Rapid release of propane from the tank into the atmosphere will cause frost to accumulate at the liquid level on the outside of the tank. This frost is produced because of the rapid drop in pressure and temperature inside the tank. On an outside tank used to supply fuel to a building or facility, a frost line may be readily visible, but because a tank or cylinder may be concealed in the vehicle, the technical rescuer may not have access to visualize this frost line to determine the amount of remaining product in the tank.

The required working pressure of an ASME tank supplying propane as fuel for passenger vehicles is designed to provide either 250 psi (1724 kPa) if constructed prior to April 1, 2001, or 312 psi (2151 kPa) if constructed on or after April 1, 2001. This is a significant difference from natural gas, which is stored under pressures of 3000 to 3600 psi (20,684 to 24,821 kPa).

Passenger vehicle propane tanks must contain an identification label on the tank itself that displays at a minimum the following information: water capacity, working pressure, serial number of the tank, and manufacturer.

NFPA 58, *Liquefied Petroleum Gas Code,* also requires that main shut-off valves on a container for liquid or vapor be accessible and operated without the use of any tools, or that there be specific equipment provided that can shut off the valve. Also, any LPG container that is permanently installed in a vehicle requires that vehicle to be clearly marked with an identification label adhered to the right lower rear section of the vehicle. This identification label should have a diamond shape with the word "Propane" in white or silver reflective lettering against a black or dark background Figure 4-10 ▶. Even with this type of labeling requirement in place, because of the inconsistencies that can occur with vehicle identification badging, it is not recommended to rely on these types of identification markings as the only sign that you are dealing with an alternative fueled

Figure 4-10 Any vehicle with a permanently installed LPG container must be clearly marked with a diamond shape with the word "Propane" in white or silver reflective lettering against a black or dark background.

vehicle. Always use precaution and monitor the environment with a gas reading meter to determine if any flammable vapors exist upon approach.

Proper labeling may be missing from a vehicle that has had aftermarket alterations. The fuel system may have been converted from a gasoline engine to an alternative fuel engine, such as propane or natural gas. There are some conversion kits that allow the user to run both types of fuel systems; when the gasoline tank runs low, a switch can be flipped so the vehicle runs on propane from an onboard aftermarket-installed tank. These types of propulsion systems will more than likely not have any identification badging on the outside of the vehicle to warn responders.

Liquid Petroleum Gas Emergency Procedures

As just mentioned with natural gas emergency procedures, it is highly recommended that the technical rescuer review emergency response guides pertaining to propane-fueled vehicles, which are provided by some vehicle manufacturers. Also, the DOT's *ERG*, particularly guide 115, is a great reference for initial planning. Some of the guides recommend steps for mitigating any potential problems or leaks before attempting to extricate any patients. This will be determined by the incident commander; quick hazard and risk analyses will have to be determined prior to taking action. Keep in mind again that propane is $1\frac{1}{2}$ times heavier than air and will accumulate in low-lying areas such as a ditch or lower road embankment, or it can accumulate in confined spaces such as passenger compartments or truck cargo areas. Propane disperses well beyond its vapor cloud and will seek out ignition sources, which can cause a flashback to the leak.

If there is a substantial breach in the tank, there will be an initial blow-off of product to reduce the overall tank pressure. There can be a freezing of the product with a slower leak.

General emergency procedures when dealing with vehicles using LPG as a fuel include the following. Note, some of these steps are not necessarily completed in succession; some of them can be completed simultaneously, depending on the number of personnel on scene who are available to be assigned to complete each step.

- Don the appropriate PPE, including SCBA, and clear the scene of all bystanders and hazards.
- Establish hazard control zones (hot, warm, and cold). Utilize the DOT's *ERG* as a reference guide to determine initial zone sizing.
- Conduct the inner and outer surveys. If possible, approach the vehicle from upwind and uphill and from the sides because like CNG tanks, LPG tanks may be stored in the trunk area. A loud hissing noise may indicate a leak or the PRD rapidly expelling product from the tank. Unlike CNG and compressed hydrogen gas, which both rise and dissipate quickly, propane is heavier than air and will linger and accumulate in lower areas or confined spaces, causing a flashback to the leak or product release point. The technical rescuer must ensure that the area is safe to work in before attempting to operate on the vehicle.
- Utilize a combustible gas meter, if carried on an apparatus, to detect possible leaks and concentrations of combustible gases or vapors in the surrounding atmosphere. This is a continuous process until the leak has been contained and stopped.
- Look for any vehicle identification badging on the outside body of the vehicle, which can indicate LPG on board. This label must have a diamond shape with the word "Propane" in white or silver reflective lettering against a black or dark background.
- Two $1\frac{3}{4}$-inch (44-mm) charged hose lines should be deployed to protect personnel and to disperse any significant release of propane product to keep it below the flammable range.
- When the scene is safe, ensure that the vehicle's ignition is turned off, the keys are removed from the ignition, and the vehicle is placed in park.
- Stabilize the vehicle from movement with cribbing. Avoid placing any cribbing under high-voltage wiring, alternative fuel supply lines or cylinders, and high-voltage battery packs.
- Attempt, if needed, any necessary component adjustments before disabling the power to the vehicle, such as activating the electric parking brake, adjusting power seats, and releasing hatchbacks.
- Disconnect the 12-volt DC battery starting with the negative line first. (Location of the 12-volt DC battery will vary on each model of vehicle.)
- Manually turn off the gas at the tanks utilizing the shutoff valve. If the tanks are located in the trunk, remember that there is the potential for a large amount of propane product to accumulate in this space.
- Manufacturers recommend removal of the main fuse from the vehicle to ensure that the electrical system is disabled. This will be an AHJ decision, as location of fuses can vary greatly among models.

Fires involving vehicles equipped with propane can present unique challenges to the rescuer. For propane to be combustible, it must fall within its flammability range, which is 2.15 to 9.6 percent. Thus, if an air mixture contains only 2 percent propane, it will not support combustion; likewise, if an air mixture contains 10 percent propane, it will also not support combustion. Continuous flame impingement on a tank containing propane will cause the overpressurization and eventual failure of the tank in an explosive reaction known as a <u>BLEVE</u>, or boiling liquid expanding vapor explosion. This reaction occurs when a pressurized liquified material starts to boil from the flame impingement and releases vapor, which in turn sets off the pressure relief valve. When the pressure relief valve can no longer compensate for the expanding vapor, which is 270 parts vapor to 1 part liquid, the tank will start to bulge and eventually fail at its weakest point, releasing a fireball of product and tank pieces. To counter this, hose streams must be directed toward the vapor space of the tank to cool the tank until the pressure relief valve is reseated; the fire is then extinguished. If there is a leak with an active fire, or the PRD has a flame coming out of it, the fire must not be extinguished or the product will just be free to release and seek out a new ignition source. The tank must be cooled, the leak must be stopped, or the main shut-off valve must be turned off before the fire is extinguished.

When attempting to extricate on an LPG-equipped vehicle utilizing power or hand tools, the same precautions should be utilized as were described for CNG-equipped vehicles. Ensure that the emergency procedures discussed in Skill Drill 4-1 are followed. This includes wearing full protective gear, including SCBA to ensure safety. Also, examine the vehicle carefully for fuel line locations before attempting any cutting. The techniques that will require extra caution will be any of the dash displacement techniques where relief cuts are made or tearing will occur when the dash is released or pushed forward. In addition, techniques that involve cutting and dropping the floorboard area under the brake and gas pedal should not be attempted. Such maneuvers could expose high-voltage wires or gas lines running under the rocker panel or channel area. This should be avoided regardless of the type of fuel system the vehicle utilizes or whether the power supply has been secured. This is not a safe practice.

Biodiesel

<u>Biodiesel</u> is a fuel used solely for diesel engines that is processed from domestic renewable resources such as plant oils, grease, animal fats, used cooking oil, and, more recently, algae. Biodiesel can be utilized by itself as a diesel fuel (B100—100 percent biodiesel), although B100 is not recommended for use in low temperatures. Biodiesel can also be blended with petroleum diesel at varying percentages, B2 (2 percent biodiesel), B5, B20. Biodiesel produces less air pollutants than petroleum-based diesel, is safe, nontoxic, and biodegradable.

The DOT does not classify biodiesel as a flammable liquid because it has a high flash point (199°F [93°C]). As compared to other fuels, biodiesel has a flash point that is much higher than that of petrodiesel (100°F [38°C]) or gasoline (−45°F [−43°C]).

Figure 4-11 An example of a biodiesel identification badge.

Biodiesel-blended fuels will act as a hydrocarbon-type fuel or a polar solvent depending on the fuel's purity and type of blend. It is best to use an aqueous film-forming foam (AFFF) because this type of foam system can be used on both polar solvent and hydrocarbon-based fires.

An example of a biodiesel identification badge is shown in Figure 4-11 ▲.

Biodiesel Emergency Procedures

Emergency procedures for vehicles utilizing biodiesel as a fuel are similar to those for a standard conventional vehicle. Because of its high flash point, biodiesel is not considered a flammable liquid. However, it will still burn and will require proper emergency procedures for handling a spill, which, again, are similar to those for handling a hydrocarbon spill. Using a foam blanket with an alcohol-resistant aqueous film firefighting foam (AR-AFFF) to suppress vapors is the best practice, including encompassing any diking procedures to control run-off.

Rescue Tips

Alcohol-resistant aqueous film firefighting foam (AR-AFFF) is best used on a biodiesel or biodiesel blended fuel spill to suppress vapors from finding any ignition source.

Hydrogen

<u>Hydrogen</u> is one of the most abundant elements on earth. It is an odorless, colorless, flammable, nontoxic gas that combines easily with other elements. Hydrogen came into the spotlight in 1937 after the tragedy of the Hindenburg airship Figure 4-12 ▶. The airship utilized hydrogen gas to create its buoyancy, but during a maneuver to dock the ship, static electricity caught the ship's outer liner on fire. The product used to coat the liner was extremely flammable and ignited, easily consuming the entire vessel. The hydrogen gas in the ship escaped almost immediately once the shell was breached. Hydrogen was proven not to be the culprit in this accident, which was the belief for many decades.

Figure 4-12 Hydrogen came into the spotlight in 1937 after the Hindenburg tragedy.

Figure 4-13 The identification label for hydrogen may have the word "Hydrogen" in white or silver reflective lettering or may have the word "Hydrogen" somewhere on the vehicle.

Hydrogen is relatively buoyant, being 14 times lighter than air; it rises and disperses at a rate of 44 miles per hour (mph) (71 kilometers per hour [kph]) at approximately 66 feet (20 m) per second; thus its contents are quickly emptied and the flammability concentrations dissipated on smaller tanks. Hydrogen has a very wide flammability range, between 4 and 75 percent in air, and its flame burns almost clear, making it very difficult to see when ignited during daylight hours. The dispersion rate of hydrogen is two times faster than helium and six times faster than natural gas. There is no chemical odor additive in hydrogen because the dispersion rate and buoyancy are too great for any odor chemical to keep up with it. The expansion volume ratio of liquid hydrogen into a vapor or gaseous state is 1 part liquid to 848 parts vapor; this is normally expressed as 1:848.

Hydrogen is unique because it can be produced domestically from multiple resources such as fossil fuels, plants/algae, or water. In the United States, 95 percent of the hydrogen used today is produced from the methane in natural gas using a high-temperature steam process called steam methane reforming. This process separates the hydrogen from the methane.

Hydrogen gas can be utilized directly on a modified internal combustion engine (ICE) or it can be used as a catalyst for producing electricity in a fuel cell to run a vehicle. Hydrogen fuel cell vehicles will be discussed later in this chapter. Vehicles fueled by hydrogen are required to be clearly marked with an identification label adhered to the right lower rear section of the vehicle. Using the same labeling identification markings found with CNG, LNG, and propane, the identification label for hydrogen may have the word "Hydrogen" in white or silver reflective lettering against a blue or dark background or may have the word "Hydrogen" somewhere on the vehicle **Figure 4-13 ▶**.

Rescue Tips

Hydrogen gas can be utilized directly on a modified internal combustion engine or it can be used as a catalyst for producing electricity in a fuel cell to run a vehicle.

Hydrogen Storage Tanks

There are several ways to store hydrogen on a passenger vehicle when used as a fuel:

- Hydrogen can be stored as a liquid, but must be cooled to −423° F (−253° C) or it will boil off as a gas.
- Hydrogen can be compressed and stored in high-pressure storage tanks with pressures of 3600 psi (24,821 kPa), 5000 psi (34,474 kPa), and 10,000 psi (68,948 kPa).
- Hydrogen can be chemically combined in hydride form with certain metals, which can store it more compactly and efficiently than in a gas form.
- Hydrogen can also be stored in the microscopic pores of carbon nanotubes.

Hydrogen storage tanks must conform to several safety standards. Because of its low-molecular structure, hydrogen has very strict storage requirements; hydrogen can cause certain metals to become brittle and fail. Currently, hydrogen storage tanks must meet the federal government's Federal Motor Vehicle Safety Standard 304 (49 CFR 571.304), *Compressed Natural Gas Fuel Container Integrity*. The International Association for Natural Gas Vehicles lists four specific pressure cylinder types for various applications and fuel storage such as CNG, LNG, and hydrogen. These are the various types of fuel tanks that a rescuer can encounter on an alternative fuel vehicle or fuel cell vehicle.

- **Type 1:** This cylinder type is composed of steel only. Only paint covers the outside of the cylinder. This is the most common type of cylinder.
- **Type 2:** This cylinder type is composed of steel or aluminum with a partial hoop wrap that goes around the cylinder. The wrapping material, which goes over the sidewall, can be made of fiberglass or carbon fiber.
- **Type 3:** This cylinder type is composed of the same material as the Type 2 cylinder; however, the wrapping encompasses the entire tank, including the domes. This type of cylinder has a metal liner, usually made of aluminum.

- **Type 4:** This cylinder type has a nonmetallic liner, usually plastic, and is fully wrapped, including the domes, with the same kind of material used for the Type 2 cylinder.

New standards for hydrogen gas vehicles are currently being drafted.

Hydrogen Emergency Procedures

Emergency procedures for vehicles that utilize hydrogen as a fuel are similar to CNG and LNG procedures. It is highly recommended that the technical rescuer review emergency response guides dealing with vehicles that utilize hydrogen as a fuel. Also, the DOT's *ERG*, particularly Guide 115, is a great reference for initial planning. Most guides recommend that any potential problems or leaks be mitigated before attempting to extricate any patients. This will be determined by the incident commander; quick hazard and risk analyses will have to be determined prior to taking action. The hazard analysis identifies situations or conditions that may injure people or personnel, or may damage property or the environment. The risk analysis assesses the risk to the rescuers compared to the benefits that might come from the rescue. Hazard and risk analyses are continual processes that are reevaluated throughout the entire incident.

General emergency procedures when dealing with hydrogen-fueled vehicle incidents include the following. Note, some of these steps are not necessarily completed in succession; some of them can be completed simultaneously, depending on the number of personnel on scene who are available to be assigned to complete each step.

- Don the appropriate PPE, including SCBA, and clear the scene of all bystanders and hazards.
- Establish hazard control zones (hot, warm, and cold). Utilize the DOT's *ERG* as a reference guide to determine initial zone sizing.
- Conduct the inner and outer surveys. If possible, approach the vehicle from upwind and uphill and from the sides because hydrogen tanks can be stored in the trunk area or just below the rear section of the vehicle. Look for any visible vapor clouds and listen for hissing noises, which indicate product release through a leak or through a PRD.
- Utilize a hydrogen-specific gas meter, if carried on an apparatus, to detect possible leaks and concentrations of combustible gases or vapors in the surrounding atmosphere. This is a continuous process until the leak has been contained and stopped.
- Look for a vehicle identification badge; in this situation, it may be the word "Hydrogen" in white or silver reflective lettering against a blue or dark background or may have the word "Hydrogen" somewhere on the vehicle.
- Two 1¾-inch (44-mm) charged hose lines should be deployed to protect personnel and to disperse any significant release of vapor from liquid hydrogen or a compressed hydrogen gas product to keep it below the flammable range.
- When the scene is safe, ensure that the vehicle's ignition is turned off, the keys are removed from the ignition, and the vehicle is in park.

- Stabilize the vehicle from movement with cribbing. Avoid placing any cribbing under high-voltage wiring, alternative fuel supply lines or cylinders, and high-voltage battery packs.
- Attempt, if needed, any necessary component adjustments before disabling the power to the vehicle, such as activating the electric parking brake, adjusting power seats, and releasing hatchbacks.
- Disconnect the 12-volt DC battery starting with the negative line first. (Location of the 12-volt DC battery will vary on each model of vehicle.)
- Manually turn off the gas at the tanks utilizing the shut-off valves.
- Manufacturers recommend removal of the main fuse from the vehicle to ensure that the electrical system is disabled. This will be an AHJ decision, as location of fuses can vary greatly among models.
- Some manufacturers recommend utilizing service disconnects as indicated in the manufacturer's emergency response guide for that particular vehicle. This will be an AHJ decision, as location of service disconnects can vary greatly among models.

Fires involving vehicles utilizing hydrogen gas as the fuel source, whether as a compressed gas or a cryogenic liquid, should not be extinguished until the leak can be isolated and eliminated or the fuel tank containing the product can be shut down. The same procedures are to be followed for PRDs that are releasing product with visible flame. Do not extinguish the flame; instead, cool the tank or eliminate the flame impingement, and the PRD will reset and self-extinguish once the tank and product are cooled. Remember that pressure relief valves are also required to be vented to the outside of the vehicle, with the relief valve discharge points directed upward or downward within 45 degrees of vertical. Emergency personnel should be aware of these discharge points. The high pressure of hydrogen storage tanks (3000 to 10,000 psi [20,684 to 68,948 kPa]) causes the contents to release quickly, and the fire at the point of product release should self-extinguish. Other exposures such as the vehicle itself, additional vehicles, or surrounding structures can be protected and/or extinguished with hose streams if needed, but until the leak can be isolated and eliminated, it is safer to let the product burn itself out.

Hydrogen Fuel Cell Vehicles

A <u>fuel cell</u> is an electrochemical device that combines hydrogen and oxygen to produce electricity to power a motor or generator, with the by-products of this process being water and heat. In a vehicle powered by a fuel cell, the electric motor is powered by electricity generated by the fuel cell. The fuel cell uses hydrogen that is stored in an onboard tank combined with the outside oxygen to produce the electricity. Hydrogen fuel cell vehicles are potentially two to three times more efficient than conventional vehicles, emitting little to no greenhouse gas emissions. The space industry has used this technology for a number of years.

The first fuel cell was developed in 1839 by Sir William Grove, who is also known as the "father of the fuel cell." He

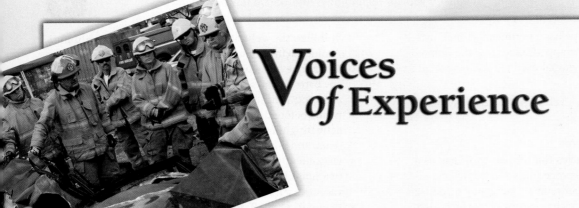

Voices of Experience

One unit responding to a car fire gave an on-scene arrival transmission of "Engine 4 on scene of a vehicle fire . . . *God Almighty*!" and the transmission stopped. Upon arrival, the battalion chief found the crew fighting a fully involved vehicle fire. The vehicle was on its roof in the middle of the intersection.

After the incident was under control and the fire extinguished, the chief discussed the incident with the captain. The captain explained that when they arrived the engine of the vehicle was on fire, but before they were able to lay a line, the car exploded, flipped over, and was immediately fully involved. Conversation with the vehicle owner led to the discovery that her husband used propane from the main tank at their farm to inflate the tires.

This incident did not involve advanced vehicle technology or an alternative powered vehicle; however, it serves as a reminder to responders to always be prepared to find dangerous fuels where you least expect them to be.

J. T. Cantrell
Pulaski County Office of Emergency Management
Little Rock, Arkansas

> **"The captain explained that when they arrived the engine of the vehicle was on fire, but before they were able to lay a line, the car exploded, flipped over, and was immediately fully involved."**

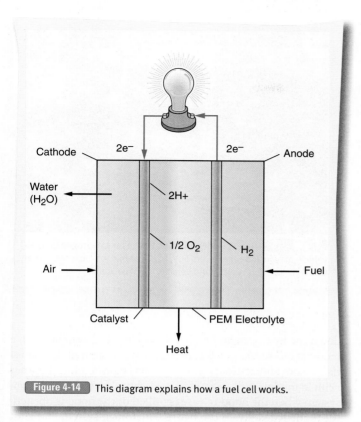

Figure 4-14   This diagram explains how a fuel cell works.

Labels in figure: Cathode, $2e^-$, $2e^-$, Anode, Water (H_2O), $2H+$, Air, $1/2$ O_2, H_2, Fuel, Catalyst, PEM Electrolyte, Heat

referred to his invention as the gas voltaic battery, which many years later was changed to the fuel cell.

Four basic elements make up a fuel cell: the anode, the cathode, the electrolyte, and the catalyst **Figure 4-14 ▲**. In a fuel cell, an electrolyte membrane called a **polymer exchange membrane**, or proton exchange membrane (PEM), is placed between an anode (negative electrode) and a cathode (positive electrode). The PEM exchanges positive electrons. The process begins with hydrogen from an onboard storage tank that enters the anode and is split into positive ions and negative electrons. The positive hydrogen ions pass through the PEM, which is only permeable to positive ions, and combine at the cathode with the oxygen supplied from outside air, creating the by-product, which is water. The water is either used in some other area of the vehicle for cooling or is omitted out the tailpipe of the vehicle. The negative hydrogen electrons are then used to provide the electrical current to power the vehicle. Heat is also created from the chemical reaction, which requires the use of a coolant (water) to keep the fuel cell at the proper temperature. A separator is used to ensure that the positive and negative elements are routed to the correct paths. This entire process encompasses just one fuel cell. Each cell produces approximately 0.7 to 1.1 volts of electricity.

Hydrogen Fuel Cell Vehicle Electrical Design

A fuel cell vehicle by definition is a hybrid vehicle system in which two separate sources of power are used individually or combined as a propulsion mechanism for the vehicle. For

example, hydrogen fuel compressed as a gas or liquefied and stored in an onboard cylinder along with electrical energy produced from a separate battery pack are examples of two separate sources of power that are used individually or combined to propel the motor.

The basic components of a fuel cell vehicle system consist of a fuel cell module pack/stack, electric motor, generator, hydrogen storage system, and battery pack. Multiple fuel cells are stacked together and placed in a series where one complete system for a vehicle can consist of over 300 to 400 individual cells producing over 400 volts of DC electricity. The vehicle will also utilize a battery pack and/or large storage capacitor (ultracapacitor) for energy reserve to make the vehicle more efficient. The battery packs are configured with the same design as the ones found in the hybrid electric vehicle system; they are comprised of several stacked cells consisting of nickel metal hydride (NiMH) or lithium ion batteries. These batteries can produce upwards of 300 or more volts of DC electricity, depending on the amount of individual cells utilized. The vehicle will also be equipped with a standard 12-volt DC lead-acid battery that can be used for the initial starting of the vehicle or can offer auxiliary power for various electrical components. This standard 12-volt DC lead-acid battery is the basic two-pole design with a black negative pole and a red positive pole. The location of the 12-volt DC battery will vary with each model of vehicle. Most fuel cell vehicles come equipped with a regenerative energy braking system that utilizes or captures the kinetic energy produced when the vehicle's brakes are applied. The energy produced from the brakes is used to recharge the battery pack or is stored in the onboard energy storage capacitor.

Hydrogen Storage System for a Fuel Cell Vehicle

A fuel cell design will normally utilize a compressed hydrogen gas system comprised of several Type 3 or 4 cylinders as opposed to a liquefied hydrogen cryogenic storage system, which requires extreme temperatures (−423° F [−253° C]) to keep the hydrogen in the liquid state. The compressed hydrogen gas storage cylinders come in pressures of 3600 psi (24,821 kPa), 5000 psi (34,474 kPa), and 10,000 psi (68,948 kPa), which then have to be regulated down to a nominal pressure so the gas can be utilized as it enters the fuel cell module. Reinforced framing material is commonly added to the existing frame of the vehicle to protect the storage cylinders against impacts **Figure 4-15 ▶**.

Hydrogen lines running from the cylinders in the rear of the vehicle to the fuel cells in the front of the vehicle are typically routed underneath the vehicle outside the passenger compartment. Some vehicle models, such as the Toyota Fuel Cell Hybrid Vehicle-Advanced (FCHV-adv), use a red color-coding system on these lines for identification purposes, but this is not a standardized practice among manufacturers.

All hydrogen storage tanks come equipped with a PRD or **temperature relief device (TRD)**, which rapidly releases the product through a small metal tube attachment when detecting excessive amounts of heat at a preset temperature. The hydro-

Figure 4-15 Reinforced framing material is commonly added to the existing frame to protect the storage cylinders against impacts.

Figure 4-16 A vehicle identification badge for a hydrogen fuel cell vehicle may include the letters FCV, FCHV, or FCX, or the words "Fuel Cell Vehicle," "Fuel Cell Hybrid Vehicle," or "Fuel Cell."

gen can take up to several minutes to release all of its contents and is identified by a loud hissing noise when activated. PPE, including SCBA, should be worn when approaching the vehicle; as mentioned earlier in the chapter, hydrogen burns clean with an almost clear flame that is difficult to see. A liquid hydrogen release will form a visible vapor cloud because the cryogenic state of the hydrogen temporarily freezes the air around the product as it is rapidly released in the atmosphere. It is not recommended to direct a water stream into this area because of the possibility of freezing/icing the PRD and blocking the product release or the release of pressure from the cylinder. Also, remember never to direct a hose stream into the liquid state of hydrogen because the warmer water will immediately cause a rapid boil-off phase, instantly vaporizing and expanding the liquid hydrogen into its gaseous state. A fog pattern can be directed to disperse any visible vapors.

Rescue Tips

Because of the possibility of freezing, it is not recommended to direct a water stream at an activated PRD of a liquid hydrogen storage cylinder that is expelling product.

There are numerous safety features built into the fuel cell vehicle, that vary among different models. Some of these features include hydrogen leak detectors placed in strategic locations throughout the vehicle; these sensors, when detecting a leak, will deactivate the hydrogen storage system, stopping the flow of hydrogen gas through the lines. Electronic control units (ECUs), in some models, continuously monitor temperatures and pressures for the entire hydrogen system including its components; any irregularity or leak detected will cause the ECU to shut down the hydrogen system and the medium- to high-voltage electrical system. Most vehicle models incorporate a type of crash detection system that deactivates and shuts

down the hydrogen and high-voltage electrical system when a moderate to severe crash is detected; the system utilizes inertia sensors and activates when any air bag is deployed. Other vehicle models will deactivate and shut down the hydrogen system when the hood release is pulled or the hood is opened. All models deactivate the hydrogen storage system and shut down the gas lines, including the medium- to high-voltage lines, when the ignition key is in the off position and/or the key is removed. Some storage cylinders come equipped with manual shut-off valves that can be accessed from underneath or from the side of the vehicle; research the various models' emergency response guides to find out the type and location of the manual shut-off valves.

Hydrogen fuel cell vehicles are generally recognizable through vehicle identification badges located on the sides, front hood, and rear of the vehicle. Examples of the letters that may be seen include FCV, FCHV, or FCX, or the words "Fuel Cell Vehicle," "Fuel Cell Hybrid Vehicle," or "Fuel Cell" **Figure 4-16**. Again these identification badges are not standardized among manufacturers.

Hydrogen Fuel Cell Emergency Procedures

Emergency procedures for fuel cell vehicles that utilize an onboard storage tank consisting of hydrogen as a catalyst to generate electricity will be similar to the procedures of CNG and LNG systems. It is highly recommended that the technical rescuer review emergency response guides dealing with vehicles that utilize hydrogen. Also, the DOT's *ERG*, particularly Guide 115, is a great reference for initial planning. Most guides recommend that any potential problems or leaks be mitigated before attempting to extricate any patients. This will be determined by the incident commander; quick hazard and risk analyses will have to be determined prior to taking action. The hazard analysis identifies situations or conditions that may injure people or personnel, or may damage property or the environment. The risk analysis assesses the risk to the rescuers compared to the benefits that might come from the rescue. Hazard and risk analyses are continual processes that are reevaluated throughout the entire incident.

General emergency procedures when dealing with hydrogen fuel cell vehicle incidents include the following. Note, some of these steps are not necessarily completed in succession; some of them can be completed simultaneously, depending on the number of personnel on scene who are available to be assigned to complete each step.

- Don the appropriate PPE, including SCBA, and clear the scene of all bystanders and hazards.
- Establish hazard control zones (hot, warm, and cold). Utilize the DOT's *ERG* as a reference guide to determine initial zone sizing.
- Conduct the inner and outer surveys. If possible, approach the vehicle from upwind and uphill and from the side or corners of the vehicle because of unexpected lunging or reversing of the vehicle. The vehicle may be silent and appear to be turned off because it is void of an internal combustion engine, which produces the noise we are accustomed to hearing. Remember that the system can still be in an all-electric "ready" mode, capable of engaging the drive or reverse motor at any moment. This is similar to an electric golf cart that is always in a "ready" mode accelerating when the pedal is depressed. Look for any visible vapor clouds and listen for loud hissing noises, which indicate product release through a leak or through a PRD.
- Utilize a hydrogen-specific gas meter, if carried on an apparatus, to detect possible leaks and concentrations of combustible gases or vapors in the surrounding atmosphere. This is a continuous process until the leak has been contained and stopped.
- Look for a vehicle identification badge; in this situation, it may be the letters FCV, FCHV, or FCX, or the words "Fuel Cell Vehicle," "Fuel Cell Hybrid Vehicle," or "Fuel Cell." Or, there may be a blue triangle with "Hydrogen" written in white on the rear trunk area.
- Two 1¾-inch (44-mm) charged hose lines should be deployed to protect personnel and to disperse any significant release of vapor from liquid hydrogen or compressed hydrogen gas product to keep it below the flammable range.
- Never assume that the vehicle's ignition is turned off because it is silent or still.
- When the scene is safe, manually turn the ignition key to the off position, remove the keys, and be sure the vehicle is in park. Various emergency response guides state that smart keys need to be a minimum of 15 feet (5 m) away from the vehicle, which removes them from operational range. This is common for most vehicles now. Place the parking brake on, if accessible.
- Stabilize the vehicle from movement with cribbing. Avoid placing any cribbing under high-voltage wiring, alternative fuel supply lines or cylinders, and high-voltage battery packs.

- Manually engage the hood release device, which, in some vehicle models, disengages the hydrogen system and the medium- to high-voltage electrical system.
- Attempt, if needed, any necessary component adjustments before disabling the power to the vehicle, such as activating the electric parking brake, adjusting power seats, and releasing hatchbacks.
- Safely disengage the 12-volt DC battery, which will shut down the flow of hydrogen gas and medium- to high-voltage electricity. Remove or cut the negative cable first and ensure that the cable does not fall back on the vehicle and make contact with the frame in any way. A 12-volt DC battery can be located in various locations in the vehicle, depending on the manufacturer and vehicle model. Traditional locations include under the engine compartment hood, inside one of the wheel wells, under the backseat, or in the trunk. *Warning*: Energy capacitors in some models can hold power for 5 to 10 minutes after the power has been disengaged.
- Manufacturers recommend removal of the main fuse from the vehicle to ensure that the electrical system is disabled. This will be an AHJ decision, as location of fuses can vary greatly among models.
- Manually shut off the cylinder tank valve if the cylinder comes equipped with one.
- Some manufacturers recommend utilizing service disconnects as indicated in the manufacturer's emergency response guide for that particular vehicle. This will be an AHJ decision, as location of service disconnects can vary greatly among models.

In a fuel cell vehicle, the medium- to high-voltage wires run along the undercarriage of the vehicle, normally on the opposite side of the hydrogen gas lines; this can vary with each vehicle model. These wires are normally protected by some framing material and/or are wrapped in a protective casing. Remember that these wires have ground-fault and short-circuit protection; if there is a break in the line, a relay will kick in and open the circuit, which will isolate and disable the voltage. With various fuel cell vehicle models, fully discharging the voltage from the capacitor can take from 5 to 10 minutes once a breach in the system is detected or the power is turned off.

When performing various extrication techniques, precaution must be observed when the vehicle is overturned on its roof. If the decision to perform a tunneling technique through the trunk area is considered, remember that the hydrogen storage tanks and battery packs can be placed in the trunk area, under the rear seat section, or under the vehicle, so an alternative method may be safer. Also, when considering a dash displacement technique, such as a dash lift technique, or the dash roll technique, remember that the hydrogen gas lines run along the undercarriage up into the motor/generator compartment, and the medium- to high-voltage wires also run on the undercarriage from the battery pack in the rear to the motor/generator compartment located at the front of the vehicle. Opening up the dash area can potentially expose the technical rescuer to the hydrogen gas lines and the medium-

and high-voltage wires. The hydrogen system and the power to medium- and high-voltage wires should have been isolated and disconnected prior to performing any extrication, but it is still not recommended to cut into any medium- to high-voltage wire or hydrogen gas lines regardless of whether the power has been neutralized or the hydrogen system has been disconnected. In addition, techniques that involve dropping the floorboard area under the brake and gas pedal should not be attempted; such maneuvers could expose high-voltage wires or gas lines running under the rocker panel/channel area, which should be avoided regardless of the type of fuel system the vehicle utilizes or whether the power supply has been secured. This is not a safe practice.

Hybrid Electric Vehicles

A **hybrid electric vehicle (HEV)** is defined as a vehicle that combines two or more power sources for propulsion, one of which is electric power **Table 4-1 ▾**. Several different types of vehicles on the roadway today are designed with a propulsion system that uses either a combination of electric power and another type of fuel (such as gasoline) or electric power only. Some of the abbreviations used to label these vehicles include PHEV (plug-in hybrid electric vehicle) and EREV (extended range electric vehicle).

Conventional or alternative fuels can be combined with electric power, which is stored in a battery pack system inside

Table 4-1   Summary of Hybrid Electric Vehicles (HEVs)

Manufacturer	Model	Type	Class	Year	Web Link to Vehicle Information
Audi	Q7 TDI Hybrid	SUV	HEV	Concept	www.audiusa.com
BMW	ActiveHybrid 7	Sedan	HEV	2010	www.bmwusa.com
	X6 Hybrid	SUV	HEV	2010	
Cadillac	Escalade Hybrid	SUV	HEV	2009	www.cadillac.com
Chevrolet	Malibu	Sedan	HEV	2009	www.chevrolet.com/hybrid
	Silverado Hybrid	Pickup	HEV	2009	
	Tahoe Hybrid	SUV	HEV	2009	
Chrysler	Aspen Hybrid	SUV	HEV	*D/NLP	www.chrysler.com/en/2009/aspen/hybrid/
Dodge	Ram Hybrid	Pickup	HEV	2010	www.dodge.com
	Durango Hybrid	SUV	HEV	*D/NLP	
	Grand Caravan Hybrid	Van	HEV	Concept	
Ford	Reflex	Coupe	HEV	Concept	www.ford.com
	Fusion Hybrid	Sedan	HEV	2009	
	Escape Hybrid	SUV	HEV	2009	
GMC	Sierra Hybrid	Pickup	HEV	2009	www.gmc.com
	Yukon Hybrid	SUV	HEV	2009	
Honda	CR-Z Hybrid	Coupe	HEV	2010	www.honda.com
	Civic Hybrid	Sedan	HEV	2009	
	Insight	Sedan	HEV	2009	
	Fit Hybrid	Sedan	HEV	2010	
	Accord Hybrid	Sedan	HEV	*D/NLP	
Hyundai	Sonata Hybrid	Sedan	HEV	2010	www.hyundaiusa.com
	Accent Hybrid	Sedan	HEV	2010	
Infiniti	M35 Hybrid	Sedan	HEV	2011	www.infinitiusa.com
Lexus	HS 250h	Sedan	HEV	2009	www.lexus.com
	GS 450h	Sedan	HEV	2009	
	LS 600h L	Sedan	HEV	2009	
	RX 450h	SUV	HEV	2009	
	RX 400h	SUV	HEV	2009	
Mazda	Tribute HEV	SUV	HEV	2009	www.mazdausa.com
Mercedes	S400 Blue Hybrid	Sedan	HEV	2009	mbusa.com
	ML 450 Hybrid	SUV	HEV	2009	
Mercury	Milan Hybrid	Sedan	HEV	2009	www.mercuryvehicles.com
	Mariner Hybrid	SUV	HEV	2009	
	Meta One	Van	HEV	Concept	

Continues

Table 4-1   Summary of Hybrid Electric Vehicles (HEVs) (Continued)

Manufacturer	Model	Type	Class	Year	Web Link to Vehicle Information
Nissan	Altima Hybrid	Sedan	HEV	2009	www.nissanusa.com
Porsche	Cayenne S Hybrid	SUV	HEV	2010	www.porsche.com
Saab	BioPower Hybrid	Sedan	HEV	Concept	www.saabusa.com
Toyota	Volta	Coupe	HEV	Concept	www.toyota.com/hsd
	A-BAT Hybrid Truck	Pickup	HEV	Concept	
	Prius	Sedan	HEV	2009	
	Camry Hybrid	Sedan	HEV	2009	
	Hybrid X	Sedan	HEV	Concept	
	Highlander Hybrid	SUV	HEV	2009	
	Sienna Hybrid	Van	HEV	Concept	
Volkswagen	Touareg Hybrid	SUV	HEV	2011	www.vw.com
Volvo	3CCC	Coupe	HEV	Concept	www.volvo.com

*D/NLP: Discontinued/No Longer Produced

Source: Reprinted with permission from *Fire Fighter Safety and Emergency Response for Electric Drive and Hybrid Electric Vehicles,* Copyright © 2010, The Fire Protection Research Foundation, Quincy, MA. This reprinted material is not the complete and official position of The Fire Protection Research Foundation on the referenced subject.

the vehicle. Hybrid technology has been around for over a century **Figure 4-17 ▼**. The first U.S. patent issued for a hybrid vehicle was in 1909 by German-born Henri Pieper. Hybrid technology is used to power advanced systems found in submarines (nuclear/electric) and trains (diesel/electric) down to the simplest forms of transportation such as the Moped (gasoline/electric). HEVs have the power of conventional vehicles, are economical, and produce lower emissions.

The battery system generally used to supply power to the HEV is a nickel-metal hydride (NiMH) battery or a lithium ion battery configured in a pack of individual cells **Figure 4-18 ▶**. Each battery cell, depending on its size, will produce 1 to 1.5 volts of DC power. These cells are then combined and encased in modules and set in series, producing the specific voltage or DC power depending on the type of HEV system design. The HEV

Figure 4-18   The battery system generally used to supply power to the HEV is configured in a pack of individual cells.

design may utilize a high-voltage (> 60 volt DC) or a medium- to low-voltage system. The NiMH battery carries a small amount of an alkaline base electrolyte composed of sodium and/or potassium hydroxide. This electrolyte is absorbed into the cell plates and should not pose a leak hazard if the case is damaged.

Full and Mild Hybrid Electric Vehicle Designs

HEVs are unofficially classified as being full or mild in terms of how they utilize the electric power that is generated. A **full hybrid** vehicle can use either its electric motor or an internal combustion engine (ICE), or a combination of both, to propel itself. In contrast to the full hybrid, the **mild hybrid** vehicle must use electric power in conjunction with the ICE for vehicle propulsion.

Figure 4-17   One of the first hybrid vehicles, designed in the early 1900s.

Table 4-2 Summary of Plug-in Hybrid Electric Vehicles (PHEVs)

Manufacturer	Model	Type	Class	Year	Web Link to Vehicle Information
Cadillac	Converj	Sedan	PHEV	Concept	www.cadillac.com
Chevrolet	Volt	Sedan	PHEV	2010	www.chevrolet.com/hybrid
Fisker	Karma	Luxury	PHEV	2010	karma.fiskerautomotive.com
Ford	Escape Plug-in Hybrid	SUV	PHEV	2012	www.ford.com
GMC	Plug-in Crossover SUV	SUV	PHEV	2011	www.gmc.com
Toyota	Prius Plug-in	Sedan	PHEV	2012	www.toyota.com/hsd
Volvo	V70 Plug-in Hybrid	Van	PHEV	2012	www.volvo.com

Source: Reprinted with permission from *Fire Fighter Safety and Emergency Response for Electric Drive and Hybrid Electric Vehicles*, Copyright © 2010, The Fire Protection Research Foundation, Quincy, MA. This reprinted material is not the complete and official position of The Fire Protection Research Foundation on the referenced subject.

Mild hybrid vehicles can be subdivided into two additional categories: the start/stop mild hybrid and the integrated motor assist mild hybrid that is used in Honda vehicles. The **start/stop mild hybrid** is not a true hybrid system by definition because the motor/generator is not used to propel the vehicle, it is designed to turn off the vehicle's ICE when the vehicle is idle, and it will turn the vehicle's ICE back on when the accelerator is activated. The **integrated motor assist (IMA) mild hybrid** system is also designed to start and stop the HEV's ICE, and, in addition, it will assist the ICE when acceleration is needed. The IMA HEV system will also power electrical devices such as the air conditioning, power steering, and various electronics in the vehicle, and it can store and utilize the vehicle's regenerative braking energy.

Mild hybrids generally work off of a low- to medium-voltage range. The voltage for these types of hybrids can range from 36 to 42 volts DC, utilizing a series of NiMH batteries or an advanced lead-acid battery known as a valve-regulated lead-acid battery. The NiMH battery in the full hybrid produces a high voltage for propulsion and for powering various components in the range of 144 to 300 volts DC, with some vehicles using a step-up inverter capable of producing power up to 650 volts DC.

All full and some mild HEVs come equipped with a regenerative energy braking system that utilizes or captures the kinetic energy produced when the brakes are applied. The energy produced by the braking system is used to recharge the battery pack or is stored in the onboard energy storage capacitor.

All HEVs come equipped with the standard 12-volt DC lead-acid battery that can be used for the initial starting of the vehicle or offer auxiliary power for various electrical components. This standard 12-volt DC lead-acid battery is the basic two-pole design with a black negative pole and a red positive pole. The location of the 12-volt DC battery will vary with each model of vehicle.

Plug-in Hybrid Electric Vehicle

A **plug-in hybrid electric vehicle (PHEV)** is simply a hybrid vehicle with the ability to recharge its battery system using a plug-in cord that can run off general house current in the range of 120 volts, also known as a Level 1 charging system **Table 4-2 ▲**. A 240-volt charging system is known as a Level 2 charging system, which can cut the charging time in half for some models. There is also a Level 3 or DC Fast Charge sys-

tem that ranges up to 480 volts, which is generally beyond the household current level and is normally seen for commercial use only. A Level 1 charging system for a PHEV can take up to 8 hours or longer, depending on the vehicle manufacturer, to fully charge the vehicle. Some Level 3 or DC Fast Charge systems can charge a battery pack to 80 percent capacity in 15 minutes, but again, this depends on the type of system setup and the vehicle manufacturer.

A PHEV gives consumers greater flexibility because the vehicle, when fully charged, can run solely on electric power over short distances. The PHEV has the same components and features as a similar HEV model of vehicle with the exception of a larger battery pack to extend the vehicle's urban driving range on battery power only. There is a potential problem for responders when the vehicle is plugged in and charging from the house current: A rescuer responding to a vehicle fire in the garage of a single-family home must disconnect the power at the vehicle or to the residence to disable the electrical feed to the vehicle before attempting to gain entry into the vehicle to complete any interior extinguishment.

Extended Range Electric Vehicle

The **extended range electric vehicle (EREV)** utilizes a series-type propulsion system that allows the vehicle to run on all battery or all electric power until it is near depletion, which occurs in the range of 40 miles or more depending on the manufacturer. An onboard gasoline engine acts as a generator, thus extending the range up to several hundred miles. The EREV can also be plugged into a power grid to fully recharge its high-voltage battery. A popular vehicle in the EREV class is the Chevrolet Volt.

■ Voltage Color Coding

In some mild hybrid systems, a low- or medium-voltage cable (36 to 42 volts DC) can be identified by a blue cable, and a high-voltage cable is always identified by an orange cable **Figure 4-19 ▶**. The color orange for high-voltage cables is an industry standard. Currently there is no industry color standard for medium- or intermediate-voltage cables, but they can appear in yellow and/or blue depending on the manufacturer.

One mild hybrid vehicle utilizes an inverter, which converts the current from 36 volts DC (blue cable in some models)

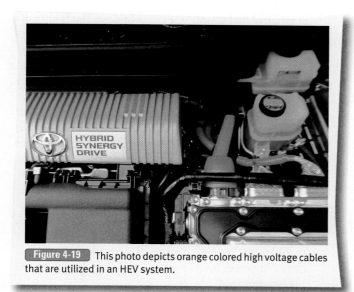

Figure 4-19 This photo depicts orange colored high voltage cables that are utilized in an HEV system.

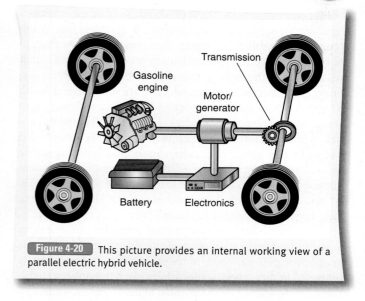

Figure 4-20 This picture provides an internal working view of a parallel electric hybrid vehicle.

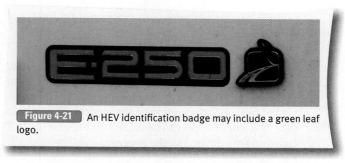

Figure 4-21 An HEV identification badge may include a green leaf logo.

into alternating current (AC), stepping the voltage up to 120 volts DC. This 120-volt DC high-voltage cable is colored orange to denote the higher voltage. Another hybrid vehicle utilizes a DC to DC inverter, stepping down the high-voltage output to 42 volts DC, which it then uses to power the electric power steering system. This particular manufacturer uses a yellow cable to indicate that the voltage is in the medium range (42 volts DC), along with the orange cable placed before the DC to DC inverter to denote the higher voltage. All cables, regardless of color, must be respected and carefully evaluated for their voltage capacity.

Rescue Tips

Be aware that some manufacturers may cover exposed voltage cables in a protective black casing, thus concealing the blue, yellow, or orange color coding that is used by some manufacturers to identify voltage capacity.

Hybrid Electric Vehicle Drive Systems

The basic components of an HEV system consist of an electric motor, generator, ICE, and battery pack. The drive system for a hybrid vehicle can be designed in series or parallel. In a **series drive system**, the electric motor turns the vehicle's transmission to provide propulsion, or it is used to charge the batteries or store power in a capacitor. The ICE does not provide propulsion to the vehicle as it is designed to do in a conventional vehicle; the ICE supplies power to the electric motor only. The more common **parallel drive system** can use the vehicle's ICE and/or the electric motor to power the vehicle's transmission to provide propulsion. Which propulsion system is utilized is dependent upon the speed of the vehicle and will vary among manufacturers **Figure 4-20 ▶**.

HEVs are generally recognizable through vehicle identification badging located on the sides or rear of the vehicle. A green leaf logo, the word "Hybrid" or letter H are some common vehicle badges that the technical rescuer may find, but be aware

that there is no standardization of vehicle identification badging among manufacturers **Figure 4-21 ▲**.

Hybrid Electric Vehicle Emergency Procedures

All hybrid vehicles have built-in safety features that shut down the medium- and high-voltage lines for various situations. Safety devices such as inertia relays will open when detecting a collision or significant impact, immediately disabling the high-voltage system. Ground faults on each wire will detect any leaks, line breaches, or short circuits, disabling the high-voltage service. Thermal detection devices will shut down the high-voltage system if it detects a temperature rise greater than a specific set temperature. The medium- to high-voltage wires run along the undercarriage of the vehicle, either in the center or on the sides. They are normally protected by framing material and/or are wrapped in a protective casing. Remember that these wires have ground-fault and short-circuit protection; if there is a break in the line, a relay will kick in and open the circuit, which will isolate and disable the voltage. With various hybrid vehicle models, fully discharging the voltage from the capacitor can take from 5 to 10 minutes once a breach in the system is detected or the power is turned off.

Turning the main engine key in the off position and removing the key is one way to disable the high-voltage system as well as eliminate the vehicle's 12-volt battery. Manufacturers recommend pulling the main fuse, which is another way to disable the high-voltage system. These procedures can be incorporated together to disable the power system. If the main fuse is not recognizable or difficult to locate, then remove all of the fuses in the box. There are also manual battery service disconnects that will shut down the high-voltage system, but these service disconnects are specific to each model and are sometimes difficult to locate because they are designed for service technicians to access, not emergency responders. Trying to access the battery pack to find a manual disconnect is not a safe or recommended practice for emergency responders. Some models disengage the high-voltage system when the hood to the engine compartment is opened, but again this is not standardized among all HEV models. To ensure a full power-down of the HEV, follow the safety procedures that are explained here.

Safety procedures that the technical rescuer can take prior to performing extrication when dealing with an HEV include the following. Note, some of these steps are not necessarily completed in succession; some of them can be completed simultaneously, depending on the number of personnel on scene who are available to be assigned to complete each step.

- Don the appropriate PPE, including SCBA if needed, and clear the scene of all bystanders and hazards.
- Establish hazard control zones (hot, warm, and cold).
- Conduct the inner and outer surveys. If possible, approach the vehicle from upwind and uphill and from the side or corners because of the possibility of unexpected lunging or reversing of the vehicle. The vehicle may be silent or appear to be turned off because the ICE has been shut down, but the system can still be in an all-electric "ready" mode (most vehicle models show this on the vehicle dash display panel), capable of engaging the drive or reverse motor at any moment **Figure 4-22**.
- Look for a vehicle identification badge; in this case, it may be a green leaf logo, the word "Hybrid" or letter H.
- One $1\frac{3}{4}$-inch (44-mm) charged hose line should be deployed for scene and personnel protection.
- Never assume that the vehicle is turned off because it is silent or still.
- When the scene is safe, manually turn the engine key to the off position, make sure the vehicle is in park, and remove the keys. Various emergency response guides state that smart keys need to be a minimum of 15 feet (5 m) away from the vehicle, which removes them from operational range. Place the parking brake on if accessible.
- Stabilize the vehicle from movement with cribbing. Avoid placing any cribbing under high-voltage wiring, alternative fuel supply lines or cylinders, and high-voltage battery packs.
- Attempt, if needed, any necessary component adjustments before disabling the power to the vehicle, such as activating the electric parking brake, adjusting power seats, and releasing hatchbacks.
- Safely disengage the 12-volt battery, which will shut down the flow of high-voltage electricity. Remove or cut the negative cable first and ensure that the cable does not fall back on the vehicle and make contact with the frame in any way. 12-volt batteries can be found in various locations in the vehicle depending on the manufacturer and model of the vehicle. The battery may be in the traditional location under the engine compartment hood, or inside one of the wheel wells, under the backseat, or in the trunk. *Warning*: Energy capacitors in some models can hold power for 5 to 10 minutes after the power has been disengaged.
- Manufacturers recommend removal of the main fuse from the vehicle to ensure that the electrical system is disabled. This will be an AHJ decision, as location of fuses can vary greatly among models.
- Some manufacturers recommend utilizing service disconnects as indicated in the manufacturer's emergency response guide for that particular vehicle. This will be an AHJ decision, as location of service disconnects can vary greatly among models.

When performing various extrication techniques, precaution must be observed when the vehicle is overturned on its roof. If the decision to perform a tunneling technique through the trunk area is considered, remember that the battery packs are normally placed under the rear seat or trunk area, so an alternative method may be safer. Also, when considering a dash displacement technique, such as a dash lift technique, or the dash roll technique, keep in mind that the medium- to high-voltage wires run along the undercarriage from the battery pack in the rear to the motor/generator located at the front of the vehicle. Opening up the dash area may expose the technical rescuer to the medium- and high-voltage wires below the vehicle. Power to these wires should have been isolated and disconnected prior to performing any extrication; however, it is still not recommended to cut into any medium- to high-voltage wire, regardless of whether the power has been neutralized. In addition, techniques that involve cutting and dropping the floorboard area under the brake and gas pedal should not be attempted. Such maneuvers could expose the rescuer to high-voltage wires or gas lines running under the rocker panel/channel area. This is not a safe practice and should be avoided regardless of the type of fuel system the vehicle utilizes or whether the power supply has been secured. There are alternative techniques that are much safer.

Figure 4-22 Most hybrid electric vehicle models show the "Ready" mode light on the vehicle dash when the power is on.

Rescue Tips

If the decision to perform a tunneling technique through the trunk area of an HEV is considered, remember that the battery packs are normally placed under or behind the rear seat or in the trunk area, so an alternative method may be safer.

All-Electric Vehicles

The electric vehicle (EV) or battery electric vehicle (BEV) is 100 percent electric and is energy efficient and environmentally friendly, emitting no air pollutants **Table 4-3 ▾**. EVs are propelled by one or more electric motors, which are powered by rechargeable battery packs. The EV does not have a tailpipe because it does not emit exhaust.

Available as of late 2010, the Nissan Leaf is the first EV produced by a major auto manufacturer **Figure 4-23 ▸**.

The EV uses regenerative braking to recharge the battery, which is a laminated lithium ion battery with a capacity of 24 kilowatt-hours. This is a high-voltage battery (approximately 400 volts DC). The vehicle can also be plugged in to be recharged. EVs most commonly charge from conventional power outlets or charging stations. Fully recharging the battery

Table 4-3   Summary of Electric Vehicles (EVs)

Manufacturer	Model	Type	Class	Year	Web Link to Vehicle Information
AC Propulsion	eBox	Sedan	EV	2009	www.acpropulsion.com
BMW	City	Coupe	EV	2012	www.bmwusa.com
BYD	E6	Sedan	EV	Concept	www.byd.com
Chevrolet	S-10 Electric	Pickup	EV	*D/NLP	www.chevrolet.com/hybrid
Chrysler	Epic Electric Minivan	Van	EV	D/NLP	www.chrysler.com
Coda Auto	Hafei Saibao 3 EV	Sedan	EV	2010	www.codaautomotive.com
Daimler	Smart For Two (ED)	Coupe	EV	2010	www.smartusa.com
Dodge	Circuit	Coupe	EV	2011	www.dodge.com
Ford	Electric Ranger	Pickup	EV	*D/NLP	www.ford.com
	Focus EV	Sedan	EV	2011	
GMC	EV1	Sedan	EV	*D/NLP	www.gmc.com
Honda	EV Plus	Sedan	EV	*D/NLP	www.honda.com
Keio	Eliica	Coupe	EV	Concept	www.eliica.com/English/
Lightning	GT	Coupe	EV	2010	www.lightningcarcompany.com
Mercedes	BlueZero	Sedan	EV	Concept	mbusa.com
Miles EV	ZX 40S	Sedan	EV	2009	www.milesev.com
Mini Cooper	Mini E	Sedan	EV	Concept	www.miniusa.com
Mitsubishi	iMiEV	Sedan	EV	Concept	www.mitsubishicars.com
Modec	Box Van	Van	EV	2009	www.modeczev.com
Mullen	L1x-75	Coupe	EV	Concept	www.mullenmotorco.com
Nissan	Altra	Sedan	EV	*D/NLP	www.nissanusa.com
	Leaf	Sedan	EV	2010	
Phoenix	Phoenix SUV	SUV	EV	*D/NLP	www.phoenixmotorcars.com
	Phoenix Pickup	Pickup	EV	*D/NLP	
Pininfarina	Blue Car	Sedan	EV	2010	www.pininfarina.com
Porteon	EV	Sedan	EV	Concept	www.porteon.net
Renault	Fluence	Coupe	EV	2011	www.renault.com
Smith	Edison Panel Van	Van	EV	2009	www.smithelectricvehicles.com
Solectria	Force	Sedan	EV	*D/NLP	www.azuredynamics.com
Subaru	R1E	Coupe	EV	Concept	www.subaru.com
Tesla	Model S	Coupe	EV	2011	www.teslamotors.com
	Roadster	Coupe	EV	2009	
Think	Th!nk City	Coupe	EV	2009	www.think.no
Toyota	RAV4 EV	SUV	EV	*D/NLP	www.toyota.com/hsd
	FT-EV	Coupe	EV	Concept	
Universal	UEV Spyder	Coupe	EV	*D/NLP	n/a
Velozzi	Supercar	Coupe	EV	Concept	www.velozzi.org
Venturi	Fetish	Coupe	EV	2009	www.venturifetish.fr
Wrightspeed	X1	Coupe	EV	Concept	www.wrightspeed.com

*D/NLP: Discontinued/No Longer Produced

Source: Reprinted with permission from *Fire Fighter Safety and Emergency Response for Electric Drive and Hybrid Electric Vehicles*, Copyright © 2010, The Fire Protection Research Foundation, Quincy, MA. This reprinted material is not the complete and official position of The Fire Protection Research Foundation on the referenced subject.

Figure 4-23 The Nissan Leaf is the first EV produced by a major auto manufacturer.

Figure 4-24 The neighborhood electric vehicle (NEV).

pack can take up to 20 hours utilizing a Level 1 charging system. The high-voltage battery is encased in steel and located in the undercarriage. A 12-volt DC battery is located under the hood to supply power to the low-voltage devices such as the lights, horns, and other accessories.

Many EVs can travel 100 to 200 miles (161 to 322 kilometers) without charging. The range of the vehicle may be affected by temperature, speed, topography, driving style, and cargo as well as the manufacturer. As with other advanced vehicles, EVs can be recognized through vehicle identification badging. For example, the Nissan Leaf has two "zero emission" vehicle

badges. One is located on the rear of the vehicle, and the other is located on the driver-side door panel.

A subclass of the EV is the **neighborhood electric vehicle (NEV)** **Figure 4-24 ▲**. NEVs are classified as battery-operated low-speed vehicles with a top speed of 25 mph (40 kph) and are approved for street use on public roadways with speeds posted of no greater than 35 mph (56 kph) **Table 4-4 ▼**. NEVs can vary on distance traveled on a single charge but generally provide up to 35 miles of travel on a full charge. A NEV utilizes

Table 4-4   Neighborhood Electric Vehicles

Manufacturer	Model	Type	Class	Year	Web Link to Vehicle Information
AEV	Kurrent	Coupe	NEV	2009	www.getkurrent.com
Aptera	2E	Coupe	NEV	2009	www.aptera.com
BB Buggies	Bad Boy Buggy	Coupe	NEV	2009	www.badboybuggies.com
BG Auto	BG C100	Coupe	NEV	2009	www.bgelectriccars.com
Commuter	Tango T600	Coupe	NEV	2009	www.commutercars.com
Dynasty	IT	Coupe	NEV	2009	www.itiselectric.com
Elbilen	Buddy	Coupe	NEV	2009	n/a
FineMobile	Twike	Coupe	NEV	2009	www.twike.us
Flybo	XFD-6000ZK	Coupe	NEV	2009	www.flybo.cn
GEM	GEM Car	Coupe	NEV	2009	www.gemcar.com
Myers	NmG	Coupe	NEV	2009	www.myersmoters.com
Obvio	828e	Coupe	NEV	2009	www.obvio.ind.br
Reva	NXR / NXG	Coupe	NEV	2009	www.revaglobal.com
Spark Electric	Comet	Coupe	NEV	*D/NLP	n/a
Venture	Pursu	Coupe	NEV	2009	www.flytheroad.com
	VentureOne e50	Coupe	NEV	Concept	xprizecars.com/2008/06/venture-vehicles-ventureone.php
Zap	Xebra	Coupe	NEV	2009	www.zapworld.com
Zenn Motors	CityZenn	Coupe	NEV	2009	www.zenncars.com

*D/NLP: Discontinued/No Longer Produced

Source: Reprinted with permission from *Fire Fighter Safety and Emergency Response for Electric Drive and Hybrid Electric Vehicles*, Copyright © 2010, The Fire Protection Research Foundation, Quincy, MA. This reprinted material is not the complete and official position of The Fire Protection Research Foundation on the referenced subject.

a standard Level 1 charging system, or a 120-volt household outlet, to recharge its battery system.

All-Electric Vehicles Emergency Procedures

During the inner survey, look for indications that the high-voltage system is on. For example, upon approaching the vehicle, look to see if the "ready" indicator is on, if the charge indicator is on, if the air-conditioning remote timer indicator is on, or if the remote-controlled air-conditioning system is active. These are all indications that the high-voltage system of the vehicle is on. Technical rescuers should refer to the manufacturer's emergency response guides for information on how to disable the high-voltage system in various EVs. The high-voltage cables for the Nissan Leaf are located in the undercarriage and under the hood. All high-voltage cables are color-coded orange on the Nissan Leaf with a "WARNING" label.

Safety procedures that the technical rescuer can take prior to performing extrication when dealing with an EV include the following. Note, some of these steps are not necessarily completed in succession; some of them can be completed simultaneously, depending on the number of personnel on scene who are available to be assigned to complete each step.

- Don the appropriate PPE, including SCBA if needed, and clear the scene of all bystanders and hazards.
- Establish hazard control zones (hot, warm, and cold).
- Conduct the inner and outer surveys. If possible, approach the vehicle from upwind and uphill and from the side or corners because of the possibility of unexpected lunging or reversing of the vehicle. The vehicle may be silent or appear to be turned off, but the system can still be in an all-electric "ready" mode (most vehicle models show this on the vehicle dash display panel), capable of engaging the drive or reverse motor at any moment.
- Look for a vehicle identification badge; in the case of the Nissan Leaf, it may be a "zero emission" badge.
- One 1¾-inch (44-mm) charged hose line should be deployed for scene and personnel protection.
- Never assume that the vehicle is turned off because it is silent or still.
- When the scene is safe, manually turn the engine key to the off position, make sure the vehicle is in park, and remove the keys (unless it is a push-button start system). Various emergency response guides state that smart keys need to be a minimum of 15 feet (5 m) away from the vehicle, which removes them from operational range. Place the parking brake on if accessible.
- Ensure that the remote heating/air-conditioning system is deactivated.

- Ensure that the charging (electric plug) is disconnected.
- Stabilize the vehicle from movement with cribbing. Avoid placing any cribbing under high-voltage wiring and high-voltage battery packs.
- Attempt any necessary component adjustments before disabling the power to the vehicle, such as activating the electric parking brake, adjusting power seats, and releasing hatchbacks.
- Safely disengage the 12-volt battery, which will shut down the flow of high-voltage electricity. Remove or cut the negative cable first and ensure that the cable does not fall back on the vehicle and make contact with the frame in any way. A 12-volt battery can be found in various locations in the vehicle, depending on the manufacturer and model of the vehicle. The battery may be in the traditional location under the engine compartment hood, inside one of the wheel wells, under the backseat, or in the trunk. *Warning*: Energy capacitors in some models can hold power for 5 to 10 minutes after the power has been disengaged.
- Manufacturers recommend removal of the main fuse from the vehicle to ensure that the electrical system is disabled. This will be an AHJ decision, as location of fuses can vary greatly among models.
- Some manufacturers recommend utilizing service disconnects as indicated in the manufacturer's emergency response guide for that particular vehicle. This will be an AHJ decision, as location of service disconnects can vary greatly among models.

Ongoing Education

The technical rescuer must adapt and change with advancing vehicle technology. Alternative fueled vehicles, HEVs, EVs, and fuel cell vehicles will become the dominant forms of transportation in the not-so-near future. Departmental/organizational standard operating procedures should be developed to reflect the emergency procedures for handling the specific vehicle types discussed in this chapter. Each member in the organization should be trained to at least a basic level of competency in dealing with advanced vehicle technology. Technical rescuers involved in vehicle extrication practices and procedures have an inherent responsibility to improve their skills and remain on the cutting edge of technology; self-motivation with continual training and education will provide the means to stay focused, while always looking for ways to improve. Stay current on the emergence of these vehicles, and download manufacturer emergency response guides from manufacturer web sites regularly.

Wrap-Up

■ Ready for Review

- Alternative powered vehicles are vehicles that utilize fuels other than petroleum or a combination of petroleum and another fuel for power.
- The Energy Policy Act of 1992 outlines a list of fuels that can be classified as an "alternative fuel" for vehicles.
- Most vehicle manufacturers identify the vehicle type or fuel type through a label known as a vehicle identification badge.
- Flexible fuel vehicles can run on gasoline alone or utilize the E85 blend of up to 85 percent ethanol and 15 percent gasoline.
- Natural gas is a fossil fuel primarily composed of methane that can be utilized as a compressed natural gas (CNG) or liquefied natural gas (LNG).
- A safety feature for high-pressure cylinders is the PRD, which is designed to rapidly release all of the gas when exposed to high temperatures, such as during a fire.
- Liquid petroleum gas (LPG), also known as propane, is produced from the processing of natural gas and is also produced as part of the refining process of crude oil. Propane is the third most common engine fuel today, after gasoline and diesel.
- Biodiesel is a fuel used solely for diesel engines that is processed from domestic renewable resources such as plant oils, grease, animal fats, used cooking oil, and, more recently, algae. Biodiesel can be utilized by itself as a diesel fuel or blended with petroleum diesel.
- Hydrogen is one of the most abundant elements on earth. As a fuel, hydrogen can be compressed and stored in high-pressure storage tanks with pressures of 3600 psi (24,821 kPa), 5000 psi (34,474 kPa), and 10,000 psi (68,948 kPa). Hydrogen can be chemically combined in hydride form with certain metals, which can store it more compactly and efficiently than in a gas form.
- A fuel cell is an electrochemical device that utilizes a catalyst-facilitated chemical reaction of hydrogen and oxygen to create electricity, which is then used to power an electric motor. The basic components of a fuel cell vehicle system are a fuel cell module pack/stack, electric motor, generator, hydrogen storage system, and battery pack.
- A hybrid electric vehicle (HEV) is a vehicle that combines two or more power sources for propulsion, one of which is electric power. A full hybrid vehicle can use either its electric motor or its internal combustion engine, or both, to propel itself. In contrast to the full hybrid, the mild hybrid vehicle cannot propel itself on electric power alone; it must use electric power and the internal combustion engine.
- The electric vehicle or battery electric vehicle is 100 percent electric and is propelled by one or more electric motors, which are powered by rechargeable battery packs. The EV does not have a tailpipe because it does not emit exhaust.

■ Hot Terms

Alternative powered vehicle A vehicle that utilizes fuels other than petroleum or a combination of petroleum and another fuel for power.

Biodiesel A safe, nontoxic, biodegradable fuel used solely for diesel engines that is processed from domestic renewable resources such as plant oils, grease, animal fats, used cooking oil, and, more recently, algae. Biodiesel can be utilized by itself as a diesel fuel or blended with petroleum diesel at varying percentages.

BLEVE Boiling liquid/expanding vapor explosion; an explosion that occurs when pressurized liquefied materials inside a closed vessel are exposed to high heat.

Compressed natural gas (CNG) A fuel utilized primarily in fleet vehicles; natural gas is compressed for high-pressure storage and distribution systems.

Department of Transportation's *Emergency Response Guidebook (ERG)* A preliminary action guide for responders operating near hazardous materials.

Electric vehicle (EV) A vehicle that is 100 percent electric, emits no air pollutants, and is propelled by one or more electric motors, which are powered by rechargeable battery packs.

Emergency response guides Booklets prepared by vehicle manufacturers to educate and assist emergency response personnel in responding to emergencies dealing with specific types and models of vehicles such as hybrid/electric, hydrogen fuel cell, and alternative fuel systems.

Ethanol A fuel comprised of an alcohol base that is normally processed from crops such as corn, sugar, trees, or grasses.

Extended range electric vehicle (EREV) A vehicle that utilize a series-type propulsion system that allows the vehicle to run on all battery or all electric power until it is near depletion, which occurs in the range of 40 miles or more depending on the manufacturer.

Flexible fuel vehicle (FFV) A vehicle capable of running on gasoline alone or utilizing the E85 blend of up to 85 percent ethanol and 15 percent gasoline.

Fuel cell An electrochemical device that utilizes a catalyst-facilitated chemical reaction of hydrogen and oxygen to create electricity, which is then used to power an electric motor or generator, with the by-products of this process being water and heat.

Full hybrid A vehicle that uses either its electric motor or its internal combustion engine, or a combination of both, to propel itself.

Hybrid electric vehicle (HEV) A vehicle that combines two or more power sources for propulsion, one of which is electric power.

Hydrogen An odorless, colorless, flammable, nontoxic gas that combines easily with other elements.

Integrated motor assist (IMA) mild hybrid A hybrid system used by Honda that is designed to start and stop the hybrid electric vehicle's internal combustion engine; in addition, it will assist the internal combustion engine when acceleration is needed.

Liquefied natural gas (LNG) A colorless, odorless, nontoxic natural gas that floats on water and is lighter than air when released as a vapor.

Liquid petroleum gas (LPG) Also known as propane, a fossil fuel produced from the processing of natural gas and also produced as part of the refining process of crude oil. Propane is the third most widely utilized fuel source behind gasoline and diesel; it is commonly used with forklifts and other similar work units.

Methanol An alcohol-based fuel similar to ethanol. It is also known as a wood alcohol because it is processed from natural wood sources such as trees and yard clippings. It may be utilized as a flex fuel in a ratio of 85 percent methanol to 15 percent gasoline, better known as M85.

Mild hybrid A vehicle that uses electric power in conjunction with the internal combustion engine for vehicle propulsion.

Natural gas A fossil fuel primarily composed of methane that can be used as a compressed natural gas (CNG) or liquefied natural gas (LNG).

Neighborhood electric vehicle (NEV) A vehicle that is classified as a battery-operated low-speed vehicle with a top speed of 25 mph (40 kph) and that is approved for street use on public roadways with speeds posted of no greater than 35 mph (56 kph).

Parallel drive system A system that can use either the vehicle's internal combustion engine or the electric motor to power the vehicle's transmission and provide propulsion.

Plug-in hybrid electric vehicle (PHEV) A hybrid vehicle that can recharge its battery system using a plug-in cord that can run off general house current in the range of 120 volts, also known as a Level 1 charging system.

Polymer exchange membrane Also known as a proton exchange membrane, a thin membrane used in a fuel cell system that is placed between the anode and cathode and through which positive electrons are passed.

Pressure release device (PRD) A safety feature built into high-pressure storage cylinders that is designed to rapidly release gas contents when exposed to high temperatures, such as during a fire.

Series drive system A system that uses the internal combustion engine alone to run an onboard generator, which in turn can either run the electric motor that turns the vehicle's transmission (providing propulsion) or be used to charge the batteries or store power in a capacitor. The internal combustion engine does not provide direct propulsion to the vehicle.

Smart key A device that uses a computerized chip that communicates through radio frequencies to unlock or lock a vehicle as well as start a vehicle remotely without the requirement of traditional keys.

Start/stop mild hybrid A vehicle that is not a true hybrid system by definition. The motor/generator is not used to propel the vehicle; it is designed to turn off the vehicle's internal combustion engine when the vehicle is idle and will turn the vehicle's internal combustion engine back on when the accelerator is activated.

Temperature relief device (TRD) A device that rapidly releases product through a small metal tube attachment when detecting excessive amounts of heat at a preset temperature.

Vehicle identification badge A type of label that vehicle manufacturers utilize to identify the type of vehicle or the fuel that is used in the vehicle.

Technical Rescuer *in Action*

As the officer on the engine, you are dispatched to a vehicle rollover with possible entrapment. Upon your arrival as you are approaching the scene, you notice one vehicle with heavy damage resting on its side. You can clearly see the undercarriage of the vehicle from your position. The undercarriage reveals a large cylinder attached at the rear and an orange cable running up the side from the rear to the engine compartment. You advise dispatch that you have a hybrid electric vehicle with an alternative fuel source. You advise on injuries and the need for the hazardous materials team to respond.

1. What can you look for on the vehicle that may indicate the type of alternative fuel in use?

A. Don't do anything until the hazardous materials team arrives.

B. There is no way to tell the type of fuel being used.

C. A product label on the tank.

D. Notify the vehicle manufacturer to find out.

2. A good source for research and information on auto extrication is the Internet. A lack of information or misinformation can cause rescuers to:

A. overreact.

B. fail to act at all.

C. save more lives.

D. A and B

3. To use the same emergency procedures on every incident involving one or more advanced vehicle systems is not practical and can be very dangerous. To avoid doing so requires:

A. an extensive library of vehicle operating manuals of the most popular models.

B. recognition of all hazards to be mitigated and neutralized before rescue extrication.

C. preplanning, study, and training.

D. B and C

4. After approaching the emergency scene from uphill and upwind, the company officer should perform:

A. a left to right survey.

B. a north to south survey.

C. a thorough and ongoing survey.

D. inner and outer surveys.

5. When approaching an HEV during an inner survey, the technical rescuer should approach the vehicle from the corner or the:

A. front.

B. rear.

C. side.

D. center.

6. How many hose lines should be deployed with these operations?

A. One 1¾-inch hose line

B. Two 1¾-inch hose lines

C. One 1¾-inch hose line with a 2A/10BC portable fire extinguisher

D. One 2½-inch hose line

7. To prevent accidental ignition, smart keys must be kept a minimum of:

A. 10 feet away from the vehicle.

B. 12 feet away from the vehicle.

C. 15 feet away from the vehicle.

D. 20 feet away from the vehicle.

8. When disconnecting a car battery:

A. cut the negative cable first, then the positive.

B. cut the positive cable first, then the negative.

C. cut both cables simultaneously to prevent an electrical arc.

D. it no longer matters which cable is cut first because advanced vehicle technology has eliminated the hazard.

9. When approaching the emergency scene, look for any visible vapor cloud and listen for a loud hissing noise; both indicate product release through a leak in the system or through the:

A. pressure release device (PRD).

B. pressure regulating valve (PRV).

C. emergency pressure equalizing unit (EPEU).

D. a loose fill cap (FC).

10. Most vehicle manufacturers will identify the vehicle type or fuel type through a label known as a vehicle identification badge. Badges can be found at various locations on the vehicle body. Which of the following statements about badges is correct?

A. Be aware that they are usually found on the exterior right side of the trunk compartment.

B. Be aware that they are usually found on the driver-side rear bumper.

C. Be aware that badge labeling is not standardized and can vary in design from one manufacturer to another.

D. Be aware that vehicle identification badges have to be placed on both sides of the vehicle similar to DOT placards.

Supplemental Restraint Systems

NFPA 1006 Standard

There are no objectives for this chapter.

Knowledge Objectives

After studying this chapter, you will be able to:

- Explain the difference between an active restraint device and a passive restraint device. (page 86)
- Explain what a supplemental restraint system consists of. (pages 86–87)
- Explain the air bag deployment process. (page 87)
- Explain the basic components that make up an air bag system. (pages 88–92)
- Describe a roll-over protection system and how it is activated. (page 92)
- Describe a seat belt pretensioning system and how it is activated. (page 92)
- Explain the emergency procedures for handling vehicles equipped with air bags. (pages 94–96)

Skills Objectives

After studying this chapter, you will be able to perform the following skill:

- Disconnect power on a vehicle to disable an air bag system. (pages 94–95, Skill Drill 5-1)

ou are on an extrication incident and the vehicle your crew is about to work on contains multiple undeployed air bags throughout the vehicle.

1. As the technical rescuer, what actions will you take?
2. Why should you determine if the vehicle has electric seats?
3. Why is it important to disconnect the 12-volt DC battery?

Introduction

"To grasp and hold a vision, that is the very essence of successful leadership." —Ronald Reagan

In 1967, the National Highway Traffic Safety Administration (NHTSA) issued **Federal Motor Vehicle Safety Standards (FMVSSs)**. FMVSSs were enacted to protect the public from unreasonable risk of crashes, injury, or death resulting from the design, construction, or performance of a motor vehicle. The first of many subsequent regulations that outlined minimum safety requirements for motor vehicles mandating compliance from vehicle manufacturers was FMVSS 209, *Seatbelt Assemblies*. FMVSS 209 includes requirements for the straps, buckles, hardware, and fasteners of seat belt assemblies. FMVSS 208, *Occupant Crash Protection*, was adopted at a later date and specified the type of occupant restraints, seat belts, and air bags required.

Supplemental Restraint Systems

In the early 1970s, fatalities from vehicle accidents started to increase dramatically; this was due largely to the occupants failing to wear seat belts while driving. To counter the growing problem of consumers disregarding the importance of proper seat belt usage, the auto industry introduced the air cushion restraint system. The air cushion restraint system was an air bag device that, at the time, was considered a replacement option to the seat belt, offering occupant protection in head-on collisions. Unfortunately, this theory did not work out as planned, as more vehicle accident fatalities started mounting. As time went by, the air cushion restraint system faded away and seat belt education and enforcement started to increase. In the 1980s, a system similar to the air cushion restraint, the **air bag**, emerged as a supplement to the seat belt. By working in conjunction with the seat belt, the air bag became known as a **supplemental restraint system (SRS)** ▸ Figure 5-1 ▸ .

Manual seat belts are classified as an **active restraint device** because the occupant has to activate the system by engaging the seat belt mechanism into the anchor unit. A vehicle air bag is classified as a **passive restraint device** because the occupant does not have to activate the device to make it function; the system is automatically activated when power is applied to the vehicle.

As the demand for vehicle air bags grew, new and stricter regulation coincided. In 1984, FMVSS 208 was amended to mandate that motor vehicles manufactured after April 1989 be equipped with a passive restraint system; this included air bags and automatic seat belts. At the time, most vehicles only offered a single-stage air bag system. In this "first-generation" air bag system, the mechanism would fire at a preset discharge rate and pressure. This inflation rate, pressure, and velocity were tested and measured to protect an average-size adult male; children, women, or smaller statured individuals were not factored into the equation. The test results determined that in order to deploy

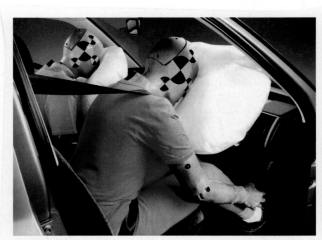

Figure 5-1 Air bags working in conjunction with seat belts are known as supplemental restraint systems.

the air bag in approximately 40 to 45 milliseconds, the velocity and pressure needed would be an inflation speed of 150 to 200 miles per hour (mph) (241 to 322 kilometers per hour [kph]). Because of the power exerted, out-of-position occupants or unbelted occupants who made contact at the inflation stage would be subject to a crushing force, sometimes causing fatalities and multiple injuries (fractures, contusions, lacerations, soft-tissue injuries, hearing impairment, burns). This caused FMVSS 208 to be amended again in 1998.

Air bags became the main focus of the 1998 FMVSS 208 amendment. The revised regulation mandated that air bags be depowered and a deactivation switch be added to passenger-side air bags for vehicles such as light-duty trucks that had no rear seating. These depowered air bags, also known as second-generation air bags, were now designed to inflate at a lesser force. The industry and the NHTSA believed that this correction would significantly reduce the casualties caused by its predecessor.

In 1996, the NHTSA proposed changes to the federal air bag requirement to encourage the introduction of "smart" air bag systems. These changes would not come into existence until 2000. In 2000, FMVSS 208 was amended again. The revision outlined the requirements of *advanced* frontal air bag systems, or third-generation air bag systems, to be phased in for all passenger vehicles and light-duty trucks by 2004, with full compliance for all passenger vehicles manufactured after September 1, 2006. These advanced **smart air bag systems** are using adaptive response features that automatically adjust the pressure in the air bag by utilizing multistage **inflators** and basing the deployment force on a number of calculated factors such as crash severity, occupant's weight, proximity to the air bag, seat belt usage, and seat position. To obtain the crash severity threshold, a vehicle was crashed into a stationary barrier at 14 mph (23 kph) or greater. The 14 mph (23 kph) was determined to be the minimum cutoff speed for air bag activation Figure 5-2 ▾.

These new requirements addressed several specific advanced features to protect occupants from air bag injuries:

- Dual-stage or multistage inflation process: The inflation process incorporates a full-force deployment and a reduced or multistage reduced force deployment option;

the first feature of this revised FMVSS 208 regulation requires the system to be equipped with a reduced deployment option when an occupant is too close to the deployment zone.

- A **suppression system**: The suppression system shuts down the air bag if an occupant classification system detects a child in the air bag deployment zone or if one of the many sensors detects a high risk potential by acquiring the occupant's weight, height, proximity to the air bag, seat belt usage, and seat position; the system sends this information to the electronic control unit, which will then shut off the air bag if a high risk to the occupant is determined.

■ Air Bag Deployment Process

A four-stage process occurs when an air bag is deployed in a crash sequence:

- The crash itself
- The crash sensor detecting deceleration
- The air bag deploying and inflating
- The occupant moving forward and striking the bag as deflation occurs

The first stage is the crash itself, which occurs when the vehicle strikes an object or is struck by an object. The second stage occurs when the crash sensor, or **accelerometer**, detects an immediate deceleration of the vehicle, sending the information to the electronic control unit, which then determines the severity of the crash. If the electronic control unit detects a deceleration equivalent to a stationary barrier crash at 14 mph (23 kph) or greater, stage three occurs. During stage three, the air bag deploys and the occupant is momentarily forced backward by the impact and sudden deceleration of the vehicle. Stage four occurs after the bag has fully inflated and the occupant has resumed moving forward, striking the bag as deflation occurs; this action causes the gas in the bag to be forced out of the small vent holes on its sides.

A vehicle crash is measured in milliseconds, with 1000 milliseconds equaling 1 second. The entire vehicle crash process, which involves dissipating all of the kinetic energy of the vehicle, takes approximately 100 to 125 milliseconds, occurring faster than the blink of an eye, which takes approximately 300 to 400 milliseconds. To understand how an occupant's body reacts inside a vehicle as the vehicle strikes an object and causes it to rapidly decelerate, we need to look at Newton's First Law of Motion. It states that an object in motion will remain in motion until it is disrupted by an external force. If a vehicle traveling at 50 mph (80 kph) suddenly stops, the vehicle may stop but the objects inside the vehicle that are not attached continue to travel at that speed until something stops their forward motion, such as the steering wheel, dash, or windshield. The air bag is engineered to absorb the force of the occupant's body, accelerating at 50 mph (80 kph) and then gradually dissipating as the gas is pushed out of the bag through its vent holes. The gas inside the air bag must be precisely set with the correct volume to prevent the occupant from bottoming out and striking the steering wheel or dash.

Figure 5-2   A barrier crash test is regularly performed by vehicle manufacturers and various testing facilities.

Figure 5-3 The air bag, initiator, and inflator.

Air Bag Components

Several components make up an air bag, including the following Figure 5-3 ▲ :

- Air bag
- Initiator
- Electronic control unit
- Inflator
- Sensors

Air Bag

The air bag itself normally consists of a strong, durable nylon or blended material that is folded in a certain manner to assist during inflation. The bag is coated with a powdered substance, normally consisting of talcum, chalk, or cornstarch, which is used as a lubricant to assist in deployment ease. Once the air bag has deployed, the powdered residue will visibly float in the air and may be a mild irritant to individuals with respiratory ailments. The air bag also comes equipped with several tethers, which are designed to manage the speed of the deployment. The air bag cover for the driver- and passenger-side air bags is composed of a plastic material that is scored and designed to tear apart and separate when the air bag inflates.

The size of the bag will vary depending on the manufacturer specifications, type of vehicle design, type of bag (driver, passenger, or side air bag), or location of the bag (whether on the driver or passenger side). Vehicles with convertible and/or removable roof systems may combine the side-impact air bags in the doors or seat backs into one larger bag to provide protection for the passenger's head and torso; such vehicles may also incorporate a rear seat air bag system along with a deployable roll-over protection system, described later in this chapter. Air bag manufacturers are continuously researching and developing better ways to protect vehicle occupants, and this technology is always evolving; it is up to you as a technical rescuer to keep current with these changes.

The most common air bags are the driver and front passenger air bags, which are mandatory in all vehicles. Other variations of air bags located within a vehicle include, but are not limited to, the following:

- **Side-impact air bags**: There are three types of side-impact air bags designed to protect the following areas of the occupant: the head, the chest/upper torso, and a combination of the head and chest/upper torso. Side-impact air bags are designed to activate immediately upon impact. All three types of side-impact air bags can be found in the door, seat backs, roof posts, or roof rails. These bags may be labeled HPS (head protection system), IC (inflatable curtain), SIPS (side-impact protection system), or ROI (roll-over inflator air bag). Some of these side-impact air bags are designed to maintain inflation for a few seconds to help protect the occupant in secondary impacts or rollovers Figure 5-4 ▼ .
- Knee air bags: designed to protect the occupant's abdomen, pelvis, and lower extremities, preventing the occupant from being pulled under the dash area.
- Seat belt air bags: These air bags are designed to protect the occupant's upper and lower torso area as well as the pelvic area; they are designed to reduce the sudden tensioning or "clothesline effect" that can occur from the automatic engagement of a standard or pretensioning seat belt system.
- Rear seat deployment air bag systems: These air bags can be deployed from the center roof area, seat belt, door, or roof post, depending on various manufacturers.
- Outside pedestrian protection system: In some rare cases (mostly European), there may be an air bag positioned in the front bumper/hood area, deploying and protecting pedestrians when they are inadvertently struck by a vehicle.

With the exception of the driver-side and front passenger-side air bag, there are no standardized locations of vehicle air bags Figure 5-5 ▶ . Vehicle manufacturers are free to install air bags in any location they believe provides the best occupant protection and is the most economically feasible. Most manufacturers offer these unique air bags as upgrades or options to purchase by the consumer.

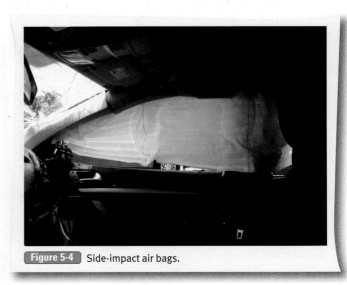

Figure 5-4 Side-impact air bags.

Figure 5-5 Locations of air bags will vary. This photo shows a knee air bag.

Initiator

Most front air bags utilize an <u>initiator</u> device such as a <u>squib</u>, which is a miniature explosive device, to ignite the propellant that produces the nitrogen or other inert gas that fills the nylon air bag **Figure 5-6 ▾**.

Figure 5-6 An initiator device, such as a squib, is utilized to activate an air bag.

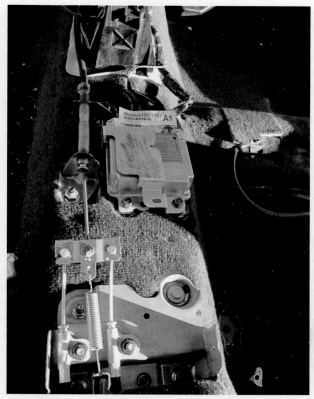

Figure 5-7 The electronic control unit (ECU), also known as the air bag control unit (ACU), is the brains of an air bag system.

Electronic Control Unit

The <u>electronic control unit (ECU)</u>, also known as the air bag control unit (ACU), is the brains of an air bag system **Figure 5-7 ▲**. This device is a small processing unit that is generally located in the center of the vehicle under and/or between the seats, but it may be located in other areas depending on the vehicle manufacturer's preference or vehicle design. In advanced air bag systems, this processor is preset with multiple crash algorithms, which calculate the deployment level needed to best protect the occupants. Three basic deployment levels are common with advanced air bag systems, each determined by the algorithmic factors. They are no deployment, low deployment, and full deployment. In each scenario, the vehicle crash sensor must first detect a crash and then instantly (within milliseconds) send a signal to the ECU. The ECU determines if the speed of the deceleration exceeds the crash threshold level of 14 mph (23 kph).

The ECU is designed to limit unnecessary deployments that occurred in first-generation air bag systems. The ECU works in conjunction with multiple sensors located throughout the vehicle, continuously monitoring changes in information. Within milliseconds, this information is sent back to the ECU. The ECU also contains an energy capacitor that acts as a backup system in lieu of any power disruption, such as a battery disconnect. The amount of time that the energy capacitor holds power will vary with each manufacturer, ranging from 30 seconds up to 30 minutes. The energy capacitor is a very important component because it holds power when the vehicle's 12-volt DC battery has been disconnected, thus ensuring that all the power sources have not been eliminated. An ECU can also simultaneously activate a seat belt pretension system, along with an air bag, to give added protection. Some ECUs have a feature that records and stores information from accidents that have occurred, acting similar to a flight recorder. This can provide valuable information to manufacturers for quality improvement measures and for crash investigation teams in determining what occurred during the crash.

Inflator

One of the most critical design features for an air bag is the ability to fill up the bag instantaneously, in milliseconds, from the onset of the collision. This is accomplished using one of two basic inflation systems—a stored compressed gas system or a gas generation system. These high-pressure inflators are critical considerations of the technical rescuer because they are generally positioned in the cut zone areas such as roof posts/pillars or roof rails. Exposing areas before any cutting or spreading operation begins is vital to the safety of the rescuer as well as the patients. The techniques and skill drills discussed in this text include precautions and procedures in exposing and operating around these inflators.

A **stored compressed gas system** can be comprised of a single stage or multistage inflation process. These types of systems commonly utilize an inert gas such as argon, helium, or another type of gas such as nitrous. The gas is stored in a steel or aluminum cylinder, normally at a pressure of 2500 to 3500 psi (17237 to 24132 kPa), but these pressures may be greater in some designs, depending on the size of bag, its use, and the manufacturer preference or vehicle design. The igniter, or squib, sets off a burst or rupture disc that acts as a seal, holding back the compressed gas. When activated, the disc breaks open, releasing the gas from the chamber, which expands and instantly fills the bag. The rapid release of the compressed gas in the ambient air could cause the cylinder to freeze, resulting in variable bag inflation. To prevent this, a heating element keeps the temperature of the gas constant to maintain the proper inflation ratio of the bag.

Multiple gas storage cylinders may be located in various areas of the vehicle, depending on the manufacturer. It is difficult to explain or list all of these cylinder locations because there can be such a variance between manufacturers, and this technology is always evolving. The best practice model for the technical rescuer in dealing with cylinder/inflator locations is to always expose the area to be cut and/or spread if it can be exposed before any operation begins. A phrase that is consistently used in this text is "expose and cut." This is not a suggestion, but an action that must be implemented at every vehicle extrication incident **Figure 5-8 ▶**.

Multistage inflators, sometimes known as hybrid inflators, are cylinders that can be comprised of two separate chambers of compressed gas—one with a large amount of product and the other with a smaller amount of product. Both chambers have

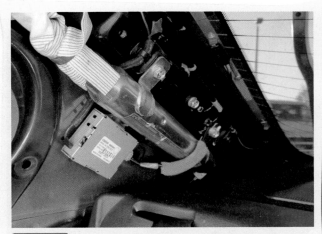

Figure 5-8   The gas that is used to inflate the air bag is stored in a steel or aluminum cylinder, normally at a pressure of 2500 to 3500 psi (17237 to 24132 kPa). These pressures may be greater in some designs, depending on the size of bag, its use, and the manufacturer preference or vehicle design. Multiple gas cylinders can be located in various areas within the vehicle, depending on the manufacturer.

initiators that can fire independently or together. Depending on the severity of the crash (>14 mph [23 kph] and <25 mph [40 kph]), weight, height, and proximity of the occupant, and whether the occupant is wearing a seat belt, the ECU may tell the inflator to release the smaller chamber, or 50% to 75% of the gas product. If the ECU determines that a full deployment is needed (crash is >25 mph [40 kph]), both chambers will fire simultaneously, filling the air bag to the appropriate inflation ratio to protect the occupant.

A **gas generation system** utilizes a chemical reaction that rapidly produces the gas, most commonly nitrogen gas. The gas instantly fills the bag at a rate of approximately 200 to 250 mph (322 to 402 kph), completely filling the bag in approximately 30 to 40 milliseconds. The most common solid fuel used for the nitrogen gas generation is sodium azide. Sodium azide is a very volatile substance; when mixed with water, it rapidly changes to hydrazoic acid in both a toxic gas and a liquid state. Sodium azide also changes into a toxic gas when it comes in contact with solid metals such as lead or copper. In the air bag system, sodium azide is detonated by the igniter or squib, which rapidly and completely decomposes the product through the combustion process. This produces the inert nitrogen gas and a small amount of the by-product sodium hydroxide, also known as lye. Additional chemicals are added to neutralize the sodium hydroxide. Most of the other by-products that are produced after the chemical reaction are contained by filters. Air bag system manufacturers are presently designing and using safer environmentally friendly nonazide gas generation propellants in driver- and passenger-side air bag systems.

Common driver-side air bag housing units that are found in most vehicles can contain approximately 2 ounces (57 grams) of sodium azide, depending on the size of the bag and its use. Because of its volatility and hazardous nature, sodium azide is utilized in pellet form for easier product containment. A driver-side air bag must produce enough gas to inflate the air bag approximately 10 inches (254 mm) into the compartment space of the front driver area. The deployment zone is measured to 10 inches (254 mm), which is the estimated safe distance that an occupant should be seated from the steering wheel. A common passenger-side air bag housing unit can contain approximately 7 ounces (198 grams) of sodium azide because of the additional deployment space as compared to a driver-side air bag; the air bag has to inflate big enough to fill the front passenger area. The deployment zone area is measured to 20 inches (508 mm), which is the estimated safe distance that an occupant should be seated from the dash area. Also, most passenger-side air bags are designed to strike and deflect off of the front windshield upon deployment, slowing the speed considerably. The technical rescuer should be aware that the fractured windshield or spidering effect left on the windshield from the air bag striking it can commonly be mistaken for the occupant striking the windshield Figure 5-9 ▶.

Side-impact air bags located in the roof rail, seats, doors, or roof posts are designed to react and deploy at a much faster rate because of the minimal distance/proximity of the occupant to the impact. A typical side-impact air bag must inflate within 10 to 15 milliseconds as compared to the standard front driver and passenger air bag, which has an inflation range of 30 to 40 milliseconds.

Rescue Tips

Because of the minimal distance/proximity of the occupant to the impact, a typical side-impact air bag must inflate within 10 to 15 milliseconds. In comparison, the standard front driver and passenger air bag has an inflation range of 30 to 40 milliseconds.

The common estimated elapsed time from bag deployment to bag deflation is approximately 100 to 150 milliseconds. Some advanced inflators are designed to keep the bags inflated for longer to assist with secondary crashes and roll-overs. These systems are currently designed for side-impact air bags where the inflators utilize compressed helium gas, releasing it in a cold state as opposed to using the heating element. This cold state allows the expanded gas to maintain its inflation ratio for longer durations to protect the occupant in the event of a roll-over, which can occur several minutes after the initial impact.

Sensors

Sensors in a vehicle air bag system are designed to measure variances in preset factors. The sensors send information back to the ECU, which determines whether or not to deploy the air bags. First-generation air bag crash detection sensors commonly used an electromechanical or magnetic sensing device consisting of a steel ball in a tube, where the steel ball was held in place by a spring or magnet. When the vehicle detected an impact deceleration at a predetermined rate, it would jar the ball, moving it through the tube, which would complete an electrical circuit and fire the air bag. Manufacturers would also include a **safing sensor** to prevent false deployments that might

Figure 5-9 A fractured windshield is commonly caused by the passenger-side air bag deflecting off of it during deployment.

occur from jarring the vehicle by driving in a pothole or over an object that would strike the undercarriage. This type of sensor has a deceleration setting lower than the crash-type sensor, and both the crash sensor and safing sensor had to be activated at the same time for the air bag to deploy.

Impact or crash sensors today are designed to detect a rapid deceleration of the vehicle through the use of a micro-electromechanical systems (MEMS) accelerometer. A MEMS accelerometer is constructed of small circuits integrated with micromechanical elements. When a crash is detected through a rapid deceleration of the vehicle, the microscopic mechanical element will move. The movement is detected by the circuit board, which sends a signal back to the ECU telling it that a deployment threshold has been reached. Again, this process occurs in milliseconds.

There are several different types of sensors that serve various functions and are designed to report a constant stream of information back to the ECU. These sensors are located throughout the vehicle, depending on their type and use Figure 5-10 ▼.

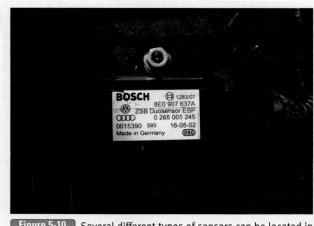

Figure 5-10 Several different types of sensors can be located in a vehicle.

Occupant Classification System

An <u>occupant classification system</u> normally consists of three different types of sensors including a suppression system, which utilizes all of these sensors for automatic deactivation:

- Seat position sensor: detects the proximity of the occupant to the air bag
- Seat belt sensor: detects if the occupant seat belt is engaged and locked in the housing unit
- Occupant weight sensor: measures the weight of the occupant, determining whether the occupant has met a preset weight threshold limit.

If an occupant measures below the preset threshold limit, such as when a child car seat is placed in the seat, then the air bag will be turned off. This is critical to the technical rescuer because the system constantly monitors the weight in the seat. When the rescuer enters the vehicle and places his or her weight on the seat, the sensor will measure that weight and send the information back to the ECU, thus arming it for possible deployment. There will also be an air bag status indication light located somewhere on the dashboard panel, which will display that the air bag for that area has been shut off or disabled. This will vary with each individual manufacturer. Some vehicles may be equipped with a manual on/off switch that is designed to disable the front passenger-side air bag, but again, this feature will vary with each individual manufacturer.

> **Rescue Tips**
>
> With an occupant classification system, as the rescuer enters the vehicle and places his or her weight on the seat, the sensor will measure that weight and send the information back to the ECU, thus arming it for a possible full deployment.

Roll-over Protection System

<u>Roll-over protection system (ROPS)</u> systems were initially designed for convertible vehicles to protect occupants in vehicle roll-over incidents by means of a deployable roll bar **Figure 5-11 ▶**. Unlike fixed roll bars, these deployable roll bars are concealed until activated. They have a reaction time of less than 0.3 second and are activated by sensor detection consisting of an <u>inclinometer sensor</u>, or tilt sensor, which detects vehicle inclination or tilt with lateral acceleration (detects how fast the vehicle's tilt is changing). A <u>gravitational acceleration sensor (G-sensor)</u> detects a vehicle's weightlessness, such as that experienced in a freefall, when the vehicle starts to roll and come down.

To activate the ROPS, the sensors must detect a significant vehicle tilt with lateral acceleration including a sustained G-force; the ECU of the ROPS then determines when to deploy the roll bars. The roll bars can extend up to 20 inches (508 mm) in some models. Exercise caution when operating around a vehicle containing an undeployed roll bar. Avoid placing any parts of the body over an undeployed roll bar. For instance, if spinal immobilization is used on a patient

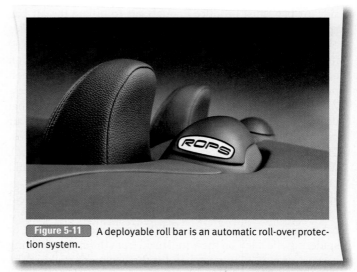

Figure 5-11 A deployable roll bar is an automatic roll-over protection system.

in the backseat, do not lean across the back of the vehicle to access the patient.

When encountering a ROPS, the technical rescuer must follow the same safety guidelines and electrical disconnection procedures that are established for vehicle air bag systems, as discussed in this chapter.

Seat Belt Pretensioning System

A <u>seat belt pretensioning system</u> is designed to automatically tighten or take up slack in a seat belt when a crash is detected **Figure 5-12 ▶**. They can be activated in conjunction with the vehicle air bags from the ECU or act independently. A seat belt pretensioner can be set up to operate at the belt buckle attachment or to operate at the anchor attachment utilizing the spool. The belt buckle will operate by pulling down and/or back on the buckle itself by means of a cable attachment or piston rod. This type of system is commonly activated by a small pyrotechnic charge or firing mechanism, which draws back on the cable or piston attachment, thus preventing forward movement of the occupant. The anchor attachment also will utilize a pyrotechnic charge or chemical charge to draw back or ratchet back the seat belt webbing on the spool, thus limiting forward movement of the occupant. To help stop the occupant less suddenly, some seat belt webbing comes equipped with a stretch design sewn into the webbing that will slowly stretch the belt at predesigned seams to gradually slow the occupant's forward movement. Seat belt assemblies can be housed in any post or column, under the seats, or in the center console. If cutting through a post at the bottom or top, the molding must be removed to reveal the pretensioning system in order to cut around the device.

The concept of dissection underlies the processes discussed in this chapter. As professionals, our job is not to rip or tear the vehicle apart; it is to dissect it section by section, fully comprehending the action taken, just as a surgeon would operate on a patient. Vehicle extrication is a step-by-step technical process, requiring continuous training to become proficient.

Voices of Experience

On Wednesday, July 14, 2010, we responded to several incidents due to flash flooding. We had assisted numerous stranded motorists. Our tactics were different depending on the situation.

During one of the assists, we came across a female stranded in her vehicle in knee-high water. We made verbal contact and determined that the best way to remove her from the hazard was to manually push her vehicle approximately 20 feet (6 meters) in reverse to higher ground. We asked her to place the vehicle in neutral with the ignition off. Three fire fighters placed themselves at the hood of the vehicle, and I placed my hand on the driver-side B-post to push the car from that point. The vehicle was a 2009 Mitsubishi Galant. Due to the height of the vehicle, my head was even with the opened window. We began to push the vehicle and almost immediately, the driver's steering wheel air bag deployed. This created a loud noise and stunned me, which caused me to stumble back off of the vehicle. I immediately heard ringing and experienced hearing loss in my left ear. The driver of the vehicle was assessed and found to have no injuries. I was assessed at the emergency department and had a follow-up with a specialist the following day. It was determined that I only had temporary hearing loss from the explosion and my hearing would return in a short period of time.

> **"Once subjected to water, the ECU can trigger the pyrotechnic device and cause the air bags to activate."**

We go through countless hours of training on new technology, and this includes the dangers present from unintentional air bag deployment. There was no impact to the vehicle at any time, and to this day, I have never heard of air bags deploying due to water damage and have not seen any such information relayed to the emergency services community. This incident caused me to do some research to see why this air bag deployed.

It was found that the ECU, which reads the air bag sensors and triggers the ignition of a gas generator propellant to rapidly inflate the air bags, is commonly located under the driver's seat or floorboards. This places the unit at a very low point in the vehicle, which subjects it to water damage. Once subjected to water, the ECU can trigger the pyrotechnic device and cause the air bags to activate. This can happen instantaneously or even days after the flooding. There have been several documented cases of this happening.

I felt compelled to share the story of this incident to hopefully avoid any further injuries. I was very fortunate to not be struck with the air bag or sustain any permanent hearing damage. We were also fortunate that the occupant of the vehicle was not injured. Manufacturer recommendations state that any time a vehicle is involved in a flood or has significant water damage, the vehicle's battery should be disconnected and the vehicle towed. We are changing our department policy to avoid this incident from happening in the future.

By sharing information, we can learn from each other's incidents and mistakes so that we can avoid any injuries or deaths in the future. Stay safe!

Keith Davis
Collingswood Fire Department
Collingswood, New Jersey

A.

B.

Figure 5-12 A seat belt pretensioning system is designed to automatically tighten or take up slack in a seat belt when a crash is detected. **A.** Belt side. **B.** Buckle side.

Emergency Procedures

Never assume that an air bag is dead just because the power has been disconnected, regardless of the amount of time that has elapsed. As discussed, a vehicle air bag system comes equipped with an energy capacitor, which can store power for up to 30 minutes in some models. Air bag inflators are always "live" until deployed, and even then they can be a multistage unit with one chamber still containing compressed gas. The best defense is recognition and identification, disconnecting the power supply, and proper distancing. Attempting to disable the inflator can potentially cause the air bags to deploy. Vehicle air bag installation, repair, and removal procedures are to be performed by licensed or certified repair technicians only.

Disconnecting Power

Several things can be done to ensure that power is disconnected to the air bag system. The first is to always remember that there is a backup energy system with storage capacitors. One problem that can be encountered when the power is disconnected to the vehicle is the inability to adjust the seats if power seats or other <u>beneficial systems</u> are installed in the vehicle. The ability to move a seat back in certain instances can greatly assist rescue efforts. The rescuer should be aware of this prior to disconnecting the power. To disconnect the power on a vehicle to disable an air bag system, follow the steps in **Skill Drill 5-1 ▸**:

1. Remove the key from the ignition. If it is a smart key, then it must remain at a preset distance from the vehicle for complete deactivation.
2. Manually activate (turn on) an electrical component of the vehicle such as the emergency warning flashers, turn signal, or radio to indicate to the rescuers whether the power is still connected.
3. If the vehicle has any beneficial systems, such as electric seat adjustments, electric windows, adjustable pedals, or other adjustable components, determine if these components can be adjusted to provide better access before disconnecting power. (**Step 1**)
4. Remove or cut the battery cables starting with the negative side. Fold the cables back onto themselves and cover the opening to avoid the cables reconnecting with a terminal or the vehicle frame. Be aware that the main 12-volt DC battery may not be located in the traditional location under the engine compartment hood; it can be found inside one of the wheel wells, under the rear/backseat, or in the trunk on some vehicles. Verify that this is the only battery in the vehicle. (**Step 2**)
5. Some vehicle manufacturers recommend removing the main fuses as an additional safety step, but location and time expended must be considered. (**Step 3**)

Recognizing and Identifying Air Bags

Vehicle air bag recognition and identification can be determined by the air bag badging or labeling system. These markings consist of acronyms that are generally located in proximity to the inflator **Figure 5-13 ▸**.

Standard air bag acronyms that can be found include:
- SRS (supplemental restraint system)
- SIR (supplemental inflatable restraint)
- HPS (head protection system)
- IC (inflatable curtain)
- SIPS (side-impact protection system)
- ROI (roll-over inflator)

Rescue Tips

Starting with vehicles manufactured in 1998, all vehicles must contain a driver- and passenger-side air bag per FMVSS 208.

Various other acronyms are used to label vehicle air bags, but the idea here is to recognize the markings and understand

Skill Drill 5-1

NFPA 1006, (10.1.5)

Disconnecting Power on a Vehicle to Disable an Air Bag System

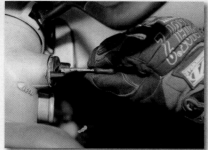

1 Remove the key from the ignition or the smart key from the vicinity. Manually activate (turn on) an electrical component of the vehicle such as the emergency warning flashers, turn signal, or radio to indicate to the rescuers whether the power is still connected. If the vehicle has beneficial systems, determine if these components can be adjusted to provide better access before disconnecting power.

2 Remove or cut the battery cables starting with the negative side. Fold the cables back onto themselves and cover the open end to avoid the cables reconnecting with a terminal or the vehicle frame. Verify that this is the only battery in the vehicle.

3 Some vehicle manufactures recommend removing the main fuses as an additional safety step, but location and time expended must be considered.

Figure 5-13 A labeling system using acronyms is generally used to indicate that an air bag system is present. The label is typically located in proximity to the inflator. This label indicates there is a Head Protection System in this vehicle.

that some type of air bag system is located in that general vicinity. These acronyms/letters may be embossed, raised, or sewn into the plastic, cloth, or leather material, depending on the manufacturer's preference. Starting with vehicles manufactured in 1998, all vehicles must contain a driver- and passenger-side air bag. All other air bags will have to be located by the rescuer. One of the assignments for the rescuer positioned inside the vehicle is to scan the entire interior of the vehicle for air bag locations. A good practice is that once the interior rescuer locates an air bag, it should be clearly marked with red tape, or other type of marking system, and prior to any cutting or spreading, this area should be exposed and relayed to all technical rescuers on scene.

■ Distancing

Once an air bag location has been identified, the next precaution is to maintain the proper distance from the deployment zone, which depends on the type of air bag system that is installed. The accepted rule of thumb for proper distancing is 10 inches (254 mm) for the driver-side air bag, 20 to 25 inches (508 to 635 mm) for passenger-side air bags, and 5 to 15 inches (127 to 381 mm) for side-impact air bags. These are only recommendations and cannot be guaranteed; each manufacturer has different components and various sizes of air bags installed in the vehicle.

Extrication Precautions

When performing extrication on any vehicle that has a potentially live air bag system, there are certain rules that the technical rescuer must adhere to prior to removal of any sections of the vehicle.

- **Rule 1:** The proper procedures for disconnecting power and proper distancing shall be conducted and maintained.
- **Rule 2:** The technical rescuer should never place anything, such as a backboard or other hard protection, between the patient and an undeployed air bag. Doing so can cause the object to be violently thrust into the patient or rescuer if the air bag deploys.
- **Rule 3:** The technical rescuer should never try to contain the air bag by tying it up with webbing or some type of bag-containment system; the shear force of the bag deploying at greater than 200 mph (322 kph) can rip any containment design apart or possibly cause the steering wheel to come apart.
- **Rule 4:** Always inspect before you dissect! Inspection is critical! The technical rescuer *must always* expose the area that is going to be cut, spread, or pushed for any air bag components or high-powered wires prior to performing any maneuver. The tendency to succumb to the adrenalin rush that comes with trying to get the patient out of the vehicle immediately must be held in check, and safety must be maintained above all things. Inadvertently cutting into a high-pressure air bag inflator (2500 to 3500 psi [17236 to 24131 kPa] or higher) can cause the cylinder to fragment violently, separating into pieces of metal projectiles and potentially causing injury to the patient or rescuer. All of the plastic molding or material coverings for roof posts, roof rail channels/liners, seat backs, and lower door posts must be removed or pulled back and examined for inflators or high-powered wiring that can be encountered with hybrid or fuel cell vehicle systems. Remember the phrase "Expose and cut"! If an inflator is located, cut a few inches in front or behind the cylinder, avoiding the wires or initiator clips; the actual nylon bag that is attached to the cylinder can be cut into and through without any problem of deployment **Figure 5-14 ▶**.
- **Rule 5:** Always consider the location of the ECU any time you are operating tools along the center console and floorboard area. It can be difficult to pinpoint the exact location of the air bag ECU; each vehicle setup may be different.
- **Rule 6:** Always be cognizant of side-impact sensors before attempting to spread or crush a door. Let's say a vehicle with a full complement of air bag components crashes its front end, causing the driver-side and front passenger-side air bags to deploy. The side-impact air bags do not deploy, and the vehicle framing partially collapses onto itself, causing the vehicle's doors to become jammed in the framing. The moment a hydraulic tool is inserted into the door frame, a side-impact sensor can potentially activate, causing any or all of the side air bags on that side of the vehicle to deploy. One possible solution is to expose the hinges of the door with a wheel well crush technique utilizing a hydraulic spreader or remove a panel section with a pneumatic air chisel and then cut the hinges with the pneumatic air chisel or hydraulic cutter, also removing the attached swing bar if one exists. Pull the door back and cut the wires that enter into the door frame hole between the hinges. This will disconnect power to the door bag only; when the door is pulled back away from the occupant, this should provide enough distance between the occupant and the air bag. This would be the only time that wires should be cut. Be aware that this is not a foolproof method; there is always the potential for the bag to deploy when the wires are compressed before they shear.

Figure 5-14 Always inspect the area for an air bag inflator before committing to cutting into a section of the vehicle.

Rescue Tips

Avoid or be very cautious of using the spreader in the center console area where the ECU may be located. Several years ago, an educational video was circulating that depicted a "live" extrication of a victim trapped in a Mitsubishi vehicle. The crew was shown attempting to maneuver the hydraulic spreader next to the center console for better leveraging; they were unaware that the air bag ECU was in the same location. When the tool was engaged, it crushed the housing of the ECU, causing a breach in the circuitry and sending a signal to fire all of the air bags in the vehicle. Both driver- and passenger-side air bags deployed instantaneously, causing serious injuries to two of the fire rescue personnel.

Wrap-Up

■ Ready for Review

- In 1967 the National Highway Traffic Safety Administration (NHTSA) issued a federal mandate titled the Federal Motor Vehicle Safety Standard 209 (FMVSS 209), *Seatbelt Assemblies*. This was the first of many subsequent regulations that outlined minimum safety requirements for motor vehicles mandating compliance from vehicle manufacturers.
- In the 1980s, the air bag became known as a supplemental restraint system (SRS) by working in conjunction with the seat belt.
- Manual seat belts are classified as an active restraint device because the occupant has to activate the system by engaging the seat belt mechanism into the anchor unit.
- A vehicle air bag is classified as a passive restraint device because the occupant does not have to activate the device to make it function; the system is automatically activated when power is applied to the vehicle.
- Smart air bag systems are using adaptive response features such as dual-stage or multistage inflation and suppression systems.
- A four-stage process occurs when an air bag deploys in a crash sequence: the crash itself, the crash sensor detecting deceleration, the air bag deploying and inflating, and the occupant moving forward and striking the bag as deflation occurs.
- Several components make up an air bag, including the air bag, initiator, electronic control unit, inflator, and sensors.
- Roll-over protection systems (ROPS) were initially designed for convertible vehicles to protect occupants in vehicle roll-over incidents. Roll bars are concealed until activated by sensors.
- Seat belt assemblies can be housed in any post or column, under the seats, or in the center console. If cutting through a post at the bottom or top, the molding must be removed to reveal the pretensioning system in order to cut around the device.
- Never assume that an air bag is dead just because the power has been disconnected; a vehicle air bag system comes equipped with an energy capacitor, which can store power for up to 30 minutes in some models.
- Eliminating potential hazards of SRS systems may include disconnecting power, recognizing and identifying air bags, distancing, and following certain rules.

■ Hot Terms

Accelerometer A sensor that detects a crash.

Active restraint device A device that the occupant must activate; for example, a seat belt is an active device because the occupant has to engage the seat belt mechanism into the anchor unit.

Air bag An inflatable plastic bag that inflates automatically to cushion passengers in the event of a collision.

Beneficial systems Auxiliary-powered equipment in motor vehicles or machines that can enhance or facilitate rescues such as electric, pneumatic, or hydraulic seat positioners, door locks, window operating mechanisms, suspension systems, tilt steering wheels, convertible tops, or other devices or systems that facilitate the movement (extension, retraction, raising, lowering, conveyor control) of equipment or machinery.

Electronic control unit (ECU) Also known as the air bag control unit (ACU), this is the brains of an air bag system, consisting of a small processing unit generally located in the center of the vehicle.

Federal Motor Vehicle Safety Standards (FMVSSs) Safety standards enacted to protect the public from unreasonable risk of crashes, injury, or death resulting from the design, construction, or performance of a motor vehicle.

Gas generation system An inflation system that completely fills the air bag to the appropriate inflation ratio to protect the occupant.

Gravitational acceleration sensor (G-sensor) A sensor that detects a vehicle's weightlessness, such as that experienced in a freefall, when the vehicle starts to roll and come down.

Inclinometer sensor A tilt sensor that detects vehicle inclination or tilt with lateral acceleration (detects how fast the vehicle's tilt is changing).

Inflator One of the most critical design features for an air bag, providing the ability to fill up the bag instantaneously in milliseconds from the onset of the collision. There are two basic inflation systems—a stored compressed gas system and a gas generation system.

Initiator A device such as a squib (a pyrotechnic device) that activates the air bag through an electrical current, which becomes instantly hot and ignites the combustible material inside the containment housing, or through ignition of a burst disc, which releases compressed gas.

Multistage inflators Also known as hybrid inflators, cylinders that can be comprised of two separate chambers of compressed gas—one with a large amount of product and the other with a smaller amount of product.

Occupant classification system A system consisting of three different types of sensors: the seat position sensor, which detects the proximity of the occupant to the air bag; the seat belt sensor, which detects if the occupant's seat belt is engaged and locked in the housing unit; and the occupant weight sensor, which measures the weight of the occupant, determining whether the occupant has met a preset weight threshold limit.

Passive restraint device A device that the occupant does not have to activate for it to function; the system is automatically activated when power is applied to the vehicle.

Roll-over protection system (ROPS) (vehicles) A system designed to protect occupants in vehicle roll-over incidents by means of a deployable roll bar.

Safing sensor A type of air bag sensor that has a deceleration setting lower than the crash-type sensor. The sensor prevents false deployments.

Seat belt pretensioning system A system designed to automatically tighten or take up slack in a seat belt when a crash is detected.

Side-impact air bag An air bag designed to activate immediately upon impact to protect the following areas of the occupant: the head, the chest/upper torso, and a combination of the head and the chest/upper torso. There are three types of side-impact air bags. All three types can be found in the door, seat backs, roof posts,

or roof rails. These air bags may be labeled HPS (head protection system), IC (inflatable curtain), SIPS (side-impact protection system), or ROI (roll-over inflator air bag).

Smart air bag system An air bag system that will automatically adjust the pressure in the air bag by utilizing multistage inflators and basing the deployment force on a number of calculated factors, such as crash severity, occupant's weight, proximity to the air bag, seat belt usage, and seat position.

Squib A pyrotechnic device.

Stored compressed gas system An inflation system comprised of a single-stage or multistage inflation process. The igniter or squib sets off a burst or rupture disc that acts as a seal, holding back the compressed gas. When activated, the disc breaks open, releasing the gas from the chamber, which expands and instantly fills the air bag.

Supplemental restraint system (SRS) A system that uses supplemental restraint devices such as air bags to enhance safety in conjunction with properly applied seat belts. Seat belt pretensioning systems are also considered part of an SRS.

Suppression system A device that shuts down the air bag if an occupant classification system detects a child in the air bag deployment zone or if one of the sensors detects a high risk potential by acquiring the occupant's weight, height, proximity to the air bag, seat belt usage, and seat position; the system sends this information to the electronic control unit, which will then shut off the air bag if a high risk to the occupant is determined.

Technical Rescuer *in Action*

You arrive on scene at an incident involving two vehicles and two victims. One of the vehicles has rolled over. Because multiple air bags have deployed, you cannot get a good visual of the victims from your vehicle. After performing a scene size-up, you carefully approach and discover that both victims are wearing seat belts and one of the vehicles has a roll-over protection system. All of the information you have learned about supplemental restraint systems begins to flood your mind.

1. You know that air bags are known as supplemental restraint devices, which are designed to be a supplement to:
   A. safe driving.
   B. wearing a supplemental restraint device.
   C. being properly positioned in the seat.
   D. wearing a seat belt.

2. Manual seat belts are classified as:
   A. pretensioning.
   B. active restraint devices.
   C. self-activating.
   D. supplemental restraint devices.

3. Air bags are classified as:
   A. pretensioning.
   B. active restraint devices.
   C. self-activating.
   D. passive restraint devices.

4. Before you get any closer, you recall that the technical rescuer's best defense against a live air bag is recognition/identification, disconnecting the power supply, and:
   A. deactivation.
   B. proper distancing.
   C. activation.
   D. pushing the bag out of the way.

5. A vehicle air bag system consists of all of the following except:
   A. the air bag.
   B. the electronic control unit.
   C. the sensor.
   D. the ring-clip.

6. The entire vehicle crash process, which involves dissipating all of the kinetic energy of the vehicle, takes approximately:
   A. 100 to 125 milliseconds.
   B. 300 to 425 milliseconds.
   C. 800 to 825 milliseconds.
   D. 2 seconds.

7. An occupant classification system normally consists of the following sensors except a:
   A. seat position sensor.
   B. seat belt sensor.
   C. crash sensor.
   D. occupant weight sensor.

8. The roll bars in a roll-over protection system can extend up to how many inches in some models?
   A. 10 inches
   B. 15 inches
   C. 12 inches
   D. 20 inches

9. The seat belt pretensioning system can be located in any of the following areas except:
   A. under the seat.
   B. in the B-post column.
   C. under the center console.
   D. in the dashboard.

10. Seat belt pretensioners activate at the belt buckle attachment or the:
    A. anchor attachment using the spool.
    B. webbing section.
    C. door.
    D. ECU.

NFPA 1006 Standard

Chapter 5, Job Performance Requirements

5.4 **Maintenance.**

5.4.1 Inspect and maintain hazard-specific personal protective equipment, given clothing or equipment for the protection of the rescuers, including respiratory protection, cleaning and sanitation supplies, maintenance logs or records, and such tools and resources as are indicated by the manufacturer's guidelines for assembly or disassembly of components during repair or maintenance, so that damage, defects, and wear are identified and reported or repaired, equipment functions as designed, and preventive maintenance has been performed and documented consistent with the manufacturer's recommendations. (pages 102–107)

(A) Requisite Knowledge. Functions, construction, and operation of personal protective equipment; use of record-keeping systems of the AHJ; requirements and procedures for cleaning, sanitizing, and infectious disease control; use of provided assembly and disassembly tools; manufacturer and department recommendations; pre-use inspection procedures; and ways to determine operational readiness. (pages 102–107)

(B) Requisite Skills. The ability to identify wear and damage indicators for personal protective equipment; evaluate operational readiness of personal protective equipment; complete logs and records; use cleaning equipment, supplies, and reference materials; and select and use tools specific to the task. (pages 102–107)

5.4.2 Inspect and maintain rescue equipment, given maintenance logs and records, tools, and resources as indicated by the manufacturer's guidelines, equipment replacement protocol, and organizational standard operating procedure, so that the operational status of equipment is verified and documented, all components are checked for operation, deficiencies are repaired or reported as indicated by standard operating procedure, and items subject to replacement protocol are correctly disposed of and changed. (pages 102–107)

(A) Requisite Knowledge. Functions and operations of rescue equipment, use of record-keeping systems, manufacturer and organizational care and maintenance requirements, selection and use of maintenance tools, replacement protocol and procedures, disposal methods, and organizational standard operating procedures. (pages 102–107)

(B) Requisite Skills. The ability to identify wear and damage indicators for rescue equipment, evaluate operation readiness of equipment, complete logs and records, and select and use maintenance tools. (pages 102–107)

Knowledge Objectives

After studying this chapter, you will be able to:

- Explain the basic categories of tools. (page 102)
- Describe the various types of PPE, their uses, and maintenance. (pages 102–107)
- Describe the various types of hand tools and how they operate. (pages 107–115)
- Describe the various types of pneumatic tools and how they operate. (pages 115–121)
- Describe the various types of electric-powered tools and how they operate. (pages 122–127)
- Describe the various types of fuel-powered tools and how they operate. (pages 127–129)
- Describe the various types of hydraulic tools and how they operate. (pages 129–132)
- Describe the various types of stabilization tools and how they operate. (pages 132–135)
- Describe how to properly organize equipment on the scene. (page 135)
- Describe the classifications of rescue vehicles and their uses. (pages 135–136)
- Describe class B foam and its benefits during vehicle extrication. (pages 136–137)
- Describe the types of signaling devices and power detection used during technical rescue incidents. (pages 137–138)
- Describe the most common stretchers used by rescue services when removing victims. (pages 138–139)

Skills Objectives

There are no skills objectives for this chapter.

you are on the scene of an extraction incident and are told that one of the hydraulic tools has failed.

1. What alternative tools are available?
2. Will you need to call for backup?

Introduction

"The mechanic that would perfect his work must first sharpen his tools." —Confucius

The technical rescuer must not only have a vast working knowledge of tools used in the field, but he or she must be proficient with them to be successful. With the multitude of tools available, it would take an entire book to cover all of them. This chapter will cover some of the more common tools used in vehicle extrication.

Tools for vehicle extrication purposes can be broken down into basic categories:

1. Hand tools
2. Pneumatic tools
3. Hydraulic tools
4. Electric- or battery-operated tools (nonhydraulic)
5. Fuel-powered tools

Note: Stabilization tools can be classified as a hand, pneumatic, or hydraulic type.

Before the technical rescuer can start to work with tools, he or she must wear full personal protective equipment (PPE) for safety. National Fire Protection Association (NFPA) 1500, *Standard on Fire Department Occupational Safety and Health Program*, and NFPA 1951, *Standard on Protective Ensembles for Technical Rescue Incidents*, describe the protective ensemble required to be worn by the technical rescuer at the scene of rescue or recovery operations.

Personal Protective Equipment

The **protective ensemble** includes the body piece, helmet, eye protection, gloves, footwear, and hearing protection. Respiratory protection may also be necessary if operating in an area known or suspected to be hazardous. To be compliant to NFPA 1951, the protective ensemble shall provide protection from exposure to physical, thermal, liquid, and body fluid–borne pathogen hazards. The selection of the specific components of

PPE will depend on the hazards at the scene, the duration of the incident, the availability of equipment, and the weather.

Rescue Tips

Always make it a priority to wear all of the required protective gear when at a call or at a training drill.

Head Protection

The helmet will protect the head and has a retention system (strap) to maintain the helmet position on the head. Helmets come in several types, including the traditional fire helmet, the lighter USAR (urban search and rescue) helmet, and the Euro-style helmet [Figure 6-1]. Both the traditional and USAR helmets are acceptable, with the Euro-style helmet being accepted if it is NFPA-compliant.

Inspect helmets for any signs of damage to the shell such as cracks, major chips, or gouges. Such flaws could compromise the integrity of the shell; the helmet should be removed from service if these defects are evident. The suspension system, chin strap, and visor should also be inspected for any signs of damage. Inspect helmet liners and hoods for unusual wear, tears, or other damage. Keep all head protection clean to facilitate inspection. In addition to these general guidelines, familiarize yourself with any additional recommendations or field maintenance guidance or procedures provided by the manufacturer of the equipment.

Body Protection

The type of body protection used will depend on the hazards present, the authority having jurisdiction (AHJ), and/or the level of comfort for the wearer while maintaining full protective compliance. Protective coveralls or coats/pants protect the upper torso, lower torso, arms, and legs. There are several types that can be worn. Most companies wear full structural firefighting clothing that meets all NFPA-required standards, but such

A.

B.

C.

Figure 6-1   Head protection. **A.** Fire fighter's helmet. **B.** USAR helmet. **C.** Euro-style helmet.

A.   B.

Figure 6-2   The technical rescuer must wear a full personal protective ensemble at all rescue or recovery operations. **A.** Structural firefighting clothing. **B.** Extrication jumpsuit.

clothing tends to be bulky and hot. Newer, lightweight turnout gear, offering the same level of protection, is now available. Some companies prefer the coverall or jumpsuit. A fully compliant extrication jumpsuit is heavier grade with reinforcements sewn on the knees and elbow areas. Extrication jumpsuits are a great alternative to turnout gear, providing easier movement and less fatigue from overheating Figure 6-2 ▲. There are many different types of jumpsuits; make sure the suit can offer the extra protection needed to operate safely at an extrication incident. Leave the lightweight mechanic-type jumpsuits at the station; these will offer little or no protection. Remember, the AHJ representing each agency will have the ultimate decision on what will be worn by personnel.

Eye and Face Protection

Exposure to debris, dirt, dust, fumes, bright light, and hazardous fluids can potentially cause temporary or even permanent damage to vision or can enable a point of entry for any bloodborne pathogens (microorganisms in the blood) or other body fluids that can transmit illness and disease in people. Face and eye protection can be provided by a respiratory face piece, or a face shield or goggles Figure 6-3 ▶. Respirators offer the most complete protection and will be discussed later in this chapter. Face shields normally only offer protection from the top of the eyes to the nose area, leaving an open bottom section where a hazardous substance can enter from underneath the shield and affect the eyes. Goggles are a good choice for areas where respirators are not needed; they form-fit against the skin and provide protection on all sides. Unfortunately, in hot or humid environments, goggles tend to hold a lot of condensation and will fog over, obscuring the wearer's vision. Safety glasses are

A.

B.

Figure 6-3 Eye and face protection. **A.** A respirator face piece provides maximum protection for the eyes and the face. **B.** Rescuers may need specialty eye protection such as a face shield or mask and goggles.

also commonly used at emergency incidents. Safety glasses should include a retainer strap (chums) and a side shield or a wraparound design to protect the sides of the eye. The disadvantage of goggles is that prescription glasses are difficult or impossible to wear under them. Some goggles and safety glasses can be made with prescription lenses, but the cost has to be considered. Some goggle systems offer clip adapters as an option to attach prescription lenses inside the goggles. NFPA 1500 states that all face and eye protection should meet the requirements of

the American National Standards Institute Z87.1 (ANSI Z87.1) standard, which establishes performance criteria and testing requirements for occupational and educational eye and face protection devices.

The inspection of eye protection should focus on looking for scratches or gouges that will impair the vision of the user, as well as broken or missing pieces. In addition to these general guidelines, familiarize yourself with any additional recommendations or field maintenance guidance provided by the manufacturer.

■ Hand Protection

Protective gloves should protect the hands and wrist, and should extend no less than 1 inch (25 millimeters [mm]) from the wrist crease. Gloves are a very important component for the technical rescuer. If the gloves are too bulky, for example, like traditional fire gloves, simple procedures, such as fully depressing the trigger on an air chisel or reciprocating saw will be extremely difficult and cumbersome. Leather work gloves from the hardware store will not work either; although they provide dexterity, they will provide little to no protection for your hands. Extrication gloves have Kevlar® or other heavy material sewn on in areas that require the most protection such as fingertips, palms, and knuckles, but do not restrict movement **Figure 6-4 ▼**. Extrication gloves are not impervious to all penetrations; a stray shard of glass or jagged metal could make it through the protective barrier. Still, extrication gloves offer the best level of protection as compared to other types of gloves. For additional protection, surgical latex gloves should be worn under the extrication gloves to offer an added barrier against any bloodborne pathogens or other biohazards. This standard practice will not compromise dexterity and provides bloodborne pathogen protection.

Inspect gloves for rips, tears, weak or missing stitching, and exposure to contaminants. They should be cleaned, decontaminated, sanitized, and dried according to the manufacturer's instructions, and you should be familiar with any additional recommendations or field maintenance guidance provided by the manufacturer.

Figure 6-4 It is critical that the technical rescuer wear proper gloves; he or she must purchase gloves with maximum dexterity and protection.

tection should be used routinely. Hearing protection, such as earplugs, earmuffs, and noise-reducing or -canceling headphones, is designed to reduce the decibel (dB) level that enters the wearer's ear canal. Sound is measured in decibels, with a normal verbal conversation between two individuals measuring at around 60 dB. Damage and/or hearing loss can occur with prolonged exposure to decibel levels greater than 85 dB. A reciprocating saw measures in the 100 to 105 dB level, and a typical hydraulic engine, depending on the manufacturer type, can measure 60 to 95 dB. Damage sustained to one's hearing can occur gradually over continual or prolonged exposure; the rescuer may not notice any damage occurring to the eardrum until it is too late Figure 6-6 ▾.

A.

B.

Figure 6-5 Boots used in rescues are typically either fire fighters' boots or safety work boots. **A.** Fire fighters' boots. **B.** Safety work boots.

Foot Protection

Footwear should consist of boots containing puncture-resistant materials to protect the entire foot, including the sides, sole, and ankle. The boot should also be equipped with an impact- and compression-resistant toecap to prevent crushing injuries to that area of the foot. Most boots used in rescue situations are either fire fighters' boots or safety work boots Figure 6-5 ▴.

When inspecting footwear, look for wear, tears, or holes in the leather or rubber, as well as damage or excessive wear to the soles, laces, or zipper. Waterproof boots in accordance with the manufacturer's directions, and familiarize yourself with any additional recommendations or field maintenance guidance provided by the manufacturer.

Hearing Protection

Because rescuers are commonly exposed to high levels of noise from the equipment used to perform extrication, hearing pro-

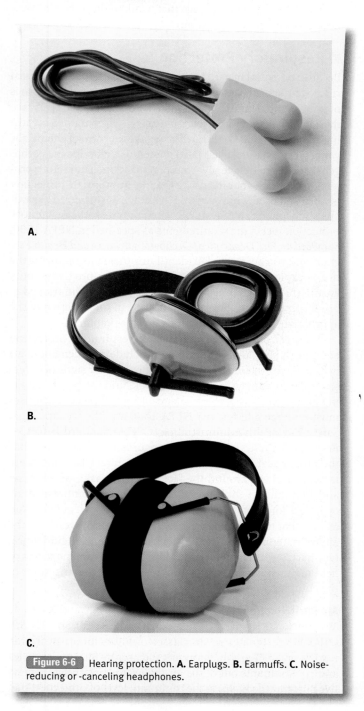

A.

B.

C.

Figure 6-6 Hearing protection. **A.** Earplugs. **B.** Earmuffs. **C.** Noise-reducing or -canceling headphones.

Earplugs come in a variety of styles and are designed to be inserted into the outer ear canal. They conform to the shape of the canal and are disposed of after use. Earplugs are inexpensive and easy to carry.

Earmuffs are designed to fit over the ears and can be more comfortable than earplugs. Some are designed to fit over the head while others are designed to be mounted onto a helmet.

Noise-canceling headphones, like earmuffs, fit over the head of the wearer and are designed to emit an electronic signal that blocks any high-decibel frequency that can be damaging to the ear. These headphones are expensive and prevent the rescuer from wearing a helmet during extrication.

Inspection of hearing protection includes ensuring cleanliness and checking for damage that would reduce the effectiveness of the earplugs or earmuffs.

Respiratory Protection

Because potential hazards in the air we breathe can pose great risk to rescuers, the selection and use of respiratory protection is of utmost importance, whether it is wearing a simple N-95 face mask or full respiratory protection using self-contained breathing apparatus (SCBA). Hazards may take the form of chemicals, vapors, fumes, dust, glass particulate/dust, bloodborne pathogens, or oxygen deficiency. Your organization's respiratory protection program should ensure that the appropriate protection is worn and used while working in known or suspected hazardous atmospheres; it is available when required; the respiratory protection meets the requirements as specified in NFPA 1500, *Standard on Fire Department Occupational Safety and Health Program*; and you are trained in identifying incidents requiring the use of respiratory protection. Additionally, this program should ensure that you are physically capable of using and trained to use respirators and that you are trained in the selection of the appropriate respirator to use for a given situation.

Inspection and maintenance of respiratory protection equipment should, as a minimum, include replacement of air cylinders, inspection of system components for signs of excessive wear or damage, an air leak check, and the proper cleaning and sanitation of the face and/or mouth piece. Each user must undergo a fit test per NFPA 1500 and the Occupational Safety and Health Administration (OSHA) standard 1910.134, *Respiratory Protection*. This test ensures that the wearer has the proper seal (i.e., the mask fits the wearer's face and there is no air leakage around the face and the mask), which is set and maintained throughout usage. Check compressed air cylinders for damage to the cylinder shell and regulator, and ensure that the cylinder has been hydrostatically tested at the appropriate intervals. In addition to these general guidelines, familiarize yourself with any additional recommendations or field maintenance guidance provided by the manufacturer of the equipment.

Air-Purifying Respirators

The National Institute for Occupational Safety and Health (NIOSH) designates seven classes of filters for air-purifying respirators (APRs), which protect the user by filtering particles and contaminants out of the air the user is breathing. Ninety-five percent (the N-95) is the minimum level of protection that

Figure 6-7   Self-contained breathing apparatus.

NIOSH will qualify for effective filtration protection. This level should be sufficient to protect the technical rescuer from glass particulates but not for any hazardous fumes, which would require full SCBA to be worn. APRs do not have a separate source of air, but rather filter and purify ambient air before it is inhaled.

Self-Contained Breathing Apparatus

Self-contained breathing apparatus (SCBA) is a respirator with an independent air supply allowing rescuers to enter dangerous atmospheres Figure 6-7 ▲. SCBA is available in 30- to 60-minute versions and protects against almost all airborne contaminants. SCBA may be a disadvantage in some situations because of the limited air supply. Also, the bulk of the frame that holds the SCBA in place may make working in confined or restricted spaces difficult, and the weight of the unit may cause excessive fatigue during long operations.

Supplied Air Respirator

In a supplied air respirator/breathing apparatus (SAR/SABA), breathing air is supplied by air line from either a compressor or stored air (bottle) system located outside the work area Figure 6-8 ▶. The apparatus has the same advantages as SCBA in regard to the face piece and self-contained air supply. System components of a SAR include the positive pressure–type respirator, the escape bottle (5–10 minutes, with 10 minutes recommended), the 300-foot (91-meter) maximum air line (depending on flow required and manufacturer), and a compressor or stored air system.

The advantages of a SAR are its air supply, which is not limited to what you take with you; its small size, which allows access into smaller spaces and provides for more maneuver-

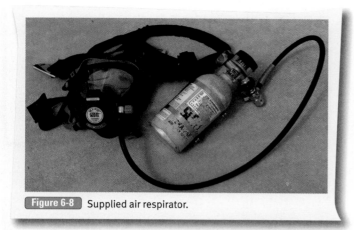

Figure 6-8 Supplied air respirator.

ability; and its lighter weight, which helps reduce fatigue during long operations. Disadvantages to using SARs include the potential for air line entanglement or damage and the possibility that the small escape bottle size may not allow much time in an emergency to evacuate to a safe place. SARs should be used in conjunction with SCBA so there is enough air for egress in the event of a primary air supply failure. Remember that all face pieces, regardless of the type of apparatus, must undergo a fit test to ensure a proper seal on each individual user.

Maintenance of PPE

NFPA 1851, *Standard on Selection, Care, and Maintenance of Protective Ensembles for Structural Fire Fighting and Proximity Fire Fighting*, covers procedures for inspections, cleaning, decontamination, repairs, storage, and record keeping for the protective ensembles for structural and proximity firefighting. This standard concerns some of the following gear pertaining to technical rescue: coats, pants, helmets, hoods, gloves, and footwear elements. A departmental standard operating procedure (SOP) should be created or implemented regarding proper maintenance of PPE. Some of the key aspects of proper maintenance include the following:

- Always inspect PPE before and after each use to check for damage that may have occurred during use or while in storage.
- Follow the manufacturer's instructions in terms of inspection, maintenance, cleaning, limitations, and repair methods, including the use of maintenance tools.
- Document all inspection and maintenance activities performed in accordance with your organization's SOPs. A maintenance log listing each type of PPE may be the best way to handle this.
- Consider utilizing/contracting an independent company that follows NFPA 1851 guidelines by specializing in cleaning and maintenance of PPE; this limits any secondary exposures that may occur when attempting to clean contaminated gear at the station. Some of these companies will also supply record-keeping options by tracking all of the maintenance and history for the gear within an agency.
- Report any problems detected immediately so PPE can be repaired and/or replaced, if deemed necessary.

Wearing the same gear that has been through several incidents and is covered in various by-products of combustion is not only foolish but dangerous to fellow personnel as well as the wearer. Secondary exposure to contaminants left on gear can affect personnel well after the call is over. Clean all gear at periodic intervals and after every exposure or excessive soiling.

Hand Tools

A **hand tool** is described as any tool or equipment that operates from human power. The technical rescuer must have a thorough working knowledge of the different types of hand tools that can be utilized at an extrication incident. Hand tools are the basis of all working tools; being able to utilize a hand tool effectively or when all other tools fail is an asset the technical rescuer needs to acquire. The first inclination of most rescuers may be to use power tools. However, in some situations, hand tools may be more efficient than power tools. At some point, each technical rescuer will experience a total electrical or mechanical failure of the powered tools he or she is using during an incident. The rescuer will then have to rely on knowledge of hand tools to get the job done; this is the true mark of a skilled professional. Again, there is a tremendous variety of hand tools, and each individual has personal favorites. This book will cover some of the more prominent hand tools common on the extrication scene.

Hand tools can be categorized as follows:
- Striking tools **Figure 6-9 ▾** **Table 6-1 ▾**

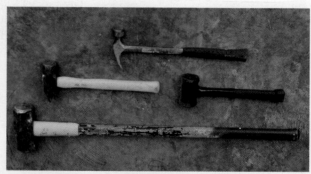

Figure 6-9 Hammer-type striking tools (from top): hammer, mallet, maul, sledgehammer.

Table 6-1   Tools for Striking
Battering ram
Hammer
Pick-head axe
Chisel
Mallet
Sledgehammer
Flat-head axe
Maul
Spring-loaded center punch

- Leverage/prying/spreading tools [Figure 6-10 ▾] [Table 6-2 ▸]
- Cutting tools [Figure 6-11 ▾] [Table 6-3 ▸]
- Lifting/pushing/pulling tools [Figure 6-12 ▸] [Table 6-4 ▸]

Figure 6-10 Leverage/prying/spreading tools.

Table 6-2   Leverage/Prying/Spreading Tools

Claw bar	Pry bar
Halligan tool	Flat bar
Kelly tool	Hydraulic spreader
Crowbar	Rabbet tool
Hux bar	

Table 6-3   Cutting Tools

Axes	Handsaws
Bolt cutters	Hydraulic shears
Chain saws	Reciprocating saws
Cutting torches	Rotary saws
Hacksaws	Seat belt cutters

A.

C.

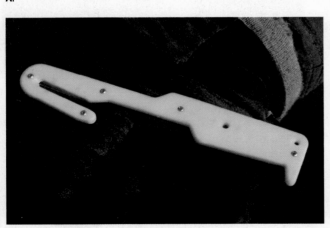

B.

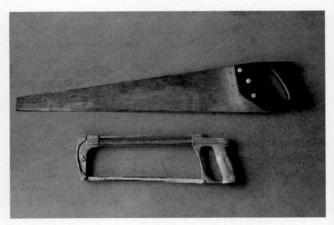

D.

Figure 6-11 Cutting tools. **A.** Trauma shears. **B.** Seat belt cutter. **C.** Bolt cutters. **D.** Handsaws.

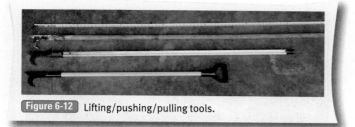

Figure 6-12   Lifting/pushing/pulling tools.

Table 6-4   Lifting/Pushing/Pulling Tools

Ceiling hook	Pike pole
Multipurpose hook	San Francisco hook
Roofman's hook	Drywall hook
Clemens hook	Plaster hook

■ Striking Tools

Striking tools are used to apply an impact force to an object. Examples include driving a nail in a shoring system, forcing the end of a prying tool into a small opening by creating a purchase or access point for another tool, or breaking a vehicle window with a center punch Figure 6-13 ▶ and using a Halligan bar to clean out the window frame of glass fragments. Included in this category are hammers, punches, and glass saws. Hammer-type striking tools have a weighted head with a long handle. They include the hammer, mallet, sledgehammer, maul, flat-head axe, and battering ram.

The **spring-loaded center punch** is a glass removal tool used on tempered glass only; it is not designed to work on laminated glass. When engaged, the center punch uses a spring-loaded plunger to fire off a steel rod with a sharpened point directly into a pinpoint area of glass, causing the glass to shatter. Two of the biggest flaws of the center punch are that the spring will normally fail from getting water in the chamber and cause it to rust, or the point of the steel rod will dull from repeated use, rendering it useless; you will notice this when the tool fails to break the glass after several attempts Figure 6-14 ▶ . The procedures for breaking tempered glass on a vehicle utilizing a spring-loaded center punch will be discussed in Chapter 9, *Victim Access and Management*.

Numerous other types of punches are available; some consist of a stationary punch design that has a hardened steel point attached to the end of a knife or the end of a pen. There is also the hammer-type punch that consists of a one- or two-sided hardened steel point head Figure 6-15 ▶ . A spring-back–type punch is about the length of a pen; the user must pull back and release a spring in the middle of the device to correctly operate it Figure 6-16 ▶ . All of these glass tools require different operating procedures. Follow the manufacturers' instructions to operate them as they were designed to be used.

Another example of a glass removal tool is a glass handsaw Figure 6-17 ▶ . The glass handsaw is a manually operated striking tool for removing glass. It is an extremely versatile tool and has several design applications primarily for removing tempered and laminated vehicle glass.

Figure 6-13   Center punches are used as striking tools.

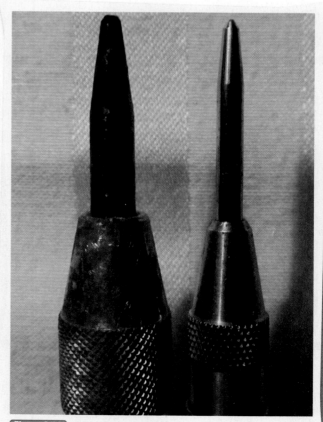

Figure 6-14   Examine the tip of your center punch to make sure it is not dull like the center punch depicted on the left in this picture.

Figure 6-15 A hammer-type punch consists of a one- or two-sided hardened steel point head.

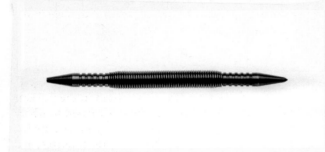

Figure 6-16 A spring-back–type punch is about the length of a pen; the user must pull back and release a spring in the middle of the device to correctly operate it.

Figure 6-17 The glass handsaw.

Figure 6-18 The glass handsaw is used here to demonstrate the proper technique of breaking tempered glass. The front guard, used against a hard surface, prevents the tool and the technical rescuer's hand from going through the window when the glass is broken.

For breaking tempered glass, the center grip handle of the glass handsaw has a slot where a center punch is stored with the tip facing inward. When ready for use, the center punch seats firmly in the handle and is then positioned with the tip facing outward. The tip of the center punch is placed against the glass with the hand guard positioned against a hard surface. Placing the front guard against a hard surface is a safety measure that prevents the tool and the technical rescuer's hand from going through the window when the glass is broken **Figure 6-18 ▲**.

Another feature that the glass handsaw offers is a notched section on the end of the tool that fits over the top lip of the glass. When the tool is turned to the side, this causes the glass to immediately fracture, eliminating the possibility of misfires from the use of a center punch. Never attempt to use the point section of the glass tool to break out tempered glass; tempered glass is designed to resist breaking from impacts such as this. Attempting to break a window comprised of tempered glass by striking it with a tool, whether it is a glass tool or a Halligan bar, is not only highly unprofessional, but can potentially cause injury to the crew, yourself, or the victim. Remember, we are technicians, trained to use our mental ability to direct and know when to use our physical ability.

These applications will be discussed in Chapter 9, *Victim Access and Management*.

■ Leverage/Prying/Spreading Tools

Two types of leverage tools are distinguished: rotating and prying **Figure 6-19 ▶**. Rotating tools are designed to turn objects and include wrenches, pliers, and screwdrivers. These types of tools are commonly used for disassembly during vehicle and machinery rescue.

Simple prying and spreading tools act as a lever to multiply the force a person can exert to bend or pry objects apart from other objects. As the name implies, they are used to pry apart objects; in addition, they are used to lift heavy objects. Common simple hand prying tools include the following equipment:

- Claw bar
- Crowbar
- Flat bar
- Halligan bar
- Kelly tool
- Pry bar

When using leverage tools, make sure you pick the right tool for the job. Remember the saying, "Work smarter, not harder."

Fire rescue and emergency personnel are ingenious in coming up with ways to make a difficult situation easier. A poor example of this would be trying to modify a tool to make it work more efficiently, which may also go against manufacturer specifications and safety guidelines. For example, a pipe is sometimes used to extend the length of a tool handle; this extension is commonly referred to as a cheater bar. This modification is meant to extend the distance from the fulcrum of the tool to where the force is being applied, thereby increasing the amount of leverage exerted. In doing so, however, the design strength of the tool may be exceeded, ultimately causing its damage or failure. These tools are susceptible to damage, and

Table 6-5	Cleaning and Inspecting Hand Tools
Metal Parts	All metal parts should be clean and dry. Remove rust with steel wool. Do not oil the striking surface of metal tools because this may cause them to slip.
Wood Handles	Inspect for damage such as cracks and splinters. Sand the handle if necessary. Do not paint or varnish; instead, apply a coat of boiled linseed oil. Check to be sure the tool head is tightly fixed to the handle.
Fiberglass Handles	Clean with soap and water. Inspect for damage. Check to be sure the tool head is tightly fixed to the handle.
Cutting Edges	Inspect for nicks or other damage. File and sharpen as needed. Power grinding may weaken some tools, so hand sharpening may be required.

it is important to take these tools out of service if they show any signs of damage, such as cracks or bends Table 6-5 ▲ .

■ Cutting Tools

Cutting tools have a sharp edge designed to sever an object. Manual cutting tools range from tools carried in the pockets of turnout coats to larger tools that must be carried to where they are needed. Cutting tools include saws, chopping/snipping shears, trauma scissors or shears, seat belt cutters, knives, and chisels. Each type is designed to work on certain types of materials and cut in a different manner. Because of their sharp edges and the force needed to make this equipment work, rescuers and victims can be injured, and cutting tools can be ruined if they are used incorrectly.

Common handsaws used in rescue include hacksaws; carpenter's handsaws; and jab, compass, keyhole, folding, and bow saws. With the exception of hacksaws, these other handsaws would rarely, if ever be used in vehicle extrication applications. Hacksaws that are used for extrication should be the heavy-duty type. They have a thicker and more durable frame so the technical rescuer can use both hands to push with force for optimum cutting performance. The heavy-duty type also offers the ability to set the blade under high-tension settings, thus preventing the blade from twisting and bending, providing a sturdy and ridged cutting action Figure 6-20 ▶ . Hacksaws are useful when metal needs to be cut under closely controlled conditions or when an electric-powered reciprocating saw cannot be utilized.

Large-tooth saws, such as the bow saw, are effective tools for cutting large timbers or tree branches; as a consequence, they can prove especially useful when a chainsaw is not available at motor vehicle collisions where tree limbs may otherwise hamper the rescue effort. The keyhole, jab, and compass saws are narrow, slender models used to cut wood, plastic, and Sheetrock quickly; their use at vehicle extrication incidents is rare. Saw blades with fine teeth are designed for cutting finished lumber and tend to cut the material more slowly. For this reason, they are not usually used in rescue work. The exception

A.

B.

Figure 6-19   Leverage tools. **A.** Rotating. **B.** Prying.

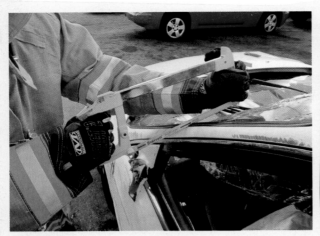

Figure 6-20 A fire fighter using a hacksaw to cut through a metal A-post of a vehicle roof.

of this statement would be the reciprocating saw blade and hacksaw blade when cutting metal; this requires fine teeth to operate more efficiently.

Chopping tools include the flat-head axe and the pick-head axe **Figure 6-21 ▼**. These tools consist of a long handle with a weighted head on the end. The head has a sharpened cutting edge that is used to strike the object. These tools should have a semi-sharpened edge so the edge will not chip if it hits an object such as a nail or a stone. Shovels may also be used as a chopping tool with the added functions of prying and material removal.

Snipping tools or shears include bolt cutters, cable cutters, insulated wire cutters, sheet metal snips, seat belt cutters, and EMS trauma scissors or shears. These types of tools operate on a leverage concept, with the fulcrum being located just behind the cutting edges. This concept allows for a concentration of the cutting force on a small area. Care must always be taken when cutting wires, including ensuring that they are not electrically energized. Seat belt cutters are normally very simple devices that utilize a razor to cut through the material. These razors are prone to dulling and rusting after they are used the first few times. It is advised to purchase seat belt cutting devices designed with blades that can be easily changed out after becoming dull or oxidized. Another cutting tool that is similar to the bolt cutter is a steering wheel ring cutter; this tool uses a ratcheting-type tensioning mechanism that builds pressure to cut through a steel steering wheel ring. A hydraulic cutter can also be used to accomplish this task **Figure 6-22 ▼**.

There are a number of knives used by rescuers **Figure 6-23 ▶**. Examples include folding knives, seat belt cutters, linoleum knives, drywall knives, and box cutters. Knives should have a retractable or folding blade that locks when open. Keep knives sharp or replace them after each use to provide for maximum readiness the next time they are needed.

Hand-operated chisels, which are used to cut wood and metal, are typically operated by striking the chisel with a hammer or mallet. They come in a variety of widths and styles and should be used only for cleaving the material for which they were designed. Wood chisels are used to remove small pieces of wood, whereas metal chisels, sometimes called cold chisels, are used to cut sheet metal or to cut off bolts and other objects. With the addition of pneumatic air chisels, hand-operated chisels would rarely, if ever be utilized in vehicle extrication applications.

A.

B.

Figure 6-21 Two types of axes. **A.** Flat-head axe. **B.** Pick-head axe.

Figure 6-22 Hydraulic cutters can be used to cut through steering wheels.

Figure 6-23 Folding knives with a serrated blade work well for a variety of fabric materials, such as seat belts, or cutting through vehicle seat fabric to expose the seat frame during rescue work.

Lifting/Pushing/Pulling Tools

Manual lifting and pushing tools include mechanical jacks and hand-operated hydraulic jacks **Figure 6-24 ▶**. Mechanical jacks can be of the screw, ratchet lever, or cam type. Jacks are used to lift or push heavy objects.

Pulling tools can extend the reach of the person using them and increase the power exerted upon an object. A variety of poles and hooks are used as pulling tools, including the pike pole. For example, the pike pole may be used to pass a ratchet strap under an elevated vehicle to a rescuer on the other side.

Another type of pulling tool is the manual winch and hoist, which uses either a chain or a cable (wire rope) and is used for pulling or lifting heavy objects **Figure 6-25 ▶**. <u>Winches</u> are used by many rescue organizations for a variety of lifting, pulling, and holding operations. Chain winches come in two varieties: chain hoist and lever hoist. The chain hoist is used primarily for lifting and can be found in models with a lifting capacity ranging from 500 to 40,000 pounds (lb) (227 to 18,144 kilograms [kg]). The lever hoist is a lifting or pulling tool with a lifting capacity ranging from 250 to 12,000 lb (113 to 5443 kg).

Cable hoists use a wire rope instead of a chain and are used primarily for pulling. The two varieties of cable hoist are the integrated cable type (come along) and the pass-through cable type.

When inspecting manual winches and hoists, look for any damage caused by overloading and friction. Check the winch for warping, bending, or cracking. Examine cables and chains

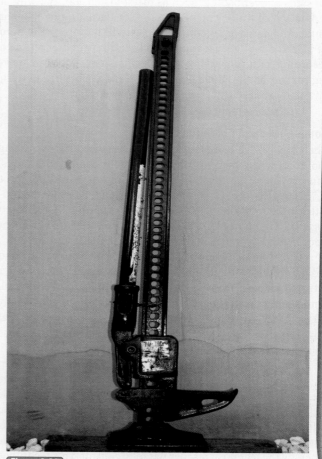

Figure 6-24 Hand-operated ratchet lever jack.

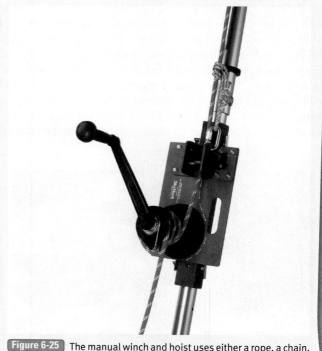

Figure 6-25 The manual winch and hoist uses either a rope, a chain, or a cable (wire rope) and is used for pulling or lifting heavy objects.

for signs of overloading, such as metal distension, twisting or separation of links, cracks, fraying, or flattening of cable or links. Any damage noted to any part will require the item to be placed out of service immediately and repaired or replaced according to the manufacturer's specifications. Lubricate moving parts properly, and keep the cable and chains free of dust, dirt, and grime.

Electrical winches are discussed later in this chapter.

Come Along and Chain Package

The <u>come along</u> is a hand-operated, ratchet lever winching tool that, when used in conjunction with chains and hooks, can provide up to several thousand pounds of pulling force **Figure 6-26 ▼**. The standard model utilized for extrication provides 2000 to 4000 lb (907 to 1814 kg) of pulling force. Used together, the come along and chain package can reap tremendous benefits in safely pulling and stabilizing operations.

The come along operates by ratcheting a wire cable around a drum. It also utilizes mechanical advantage in either a 2:1 ratio by doubling the wire cable around a pulley and then back onto itself, or 1:1 ratio/single line pull. Cable wire is normally 20 to 25 feet (6 to 8 meters [m]) in length, so keep in mind that doubling the cable to give you the extra pulling force will cause you to sacrifice the cable by half its length.

Another feature of the come along is the handle. The handle is designed to fail and bend at a certain force well before the tool fails. The failure rating should be listed on the handle **Figure 6-27 ▶**. This is an excellent safety feature that gives you full control over the object you are pulling. For example, pulling a steering column off a victim requires a very precise technique whereby the technical rescuer has total control. This technique has been rumored to be dangerous because of the possibility of the entire steering column becoming dislodged and projecting into the victim or rescuer. With the proper setup and safety checks, and by not overextending the lift, this technique is as safe as any other extrication technique with the benefit of having full control and feel over the entire length of the pull. The same technique using hydraulics should not be attempted because it will not provide the benefit of feel and control over the movement. This technique utilizing the come along will be discussed in Chapter 10, *Alternative Extrication Techniques*. Come alongs can also be used for stabilization purposes. For example, if a vehicle is on an embankment, it can be secured

to a tree, pole, or apparatus using a come along. Remember to always inspect the tool and cable for any signs of damage before and after every use.

The chain package that comes with the come along kit is rated for that system only and should not be used with any other system. Chains must be marked with a grade, which determines its <u>working load limit (WLL)</u> **Figure 6-28 ▼**. The WLL is the maximum force that may be applied to an assembly in straight tension. Load rating grades are embossed into chain links approximately every 12 to 18 inches (305 to 457 mm). Grade 80 (or System 8) and grade 100 (or System 10) are the most utilized chain types for overhead lifting and rescue.

These chains can come with an assortment of attachments for the various types of anchoring requirements encountered. The assortment of hook attachments can include the master link, which is normally an "O" (or oblong) ring, the slide hook, and the grab hook (the chain shortener), along with the basic towing company attachments, which include the "J" hook (short and long), "R" hook, and the "T" hook. These also come attached together in a cluster package. The cluster attachments are

Figure 6-26   The come along and chain package.

Figure 6-28   Load rating grades, which demonstrate a chain's working load limit (WLL), are embossed into chain links approximately every 12 to 18 inches (305 to 457 mm). This is a Grade 70 chain.

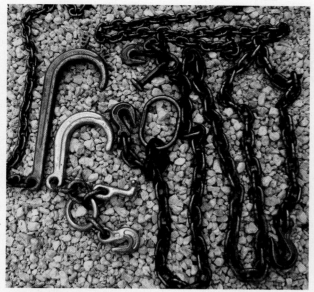

Figure 6-29 Various hooks are included with the come along and chain package, including the cluster hook, slide hook, grab hook, and master link or oblong ring.

Figure 6-30 A chain sling.

inserted in the various holes or openings found in the undercarriage or frame of the vehicle as anchor points **Figure 6-29 ▲** .

Hooks

The <u>slide hook</u> is exactly as its name states. It allows the chain links to pass freely through the throat of the hook to tighten around an object (see Figure 6-29). The slide hook should never be tip loaded by inserting the tip of the hook in a hole or opening; the hook will bend and possibly dislodge under extreme force. The <u>grab hook</u> is utilized by inserting a link of the chain into the slot of the hook. This is also known as a chain shortener; as the name suggests, it is designed to take up the slack needed to make the chain the appropriate size for the task at hand (see Figure 6-29). The <u>"O" ring</u>, or <u>oblong ring</u>, is an attachment designed to join chains together or join a chain to a come along utilizing a hook (see Figure 6-29).

Chain Sling

The <u>chain sling</u> is composed of a single chain or multiple chains—in various lengths, commonly of 5 and 9 feet (1.5 and 2.7 m)—attached to a ring, either round or oblong **Figure 6-30 ▶** . Chain slings are used for lifting heavy loads. Chain slings utilize grade-80 or grade-100 chain and can come in a variety of attachments such as single-chain, double-chain, triple-chain, quad-chain, basket, and adjustable, as well as a variety of sling fittings such as sling hooks, grab hooks, foundry hooks, self-locking hooks, and clevis hooks.

Pneumatic Tools

Tools utilizing air under pressure to operate are known as <u>pneumatic tools</u>. A wide variety of pneumatic tools are available that can nail, cut, drill, bolt, lift heavy loads, and stabilize. Several pneumatic tools the technical rescuer will encounter are air chisels, air impact wrenches, air shores, cut-off tools, and rescue-lift air bags.

The compressed air for pneumatic tools is supplied from air compressors, SCBA cylinders, or vehicle-mounted systems. Most of these tools operate at forces between 90 and 250 pounds per square inch (psi) (621 and 1724 kilopascals [kPa]) and use adjustable regulators to provide the proper operating pressure. The standard unit for measuring pressure is psi.

<u>Air compressors</u> are used to provide power to pneumatic tools or to provide breathing air. Those found on rescue units may be portable or fixed **Figure 6-31 ▶** . Breathing air compressors are generally larger because of the filtration required to meet ANSI/Compressed Gas Association (CGA) G7.1, *Commodity Specification for Air*. A minimum air quality of Grade D is used for air-supplying respirators and a minimum air quality of Grade E is used for dive operations.

Maintenance tasks for air compressors include draining water from the filters and tank, and filter replacement. Oil levels in the compressor (if applicable) and engine fluid levels should be checked after each use. Always familiarize yourself with the manufacturer's provided recommendations.

Rescue Tips

The basic toolbox essentials include various pliers, screwdrivers, a hammer, and a socket set. These are essential tools needed for potential problems requiring quick fixes.

Rescue Tips

Conspicuously label all air tools with the operating pressure so that it is readily known to the user. That way there is no question, especially if a rescuer is not intimately familiar with a given tool.

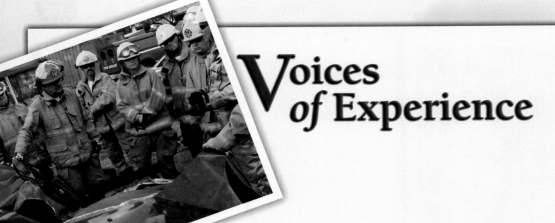

Voices of Experience

In my experience, there are some very predictable actions that must take place on the scene of a motor vehicle collision. By having defined expectations, the emergency scene is easier to choreograph and the responders will operate in a safe and efficient manner.

One of the first priorities to consider when beginning the extrication evolution is the stability of the vehicle. Regardless of its resting position, additional stabilization must be provided. Too often in the past, we have seen vehicles begin to roll away as operations begin. Cribbing, step chocks, or even the wheel chocks used for the apparatus will make the vehicle more stable.

> *"I have opened a car door by trying the handle while others were assembling pneumatic tools."*

Secondly, have tools ready. Make certain a crew member brings a prying tool, such as a Halligan bar, a striking tool, or a sledgehammer to the vehicle. Initial access must be made to the trapped victim. Although the end result may be the removal of doors and roofs with hydraulic equipment so that an immobilization device can be utilized, this is not the starting point. In short order, a rescuer needs to be put in contact with the victim or victims so that initial assessments and treatments can be made. This can be easily accomplished if you start with the right tools. We have popped many doors with simple prying tools and little effort. I have opened a car door by trying the handle while others were assembling pneumatic tools. If nothing else, the tools can be used to break and clear the glass from a window that is remote to the victim so that a rescuer can enter the vehicle and establish contact. Do not forget the old adage, "Try before you pry."

I remember one teachable moment in particular. I was a new lieutenant and watched as my crew stood outside of a vehicle yelling through the glass at a victim, all for the lack of one tool among four people. From that point forward, I have discussed each person's role in the extrication event prior to it occurring.

In summary, it is imperative that when responders arrive on scene they have a consistent plan of attack that they have prepared and trained for. Having the crew members taking actions as simple as arriving at the vehicle prepared with step chocks and a pry bar can set the entire operation up for success. Remember that "an empty-handed rescuer is a well-informed observer!" Be prepared, be smart, and be safe!

Mike Stanley
Aurora Fire Department
Aurora, Colorado

Figure 6-31 A general-use air compressor.

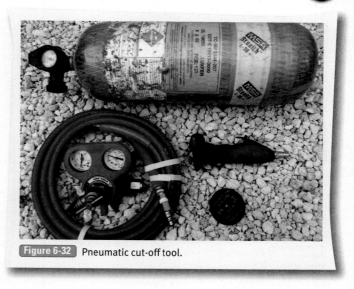

Figure 6-32 Pneumatic cut-off tool.

Pneumatic Cut-off Tool

The **pneumatic cut-off tool** utilizes a small carbide disc, normally 3 inches (76 mm) in diameter, which rotates at a high rpm to cut through most metals Figure 6-32 ▲. The disc throws off a tremendous amount of sparks, but it is a great tool to cut through hardened steel such as steel found on padlocks, ½-inch (13-mm) or less reinforcing bars, or steel found on fence posts. To provide an example of the versatility of this tool, there was one incident where the police department unsuccessfully tried to remove handcuffs from a prisoner using heavy duty bolt cutters. After covering the prisoner with a damp blanket and placing a wide flat crescent wrench under the ring of the handcuff, two sections of the metal handcuff were removed in less than 30 seconds using the pneumatic cut-off tool.

Most models of this tool operate at no greater than 90 psi (621 kPa) and will rapidly deplete an air bottle within minutes. In some models, operating the tool at greater than 90 psi (621 kPa) will potentially damage the tool; always check the manufacturer's specifications for the tool's proper operating pressure. The pneumatic cut-off tool is a great tool to carry on any apparatus; it has many applications, it is economical, and it can be stowed almost anywhere because of its small size.

Pneumatic Chisel

Pneumatic chisels, often referred to as air chisels, air hammers, or impact hammers, are used to cut sheet metal and/or hardened steel such as the kind that can be found in vehicle door hinges Figure 6-33 ▶. Pneumatic chisels do not normally create sparks when operating. For this reason, pneumatic chisels are a common tool used in vehicle extrication incidents.

The tool utilizes an adjustable regulator with a common range setting of 0 to 300 psi (0 to 2068 kPa). The gauge on the regulator displays the bottle pressure and the operating pressure of the air chisel. The regulator can attach to either a 2216-psi or 4500-psi (15,269-kPa or 31,026-kPa) air bottle, depending on which it is rated for. A high-powered air chisel can have a general operating pressure in the range of 150 to 225 psi (1034 to 1551 kPa). This tool should never be operated at 300 psi (2068 kPa); doing so will potentially damage the tool and cause more

■ Pneumatic Cutting Tools

Pneumatic cutting tools include saws, shears, and chisels. The pneumatic saws used in rescue work are either the reciprocating or rotating type. Pneumatic reciprocating saws operate just like the electric variety. Models are available with saw speeds ranging from 1600 to 10,000 revolutions per minute (rpm). Pneumatic cut-off tools, also known as whizzer saws and die grinders, use a circular, composite cut-off blade that rotates at speeds as high as 25,000 rpm. Such saws are normally used when there is a space limitation, but they can be a problem because of the sparks they generate. The pneumatic cut-off tool is discussed in the following section.

Pneumatic shears are used for cutting sheet metal; because of their limited application, they are rarely used in rescue work.

Figure 6-33 Pneumatic chisels are used to cut sheet metal or hardened steel.

blades to break. Check the manufacturer's recommended pressure settings for the various applications encountered.

The pneumatic chisel can be an extremely effective tool in the hands of a skilled technician. It is a precision tool requiring a tremendous amount of training to become proficient in its use. The technical rescuer has to fully understand all the intricacies of the tool and be able to apply it practically. Some of the intricacies of operating a pneumatic chisel include knowing the proper way to hold the tool and blade, maneuvering the angle and depth of the cut, determining what type of material to cut using the chisel, how to avoid breaking a bit, how to properly dislodge a stuck bit, when to increase and decrease pressure settings, and how to conserve air.

Rescue Tips

To develop proper control of the pneumatic chisel, a good practice tip is to engrave each other's names into the sides of old junk cars.

There are numerous air chisels available. However, for vehicle extrication, you want an air chisel industrialized and able to take the abuse of emergency work. Avoid the automotive body shop types; these are commonly underpowered and are not reliable at an emergency incident. Several blade attachments come with the industrialized kits. Some of the more common blades used for vehicle extrication are the flat blade (long and short shaft), which can be utilized for all cutting requirements, whether cutting hardened steel or lighter sheet metal, and the panel cutter or T-blade, which is utilized for lighter gauge steel such as sheet metal found in vehicle roofs. If the rescuer is proficient with the flat blade, then the panel cutter or T-blade would never have to be used.

Vehicle roofs are composed of both light- and heavy-gauge steels. The outer covering is made up of light-weight sheet metal, supported underneath by heavier gauge steel rib supports. The technical rescuer may be able to cut through the surface of the roof fairly quickly with a panel cutter or T-blade, but would

then have to switch out to a flat blade to cut the remaining steel ribs or framing that support the roof. This switch will dramatically delay the entry time and possibly cause the rescuer to have to change out air bottles because of the inefficient use and operation of the tool. This is a perfect example of why proficiency training using this tool is critical. When choosing which type of air chisel to carry on the apparatus, keep in mind that not all blades are designed alike; some blades are made of a thicker gauge high-quality alloy steel that can withstand higher operating pressures and resist accidental breakage of the tip **Figure 6-34 ▾**.

The blade retainer is also an important feature to consider when choosing the right tool. The blade retainer prevents the blade from moving and should be the quick change type, where the technical rescuer simply depresses the retainer sleeve to change out a blade. The screw-on type retainers or tools equipped with a collet are cumbersome and can slow down an operation when a blade change needs to be made quickly. Some blades also come with grooves milled at the insertion point, allowing the technical rescuer to lock the blade in place and prevent it from turning.

Rescue Tips

When operating the pneumatic chisel using a flat blade, always remember to keep half the blade showing when cutting through a section of metal to avoid burying the tip.

Figure 6-34 Not all blades are alike. The blade on the right is a thicker gauge high-quality alloy steel that can withstand higher operating pressures and resist accidental breakage of the tip. The blade on the left is thinner and has a smaller diameter, making it less forgiving than the thicker blade and prone to breaking at higher psi settings.

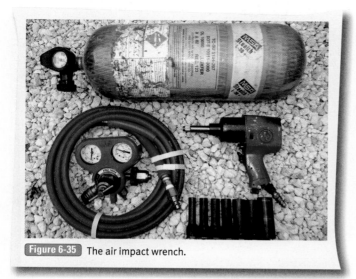

Figure 6-35   The air impact wrench.

Figure 6-36   Shoring is used where the vertical distances are too great to use cribbing or the load must be supported horizontally, such as in a trench, or diagonally, such as in a wall shore where a vehicle may have breached a structure and needs temporary structural support.

◼ Pneumatic Rotating Tools

Air Impact Wrench

The <u>air impact wrench</u> is a pneumatic rotating tool used to remove nuts and bolts of various sizes, including those found in door hinges, seats, and wheels Figure 6-35 ▲ . The tool has a general air pressure setting of 90 psi (621 kPa) and uses both metric and standard sockets. Remember to always use the sockets that come with the tool set. Never mix hand torque sockets with high-pressure impact sockets; hand torque sockets are not designed to handle the high pressure of an impact wrench and can fail and fragment, causing possible injury. Impact sockets are higher gauge steel and normally have a black finish on the outside. There are several sizes of impact wrenches, but a ½-inch (13-mm) drive model with a torque range of 180 to 325 foot pounds (244 to 441 Newton-meters) should be adequate to handle most vehicle applications. When dealing with heavier equipment such as large machinery or heavy trucks, a ¾-inch (19-mm) drive impact wrench will deliver a much higher torque range of 750 to 1200 foot pounds (1017 to 1627 Newton-meters). Utilizing an impact wrench to remove a vehicle door by taking off the hinge bolts is a very fast and easy process, but beware—most doors come equipped with a swing bar that has to be cut or spread off in order to fully remove the door.

◼ Pneumatic Lifting Tools

Air Shores

<u>Shoring</u> is used where the vertical distances are too great to use cribbing or the load must be supported horizontally, such as in a trench, or diagonally, such as in a wall shore where a vehicle may have breached a structure and needs temporary structural support Figure 6-36 ▶ . Shoring is defined as the temporary support of structures during activities such as construction, demolition, or reconstruction in order to provide stability to protect property as well as workers and the public. This working definition is obviously related to supporting structural walls but can be used interchangeably with adding structural support to unstable vehicles.

<u>Air shoring</u> is a type of strut system that comes in a variety of lengths and may be extended by the use of compressed air. The shoring is then locked in place by pins, notches, and screw collars, with ratchet straps added to the base to add tensioning. Struts and tensioning devices will be discussed later in this chapter.

Rescue Tips

Training must be provided to ensure that all equipment is used and maintained in accordance with the manufacturer's instructions.

Rescue-Lift Air Bags

<u>Rescue-lift air bags</u> are pneumatic-filled bladders used to lift an object or spread one or more objects away from each other to assist in freeing a victim Figure 6-37 ▼ . These inflatable devices can move or lift a tremendous amount of weight when inflated.

Figure 6-37   Rescue-lift air bag.

They are commonly composed of rubber with synthetic fiber linings and, depending on the various manufacturers, come in a wide arrangement of sizes and shapes, each with its own lifting capacity rating. The lifting capacity is measured in metric tonnage, with 1 metric ton equaling 2205 lb. Rescue-lift air bags are not used to stabilize a vehicle by themselves; rather, cribbing is always required in conjunction with rescue-lift air bags. Cribbing techniques are discussed in Chapter 8, *Vehicle Stabilization*. There are several rescue-lift air bag classifications used in the field. These include low-pressure rescue-lift air bags, medium-pressure rescue-lift air bags, high-pressure rescue-lift air bags, high-pressure flat-form rescue-lift air bags, and new technology (NT) locking rescue-lift air bags.

General rules when using rescue-lift air bags include the following:

- Never stack more than two bags on top of one another because this can cause one or more of the bags to come firing out when inflated just like a wet seed squeezed between your fingers. (Note: This rule does not pertain to flat-bag systems or NT-type bag systems).
- Always ensure that the valves and hoses are facing outward.
- Never place any objects such as a piece of flat wood or cribbing on top of or between the bags; doing so will only split the wood, potentially throwing pieces everywhere or causing the entire piece of cribbing to come out under extreme force. Place a piece of plywood below the bag to protect it from the ground.
- Do not use a rescue-lift air bag to pull a steering column.
- Do not use a rescue-lift air bag as the sole means to stabilize a vehicle.
- Do not overinflate the bag; overinflation will lessen the surface contact the bag has with the object being lifted. For example, think of two footballs stacked on top of one another. The surface-to-surface contact is minimal, with virtually zero stability; this is the same idea as overinflating two rescue-lift bags.
- Lift only as high as necessary.
- Always maintain the proper operating pressures. Once the bag is fully inflated, remember to utilize the shut-off valves on each bag to prevent overinflation or to isolate a bag.
- Stacking two bags on top of one another will increase the height of the lift but will not increase the lifting capacity. When stacking bags, always place the larger bag on the bottom and remember that the lifting capacity will always be determined by the lowest-pressure bag. To maximize each bag's lifting capacity, place the bags next to each other. To better illustrate this, consider trying to lift a 2-ton (1.81-metric ton) cement pole. If you take a 1-ton (0.91-metric ton) bag and then place another 1-ton (0.91-metric ton) bag on top of it and inflate both, the resulting lift capacity will only be 1 ton (0.91 metric ton), and the pole will not move. Now, if you take those same 1-ton (0.91-metric ton) bags and place them side by side and inflate them both, the resulting lift capacity will be the total of both bags, which is 2 tons (1.81

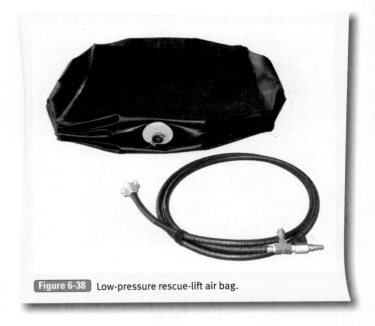

Figure 6-38 Low-pressure rescue-lift air bag.

metric tons), thus lifting the cement pole. Remember as a safety measure to always support your lifts with cribbing to compensate for the possibility of a bag failure. Cribbing in place will secure a load in the event of any critical failures.

Always examine the manufacturer's label on the bag to determine a bag's lifting and height capacity. If for some reason the label is gone, a good rule of thumb is to multiply the dimension of the bag with the operating pressure to get the lift capacity. For example, a 20- by 20-inch (508- by 508-mm) bag yields 400 square inches (258,064 square mm). Multiply 400 by the operating pressure of 114 psi (786 kPa), and it yields a lift capacity of 45,600 lb (20,684 kg) or approximately 20 tons (18.14 metric tons). Rescue-lift air bags are rated in tonnage, and 1 metric long ton equals 2240 lb.

This is a rough estimate using the full surface area of the bag; the height capacity of the bag is not added into the equation, although increasing the height of the lift while diminishing the surface contact lessens the lifting capacity.

Low-Pressure Rescue-Lift Air Bags

Low-pressure rescue-lift air bags provide a very high lift with a maximum working air pressure of approximately 7 psi (48 kPa) **Figure 6-38 ▲** . The flat design of the bag when inflated provides a large surface contact with the item being lifted. These large cushion bags are more commonly found on big tow units because of their ability to upright overturned heavy vehicles such as semi-trucks and trailers. Disadvantages of the low-pressure rescue-lift air bags are their lower lifting capacities and their thinner sidewalls, which make them less stable when inflating.

Medium-Pressure Rescue-Lift Air Bags

Medium-pressure rescue-lift air bags are not as common as the low- and high-pressure rescue-lift air bags; they have a more rugged design and utilize a working air pressure of approximately 15 psi (103 kPa), depending on the various manufactur-

Figure 6-39   Medium-pressure rescue-lift air bag.

Figure 6-41   Flat form rescue-lift air bags.

ers **Figure 6-39 ▲** . The medium-pressure bags commonly have two to three cells and are suitable for aircraft, medium or heavy truck, or bus rescue, and recovery work.

High-Pressure Rescue-Lift Air Bags

High-pressure rescue-lift air bags are the most commonly used bags among rescue agencies. These bags utilize a working air pressure of approximately 100 to 145 psi (689 to 1000 kPa), depending on the various manufacturers **Figure 6-40 ▼** . The high-pressure kits come with hoses, a regulator, a master control module, and various other attachments. For safety purposes, a good system comes with different-colored hoses for easy recognition of which bag is in use, a master control module with a dead-man release to prevent overinflating the bag if the control valve is accidentally dropped, and an in-line shut-off/pressure relief valve to isolate a bag and disconnect a line.

A rescue-lift air bag operation will generally require multiple personnel to safely accomplish the task at hand. For example, to safely remove a victim trapped under a vehicle with his or her upper torso and head exposed will require, at a minimum, five personnel to operate safely: one at the head of the victim, two on the sides for cribbing support, one at the controls, and one officer in charge of the lift and overall safety of the operation. Remember the safety rule of thumb, "Lift an inch, crib an inch." Also, remember to lift up equally across the entire vehicle as opposed to lifting on one side, which will force the opposite side downward, potentially crushing the victim.

High-Pressure Flat Form Rescue-Lift Air Bags

Flat form rescue-lift air bags are designed to retain their flat profile in the center as they are inflated **Figure 6-41 ▲** . This is in direct contrast to the traditional high-pressure bags that round out and decrease surface contact when inflated. The flat design eliminates rolling and shifting and can utilize up to three bags stacked on top of one another. The dimpled surface allows the bags to interlock with one another while side straps assist with alignment and guard against the possibility of a bag ejection. The flat bags maintain an operating pressure of approximately 116 to 118 psi (800 to 814 kPa) with various sizes and lifting capacities.

New Technology High-Pressure Rescue-Lift Air Bags

New technology (NT) high-pressure rescue-lift air bags utilize a unique lifting system where each round-shaped bag can be locked together with a threaded connector, creating one bag with multiple cells that offers a distinct height advantage over traditional flat bags **Figure 6-42 ▶** . The system utilizes a working air pressure of approximately 147 psi (1014 kPa) with the option of various height adjustments with the attachment of additional bags. Always reassess the bags before inflation to ensure that they are properly threaded and locked together; these types of systems can detach violently and cause injury or death if not secured correctly.

Figure 6-40   High-pressure rescue-lift air bag.

Near Miss REPORT

Report Number: 08-533
Report Date: 10/27/2008

Synopsis: The tool had overpressurized due to an incorrect connection on the return hose couplings.

Event Description: During a routine training meeting, we removed our extrication spreader from our rescue truck and placed the tool on the ground. Our gas-powered power unit was started and left at idle. Our mission was to open and close the tool to check the operation.

My fire fighter was in full PPE, with the shield down on his helmet, when starting the tool. When the tool opened to approximately 4 inches (102 mm), the spreader ruptured the length of the housing. The fire fighter received a blast of mineral oil–based hydraulic fluid in the face, knocking his helmet off and hitting him in the eyes and cheek.

When the tool was analyzed, the report stated that the tool had overpressurized due to an incorrect connection on the return hose couplings. The tool had been run just the night before in the same manner. The dealer has since replaced all the couplings with bleeder-style couplings. The report also stated that the relief valve had opened but could not keep up to the pressure, and that is why the tool ruptured.

The fire fighter received emergency care with a follow-up doctor and eye doctor appointment. No long-lasting effects were reported. Had this been an actual incident, the pump would have been at full throttle and would have been much worse.

Lessons Learned:

- The basic lesson learned was to have full PPE and also goggles when operating high-pressure tools.
- A second lesson, which we felt we had in place, is to have a repair/maintenance plan that covers the entire tool with your dealer. We now know of more "areas" that need to be serviced, which are now serviced annually by our request.
- A third lesson is to check the couplings *each and every* time before using the tools. Checking them after you remove the tools from storage will verify that the couplings are not knocked loose when preparing to use the tools.

Figure 6-42 NT high-pressure rescue-lift air bag.

Electric Tools

Electric-powered tools utilize a standard household current or generator to operate. **Electrical generators** may be portable or fixed and are primarily used to power scene lighting and to run power tools and equipment **Figure 6-43 ▶**. Generators range in capacity from less than 1000 watts (1 kilowatt [kW]) to 75,000 watts (75 kW), and larger. Depending on size, they provide output as 120 or 240 volts. Maintenance tasks for generators include checking for evidence of leaks or damaged parts. For those powered by an attached engine, check engine fluids and ensure proper operation. Familiarize yourself with any specific recommendations provided by the manufacturer of the equipment. One obvious rule and general disadvantage of utilizing portable generators is that they cannot be operated inside a structure because of the lethal carbon monoxide gas release and build-up. Carbon monoxide gas is known as a "silent killer" because unsuspecting victims normally pass out well before they realize they are in danger and need to escape.

A.

B.

Figure 6-43   Generators. **A.** Portable. **B.** Fixed.

Figure 6-44   Heavy-duty junction boxes are used at incidents where multiple outlets are needed.

Rescue Tips

Some tools utilize a battery as an electrical source. Battery-operated tools work very well in some applications but in comparison to an electric-powered tool, they do carry certain limitations. Batteries can only sustain a certain amount of energy and can quickly drain with continuous use. These tools may not have the reliability of continuous power that a high-amperage tool offers. Reliability is the most important feature of any tool at an incident where time is critical. Any loss of power or insufficient power will have a negative impact on the operation. However, if a generator fails and there is no power to supply a tool, it would be a good idea to have a battery-operated tool on the apparatus. If battery-operated tools are used in extrication, lithium-ion batteries appear to offer more power and last longer in continuous use than nickel-metal hydride (NiMH) batteries or nickel-cadmium (NiCd) batteries Figure 6-45 ▼ .

Adapters may be used to convert a battery-operated tool to a general current tool. Adapters are an excellent backup accessory if the batteries run out of power or malfunction.

Figure 6-45   Lithium-ion batteries appear to offer more power and last much longer in continuous use than nickel-metal hydride (NiMH) batteries or nickel-cadmium (NiCd) batteries.

Junction boxes are electrical enclosures commonly found when the need for multiple outlets is required Figure 6-44 ▶ . These boxes should be heavy duty and supplied with ground-fault circuit interrupters (GFCIs) and should comply with NFPA 70E, *Standard for Electrical Safety in the Workplace*. For long-term incidents where the generator is large and does not have plug and breaker distribution capabilities, portable distribution panels are used. These panels accept full power from the generator and provide circuits to break this power down to individual outlets. These panels should be certified according to Underwriters Laboratories (UL) Standard 1640, *Portable Power Distribution Units*, be weatherproof, and provide protection from accidental contact with live electrical connections.

One of the most utilized electric-operated tools common to the extrication incident is the reciprocating saw. Note that some hydraulic tools and saws also operate via electric-powered energy.

■ Electric Cutting Tools

Electrically operated cutting tools include the reciprocating saw, circular saw, snipping devices such as the rebar cutter, and specialty items such as the plasma cutter.

Rebar cutters are used not only for cutting reinforcing bars but also for any other round metal that is ⅝ inch (16 mm) or less in diameter. Rebar cutters operate by placing the metal to be cut between a cutting edge and a ram, which then pinches the metal to its fracture point. This tool is more commonly associated with hydraulics than electric power and is better known as a confined-space cutter.

The plasma cutter is a device that has a nozzle from which inert gas or compressed air is blown at a high speed; at the same time, an electric arc forms within the gas, turning some of the gas to plasma. The plasma is hot enough to cut through metal. Rescue services use this tool when fine cutting and minimal sparks are required.

Electric Reciprocating Saw

The reciprocating saw is a type of saw in which the cutting action utilizes a back-and-forth motion (reciprocating), or a push and pull of the blade. Electric versions normally range from 6 to 15 amps, depending on the manufacturer, with the battery versions normally ranging from 18 to 36 volts, depending on whether it is a lithium-ion, NiMH, or NiCd battery.

Several operational options are offered by various manufacturers, including the ability to change the angle of a cut to better adapt to a particular situation. Some of the angle features include the head of the saw rotating up to a 90-degree angle, the handle rotating 360 degrees, or the handle bending to a 90-degree angle Figure 6-46 ▼. Another option is an orbital action that forces the blade to lift up and rotate over after it cuts through the material. This is best utilized for wood materi-als where it is necessary for the wood fragments to be removed to provide a more efficient cutting action. When cutting metal material, it is recommended to disable this orbital feature and maintain a straight cutting action.

When choosing a blade for the reciprocating saw, consider blade type and blade thickness. Will you be cutting wood or metal? Bi-metal blades are the best choice for cutting metals found on vehicles. Bi-metal blades have a high-speed steel cutting edge with hardened teeth welded to tough, thick, flexible spring-back steel. These blades can handle the toughest jobs and can be used multiple times before having to be discarded. The more thinly constructed blades are not designed to perform the high-speed and rigorous applications of vehicle extrication. Thinner blades will heat very quickly, warp, and dull, making the blade useless.

Another consideration when choosing a blade is the TPI rating. The **TPI rating** refers to the number of teeth on the blade per inch Figure 6-47 ▼. For example, a TPI rating of 18 indicates a large number of small teeth producing a very fine cut, whereas a TPI rating of a 5 indicates much larger teeth producing a very coarse cut. For vehicle extrication purposes, the best options for a reciprocating blade are a length of 6 inches (152 mm) or 9 inches (229 mm), and a TPI rating of 9 to 14.

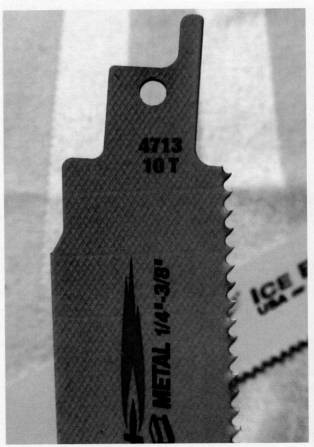

Figure 6-47 The TPI rating of a blade indicates how many teeth per inch the blade has. This blade is 10 TPI.

Figure 6-46 Several angle options are offered by manufacturers of reciprocating saws.

Last, consider the type of metal you will be cutting. Are you dealing with ferrous metals or nonferrous metals? <u>Ferrous metals</u> contain iron, cast iron, low- and medium-alloyed steels, and specialty steels, such as tooled steels and stainless steels. <u>Nonferrous metals</u> are metals or alloys free of iron, such as aluminum, copper, nickel, lead, zinc, and tin, to name a few. Why is this important to know? Nonferrous metals such as aluminum will melt under the high speed of the blade and cause the aluminum particles to actually weld themselves onto the teeth of the blade, rendering it ineffective. Some of the various hybrid vehicles and advanced technology vehicles are made of aluminum frames. A bi-metal blade or universal blade is normally effective in cutting nonferrous metals. Also, with this advanced technology in vehicles comes advanced high-strength steels, metals, and alloys such as boron or titanium; reciprocating saw blades, regardless of the type, will have very little to no success trying to cut through them. A better practice is to avoid these types of steels by cutting around them or choosing an alternate method when they are encountered. When in doubt, read the package to determine what the blade is designed to cut.

When operating the saw, some manufacturers recommend using a spray bottle containing a lubricant such as soapy water to spray on the blade while cutting in order to reduce friction and heat. These efforts may reduce wear on the blade. Remember that utilizing personnel and resources in the most efficient manner is the goal of the incident commander in charge of the operation; holding a spray bottle may not be the best choice or best example of multitasking when other objectives must be met and personnel on scene are limited.

Be very cautious when changing a blade immediately after using the tool. The blade will be extremely hot. The best method for changing a blade is to face the blade toward the ground and engage the quick-release mechanism, letting the blade fall to the ground without touching it. Keep plenty of blades on hand, and change them as often as needed.

Electric Circular Saw

The <u>circular saw</u> moves in a circular motion [Figure 6-48 ▾]. These saws come in a variety of sizes and are used primarily for cutting wood, although special blades are available that will cut metal or masonry. The 18- to 24-volt battery-powered metal-cutting circular saw with a 6½- to 7-inch (165- to 178-mm) diameter carbide tip blade that is designed for the tool is excellent for cutting fence post and rebar material that may have become impaled in the victim. The battery-powered option offers the versatility to access tight spaces.

The electric-powered version is a powerful tool that can cut through steel plate up to a ½-inch (13-mm) thickness depending on the model type. Blade diameters range from 7¼ to 9 inches (184 to 229 mm), depending on the model type. The disadvantage of the electric saw compared to the battery version is the weight; the electric saw can weigh up to 20 lb (9.1 kg), whereas the battery-operated tool is approximately 10 lb (4.5 kg), although it will not last as long.

The advantage of a metal-cutting circular saw is that there are minimal sparks and vibrations caused when the saw blade passes through the material. It is a very clean and fast cut. In comparison, a reciprocating saw will cause significant vibrations, and a hydraulic cutter will create a tremendous amount of torque and twisting of the material.

■ Electric Lifting/Pulling Tools

Electrically operated lifting and pulling devices include winches and hoists. Winches are used by many rescue organizations for a variety of lifting, pulling, and holding operations. There are two common types of winches—electrically operated and hydraulically operated. The electric winch is typically driven by an electric motor that draws its power from the vehicle's battery; it uses steel cable and can come in various pulling capacities ranging from 2000 to 12,000 lb (907 to 5443 kg) or more. NFPA 1901, *Standard for Automotive Fire Apparatus*, requires chassis-mounted winches to be rated with at least a 6000-lb (2722-kg) pulling capacity and to be remotely operated with at least a 25-foot (7.6-meter) cord. Any winch is only as strong as what it is attached to, so winches are normally connected in some fashion to the frame of the vehicle [Figure 6-49 ▸]. Bumper-type winches are attached in or on the bumper of the vehicle and are mounted permanently. Tow hitch-type winches attach into the tow hitch receiver of a vehicle and are designed to be removed when not in use.

Winch pulling capacity should be 1.5 times the gross vehicle weight of the vehicle to be pulled. If the vehicle weighs 8000 lb (3629 kg), the winch capacity should equal 12,000 lb (5443 kg). Unfortunately, the rated capacity of a winch will only pull the maximum capacity for a short duration, normally until the first layer of cable wraps around the drum. After this occurs, the capacity significantly drops due to the change in gear ratios. Electrically driven winches are easy to set up on a vehicle using universal mounting kits, and there are more features available as compared to the hydraulic type. Two disadvantages of the electric-driven winch are that it is prone to overheating if used for long durations and that it can quickly drain an electrical system if the vehicle is not running.

When using a winch, always wear full protective gear, including helmet, eye protection, and gloves, and always operate from a safe distance using the remote controls. Some manu-

Figure 6-48 A metal cutting circular saw.

A.

B.

Figure 6-49 Electric lifting and pulling tools. **A.** Winches can be attached to the bumper of a vehicle. **B.** Winches can also be attached to the tow hitch receiver.

facturers recommend placing a heavy blanket on the wire rope at the midpoint between the winch base and the anchor to absorb some of the stored potential energy should the wire rope break. Be aware that there are some possible problems that can occur with this blanket technique; the blanket can move and may be drawn into the winch. Also, setting or removing a blanket can place an added impact load on the line, causing it to fail. Always proceed with caution.

When inspecting a winch, look at the cable, hook, and gears closely for indications of damage or overloading. Periodically inspect the connections to the vehicle frame for rust, bolt tightness, or weld cracks. Periodic lubricating and other inspection/maintenance procedures should be in accordance with any recommendations or field maintenance guidance provided by the manufacturer.

■ Electric Lighting

Most rescue vehicles carry an assortment of portable and mounted lights. Typically ranging from 300 to 1000 watts (0.3

to 1 kW) or more per light, they are found in a variety of styles. Portable lights are meant to provide light where fixed lights cannot. Portable lights are usually adjustable for elevation and have a broad base to prevent them from being tipped over. There are also portable light stands containing one or more lights. These stands are adjustable in height and are very useful for providing large-area or higher height lighting **Figure 6-50 ▶**. Hand lights and helmet lights are also used by rescuers.

Fixed lights are mounted on the rescue vehicle and, like portable lights, come in a variety of styles. Some are mounted into the body of the vehicle for perimeter lighting and are not adjustable. Others are mounted so they can be adjusted for direction and elevation, some being a fixed height and some being on adjustable poles. More elaborate systems, referred to as light towers **Figure 6-51 ▶**, consist of telescoping masts with a bank of lights with up to 6000 watts (6 kW) or more of lighting. These masts, which can be vehicle or trailer mounted, can be 30 to 40 feet (9 to 12 meters) in height and provide a significant amount of overhead lighting. Because of the height of these towers, it is common to see stabilizers installed on

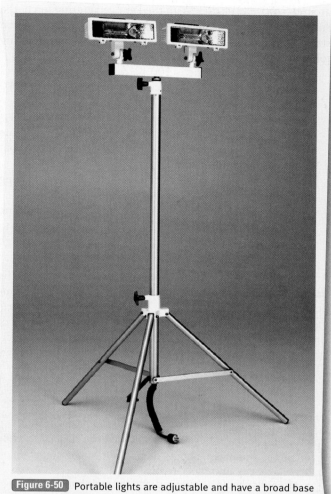

Figure 6-50 Portable lights are adjustable and have a broad base to prevent tipping.

Figure 6-51 Light towers mounted to vehicles offer a significant amount of overhead lighting.

apparatus lighting, and personal handheld lighting. LED lighting draws less current than conventional lights and produces light from passing an electrical current through a semiconductor diode. The advantage of LEDs over halogen lights or regular incandescent lights is that LEDs do not have a filament to heat or gas to ionize; LEDs function in a solid state. Another advantage of LEDs is that they do not contain mercury and are shock resistant. LED lighting ranges from 80 to over 100 lumens per energy watt. Conventional incandescent light bulbs are approximately 15 lumens per energy watt, and halogen lamps are approximately 20 lumens per energy watt. A lumen measures a light's intensity and is equal to 1 foot (0.3 m) of candle power upon 1 square foot (0.09 m²) of area.

Rescue Tips

Be sure the lights you use do not exceed the rated capacity of the generator supplying the power. Overloading the generator could damage the generator, lights, or other attached equipment.

Fuel-Powered Cutting Tools

Cutting tools include chainsaws, rotary saws, cutting torches, and exothermic torches. One of the major advantages of fuel-powered tools is the high power they can generate. Disadvantages include that they can be heavy to carry, depending on the type of tool, some require a fuel mixture of gas and oil (two-stroke engine only), and some can be difficult to start cold. A periodic maintenance schedule and thorough inspection before and after every use are crucial for consistent and reliable operation of fuel-powered tools.

Chainsaws

Fuel-powered <u>chainsaws</u> are available for cutting wood, concrete, and even light-gauge steel. Standard steel chains are

the vehicle or trailer to prevent tipping during high-wind situations.

Make sure the lights you are using do not exceed the rated capacity of the generator supplying the power. Consider not only the possible power usage of the lights but also any other needs you may have. Overloading your generator could cause damage to the generator, lights, or other attached equipment.

After use, check all lights for any physical damage such as wires pulled from the sockets or disfigurement caused by overloading or other injury. Bulbs are susceptible to shock damage and are usually easy to change, so keep a stock for replacement as needed. Never install a bulb with a higher wattage than the light's rating, and never touch one with your bare hands, as the residue left may make the bulb fail prematurely. Keep the lens and reflector clean and the lens protector in place to ensure maximum light output. Additionally, familiarize yourself with any additional recommendations or field maintenance guidance provided by the manufacturer.

Light-emitting diodes, better known as LEDs, are the future of lighting for emergency services in the area of scene lighting, used to cut wood, carbide-tipped chains can cut wood and light-gauge metal, and diamond chains are used for cutting concrete. Chainsaws are a great option to have for an incident where a vehicle has struck a tree and the tree itself or a tree limb is impeding access and has to be removed. The chainsaw is the fastest and most efficient tool to get the job done. Always remember to utilize the proper safety procedures when cutting tree branches that are possibly under tension to avoid any spring-back injuries.

All chainsaws used by rescue personnel should be equipped with a chain brake (a feature built into the chainsaw that stops the blade when manually engaged). Always wear appropriate PPE (including eye protection, hearing protection, and gloves) when working with these tools, and never operate them in enclosed spaces. Chaps are protective garments worn for lower body protection. Chaps for chainsaw use should meet the requirements of the ASTM (American Society for Testing and Materials) Standard for Leg Protection for Chain Saw Users.

■ Rotary Saws

Fuel-powered <u>rotary saws</u> (K-12 saws) are available for cutting wood, concrete, and metal. There are two types of blades used on rotary saws: a round metal blade with teeth and an abrasive disc. These discs are made of composite materials and are designed to wear down as they are used. Different styles of discs are available for concrete, asphalt, and metal. It is important to match the appropriate saw blade or saw disc to the material being cut. The application of rotary saws in vehicle extrication would be extremely limited; these saws tend to throw a tremendous amount of sparks when cutting through metal and are considered a fire safety hazard. However, if, for example, a concrete pole is impeding access at the incident and must be removed or cut down, the rotary saw would be the best option for this task.

■ Cutting Torches

<u>Cutting torches</u> produce an extremely high-temperature flame and are capable of heating steel until it melts, burns, and oxidizes, thereby cutting through the object. These tools are sometimes used for rescue situations such as cutting through heavy steel objects. The head of a cutting torch releases pure oxygen and generally is positioned at a 90-degree angle. The release of oxygen is controlled by a lever on the outside of the torch body.

Because these torches produce such high temperatures (5700°F [3149°C] or more), operators must be specially trained before using them. The most common type of cutting torch uses oxygen and acetylene, also known as oxyacetylene, to create the flame, but many rescue services have begun to use oxygen/gasoline torches, also known as Petrogen. These two torch systems can only be used to cut steel because they apply an oxidizing principle to cut.

Petrogen torches can cut through as much as 14 inches (356 mm) of steel with a flame temperature around 5200°F (2871°C). The oxyacetylene cutting torch can cut through as much as 12 inches (305 mm) of steel with a flame temperature

around 5700°F (3149°C). With the Petrogen cutting torch, the sparks produce little heat and weight, although there can be some molten metal. Other advantages of the Petrogen cutting torch include:

- The gasoline flame is 100 percent oxidizing, producing a clean burn and a cleaner cut; acetylene is only 70 percent oxidizing.
- Gasoline is readily available and is not as volatile as acetylene; with gasoline, there is no potential for a fuel line backflash because the liquid fuel prevents any flame from going back into the torch.
- Petrogen torches use much less fuel by weight, making them more economical.
- Petrogen can cut twice as fast as acetylene when cutting through 2- to 4-inch (51- to 102-mm) steel, and up to four times faster than acetylene when cutting through 8- to 10-inch (203- to 254-mm) steel.

An alternative to the Petrogen and oxyacetylene systems is the oxygen/propane torch, which offers a slightly cooler flame temperature (approximately 5112°F [2822°C]). Oxygen/propane is just as economical as Petrogen. Some of the disadvantages of the oxygen/propane torch are that it does not offer the precision cutting that an oxyacetylene torch can and that it throws a tremendous amount of slag when cutting.

Rescue Tips

All power equipment should be left in a "ready state" for immediate use at the next incident.

- All debris should be removed, and the tool should be clean and dry.
- All fuel tanks should be filled completely with fresh fuel.
- Any dull or damaged blades should be replaced.
- Teeth on chainsaw blades should be inspected for damage or missing teeth. Check the manufacturer's recommendations for the number of teeth allowed to be damaged or missing before total chain replacement is necessary.
- Belts and chains should be inspected to ensure that they are tight and undamaged.
- All guards should be securely in place.
- All hydraulic hoses should be cleaned and inspected.
- All power cords should be inspected for damage.
- All hose fittings should be cleaned, inspected, and tested to ensure a tight fit.
- Tools should be started to ensure that they operate properly.

Tools with damage to the plug or cord should be taken out of service immediately and repaired to avoid any possibility of electrocution. Also inspect tools for cracked or damaged housings, improperly operating trigger mechanisms, or improperly operating or damaged chucks.

It is very important to read the manufacturer's manual and follow all instructions on the care and inspection of power tools and equipment. It is also important to keep a record of any maintenance performed on power equipment, and to repair and report deficiencies with equipment.

Another type of cutting torch is the exothermic torch. It operates by igniting a combustible metal contained within a tube where oxygen has been forced down the center of the tube. While these torches can cut very heavy steel (even underwater), they also produce a tremendous amount of sparks and slag. A benefit of an exothermic cutting torch is that it can burn through almost anything, including ferrous and nonferrous metals, stainless steel, concrete, glass, and cast iron, to name a few materials. Also, as mentioned previously in the electric cutting tools section, is the plasma cutter. This cutting torch utilizes inert gas or compressed air that is blown from its nozzle at a high speed. At the same time, an electric arc forms within the gas, turning some of the gas to plasma. This precision cutting tool requires no preheating time (as compared to a

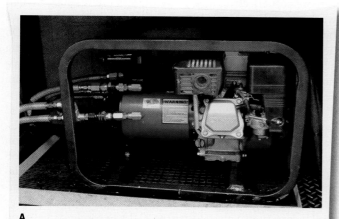

A.

B.

Figure 6-52 Hydraulic pumps. **A.** Built into the apparatus. **B.** Portable.

Petrogen torch) and comes in a portable power pack for better maneuverability.

Hydraulic Tools

Hydraulic tools operate by transferring energy or force from one area to another by using an incompressible fluid such as high-density oil. Hydraulic tools can also be operated by electric and/or battery, gasoline, or pneumatic power; they most commonly operate using a gasoline-powered engine and a hydraulic pump. This text focuses on gasoline-powered hydraulic operation, as this is the most common application for these tools. Many rescue vehicles carry hydraulic pumps used to power a variety of hydraulic rescue tools. The pumps may be built into the vehicle or they may be portable, and they may be powered by the vehicle's engine or have their own engine **Figure 6-52 ◄**.

Maintenance of hydraulic pumps includes checking the hydraulic oil after every use, checking for leaks, inspecting the equipment for any damaged or improperly functioning parts, and checking the fittings for proper operation and cleanness. For units powered by an attached engine, engine fluids should also be checked. It is important for you to be familiar with the manufacturer's maintenance recommendations. Some hydraulic tool manufacturers offer service technician classes that will certify you to perform general repair work on the unit and tools. Check with your tool manufacturer to see if this is an available option; it is a great way to fully understand the inner working of a hydraulic system.

The major advantage of hydraulic tools over any other tool is the power and speed of operation. Hydraulic tools can be superior to any other tool when utilized properly and in the right application. Disadvantages of hydraulic tools are their overall weight and their limited maneuverability in tight spaces. The standard hydraulic tool tends to be very heavy (generally in the range of 35 to 50 lb [16 to 23 kg] for a hydraulic spreader), and the ability to get into very tight spaces is limited because of its relative size. Hydraulic tool companies today are designing tools to be smaller and lighter without sacrificing power.

There are generally four types of hydraulic rescue tools used today **Figure 6-53 ▼**:

Figure 6-53 The four basic types of hydraulic rescue tools utilized today. **A.** Hydraulic spreader. **B.** Hydraulic cutter. **C.** Hydraulic ram. **D.** Hydraulic combination tool (spreader and cutter).

- Hydraulic spreader
- Hydraulic cutter
- Hydraulic ram
- Hydraulic combination tool (spreader and cutter)

Hydraulic rescue tools work through a very simple mechanical process: The hydraulic pump, powered by an engine, forces fluid into the cylinder of the particular tool you are operating. This in turn puts a gradually building pressure on the piston and piston rod inside the tool, forcing the tool to either open or close, based on the operator's positioning of the control valve. This is a basic explanation of how a hydraulic system operates; it can get more involved by adding force multiplication factors such as changing the size of the cylinder and piston, pump capacity, type of system (e.g., single stage and two stages), and so on. We will leave these factors to the engineers.

Most power units operate on either a 5000-psi or 10,500-psi (34,474-kPa or 72,395-kPa) capacity utilizing a two-stage pump system. The first stage of the pump normally operates at a maximum volume flow rate with a lower pressure building up to around 2000 to 3000 psi (13,790 to 20,684 kPa) or greater (this will vary with each manufacturer). A sensing unit then kicks in the second stage, which delivers a maximum pressure for generating a maximum force, which is either 5000 psi or 10,500 psi (34,474 kPa or 72,395 kPa), depending on the unit you are operating.

There are several different types of hydraulic fluid, including mineral (petroleum base) oil, phosphate ester, water-based ethyl glycol compounds, polyol ester (not to be confused with the more commonly used phosphate ester), and vegetable-based oil. The two most prevalent fluids required for rescue tools are phosphate ester and mineral base oil Figure 6-54 ◄ .

Phosphate ester has excellent fire-resistant qualities, but it is considered an irritant, requiring full PPE to always be worn when working around this fluid. This fluid burns like pepper spray, requiring a saline flush or a visit to the emergency department for severe exposure.

Mineral base hydraulic fluid, as compared to phosphate ester, is much easier to work with, is less of an irritant, and is more cost effective; it is, however, less fire resistant. When working with a mineral base hydraulic fluid or any other hydraulic fluid, it is recommended that you wear full PPE. Do not accidentally wipe your eyes with your gloves after handling a hydraulic rescue tool. This common mistake may occur on a hot summer day when you inadvertently wipe the sweat from your forehead.

It is a requirement of NFPA 1936, *Standard on Powered Rescue Tools*, that all internal components, including seals, valves, and fittings, of a powered rescue tool system function properly at a maximum hydraulic fluid temperature of 160°F (71°C). NFPA 1936 also lists safety and performance guidelines for powered rescue tools. Most rescue tool manufacturers are utilizing third-party certification organizations, such as the Underwriters Laboratories Inc. (UL), to conduct safety and performance tests and certify their product to NFPA 1936.

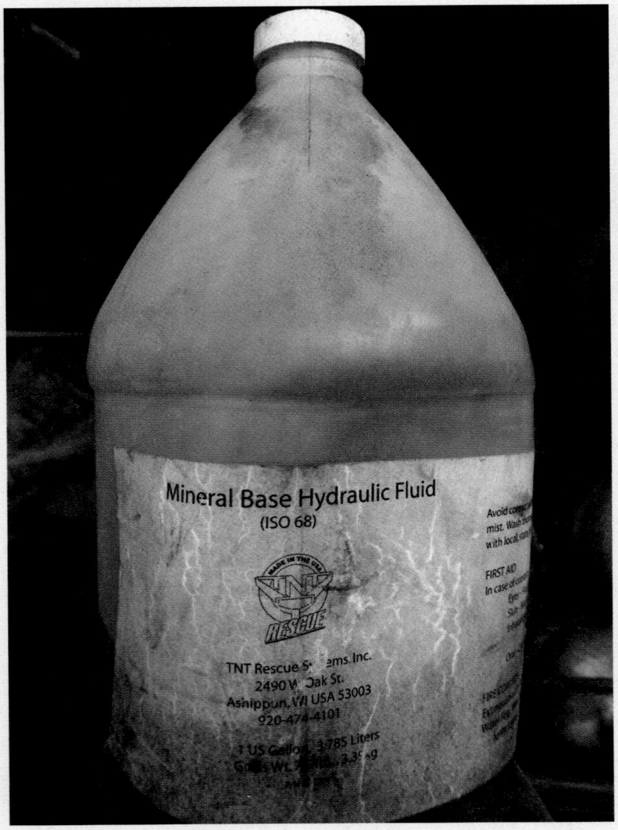

Figure 6-54 The two most prevalent fluids utilized in hydraulic rescue tools are phosphate ester and mineral base oil. This photo shows mineral base oil.

Rescue Tips

Powered rescue tools must have a <u>**dead-man control**</u> feature designed to return the control of the tool to the neutral position automatically in the event the control is released.

The following is an overview of the four common standard hydraulic-powered rescue tools. The definitions of the following hydraulic tools are taken from NFPA 1936.

■ Hydraulic Spreader

The hydraulic spreader was one of the first hydraulic tools designed to be used as a rescue tool. A <u>hydraulic spreader</u> is a powered rescue tool consisting of at least one movable arm that opens to move or spread apart material, or to crush or lift material Figure 6-55 ▶ . Spreaders range in size and force, depending on the manufacturer and whether the tool is operating at 5000 or 10,500 psi (34,474 or 72,395 kPa). A typical range of force can measure from 14,000 to 28,000 lb (6350 to 12,701 kg) of spreading force, with some manufacturers listing their tools with up to 59,000 lb (26,762 kg) or greater spreading force. Newer models are being designed and redesigned each year that are lighter and more powerful; it is very difficult to keep up with all of the changes in the technology of the "latest and

Figure 6-55 The hydraulic spreader.

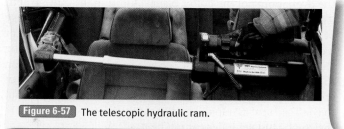

Figure 6-57 The telescopic hydraulic ram.

greatest" rescue tools on the market. The more power and speed the better, but a tool's effectiveness always comes down to the individual using it.

Hydraulic Cutter

The hydraulic cutter is a powered rescue tool consisting of at least one movable blade used to cut, shear, or sever material Figure 6-56 ▼ . There are two types of blade designs commonly used—the curved, or "O," cutter and the straight blade cutter. Both of these blades perform a different cutting action when in operation. Hydraulic cutters do not necessarily cut metal; they are designed to compress the metal until it reaches its fracture point.

The curved blade, when cutting, actually draws material in toward its center notch; this is where the majority of curved blade cutters carry the most cutting force. The straight blade cutter has a tendency to push material outward as it cuts, but it can give you a deeper cut than the curved blade cutter.

Another important feature of the blade is the manufacturing of the steel. Blades utilizing a high-grade steel seem to resist breaking or cracking more than other types of steel. The quality of a steel depends on the grade process and carbon content, which will determine the hardness of the blade and the ability to hold a cutting edge. This is significant when cutting into hardened steel such as a Nader pin or door hinge. Blades made of low-grade steel can at times fracture at their weakest point when attempting to cut through hardened steel; this is why some manufacturers do not recommend cutting Nader pins

and hinges. Most blades made of high-grade steel do not seem to have these limitations and can cut through most hardened steel with the exception of boron, titanium, and other exotic metals, which require a higher cutting force and special design from the tool itself, not just the blades.

Cutting forces on high-pressure systems (10,500 psi [72,395 kPa]) can range from 40,000 to 80,000 psi (275,790 to 551,581 kPa), with some newer models producing a cutting force upwards of 250,000 psi (1,723,689 kPa) or greater. Remember to check with the manufacturer for the cutting force of the tool your organization utilizes.

Hydraulic Ram

The hydraulic ram is a powered rescue tool with a piston or other type of extender that generates extending forces or both extending and retracting forces Figure 6-57 ▲ . Each manufacturer offers various lengths of units, but generally, individual units come in 20 inches (508 mm), 30 inches (762 mm), and 60 inches (1524 mm), or in a telescopic type where one unit can extend from 20 to 60 inches (508 to 1524 mm). Some manufacturers offer various options such as interchangeable tips or extension bars that screw on.

Hydraulic Combination Tool

A hydraulic combination tool is a powered rescue tool capable of spreading and cutting Figure 6-58 ▼ . This tool is not as effective as a dedicated cutter or spreader because of the limited range it can provide during a cut or a spread. Operating the tips of the spreader in the completely closed position will impede

Figure 6-56 The hydraulic cutter.

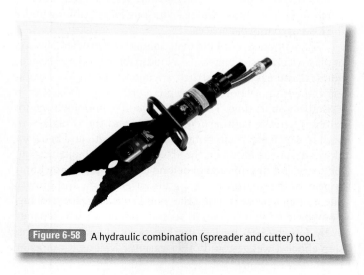

Figure 6-58 A hydraulic combination (spreader and cutter) tool.

the overlapping of the cutter blades, limiting the cutting ability that a dedicated hydraulic cutter can offer. An advantage of a combination tool is the ability to rapidly apply a spread or a cut on a vehicle without having to switch tools, thus saving valuable time. Some units also come in a hand pump version that has the hydraulics in a self-contained unit that is void of any hoses, combustion engines, or pumps. This is very useful in confined spaces, rapid deployment situations, or areas where the combustion engines cannot be used.

Stabilization Tools

Balance goes hand in hand with stabilization. The main objective in stabilization is to gain a balanced footprint by expanding the vehicle's base and lowering its center of gravity prior to performing any work on the vehicle. To properly stabilize a vehicle or any object, one must first understand the term *center of gravity* as it relates to the object being stabilized. **Center of gravity** is the area or point where the entire weight of the object is concentrated and where the load is being forced downward by the earth's gravitational pull. A vehicle on its side will have a narrow base and a high center of gravity; therefore, it will be very unstable.

There are a multitude of tools utilized to stabilize or shore up a vehicle. This section will focus on stabilization tools that use cribbing, struts, jacks, pneumatic shores, and ratchet straps.

Cribbing

Cribbing consists of short lengths of sturdy timber or composite material—usually 2 by 2 inches (51 by 51 mm), 4 by 4 inches (102 by 102 mm), or 6 by 6 inches (152 by 152 mm), in lengths of 18 to 24 inches (457 to 610 mm)—used in various configurations to stabilize loads in place. There are several cribbing designs utilized for extrication, such as step chocks, wedges, shims, and the basic box crib design **Figure 6-59 ▶**. **Step chocks** are specialized cribbing assemblies made of wood or plastic/composite blocks in a step configuration. They are typically used to stabilize vehicles. **Wedges** are objects used to snug loose cribbing under the load or to fill the void space. **Shims** are very similar in design to a wedge, but their profile is much smaller. They are also used to snug loose cribbing under the load or to fill void spaces.

As mentioned, cribbing materials are constructed of wood or plastic/composite materials. Wood cribbing is the most economical and easy to construct; anyone handy in woodworking can cut up sections and/or put a cribbing set together in a relatively short time. The types of wood commonly used for cribbing are the Southern Yellow Pine and the Douglas Fir, both of which are a soft, durable wood. Strength and behavior of wood may be based on the species of wood, moisture content, size and distribution of defects (such as how many knots appear in the board), cracks, grain orientation, and growth rate. Composite plastic is another option to choose for cribbing. Composite cribbing comes in several options, such as separate 4- by 4-inch (102- by 102-mm) sections in various lengths, step chocks, wedges, and other various designs. Composite

A.

B.

C.

D.

Figure 6-59 Cribbing designs. **A.** Step chocks. **B.** Wedges. **C.** Box crib. **D.** Shims.

cribbing costs significantly more up front but will last much longer than wood because of its resistance to damage from abuse, exposure to oils, and other various substances. Another option to wood and composite cribbing is the steel collapsible step chocks, which can be adjusted to heights from 5 to 16½ inches (127 to 419 mm), depending on the manufacturer. This multiple adjustment feature is convenient when you are dealing with vehicles such as a sport utility vehicle (SUV) or truck that sits high off the ground. Cribbing techniques are discussed in more detail in Chapter 8, *Vehicle Stabilization*.

■ Struts

Struts are structural supports or shores used as a "buttress" to stabilize and reinforce an object ▮Figure 6-60 ▾▮. Struts can be made of steel, aluminum, and wood, and can be composite/Kevlar wrapped. Some can also be operated pneumatically or hydraulically. With the exception of the wood type, struts are normally telescopic devices, either tubular or square, that slide to various lengths to accommodate a multitude of stabilization scenarios. The more common lengths used for vehicle stabilization can be anywhere from 3 feet (0.91 m) to 8 feet (2.5 m), with the struts being held in place by supporting pins. Struts can vary in support loads, depending on whether they use a single, double, or multiple pin system, and whether they are fully extended or compact. Depending on the manufacturer, most struts can generally support a WLL of 4000 to 18,000 lb (1814 to 8165 kg). This WLL can vary greatly depending on the type of strut system and the design.

Wood struts for vehicle stabilization are commonly comprised of 4- by 4-inch (102- by 102-mm) sections in various lengths ranging from 3 to 6 feet (0.91 to 2 m) or greater. Wood struts can be utilized in their simplest form by themselves or with a wood stabilization system such as the steel ground pad, base, plate, and cap.

Figure 6-60 Struts are used for structural support to stabilize and reinforce a vehicle.

Figure 6-61 The Rescue 42 Stabilization System is an example of tension buttress stabilization that offers several different tip options and a jacking attachment to utilize for lifting.

Most strut stabilization systems use a strap in a ratchet or jacking device to add tension to the object being stabilized; this method is known as **tension buttress stabilization**. Attaching a strap from the base of the strut to the vehicle and adding tension locks the vehicle in place using a diagonal force that lowers the vehicle's center of gravity by increasing the vehicle's entire footprint or floor span. This is the same manner in which the outriggers on an aerial apparatus function. There are multiple variations and applications to the tension buttress stabilization system, with each manufacturer adding various features to enhance its system ▮Figure 6-61 ▴▮. Some of these features are interchangeable heads offering different options for the varying penetration or anchoring problems that you will encounter. Some manufacturers provide a jacking device, which can add a lifting option to the strut system. The lifting capacities of each product's jacking device will vary, and each product will offer its own tip attachments or base attachments, such as a ratchet strap or cam buckle strap. It is best to contact the various manufacturers and physically try all of the strut systems in practical training evolutions before purchasing a system.

■ Jacks

Manual lifting and pushing tools include mechanical jacks and hand-operated hydraulic jacks ▮Figure 6-62 ▸▮. Mechanical jacks can be screw type, ratchet lever, or cam type. Screw-type jacks include bar screw jacks and house or trench screw jacks. These types of jacks operate on the inclined plane concept (screw) and may be capable of lifting very heavy loads. In most cases, however, these types of jacks are used to hold or support an object in place—for example, a vehicle that is elevated and resting on an object or another vehicle. Screw-type jacks should be used in conjunction with other stabilization devices to maintain a secure environment. Ratchet lever jacks are commonly known as Hi-Lift® or farmer's jacks and consist of a beam and a ratcheted lifting mechanism. This lifting mechanism consists of a long jacking pole or ratchet lever that offers leverage for lifting the vehicle. This lifting mechanism rides on the beam and can be controlled in either direction with a directional latch.

Make sure the load being carried by the jack does not shift, as this can cause the jack to tip and fall over. Always stand to the side when jacking or ratcheting the tool under a load. This

Figure 6-62 Hand-operated jack.

Figure 6-63 Ratchet straps are a great and cost-effective tool for tie-down stabilization.

is especially critical with the jacking pole or ratchet lever jack, which can become extremely hazardous when ratcheting up and even more so when ratcheting down to release a load. If the jack is not released in a safe, controlled manner, the jacking pole can fire up at the rescuer, causing severe injury or death to anyone struck or caught in the path of the pole. Always lock the safety latch and pole upright in place when the proper lifting height is reached. Some support arms on the ratcheting jack are modified by rescue personnel by welding an extension to it and/or expanding the base to increase its overall footprint. Before modifying the tool, make sure the manufacturer supports doing so because of obvious safety issues and the possibility of voiding the manufacturer's warranty.

Bottle jacks are manual, hydraulically or pneumatically operated vertical lifting shafts that commonly come in 3- to 50-ton (2.7- to 45-metric ton) lifting options with a lifting range of 4 to 18 inches (102 to 457 mm) including the base; these figures will vary greatly among manufacturers. These types of jacks need to be utilized in conjunction with other stabilization devices, such as cribbing.

Cam-type jacks (sometimes referred to as jack wrenches) use 2- by 2-inch (51- by 51-mm), 4- by 4-inch (102- by 102-mm), or 6- by 6-inch (152- by 152-mm) timbers side by side, which are held together by steel clamps. The cam mechanism is used to lengthen the timbers by grabbing one timber and lifting or pushing the other. These types of jacks can support a maximum safe working load of up to 16,000 lb (7257 kg) and can be up to 18 feet (5.5 m) long, depending on the size of the lumber used. These types of jacks are not utilized for vehicle stabilization.

When using any type of jack, be sure it is on a flat, level, hard surface that will not give way once weight is placed on the jack. If the surface underneath the jack is soft, place a flat board or steel plate under the jack to distribute the weight. Inspect all jacks on a regular basis for any damage that may have occurred to them, including any signs of overloading, cracks, damaged parts, or improper operation.

■ Ratchet Straps

A **ratchet strap** is a mechanical tensioning device that uses a manual gear-ratcheting drum to put tension on an object utilizing a webbing material **Figure 6-63 ▲**. Ratchet straps by themselves are a great and cost-effective tool for cargo or industrial tie-down stabilization. Ratchet straps are not designed to lift and are best utilized in locking vehicles together, a technique known as "marrying." When a vehicle is on top of another vehicle, the objective is to stabilize both vehicles, which is accomplished by marrying the vehicles to form one unit; the proper use of ratchet straps is very effective in this objective. Other uses of ratchet straps include tying down objects such as trees and poles on the vehicle or tying down objects that need to be secured to prevent them from entering the area of operation.

Ratchet straps come in lengths of 6, 10, 12, 15, 20, and 30 feet (1.8, 3.0, 3.7, 4.6, 6.1, and 9.1 m). The most commonly used types are composed of nylon or polyester webbing and are available in 1- to 4-inch (25- to 102-mm) widths. The 3- and 4-inch (76- and 102-mm) widths can provide breaking strength (stress on the material at the time of rupture) of up to 20,000 lb (9072 kg) with a WLL of up to 6000 lb (2722 kg). Some manufacturers offer abrasion-resistant webbing, which dramatically improves wear and tear resistance.

There are several end attachment options that can be added to make it much easier to secure strapping. Some of these options include webbing loops, flat hooks, triangle/delta rings, "O" rings, chain anchors with grab hook/chain shorteners, and double wire hooks. The most practical choice may be to custom design an end attachment that can give you several latching options. For example, an "O" ring with a double wire hook and a chain rated for 2 feet (0.6 m) with a D-ring and a locking gate may be an effective option. To maintain safety and proper rating, attachments should be accomplished through a manufacturer, not your local hardware store. Remember to check with the manufacturer to maintain safety specifications on any of the various design options.

Organization of Equipment

Proper organization of equipment includes tool staging at an incident and the proper setup and staging of tools on the apparatus. The proper staging of tools on the scene and on the apparatus can be an incredible benefit to completing the operational tasks more expediently and efficiently.

Tool staging at an incident may involve laying a tarp out at the edge of the secure work area and organizing the tools on the tarp. This will allow rescuers to locate the appropriate tools quickly.

Proper tool staging on the apparatus should ultimately begin at the planning stages for the building and design of the apparatus itself, but it can also be accomplished later, with a little ingenuity and common sense. A simple redesign or rearrangement of the tools in the compartments can consist of several options, such as acquiring hose reels and placing them on opposite end-side compartments preattached to the tools with one reel dedicated to the cutter and the other reel dedicated to the spreader. This organization allows the rescuer to open the compartment door, pull off the tool, and go. Another option is to place your step chocks and cribbing in a compartment that is readily available, where the rescuer does not have to step up onto the bed of the apparatus or remove other items to retrieve them. Cribbing ends can also be color coded by spray painting to depict various sizes and types. The proper organization of equipment will be different for each agency based on the type of apparatus and the custom preference of the users. The question to ask is simple: How can the apparatus be set up to improve the overall operations of the department?

Rescue Vehicles

Many different types of emergency vehicles may respond to rescue incidents. Some are single, self-contained units; others comprise truck/trailer combinations. Some are specialized vehicles dedicated to a single purpose; others are designed to support multiple specialties. In addition, some types of units are designed to perform both firefighting and rescue functions.

Specialized rescue vehicles are commonly classified into four categories: light, medium, heavy, and special purpose/multipurpose **Figure 6-64 ▼**. This distinction has more to do with the function of the vehicle than its size. For example, a large vehicle might carry equipment and support items only for basic vehicle extrication. Because of its minimal rescue capability, this vehicle may meet the criteria of only a light rescue. The design and equipping of any rescue vehicle should comply with NFPA 1901, *Standard for Automotive Fire Apparatus*, in addition to the needs determined by your organization.

■ Light Rescue Vehicles

Light rescue vehicles are equipped for basic rescue tasks, so they typically carry only hand tools and basic extrication and medical care equipment. Because these vehicles tend to be smaller, they are ideal for quick response; they can get to a scene quickly, handle small incidents, or stabilize the scene until additional equipment and personnel arrive. Typically built on a 1- to 1½-ton (0.9- to 1.4-metric ton) chassis, these vehicles can have a standard or crew cab, so they may be able to carry as many as five rescuers. Additionally, their light weight and small size make them ideal for going into areas where larger apparatus will not fit.

One of the most common uses for light rescue vehicles is for response to extrication incidents. The equipment typically carried on such units includes pry bars, bolt cutters, an air chisel, jacks, stabilization equipment and cribbing, hand and power saws, lighting, and portable hydraulic rescue tools. Most engines and ladder trucks that carry some rescue equipment fall into this category.

■ Medium Rescue Vehicles

Medium rescue vehicles are designed to handle most rescue situations likely to be encountered by the responding department. They may carry basic to advanced equipment applicable to a variety of specialties, but they normally are not equipped to the advanced level in more than one or two areas. For example, such units might provide advanced capabilities for confined-space and rope rescue, along with basic equipment for trench and structural collapse rescue. Other capabilities may include built-in electrical power generation, a hydraulic rescue tool pump, and an air compressor (for breathing or regular air).

A.   **B.**   **C.**

Figure 6-64   Rescue vehicles. **A.** Light rescue vehicle. **B.** Medium rescue vehicle. **C.** Heavy rescue vehicle.

Many medium rescue units are designed to serve a single purpose. This purpose could relate to any of the specialty areas, although the most commonly encountered are units devoted to trench and excavation, confined-space, rope, or ice/water rescue. Other specialty units, though not as common, focus on structural collapse, heavy extrication, and mine/tunnel rescue.

Heavy Rescue Vehicles

Heavy rescue vehicles are, by definition, the most heavily equipped vehicle in the rescue vehicle class. They are designed to handle almost any rescue incident a responding organization may encounter. These vehicles tend to be capable of advanced capabilities in multiple areas, and they carry a wide variety of specialized rescue tools. While many of these units are supplied by municipalities, regional and national heavy rescue organizations tend to own units on the high end of this capability spectrum. These teams are often referred to as urban search and rescue (USAR) teams, and they specialize in structural collapse. Many teams also have advanced capabilities in trench rescue, confined-space rescue, rope rescue, heavy extrication, and water rescue, just to name a few of the disciplines.

Special-Purpose/Multipurpose Vehicles

Special-purpose and multipurpose vehicles are also quite common. As mentioned earlier, many organizations have engines, ladder trucks, and vans that carry a limited amount of technical rescue equipment while also supporting fire and rescue operations. Special-purpose vehicles include small, off-road vehicles and boom trucks **Figure 6-65 ▶**. The off-road vehicles are growing in popularity because of their small size, all-wheel-drive capability for accessing difficult locations, ability to carry both victims and equipment, and basic firefighting capability when equipped with an optional skid pump/tank unit.

Special Equipment

Foam

Spilled fuel is a common occurrence at motor vehicle collisions. In a flammable liquid fire, the liquid itself does not burn. Only the flammable vapors evaporating from the surface of the liquid and mixing with air can burn. Depending upon the temperature and the physical properties of the liquid, the amount of vapors being released from the surface of a liquid will vary. For example, gasoline produces flammable vapors down to a temperature of −45°F (−43°C). Some vapors are lighter than air and rise into the atmosphere. Other vapors are heavier than air and flow across the surface and along the ground, collecting in low spots.

Class B foam is used to fight Class B fires—flammable and combustible liquids. There are several different types of Class B foam formulated to be effective on different types of flammable liquids. Some liquids are incompatible with different foam formulations and will destroy the foam before the foam can control the fire. Class B can be divided into two categories:

A.

B.

Figure 6-65 Rescue vehicles. **A.** Off-road vehicle. **B.** Boom truck.

hydrocarbons and polar solvents. The hydrocarbon consists of the gasoline-type (petroleum-based) fuels, while the polar solvents consist of alcohol-based fuels that are miscible; they will mix with water such as ethanol. Alcohol-resistant foam (ARF) is a type of foam utilized for alcohol-based fuel fires. Foams designed for use on hydrocarbon-type fires such as gasoline will break down when applied to a polar solvent such as ethanol. Each fuel type requires a different type of foam to suppress vapors while not breaking down. Beware of fuel blends such as E85 (flex fuel), which is a mixture of 15 percent unleaded gasoline and 85 percent ethanol; it will require a type of ARF.

The use of foam is increasing as many new types of foam have become available and efficient systems for applying them have been developed. For example, foams have been developed for use in neutralizing hazardous materials and decontamination. Many fire departments are using several different types of foam for a variety of situations.

Firefighting foam is produced by mixing foam concentrate with water and air to produce a solution used as an effective extinguishing agent. Each type of foam requires the appropriate type of concentrate, the proper equipment to mix the concentrate with water in the required proportions, and the proper application equipment and techniques. You need to become familiar with the specific types of foam used by your organization and the proper techniques for using them.

Foam Equipment

Foam equipment includes the proportioning equipment used to mix foam concentrate and water to produce the foam solution, as well as the nozzles and other devices used to apply the foam. There are many different types of proportioning and application systems. Most engine companies carry the necessary equipment to place at least one foam attack line into operation. Structural firefighting apparatus can also be designed with built-in foam proportioning systems and on-board tanks of foam concentrate to provide greater capabilities. Many fire departments specify both Class A–integrated and Class B–integrated foam systems on new apparatus or have special foam apparatus available for situations when large quantities of foam are needed. There are also quick-attack foam systems that come available in a portable package design for fast deployment and operation.

■ Signaling Devices

There are many devices available to assist rescuers in getting the job done. For example, portable or fixed communication devices such as cellular telephones help rescuers communicate with one another. Marking kits that include paint, chalk, pens/pencils, or crayon may be used to mark a scene and alert other rescuers of a potential hazard. Pickets or stakes may be used to close off the perimeter of the scene in order to keep bystanders at a safe distance. Preplans and maps help rescuers navigate through or around the scene. Traffic control devices halt traffic by changing traffic signals to allow emergency vehicles to pass through safely and quickly. There are many signaling devices that can be used to your advantage! Visual and audio devices are two of the most commonly used signaling devices in technical rescue.

Visual Devices

Often we cannot find the victims we are looking for. Specialized cameras are available to help find victims. The most common of these cameras is the thermal imaging camera (TIC) Figure 6-66 ▶ . Many fire departments carry this camera to detect hidden fire or hotspots, but they are equally useful in finding victims lightly covered in debris or dust or those who cannot be seen because of darkness.

TICs are very useful during a late night call on remote highways where there are no street lights. Victims who have been ejected a distance from the scene and might not otherwise be found can now be located immediately utilizing a TIC by scanning the area systematically.

Audio Devices

There may be other times when we cannot hear the victims we are trying to locate. There are two types of acoustic listening devices frequently used by rescue services Figure 6-67 ▶ . The

Figure 6-66 Thermal imaging cameras are the most common specialized cameras in rescue situations.

first device uses a microphone attached by an intrinsically safe wire that can be pushed or lowered into a space to listen for victims. Variations of this device allow for two-way communications when desired. The second type of device uses highly sensitive probes that can allow the victim's location to be triangulated when the readings are evaluated by a trained technician at the receiving station. Audio devices are commonly used for confined-space or structural collapse rescue.

■ Power Detection

Electrical hazards on a rescue scene are always a concern, and efforts should always be made to identify and isolate any hazards. Because it is often impossible to tell visually whether a wire or machinery is energized, every rescue organization should have an AC power locator Figure 6-68 ▶ . This tool detects the presence of electrical frequencies below 100 Hz. It does

Figure 6-67 Acoustic listening devices like this one can help find victims who cannot be seen.

These sections are fitted around a patient who is lying on the ground or another relatively flat surface. The parts are reconnected, and the patient is lifted and placed on a long backboard or stretcher. This type of stretcher is not designed to be used by itself for immobilization. When inspecting stretchers, ensure that there are no bent rails or supports, cracked welds, rips, tears, damaged buckles, or missing pieces.

Immobilization and Combination Devices

The simplest immobilization devices are the full and half, or short, backboards, which are often used in conjunction with stretchers (Figure 6-70 ▶). Other specialized devices incorporate a half board into a vest-style arrangement, which can be used, depending on the manufacturer, either alone or in conjunction with a harness for lifting a victim vertically. When inspecting immobilization devices, ensure that there are no cracks,

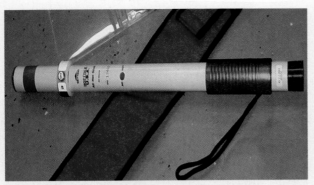

Figure 6-68 An AC power locator detects the presence of electrical frequencies below 100 Hz.

not, however, detect DC power or AC power contained in solid metal enclosures such as grounded metal conduit. There are other meters available that can detect DC power or provide voltage readings if necessary. These meters are not suitable for emergency response because they require probes to be placed directly on the wire or object to be tested.

At a motor vehicle collision, if a wire is suspected to be down and resting on the vehicle or in close proximity, then the technical rescuer must ensure that the proper jurisdictional power utility company is notified immediately and the proper action is taken. This normally consists of shutting down an area grid. Never attempt to see if a wire is hot or live by using a hot stick or pole; always assume that a wire is hot or live until the power company can determine otherwise. Take the appropriate protective safety measures, and move a safe distance from the area. Live wires are known to jump several feet when energized. What appears dead to the rescuer may all of a sudden come alive when the power company's computer system automatically, by program, reenergizes the line to detect the break areas.

Victim Packaging and Removal Equipment

When victims are found, it is important they not be further injured while removing them to definitive medical care. This is accomplished by the use of stretchers and immobilization devices appropriate to the situation and, when necessary, high-point devices that will allow for the safe movement of the stretcher in a vertical environment. The equipment used is known as victim packaging and removal equipment.

Stretchers and Litters

The most common stretchers used by rescue services when removing victims are the collapsible, scoop, and basket stretchers (Figure 6-69 ▶). Each type has advantages and disadvantages. The collapsible and basket stretchers allow for the incorporation of a backboard and may be rigged for vertical lifting. Basket stretchers can also be used for any water rescue environment when flotation devices are attached to provide buoyancy. The scoop stretcher is designed to be split into two or four pieces.

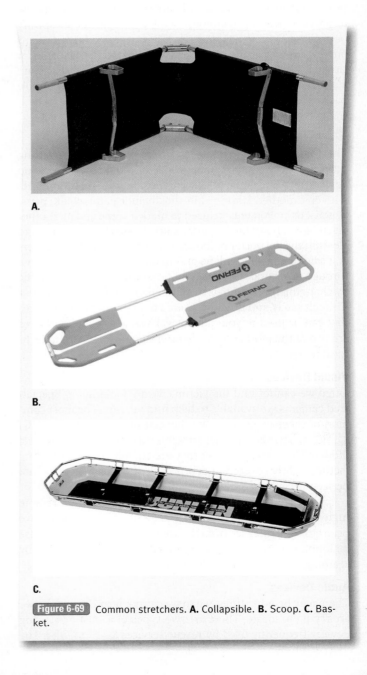

A.

B.

C.

Figure 6-69 Common stretchers. **A.** Collapsible. **B.** Scoop. **C.** Basket.

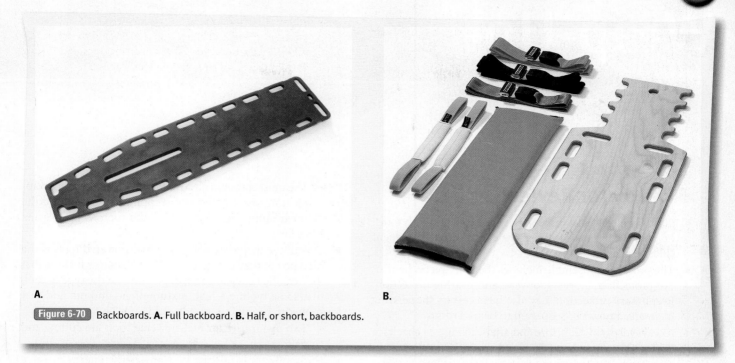

Figure 6-70 Backboards. **A.** Full backboard. **B.** Half, or short, backboards.

splinters, gouges, worn straps, damaged buckles, or missing pieces.

Research Tools

The Internet is a research tool readily available to the technical rescuer. The information and research capabilities for vehicle and machinery rescue are endless; the Internet provides an enormous amount of material on vehicle construction and design, loca-

tions of key safety and power components, physical tools, tool manufacturers—the list goes on and on. Truly the greatest tool that a technical rescuer possesses is his or her mind. The old proverbial saying that "a mind is a terrible thing to waste" speaks volumes in the field of emergencies services. It is up to you to take the initiative and apply what was gifted to you by acquiring this information through diligent researching and always being open to learning as well as by teaching and sharing what you have learned.

Wrap-Up

■ Ready for Review

- Before the technical rescuer can start to work with tools, he or she must wear full personnel protective equipment (PPE) for safety.
- The protective ensemble includes the body piece, helmet, eye protection, gloves, footwear, and hearing protection. Respiratory protection may also be necessary if operating in an area known or suspected to be hazardous.
- Always inspect PPE before and after each use to check for damage that may have occurred during use or while in storage. Follow the manufacturer's instructions regarding inspection, maintenance, cleaning, limitations, and repair methods, including the use of maintenance tools. Lastly, document all inspection and maintenance activities performed.
- The technical rescuer must have a vast working knowledge of tools used in the field and be proficient with them to be successful.
- Tools for vehicle extrication purposes can be broken down into basic categories: hand tools, pneumatic tools, electric- or battery-operated tools (nonhydraulic), fuel-powered tools, and hydraulic tools.
- Hand tools operate using human power. They can be classified in four common categories: striking tools, leverage/prying/spreading tools, cutting tools, and lifting/pushing/pulling tools.
  - Striking tools are used to apply an impact force to an object.
  - Leverage/prying/spreading tools are used to pry objects apart or to lift heavy objects. Two types of leverage tools are distinguished—rotating and prying.
  - Cutting tools have a sharp edge designed to sever an object.
  - Lifting/pushing/pulling tools can extend the reach of the fire fighter and increase the power that can be exerted upon an object.
- The working load limit (WLL) is the maximum force that may be applied to an assembly in straight tension.
- Pneumatic tools are tools utilizing air under pressure to operate.
  - The most common pneumatic tools are air chisels, air impact wrenches, air shores, cut-off tools, and rescue-lift air bags.
- Electric-powered tools utilize a standard household current or generator to operate.

- The most common electric-powered tools are the reciprocating saw, circular saw, snipping devices such as the rebar cutter, and specialty items such as the plasma cutter.
- One of the major advantages of fuel-powered tools is the high power they can generate. Disadvantages include that they can be heavy to carry, depending on the type of tool, that some require a fuel mixture of gas and oil, and that some can be difficult to start cold.
  - Two major uses for fuel-powered tools are cutting and nailing.
- Hydraulic tools operate by transferring energy or force from one area to another by using an incompressible fluid such as high-density oil. Hydraulic tools can also be operated by electric, gasoline, or pneumatic power; they most commonly operate using a gasoline-powered engine and a hydraulic pump.
  - The most common hydraulic tools are the hydraulic spreader, hydraulic cutter, and hydraulic ram.
- The main objective in stabilization is to gain a balanced footprint or base for the vehicle prior to performing any work on the vehicle. Stabilization tools include cribbing, struts, jacks, and ratchet straps.
- The proper staging of tools on the scene and on the apparatus can be an incredible benefit to completing the operational tasks more expediently and efficiently.
- Specialized rescue vehicles are commonly classified into four categories: light, medium, heavy, and special purpose/multipurpose.
- Special equipment used for vehicle extrication includes Class B foam. Class B foam is used to fight Class B fires—flammable and combustible liquids.
- Visual and audio devices are two of the most commonly used signaling devices in technical rescue.
- It is often impossible to tell visually whether a wire or machinery is energized; every rescue organization should have an AC power locator.
- Stretchers and immobilization devices allow for the safe movement of the patient.

■ Hot Terms

Adapter A device used to convert a battery-operated tool to a general current tool.

Air compressor A piece of equipment used to provide power to pneumatic tools or to provide breathing air.

Air impact wrench A pneumatic tool used to remove bolts/nuts of various sizes.

Air shoring Shoring extended by the use of compressed air; shoring is used where the vertical distances are too great to use cribbing or the load must be supported horizontally.

Center of gravity The location where entire weight of the object is concentrated and where the load is being forced downward by Earth's gravitational pull.

Chainsaw A gasoline-powered saw capable of cutting wood, concrete, and even light-gauge steel. Standard steel chains are used to cut wood, carbide-tipped chains can cut wood and light-gauge metal, and diamond chains are used for cutting concrete.

Chain sling A sling composed of a single chain or multiple chains in various lengths, commonly in lengths of 5 and 9 feet (1.5 and 2.7 m); it is attached to a ring, either round or oblong, and used for lifting heavy loads.

Circular saw An electric- or battery-powered saw that moves in a circular motion; these saws come in a variety of sizes and are used primarily for cutting wood, although special blades are available that will cut metal or masonry.

Class B foam Foam used to extinguish flammable and combustible liquid (Class B) fires.

Come along A ratchet lever winching tool that can provide up to several thousand pounds of pulling force, with the standard model for extrication being 2000 to 4000 lb (907 to 1814 kg) of pulling force.

Cribbing The most common stabilization tool that gives the user several height options when trying to rapidly stabilize a vehicle.

Cutting torch A tool that produces an extremely high-temperature flame capable of heating steel until it melts, burns, and oxidizes, cutting through the object. This tool is sometimes used for rescue situations such as cutting through heavy steel objects.

Dead-man control A control feature designed to return the control of the hydraulic tool to the neutral position automatically in the event the control is released.

Electrical generators Generators that utilize a general current or generator to operate. Primarily used to power scene lighting and to run power tools and equipment; may be portable or fixed.

Electric-powered tools Tools that utilize a general current or generator to operate.

Ferrous metals Metals that contain iron, cast iron, low- and medium-alloyed steels, and specialty steels, such as tooled steels and stainless steels.

Flat form rescue-lift air bags Pneumatic-filled bladders designed to retain their flat profile in the center as they are inflated to lift an object or spread one or more objects away from each other to assist in freeing a victim.

Grab hook A device designed to take up the slack needed to make a chain the appropriate size for the task at hand; it is utilized by inserting a link of the chain into the slot of the hook. The grab hook may also be referred to as a chain shortener.

Hand tool Any tool or equipment operating from human power.

High-pressure rescue-lift air bags The most commonly used bags among rescue agencies, these bags utilize a working air pressure of approximately 100 to 145 psi (689 to 1000 kPa) to lift an object or spread one or more objects away from each other to assist in freeing a victim. The high-pressure kits come with hoses, a regulator, a master control module, and various other attachments.

Hydraulic combination tool A powered rescue tool capable of both spreading and cutting.

Hydraulic cutter A powered rescue tool consisting of at least one movable blade used to cut, shear, or sever material.

Hydraulic ram A powered rescue tool with a piston or other type of extender that generates extending forces or both extending and retracting forces.

Hydraulic spreader A powered rescue tool consisting of at least one movable arm that opens to move or spread apart material, or to crush or lift material.

Hydraulic tools Tools that operate by transferring energy or force from one area to another by using an incompressible fluid such as high-density oil.

Junction boxes Electrical enclosures commonly found when multiple outlets are required.

Low-pressure rescue-lift air bags Air bags with a very high lift with a maximum working air pressure of approximately 7 psi (48 kPa); they are used to lift an object or spread one or more objects away from each other to assist in freeing a victim.

Medium-pressure rescue-lift air bags Air bags that have a rugged design and utilize a working air pressure of approximately 15 psi (103 kPa) used to lift an object or spread one or more objects away from each other to assist in freeing a victim. These are not as common as the low- and high-pressure rescue-lift air bags.

New technology (NT) high-pressure rescue-lift air bags Air bags that utilize a unique lifting system where each round-shaped bag can be locked together with a threaded connector, creating one bag with multiple cells that offers a distinct height advantage over traditional flat bags.

Nonferrous metals Metals or alloys free of iron, such as aluminum, copper, nickel, lead, zinc, and tin.

"O" ring or oblong ring An attachment designed to join chains together or join a chain to a come along utilizing a hook.

Pneumatic chisel A pneumatic tool used to cut through various types and sizes of metal.

Pneumatic cut-off tool A pneumatic tool utilizing a small carbide disc, normally 3 inches (76 mm) in diameter, which rotates at a high rpm to cut through most metals.

Pneumatic tools Tools that use air under pressure to operate.

Protective ensemble Personal protective gear, including the helmet, primary eye protection, coat, pants, coveralls, gloves, footwear, and hearing protection.

Ratchet strap A mechanical tensioning device with a manual gear-ratcheting drum to put tension on an object utilizing a webbing material.

Reciprocating saw A power-driven saw in which the cutting action occurs through a back-and-forth motion (reciprocating) of the blade.

Rescue-lift air bags Inflatable devices used to lift an object or spread one or more objects away from each other to assist in freeing a victim. They come in various sizes and types, such as low-pressure bags, medium-pressure bags, high-pressure bags, high-pressure flat bags, and NT (new technology) locking bag.

Rotary saw A fuel-powered saw capable of cutting wood, concrete, and metal; two types of blades are used on rotary saws: a round metal blade with teeth and an abrasive disc. The application of rotary saws in vehicle extrication is limited.

Self-contained breathing apparatus (SCBA) A respirator with an independent air supply that allows rescuers to enter dangerous atmospheres.

Shims Objects that are smaller than wedges used to snug loose cribbing under a load or to fill void spaces.

Shoring A stabilization technique used where the vertical distances are too great to use cribbing or the load must be supported horizontally, such as in a trench, or diagonally, such as in a wall shore.

Slide hook A hook that allows chain links to pass freely through the throat of the hook to tighten around an object.

Spring-loaded center punch A glass removal tool used on tempered glass that, when engaged, uses a spring-loaded plunger to fire off a steel rod with a sharpened point directly into a pinpoint area of glass, causing the glass to shatter.

Step chocks Specialized cribbing assemblies made out of wood or plastic blocks in a step configuration. They are typically used to stabilize vehicles.

Struts Structural supports used as a "buttress" to stabilize and reinforce an object. Struts can be made of steel, aluminum, composite, and wood.

Supplied air respirator/breathing apparatus (SAR/SABA) A respirator in which breathing air is supplied by air line from either a compressor or stored air (bottle) system located outside the work area.

Tension buttress stabilization A strut stabilization system that uses a strap in a ratchet or jacking device to add tension to the object being stabilized, locking the vehicle in place by using a diagonal force that lowers the vehicle's center of gravity by increasing the vehicle's entire footprint.

TPI rating A rating that indicates how many teeth per inch a blade has.

Wedges Objects used to snug loose cribbing under a load or to fill a void between the crib and the object as it is raised.

Winch Chain or cable used for a variety of lifting, pulling, and holding operations.

Working load limit (WLL) The maximum force that may be applied to an assembly in straight tension.

Technical Rescuer *in Action*

Your unit responds to a report of a motor vehicle collision. Upon your arrival, you don PPE. You notice that multiple cars are involved. The incident commander assigns you the task of performing extrication on vehicle number one.

As you perform your size-up of the vehicle, you note that it is a late-model sedan resting on four wheels. The vehicle has suffered major damage from a lateral impact accident. There are two people trapped inside the vehicle, and both will require emergency medical care. After the inner/outer survey has been completed, you confirm you have access from all sides of the vehicle.

1. At this point, what task should you perform first?
   A. Determine if there are any hazards surrounding the vehicle.
   B. Determine if the hydraulic hoses will reach the vehicle.
   C. Stabilize the vehicle.
   D. Determine what the weight of the vehicle is.

2. When referring to reciprocating saws, TPI stands for:
   A. total point interaction.
   B. two plus an inch.
   C. teeth per inch.
   D. total percentage of interference.

3. In this scenario, if you were to break the tempered side glass in the vehicle to access the victim, you could use a striking tool such as a:
   A. spring-loaded center punch.
   B. Kelly tool.
   C. steel chisel.
   D. pneumatic window punch.

4. In this scenario, you try to utilize a pry bar and find that you do not have enough leverage to pry open the door. You should then:
   A. extend the length of the handle with a pipe.
   B. sit on the pry bar and use body weight.
   C. use bystanders to assist in adding more force.
   D. utilize a different tool or tactic.

5. The best tool to use to cut through a wide C post would be a:
   A. hacksaw.
   B. reciprocating saw.
   C. K-12 rotary saw.
   D. bow saw.

6. A come along is designed to be utilized with a rated chain package.
   A. True
   B. False

7. In reference to vehicle extrication, the acronym WLL stands for:
   A. working leverage limit.
   B. working load limit.
   C. weight load limit.
   D. weight leverage limit.

8. To power pneumatic tools, the rescuer must have a supply of:
   A. compressed fluid.
   B. compressed air.
   C. 220-volt electricity.
   D. rechargeable batteries.

9. Ferrous metals contain:
   A. iron.
   B. aluminum.
   C. tin.
   D. lead.

10. Stabilization of a vehicle on its side should be relative to the vehicle's:
   A. position.
   B. height.
   C. weight.
   D. center of gravity.

NFPA 1006 Standard

Chapter 5, Job Performance Requirements

5.1 **General Requirements.** The job performance requirements defined in Sections 5.2 through 5.5 shall be met prior to being qualified as a technical rescuer. (pages 147–168)

5.2 **Site Operations.**

5.2.1 Identify the needed support resources, given a specific type of rescue incident, so that a resource cache is managed, scene lighting is provided for the tasks to be undertaken, environmental concerns are managed, personnel rehabilitation is facilitated, and the support operation facilitates rescue operational objectives. (pages 147–155)

(A) Requisite Knowledge. Equipment organization and tracking methods, lighting resource type(s), shelter and thermal control options, and rehab criteria. (pages 147–155)

(B) Requisite Skills. The ability to track equipment inventory, identify lighting resources and structures for shelter and thermal protection, select rehab areas, and manage personnel rotations. (pages 147–155)

5.2.2 Size up a rescue incident, given background information and applicable reference materials, so that the type of rescue is determined, the number of victims is identified, the last reported location of all victims is established, witnesses and reporting parties are identified and interviewed, resource needs are assessed, search parameters are identified, and information required to develop an incident action plan is obtained. (pages 155–162)

(A) Requisite Knowledge. Types of reference materials and their uses, availability and capability of the resources, elements of an action plan and related information, relationship of size-up to the incident management system, and information gathering techniques and how that information is used in the size-up process. (pages 155–162)

(B) Requisite Skills. The ability to read technical rescue reference materials, gather information, relay information, and use information gathering sources. (pages 155–162)

5.2.3 Manage incident hazards, given scene control barriers, personal protective equipment, requisite equipment, and available specialized resources, so that all hazards are identified, resource application fits the operational requirements, hazard isolation is considered, risks to rescuers and victims are minimized, and rescue time constraints are taken into account. (pages 162–166)

(A) Requisite Knowledge. Resource capabilities and limitations, types and nature of incident hazards, equipment types and their use, isolation terminology, methods, equipment and implementation, operational requirement concerns, common types of rescuer and victim risk, risk–benefit analysis methods and practices, and types of technical references. (pages 162–166)

(B) Requisite Skills. The ability to identify resource capabilities and limitations, identify incident hazards, assess victim viability (risk–benefit), utilize technical references, place scene control barriers, and operate control and mitigation equipment. (pages 162–166)

5.2.4 Manage resources in a rescue incident, given incident information, a means of communication, resources, tactical worksheets, personnel accountability protocol, applicable references, and standard operating procedures, so that references are utilized, personnel are accounted for, deployed resources achieve desired objectives, incident actions are documented, rescue efforts are coordinated, the command structure is established, task assignments are communicated and monitored, and actions are consistent with applicable regulations. (pages 152–155)

(A) Requisite Knowledge. Incident management system; tactical worksheet application and purposes; accountability protocols; resource types and deployment methods; documentation methods and requirements; availability, capabilities, and limitations of rescuers and other resources; communication problems and needs; communications requirements, methods, and means; types of tasks and assignment responsibilities; policies and procedures of the agency; and technical references related to the type of rescue incident. (pages 152–155)

(B) Requisite Skills. The ability to implement an incident management system, complete tactical worksheets, use reference materials, evaluate incident information, match resources to operational needs, operate communications equipment, manage incident communications, and communicate in a manner so that objectives are met. (pages 152–155)

5.2.5 Conduct a search, given hazard-specific personal protective equipment, equipment pertinent to search mission, an incident location, and victim investigative information, so that search parameters are established, victim profile is established, all people either involved in the search or already within the search area are questioned upon entry and exit and the information is updated and relayed to command, the personnel assignments match their expertise, all victims are located as quickly as possible, applicable technical rescue concerns are managed, risks to searchers are minimized, and all searchers are accounted for. (pages 158–160)

(A) Requisite Knowledge. Local policies and procedures and how to operate in the site-specific search environment. (pages 158–160)

(B) Requisite Skills. The ability to enter, maneuver in, and exit the search environment and provide for and perform self-escape/self-rescue. (pages 158–160)

5.2.6 Perform ground support operations for helicopter activities, given a rescue scenario/incident, helicopter, operational plans, personal protective equipment, requisite equipment, and available specialized resources, so that rescue personnel are aware of the operational characteristics of the aircraft and demonstrate operational proficiency in establishing and securing landing zones and communicating with aircraft personnel until the assignment is complete. (pages 166–168)

(A) Requisite Knowledge. Ground support operations relating to helicopter use and deployment, operation plans for helicopter service activities, type-specific personal protective equipment, aircraft familiarization and hazard areas specific to helicopters, scene control and landing zone requirements, aircraft safety systems, and communications protocols. (pages 166–168)

(B) Requisite Skills. The ability to provide ground support operations, review standard operating procedures for helicopter operations, use personal protective equipment, establish and control landing zones, and communicate with aircrews. (pages 166–168)

Chapter 10, Vehicle and Machinery Rescue

10.1.1 Plan for a vehicle/machinery incident, and conduct an initial and ongoing size-up, given agency guidelines, planning forms, an operations-level vehicle/machinery incident or simulation, so that a standard approach is used during training and operational scenarios, emergency situation hazards are identified, isolation methods and scene security measures are considered, fire suppression and safety measures are identified, vehicle/machinery stabilization needs are evaluated, and resource needs are identified and documented for future use. (pages 148–155)

(A) Requisite Knowledge. Operational protocols, specific planning forms, types of vehicles and machinery common to the AHJ boundaries, vehicle/machinery hazards, incident support operations and resources, vehicle/machinery anatomy, and fire suppression and safety measures. (pages 148–155)

(B) Requisite Skills. The ability to apply operational protocols, select specific planning forms based on the types of vehicles/machinery, identify and evaluate various types of vehicle/machinery within the AHJ boundaries, request support and resources, identify vehicle/machinery anatomy, and determine the required fire suppression and safety measures. (pages 148–155)

10.1.2 Establish "scene" safety zones, given scene security barriers, incident location, incident information, and personal protective equipment, so that action hot, warm, and cold safety zones are designated, zone perimeters are consistent with incident requirements, perimeter markings can be recognized and understood by others, zone boundaries are communicated to incident command, and only authorized personnel are allowed access to the rescue scene. (pages 162–163)

(A) Requisite Knowledge. Use and selection of personal protective equipment, traffic control flow and concepts, types of control devices and tools, types of existing and potential hazards, methods of hazard mitigation, organizational standard operating procedure, and types of zones and staffing requirements. (pages 162–163)

(B) Requisite Skills. The ability to select and use personal protective equipment, apply traffic control concepts, posi-

tion traffic control devices, identify and mitigate existing or potential hazards, and apply zone identification and personal safety techniques. (pages 162–163)

10.1.3 Establish fire protection, given an extrication incident and fire control support, so that fire and explosion potential is managed and fire hazards and rescue objectives are communicated to the fire support team. (pages 163–166)

(A) Requisite Knowledge. Types of fire and explosion hazards, incident management system, types of extinguishing devices, agency policies and procedures, types of flammable and combustible substances and types of ignition sources, and extinguishment or control options. (pages 163–166)

(B) Requisite Skills. The ability to identify fire and explosion hazards, operate within the incident management system, use extinguishing devices, apply fire control strategies, and manage ignition potential. (pages 163–166)

▌Knowledge Objectives

After studying this chapter, you will be able to:

- Understand the importance of standard operating procedures that outline universal safety procedures and best practice models for site operations. (pages 147–148)
- Discuss time management and the significance of the "Golden Period." (page 148)
- Discuss the protective ensemble required to be worn by the technical rescuer at the scene of rescue or recovery operations. (page 149)
- Define defensive apparatus placement and provide examples of this method. (pages 151–152)
- Describe various personnel resources and their roles at a vehicle extrication incident. (pages 152–154)
- Discuss methods used for personnel rehabilitation. (pages 154–155)
- Discuss types of equipment resources and the impact that these resources may have on an operation. (page 155)
- Explain the basis of scene size-up and how it is applied to the vehicle extrication incident. (pages 155–162)
- Describe how to conduct a proper inner and outer survey. (pages 158–160)
- Explain how to establish scene safety zones. (pages 162–163)
- Describe the specific hazards that may be encountered at a vehicle extrication incident. (pages 163–166)
- Discuss air medical operations, how to establish landing zones, and landing zone safety. (pages 166–168)

▌Skill Objectives

After completing this chapter, you will be able to perform the following skills:

- Respond to a motor vehicle collision. (pages 157–158, Skill Drill 7-1)
- Extinguish a vehicle fire. (page 164, Skill Drill 7-2)

ou are the technical rescuer on an extrication incident where you arrive as the company officer on scene. Upon arrival, you realize that the first arriving crew did not follow standard operating procedures and properly stabilize the vehicle. Even from your location, you can see that the vehicle is rocking back and forth as the crew works to extricate the victim. What actions should you take?

1. Should you immediately halt the operation?
2. Should you pull the crew back and have them stabilize the vehicle?
3. Should you allow the crew to continue to extricate the victim?

Introduction

"There are no secrets to success. It is the result of preparation, hard work, learning from failure." —Colin Powell

Chapter 7, *Site Operations*, Chapter 8, *Vehicle Stabilization*, and Chapter 9, *Victim Access and Management*, outline a successive three-phase process that the technical rescuer should follow at every extrication incident. This chapter will discuss the first step of this process, arrival at the scene and site operations Figure 7-1 ▾ .

Safety

Ensuring that proper safety procedures are followed in any operation, whether it is responding to an emergency incident, working on an emergency incident, conducting training, or

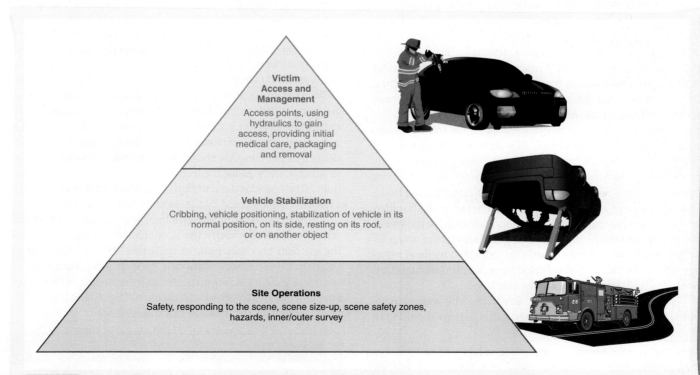

Victim Access and Management
Access points, using hydraulics to gain access, providing initial medical care, packaging and removal

Vehicle Stabilization
Cribbing, vehicle positioning, stabilization of vehicle in its normal position, on its side, resting on its roof, or on another object

Site Operations
Safety, responding to the scene, scene size-up, scene safety zones, hazards, inner/outer survey

Figure 7-1 Vehicle extrication is a technical process that requires structured successive steps to produce favorable results. This chapter will discuss the first step of this process, arrival at the scene and site operations.

simply checking out or inspecting equipment, is paramount for any organization. Standard operating procedures (SOPs) outlining universal safety procedures and best practice models need to be implemented and followed by every member of every agency. The goal is not to penalize personnel by implementing a rigid policy that is nonflexible, but to strive to conduct daily emergency and nonemergency operations in the safest and most efficient manner. An excellent reference in this area of safety is NFPA 1500, *Standard on Fire Department Occupational Safety and Health Program*. The following is an excerpt from NFPA 1500:

> **A.4.3.1** It is the policy of the fire department to provide and to operate with the highest possible levels of safety and health for all members. The prevention and reduction of accidents, injuries, and occupational illnesses are goals of the fire department and shall be primary considerations at all times. This concern for safety and health applies to all members of the fire department and to any other persons who could be involved in fire department activities.*

This statement on safety is clear and precise and should be adopted by all emergency response agencies regardless of the fact that it was adopted for a fire department. Throughout this text, safety is addressed with every technique and procedure, but it is ultimately the responsibility of personnel to be cognizant of any situation that may occur where safety can and will be jeopardized. All personnel must know how to adjust, adapt, and conform to the best practice model, avoiding or eliminating any and all potential injuries.

Time Management

At any emergency scene, time is a major factor. Spending more time on scene negatively impacts your patient's chances of survival. The late Dr. R. Adams Cowley, who founded the R. Adams Cowley Shock Trauma Center in Baltimore, Maryland, theorized that a victim of a critical traumatic injury has the greatest chance of survival if he or she is treated at a trauma facility within 60 minutes from the moment the injury occurs. Dr. Cowley believed that this "Golden Hour" was the difference between life and death. After the first 60 minutes, the body has increasing difficulty in compensating for shock and traumatic injuries. This general guideline was utilized as a standard of care for trauma victims for many years, but this critical period is now known to vary from one patient to another. Because many injured patients require definitive care in less than an hour, Cowley's concept is now commonly referred to as the "Golden Period". The <u>Golden Period</u> is the time during which treatment of shock and traumatic injuries is most critical and the potential for survival is best accomplished through rapid medical intervention. Today the "Golden Period" is a benchmark standard for emergency services all over the country.

A basic vehicle extrication involving the assessment, removal, stabilization, and packaging of a patient should take

---

* Source: Reproduced with permission from NFPA 1500-2007: *Fire Department Occupational Safety and Health Program*, Copyright © 2006, National Fire Protection Association. This reprinted material is not the complete and official position of the NFPA on the referenced subject, which is represented only by the standard in its entirety.

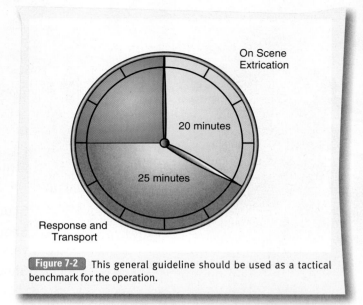

Figure 7-2   This general guideline should be used as a tactical benchmark for the operation.

no longer than 20 minutes; this does not include response or transport times. This general guideline should be used as a tactical benchmark for the operation **Figure 7-2**.

To maintain a realistic scenario, and in keeping within the Golden Period, the total incident time should be approximately 45 minutes for agencies located in urban areas. It is understood that some rural agencies may respond to extrication incidents that well exceed the 60-minute benchmark. The idea is to strive for that 20-minute on-scene extrication benchmark regardless of whether the agency area consists of urban or rural response. For example, a standard 45 minute vehicle extrication incident for a fire rescue agency in an urban response area can be broken down as follows: Once again, no more than 20 minutes should be dedicated to extrication time. This means that it should take no more than 20 minutes from the moment of arrival until the patient is extricated and placed on a backboard. This leaves approximately 25 minutes for apparatus response and transport. Although there are going to be incidents that are complex and require extended extrication time, in general, the goal should be 20 minutes for a standard, one victim extrication. Although there will be incidents that are complex and require extended extrication time, in general, the goal for the technical rescuer should be no more than 20 minutes for the standard or basic one-patient extrication. This time is monitored by the incident commander (IC).

Some dispatching agencies have a standard protocol that will automatically notify the IC at 10-minute intervals until the incident becomes static or mitigated. This is known as an <u>incident clock</u>. Check with your local dispatching agency to see if an incident clock is instituted in its protocols. NFPA 1500, *Standard on Fire Department Occupational Safety and Health Program*, references the incident clock guidelines.

Rescue Tips

The technical rescuer should utilize an incident clock for on-scene time management.

Responding to the Scene

Personal Protective Equipment

Before the technical rescuer can start the operational process, he or she must be operationally ready, not only in the area of mental and physical readiness, but particularly in the area of personal protection. As discussed previously, NFPA 1500, *Standard on Fire Department Occupational Safety and Health Program*, and NFPA 1951, *Standard on Protective Ensembles for Technical Rescue Incidents*, describe the protective ensemble required to be worn by the technical rescuer at the scene of rescue or recovery operations.

The protective ensemble includes the body piece, helmet, primary eye protection, gloves, and footwear. Reflective safety vests are utilized in hazardous areas that require high visibility, such as vehicle accident scenes on roadways. To be compliant with NFPA 1951, the protective ensemble must provide protection from exposure to physical, thermal, liquid, and body fluid-borne pathogen hazards. A self-contained breathing apparatus (SCBA) is important due to the necessity of respiratory protection from a hazardous environment or hazardous conditions such as smoke. The selection of body protection, as well as any other personal protective equipment (PPE), will be made based on the known and potential hazards at the scene of the vehicle extrication.

Dispatch Information

Emergency dispatch centers across the country can vary slightly in terminology, but per the federally mandated National Incident Management System (NIMS), common language is the goal for greater continuity and interoperability between agencies. The following is an example of an incident being dispatched utilizing common language.

A hypothetical call being dispatched may present as the following:

> Dispatch: "Engine 10, Rescue 10, respond to a passenger vehicle rollover with possible entrapment of one victim at the intersection of 12th and Maple. Police are on scene."

The technical rescuer needs to dissect the transmission from dispatch and listen for key elements of the call, such as "possible entrapment" and "vehicle rollover." The goal is to adhere to common terminology and strive to utilize plain language whenever possible.

It must be noted that emergency vehicles are not exempt from observing all traffic laws and must manage their response speeds accordingly. NFPA 1500 requires that the agency establish specific rules, regulations, and procedures relating to the operation of fire department vehicles in an emergency mode, including guidelines to establish when emergency response is authorized and when emergency response is not authorized. It is at the discretion of the authority having jurisdiction (AHJ) to make that determination.

Other important elements to listen for or consider during dispatch may include the time of day, location of the incident, speed of travel (residential roadway vs. major highway), and weather, which can vary greatly in different regions of the country. This key information should motivate the company officer to step into action and start preassigning crew members en route to the incident. It is better to preassign crew members while en route to the scene than to wait until arrival; waiting will only cause confusion and freelancing to occur. As discussed in Chapter 2, *Rescue Incident Management*, your organization should have SOPs or standard operating guidelines (SOGs) in place so there is no confusion on what is needed for each assignment. **Figure 7-3 ▼** shows a sample SOP for vehicle extrication.

Standard Operating Procedures/Guidelines

ANYTOWN FIRE RESCUE

SUBJECT: (DRAFT) Vehicle Extrication Operation Procedures/Guidelines SOG # 5-15

PURPOSE: To provide a clear understanding of the policy and procedures regarding the safe, efficient, and organized approach to any incident requiring vehicle extrication.

SCOPE/APPLICATION: To maintain safety and continuity within the organization, these guidelines are applicable to all personnel that respond, train, or otherwise perform vehicle extrication for the Anytown Fire Rescue Department.

DATE: January 25, 2011

Figure 7-3 Standard operating procedures. A sample standard operating procedure for vehicle extrication.

(continues)

GUIDELINES:

SAFETY: All personnel shall don the appropriate PPE for all vehicle extrication practices and procedures. PPE includes but is not limited to all of the following NFPA 1006 approved equipment: Helmet, Eye Protection or Goggles, Gloves, Jacket, Pants, Heavy Jumpsuit, Boots with steel toe protection. SCBA shall be donned in any IDLH environment or when deemed necessary by command or the officer-in-charge of the operation.

- For all emergency incidents that require vehicle extrication, the National Incident Management System shall be utilized for scene management and control of the incident. Please review SOG # 5-26 Incident Management System.
- For an emergency incident that is deemed a significant event, the responding unit for the district shall notify Air Rescue and place them on stand-by mode until further evaluation can be conducted. Please review the agency protocol for all air rescue operations.
- A minimum 1¾ charged hose line shall be deployed at all vehicle accidents requiring extrication.
- For an emergency incident that is deemed a significant event, an Incident Safety Officer shall be assigned.

STAGES OF EXTRICATION: The following Vehicle Extrication procedures shall be divided into three successive stages.

STAGE 1: Stabilize the Scene

FIRST ARRIVING UNIT: Upon approach of the scene, the officer shall give a general size-up of the scene stating the number of vehicles involved, type of vehicles (passenger type, cargo van, commercial truck, etc …) position of the vehicles (upright, on roof, on side, on top of another vehicle), the extent of damage (minor, moderate, heavy), and patient/s status. For example: "*Engine 14 arrival. We have two (2) passenger vehicles, one upright and one on its side, with heavy damage to both. We'll advise on injuries.*" If extrication is needed, the officer shall establish command and request a tactical channel. The second arriving officer shall assume command if command is involved in the extrication. Start additional units if needed (Engine, Rescue, TRT, Haz-Mat, Heavy Tow Unit, etc …).

I. SCENE SIZE-UP: INNER /OUTER SURVEY

A. Size-up begins at dispatch
B. Defensive apparatus placement
C. Hot /Action Zone (full PPE required)
D. Outer Scene Survey (simultaneously with inner survey)
E. Inner Scene Survey (simultaneously with outer survey)
F. Inner/Outer Scene Survey Report

STAGE 2: Stabilize the Vehicle

All vehicles requiring extrication shall be properly stabilized before any operations are performed on the vehicle/s.

II. VEHICLE POSITION

A. Vehicle Upright
B. Vehicle on its Roof
C. Vehicle on Side
D. Vehicle on top of Vehicle

STAGE 3: Stabilize the Patient

Stabilization of the patient/s require creating a large access area in the vehicle by systematically removing the vehicle from the patient/s, immobilizing the patient/s, extraction of the patient/s from the vehicle, and transporting the patient/s to the appropriate medical care facility.

III. PATIENT ACCESS AND REMOVAL

A. Patient Assessment (Interior Rescuer)
B. Patient Access

IV. INCIDENT TERMINATION:

A. Assess all personnel for CISM (Activate CISM team through FireCom)
B. Inspect all equipment utilized for possible damage
C. Replace all damaged disposable items: Reciprocating saw blades, air chisel blades, etc …
D. Replace fuel in all gas-powered units utilized at incident
E. Place all units in service

V. POST INCIDENT ANALYSIS

A. Formal Post Incident Analysis (PIA)
B. Informal Post Incident Analysis (PIA)

Figure 7-3 A sample standard operating procedure for vehicle extrication. *(Continued)*

An example of a typical preassignment model is as follows:

Company officer: Conducts the inner survey and makes sure everyone understands their assignments

Rescuer 1: Conducts the outer survey

Rescuer 2: Acquires the cribbing and tools off of the apparatus and sets up a tool staging area

Driver engineer: Blocks off traffic, places the apparatus in a defensive position, pulls off a minimum 1¾-inch [44-mm] hose line for protection, sets up traffic-diverting cones and signage, and assists Rescuer 2 with setting up the tools

This model is only an example, so keep in mind that these assignments can change upon arrival based on the nature and complexity of the incident and the number of personnel available. It is always better to be prepared and have a basic plan in place than to make one up as you go.

■ Traffic

<u>Defensive apparatus placement</u> is a component of site operations. The main goal is to block and protect the scene from the flow of traffic. Whenever possible, place emergency vehicles in a manner that will ensure safety and not disrupt traffic any more than necessary.

Apparatus should be staged in a defensive position so that it provides a barrier against motorists who fail to heed emergency warning lights. Optimal defensive apparatus placement depends on the flow of traffic in relation to the accident and area maneuverability. Consider whether you can get apparatus around the wreckage to properly stage the units.

As you approach the scene, with the accident in the same direction of travel, position the apparatus at a 30- to 45-degree angle to the wreckage, blocking off lanes to guide traffic safely around or divert it away from the designated safety or operational zones [Figure 7-4 ▾]. Also, if equipped, attempt to keep a side-mounted pump panel facing toward the incident away from oncoming traffic to protect the pump operator/driver engineer. Confirm that the apparatus's wheels face away from the accident; this will ensure that the apparatus will not be pushed into the wreckage if struck in the rear by another vehicle. If scene lighting is used, make sure the lights do not shine in the eyes of

Figure 7-5   Traffic cones can be placed to direct motorists away from the crash for scene protection.

approaching drivers. It is imperative that these safety measures be strictly followed. Passersby are normally so consumed with getting a view of the wreckage that they do not see you or the apparatus.

Even with the required reflective gear on, never assume that a passerby sees you or your crew; always take precautions. PPE must be bright to help ensure visibility during daylight hours; PPE that is used at night should be equipped with reflective material to increase visibility in the darkness. PPE must be worn at all motor vehicle collisions. In addition, the driver must also set up traffic cones or some type of signaling device to warn and divert all oncoming traffic [Figure 7-5 ▴].

Most apparatus do not carry the necessary amount of cones to properly divert traffic in a safe and appropriate manner. It may be necessary to reevaluate the amount carried on each unit based on the response area. Also, if the accident occurs at an intersection with multiple lanes and direction of travel, it is recommended that additional units be dispatched at the onset of the call to assist with lane closures and scene protection. Traffic control should be utilized at every incident. Some agencies have personnel dedicated to the traffic control function, such as fire, police, or special auxiliary police.

There are invaluable reference documents published on the subject of traffic management. For example, the U.S. Department of Transportation's (DOT) *Manual on Uniform Traffic Control Devices (MUTCD)* is a guideline for the safe operation and proper utilization of traffic control and management devices. This manual seeks to ensure uniformity of traffic control devices and traffic management procedures by providing minimum standards and guidance nationwide on emergency traffic control devices such as warning signage and traffic cone types and placement.

The U.S. Fire Administration's *Traffic Incident Management Systems* is another valuable resource that aims to enhance responder safety and traffic management at roadway emergency scenes.

The DOT's *Emergency Response Guidebook (ERG)* offers guidance for rescuers who may potentially operate at a hazardous materials incident such as a vehicle accident involving a cargo truck that was transporting hazardous materials [Figure 7-6 ▸]. This reference guides is intended to help rescuers decide which preliminary actions to take. The guide provides information on

Figure 7-4   Many fire departments place an apparatus at a 30- to 45-degree angle to the crash.

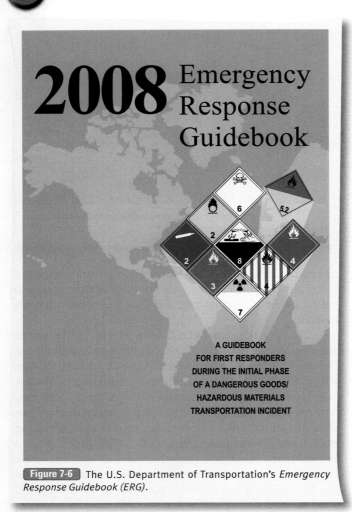

Figure 7-6 The U.S. Department of Transportation's *Emergency Response Guidebook (ERG)*.

approximately 4000 chemicals that may be encountered at an incident scene.

> ### Rescue Tips
>
> For the safety of all personnel, when placing any advance warning or traffic-diverting devices such as cones or signs, rescuers should never turn their back to oncoming traffic; they should always start with the farthest device and work backwards toward the incident.

These resources contain an enormous amount of information on traffic control devices, traffic management utilizing the NIMS model, and hazard recognition and mitigation. Some of the topics found within these resources include:

- Incident traffic size-up in determining the scope and magnitude of the incident by categorizing the scene in relation to the estimated elapsed time needed for roadway closures (minor: > 30 minutes; intermediate: 30 minutes to 2 hours; major: > 2 hours)
- Risk analysis and assessment
- Establishing traffic control zones (different from operational cold, warm, hot hazard zones)

- Proper traffic cone size and color—must be orange and no less than 18 inches (457 millimeters [mm]) in height for daylight and for low-speed roadways and 28 inches (711 mm) at night or when operating on high-speed highways with the proper type and sizing of retro-reflective material;
- Traffic cone spacing requirements—the recommendation being 15 feet (4.6 meter [m]) intervals
- Portable electronic message signs and the utilization of variable message signs that hang over the roadways at various areas in most major highways today
- Stationary prewarning signs (a requirement of NFPA 1500), including proper color, size, lettering, wording (e.g., "Emergency Scene Ahead"), spacing, and distance from the incident
- Flares—incendiary type, chemical light stick type, or light-emitting diode (LED) type
- Directional arrow panels and barricades
- Flagger position—training requirements, equipment types, and role and responsibilities (Any individual who manually directs traffic is known as a flagger.)

Many of these reference materials are readily available through Internet downloads. Technical rescuers must take advantage of these invaluable resources to better prepare and plan operationally for themselves as well as their respective agencies.

> ### Rescue Tips
>
> Traffic control should be utilized at every incident.

Crowd Control

Bystanders who act as "Good Samaritans" want to get involved and help, but most often they can be a dangerous liability that can cause havoc or injury to the victim, to themselves, or to the rescue workers. As you arrive on scene, you may see these individuals trying to pull victims from the vehicles. They may also start yelling that the vehicle is going to explode or that the people in the vehicles are dying and you need to move faster. This situation needs to be addressed and mitigated immediately and professionally. Do not get distracted by these individuals and lose focus on the operation. These individuals need to be removed from the scene either voluntarily or by intervention. Call immediately for law enforcement to assist.

Crowd control is an absolute necessity in protecting responders from individuals who might attempt to enter the emergency site. Adequate crowd control provides the necessary space for fire and rescue personnel to operate without being concerned about individuals interfering or bystanders being injured **Figure 7-7 ▶**.

Personnel Resources

The size of an incident upon conducting the scene size-up will dictate whether a large management staff is necessary for oversight and planning. Of course, as an incident grows, the need

Figure 7-7 Crowd control is essential to providing rescuers with adequate space to operate.

for additional resources and personnel can grow as well. Recognition of the need to bring in additional resources to assist in the incident is a true sign of leadership and proper scene management. Not calling out for additional help and resources not only displays a lack of leadership skills but will most certainly spell disaster for that crew and the victims. A standard rule is to always call for additional support early in the incident; these units can always be canceled or returned if the incident is determined later not to be significant or was mitigated appropriately by your crew alone. A unified command system may need to be established that incorporates multiple jurisdictions, law enforcement, the DOT, utility companies, or public works departments. SOPs should address the levels of response and indicate what positions will be staffed based on the complexity of the incident.

Law Enforcement Personnel

As discussed previously, uncontrolled incident scenes pose serious threats to everyone at the site. Law enforcement should be requested immediately in order to control crowds and traffic and/or to establish perimeter control **Figure 7-8 ▼**. In most agencies, law enforcement is on the same dispatching system as

fire rescue, and procedurally, they receive the notification call before fire rescue is dispatched. The most important point is that there must be clear communication between law enforcement and the fire department/emergency medical service agencies; this communication is the key to successful operations and general work cohesiveness. Remember that the overall safety of response personnel is the primary responsibility of the responding agency.

EMS Personnel

Although the common functional description of emergency medical services (EMS) is prehospital patient care and treatment, in some parts of the county, EMS performs extrication as well as patient care **Figure 7-9 ▼**. In other EMS agencies, vehicle extrication, victim access, and disentanglement are handled by another service such as a fire department. Some jurisdictions combine both fire and EMS systems, creating a fire rescue service. Systems can vary greatly from one jurisdiction to another, but with the utilization of a tiered classification system to identify what these agencies are capable of providing, confusion is alleviated when mutual aid is requested. A tiered fire rescue/EMS–based provider service describes the capability of the agency to provide the response, treatment, and transport of patients. The tier system is basically broken into three levels. A three-tiered system will provide three independent types of service, with various agencies or combinations of agencies providing services. A tiered system may consist of the following common components but is not limited to the types of agencies described here:

- *Three-tiered system.* The fire apparatus responds as a first responder unit to provide basic patient care and support, EMS personnel respond as a separate entity for definitive patient treatment, and a private ambulance service responds for transport of the patient to the designated medical facility.
- *Two-tiered system.* A two-tiered system consists of the fire apparatus responding as a first responder unit to provide basic patient care and support, and EMS personnel respond as a unit capable of treatment and transport.

Figure 7-8 Law enforcement personnel should be requested for crowd and traffic control at all roadway incidents.

Figure 7-9 The number of victims and types of injuries can help you determine whether ALS or BLS providers are needed.

■ *Single-tiered system.* A single-tiered system consists of one agency responding with a fire apparatus staffed with dual-certified fire fighter paramedics and a rescue unit for transport. Some fire apparatus now come equipped with patient transport capabilities. The single-tiered fire rescue system is the preferred level of medical service from an overall operational viewpoint.

As mentioned earlier, in some jurisdictions, EMS and fire service agencies are separate entities, with basic life support (BLS) being the minimum level of patient care for EMS agencies and emergency medical responder training being the minimum level of patient care for fire service agencies. Alternately, the fire service agency may not provide any medical responsibilities at all, which is considered a two- or three-tiered system, depending on the transport capabilities of the EMS unit. Two- and three-tiered systems can cause a lot of confusion and delays when trying to determine the level of care needed on scene. Remember the Golden Period described earlier in the chapter? Any delays in treatment directly impact patient survivability. Two- or three-tiered fire service agencies should incorporate EMS (preferably an advanced life support [ALS] unit) and ambulance transport services into their response and operational procedures, including training drills to improve work relations, continuity, and efficiency. ALS is the greatest level of care for the patient next to an emergency department physician responding to the incident. Unfortunately, in some areas, the only medical service available in the district may be a BLS-certified service. This does not mean that the rescuers cannot function, but it does raise the question of whether the injuries the rescuers have identified can be treated by BLS providers or whether ALS support will be required, again, causing a potential delay during a critical time. Clearly, the number of victims and types of injuries will help responders make this decision.

For example, if responders face a scenario in which patients have suffered amputations, crush injuries, severe cervical spine injuries, or other serious trauma, it makes sense that the rescuers would request ALS units. However, the National Standard Curriculum for EMTs does outline BLS-level care for these types of injuries; EMTs should provide patient care within their scope of practice and request ALS assistance as appropriate. If necessary, ALS resources may be requested from a neighboring community where a mutual aid agreement has been established. Without question, it is always best to obtain ALS medical support as early as possible when a serious incident occurs, but as rescuers, we must be reminded that not all municipalities and rural areas provide ALS medical service. For this reason, it is important that BLS service providers are trained to immediately recognize any injuries that require an ALS unit response. ALS units can always be canceled if their services are not needed.

Hazardous Materials Personnel

Hazardous materials incidents could require specialized teams, depending on the complexity of the incident. For example, an accident involving a tanker or semi-truck containing a known or unknown hazardous product would require hazardous materials personnel or a hazardous materials team to respond. In incidents involving hazardous materials, rescuers must allow a proper size-up and evaluation. All too often in incidents involving hazardous materials, rescue personnel are unnecessarily exposed to dangerous agents because they rush into the incident site before they have gathered the necessary information pertaining to the material or agent. Although awareness-level responders' activities are limited in such scenarios, they can certainly assist operations- and technician-level responders by looking through a set of binoculars and identifying placards, product labels, numbers, and other information. They can also assist in the response by preventing others from entering the incident site by sealing the site perimeter and referring to the DOT's *ERG* to identify the product, evacuation distances, the product's flammability and incompatibilities, and other pertinent data.

Additional Personnel Resources

Other valuable resources whose assistance might be requested during major incidents are state and county emergency services, state and county health departments, area hospitals, emergency flight services, and the state's National Guard civil support team. Other resources include K-9 organizations, urban search and rescue (USAR) teams, FEMA task force teams, and industry hazardous materials response teams. The mention of these types of resources may seem extreme in regard to vehicle extrication, but one major disaster such as the 6.9-magnitude earthquake that struck California in 1987 justifies the discussion of these services in relation to vehicle extrication. During this major earthquake, the two-tiered elevated Nimitz Freeway, which flows traffic from Oakland into San Francisco, collapsed onto itself, trapping and crushing an estimated 250-plus vehicles Figure 7-10 ▼. This operation required the deployment of massive resources for the freeway alone, not to mention the rest of the region that also required a response. All of these resources can play a vital role in such incidents.

■ Personnel Rehabilitation

Establishing a rehabilitation group is critical in any prolonged extrication incident. Weather can be detrimental to personnel operating on scene, whether it is in 100 percent humidity, which occurs on a daily basis in southern Florida, or it is below freezing, which is a frequent occurrence in South Dakota. A

Figure 7-10   The Nimitz Freeway after a 6.9-magnitude earthquake struck in 1987.

properly equipped rehabilitation unit providing shelter and thermal control options should be established early in the incident with personnel rotation mandated by the IC. A great resource for establishing a rehabilitation response protocol is NFPA 1584, *Standard on the Rehabilitation Process for Members During Emergency Operations and Training Exercises*. Some of the issues covered in this guide include:

- Immediate recognition and treatment of heat- or cold-related emergencies such as frostbite, heat exhaustion, or heatstroke
- Immediate shelter from potentially detrimental climate conditions
- Active or passive cooling/warming techniques based on the type of climate exposure
- Rehydration—basic fluid and electrolyte replacement
- Medical monitoring and base readings
- Personnel accountability with a release from and return to duty procedure/policy
- Transport capabilities

■ Equipment Resources

When responding to vehicle and machinery incidents, it is vital to know what additional specialized equipment is available to assist in managing an incident. All rescuers realize that vehicle and machinery incidents can require a great deal of time on scene, depending on the complexity of the incident and the factors involved. Although the availability of specialized equipment sometimes expedites rescue operations, bringing specialized equipment into the operation also adds another layer of safety concerns for the response process. This is a reminder that the need for constant and continual reevaluation of the operation should occur regardless of what specialized equipment is brought in or what the scope or size of the incident is.

Depending on the state or county's geography, assistance from other resources may be necessary during an incident, such as heavy equipment providers or operators, portable lighting companies, hardware stores, building suppliers, farm equipment sales, farm equipment mechanics, tow agencies or wrecker services, 18-wheeler operators, physicians, 4-wheeler sales, portable generator sales, fast-food restaurants, fire equipment dealers, and rescue equipment dealers Figure 7-11 ▼ . All

rescuers and other responding agencies should coordinate their training; at a minimum, a meet-and-greet session to open up dialogue and establish a working relationship with each other is valuable. Obviously, training together is the best scenario; doing so enhances all parties' equipment familiarity, identifies equipment limitations, and provides for better and more diverse service delivery for those unique or specialized incidents that can occur. Rescuers and other responders should train with the equipment that is available to them and know the equipment's limitations; these training opportunities are more tools for the toolbox.

A variety of specialized equipment may be necessary when dealing with vehicle and machinery incidents. In some incidents, more than one wrecker may be required. A heavy tow unit equipped with an articulating boom may also be useful in lifting, stabilizing, or displacing larger vehicles when necessary.

The need for this type of equipment should be determined as a part of the department's hazard assessment and analysis of its response area and included in the agency's SOPs/SOGs.

■ Communication and Documentation

Effective communication and documentation are important keys to providing the best level of service to the community. The adoption of the NIMS will lead to use of plain language and a structured incident action plan, as described in Chapter 2, *Rescue Incident Management*. Also, the utilization of tactical worksheets at large-scale incidents can be the difference for successful outcomes versus disastrous outcomes. Tactical worksheets are basically accountability sheets. They describe what engine company or rescue unit is assigned to the particular tactic. For example, Engine 23 may be assigned as extrication group 1 at vehicle 1, and Engine 34 may be assigned as extrication group 2 at vehicle 2.

Proper documentation during site operations always pays dividends for an agency when legal issues such as lawsuits occur. Proper documentation also provides justification of service when applying for agency accreditation such as with the Center for Public Safety Excellence (CPSE) or when old or outdated equipment needs to be replaced or better equipment purchased. Lastly, proper documentation can provide an avenue to improve the agency through recognition of operational deficiencies by incorporating a quality management and assurance program. Review Chapter 2, *Rescue Incident Management*, for guidelines on proper communication and documentation of the incident utilizing the NIMS model.

Scene Size-up

__Scene size-up__ is the systematic and continual evaluation of information presented in either visual or audible form. It is important to remember that size-up begins at the time the incident is dispatched, not at the time the unit arrives at the scene. The technical rescuer needs to start gathering information, formulating, and strategizing a plan while en route to the incident.

Figure 7-11 Assistance from other resources can be crucial during an incident.

Voices of Experience

"**We cannot put limitations or assumptions on the outside perimeter of an outer survey.**"

I can recall an incident that my crew and I responded to years ago where we experienced tunnel vision. We were responding to a motor vehicle collision with possible entrapment. It was approximately 1 AM. As we approached the scene, we could see the intersection of the reported accident. There were law enforcement lights flashing on the opposite side of us, behind the collision. These lights clearly caught our focus. As we approached within 500 feet (152 m) from the intersection, we drove past a vehicle that appeared to be disabled on the side of the road. It was so far from the accident scene that we assumed it was a disabled vehicle with no relation to the collision at the intersection. Unfortunately, this was our mistake; we soon discovered that vehicle was actually the cause of the accident and had been trying to flee the scene. The vehicle had traveled 500 feet (152 m) from the incident before it succumbed to the damages and stalled out.

This incident portrays the importance of a thorough scene size-up and outer survey. We cannot put limitations or assumptions on the outside perimeter of an outer survey. It is essential that we take the time to look at the big picture and the entire scene, especially upon approach of the scene.

Dave Sweet
Broward Sheriff's Office Fire Rescue Department
Broward County, Florida

Immediately upon arrival, the first company officer will size up the scene and establish command. A rapid and accurate visual size-up is needed to avoid placing rescuers in danger and to determine which additional resources, if any, are needed. Size-up at a vehicle and machinery incident can include the following evaluations:

- Recognition and mitigation of <u>**immediate danger to life and health (IDLH)**</u> hazards such as exposed utilities, water, mechanical hazards, hazardous materials, electrical hazards, explosives, and other hazards, including environmental factors
- Exposure to traffic
- Incident scope and magnitude
- Risk–benefit analysis
- Number, size, and type(s) of vehicles/machines involved (common passenger vehicle, hybrid vehicle, sport utility vehicle, etc.)
- Number of known or potential victims
- Identification of witnesses/bystanders
- Stability of vehicles/machines involved
- Access to the scene
- Necessary resources and their availability

With this information, a decision can be made to call for any additional resources, and safe and effective actions can be taken to stabilize the incident. In any event, responders should not rush into the incident scene until an assessment can be made of the situation.

When responding to a motor vehicle collision, follow the steps in **Skill Drill 7-1 ▶** :

1. Don PPE, which may include SCBA.
2. Size up the scene upon approach from inside the apparatus. Position the emergency vehicle so as to protect the scene (defensive apparatus placement).
3. Transmit an initial scene size-up report to dispatch.
4. Establish command.
5. Assign personnel and tasks.
6. Look for obvious IDLH hazards. (**Step 1**)
7. Exit the apparatus and perform the inner and outer surveys, which encompass a 360-degree walk-around of the vehicle(s) and the entire scene.
8. Assess the hazards.
9. Determine the type of vehicle involved and if the vehicle is running. Identify whether it is a conventional, hybrid, or alternative fueled vehicle.
10. Determine the number of victims involved.
11. Determine the severity of the victims' injuries.
12. Determine the level of entrapment.
13. Determine interior hazards such as supplemental restraint system (SRS) air bag components, including their locations.
14. Assess the resources available, and call for additional units if needed.
15. Provide an updated report.
16. Establish a secure work area with operational zones (hot, warm, and cold zones).
17. Establish a staging resource area.
18. Direct the staging/placement of arriving apparatus. (**Step 2**)

■ Scene Size-up Report

As the apparatus arrives on scene, the company officer will need to give a size-up report to dispatch. The main reason for the report is to give an update to the units responding so they can maintain or adjust their response. The units responding should tailor their response to what the report states. This report should be precise and detailed, but not lengthy. It should include information such as the number of vehicles involved, type of vehicles, position of the vehicles (upright, on roof, on side, on another car), extent of damage (minor, moderate, heavy), and, if known, patient status and level of entrapment. A typical size-up report may go as follows: "Engine 14 and Rescue 14 arrival. We have two passenger vehicles, one upright and one on its side, with heavy damage to both. We'll advise on injuries." If extrication is needed, the company officer will establish command and request a tactical or operational channel. A tactical or operational channel is a separate working channel designated to the incident so normal air traffic from incoming calls will not interfere with the operation.

After the initial size-up report is given and a closer evaluation of the scene has been accomplished, an update report can be conveyed to dispatch explaining the level of entrapment. To keep it simple, use the same terminology that was used to describe the damage to the vehicle. There are generally three categories of entrapment: minor, moderate, and heavy **Table 7-1 ▶** . These are suggestive guidelines only; levels of entrapment or terminology may vary from one agency to another. Consistently using the same terms to describe the condition of the vehicle(s) and the level of entrapment will convey a crystal-clear message for

Skill Drill 7-1

Responding to a Motor Vehicle Collision

1 Don PPE. Size up the scene upon approach from inside the apparatus. Position the emergency vehicle so as to protect the scene. Transmit an initial scene size-up report to dispatch. Establish command and assign personnel and tasks. Look for obvious IDLH hazards.

2 Exit the apparatus and perform a 360-degree walk-around of the entire scene utilizing the inner and outer surveys. Assess the hazards. Determine the type of vehicle involved and if the vehicle is running. Determine the number of victims involved, the severity of the victims' injuries, and the level of entrapment. Determine interior hazards such as airbags. Assess the resources available, and call for additional units if needed. Provide an updated report. Establish a secure work area and operational zones (hot, warm, and cold zones). Establish a staging resource area. Direct the placement of arriving apparatus.

Table 7-1 Entrapment Classifications

Minor Entrapment	There is no vehicle metal or material that is impinging on the patient. The patient has either no injuries or minor injuries. The extrication process to gain access to the patient would require a basic door release/removal procedure.
Moderate Entrapment	There is some vehicle metal or material that is impinging on the patient. The patient has sustained injuries that require total immobilization. The extrication process to gain access and extricate the patient will involve two or more procedures such as a door and roof removal.
Heavy Entrapment	There is a large amount of vehicle metal or material impinging on the patient. The patient has sustained substantial trauma and has multiple points of entrapment. The extrication will require an extended time of greater than the standard 20-minute objective. The extrication process to gain access and extricate the patient will generally require three or more procedures such as a door removal, roof removal, and a dash lift.

incoming units about the severity of the incident. Responding units should act according to this report. Remember that these guidelines are not set in stone, as any guideline can be modified to fit your agency's trauma criteria or terminology preference; these entrapment classifications are a simple tool that can be used to keep everyone on the same page when responding to an incident. Once decided upon, entrapment classifications should be considered to be incorporated in departmental response protocols or SOPs.

■ Inner and Outer Surveys

Both inner and outer surveys should be completed before any operations begin [Figure 7-12 ▸]. The inner and outer surveys are 360-degree inspections of the scene that are completed by two personnel. One rescuer walks in a clockwise direction around the scene and the other in a counterclockwise direction, ensuring that every area within the hot/action zone has been investigated by surveyors. These surveys provide the company officer and crew with additional information about hazards, the number of patients, the level of entrapment, the need for additional resources, and the information necessary to complete an incident action plan, or objectives for the incident strategy. Lastly, while conducting the inner and outer surveys, if any IDLH hazards are found, then they need to be called out imme-

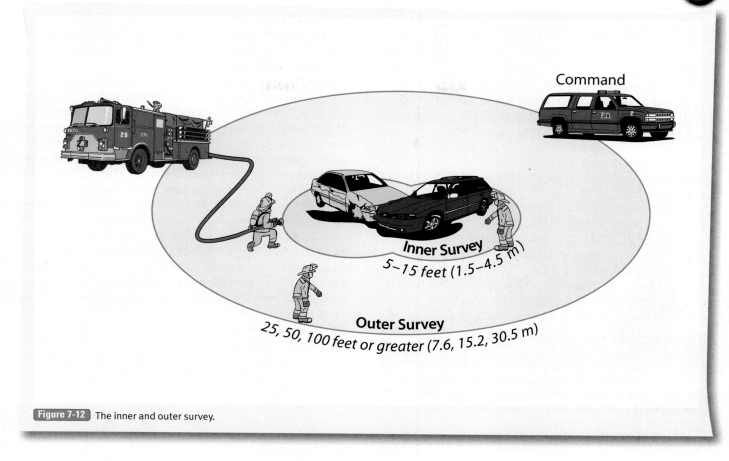

Figure 7-12 The inner and outer survey.

diately and all members ordered to "freeze" until the hazard has been mitigated, or made safe to continue.

The Inner Survey

The <u>inner survey</u> is a four-point inspection of the vehicle's front, driver's side, rear, and passenger's side including the undercarriage on all sides. It should be conducted by the first arriving company officer or the most experienced personnel. Having the company officer conduct the inner survey is more productive than assigning someone else to this. The company officer can see firsthand what the incident is presenting and immediately start to formulate a plan. Also, any IDLH hazards that are found need to be called out immediately and all members ordered to "freeze" until the hazard has been mitigated, or made safe to continue.

To begin an inner survey, the company officer will need to avoid touching the vehicle before clearing any electrical hazards. As an added safety measure, one recommendation is for the officer to lace his or her hands behind his or her back to prevent inadvertently touching the vehicle before clearing any electrical hazards. Obviously, on uneven ground, the rescuer may need the use of his or her hands and arms for balance and will need to adjust appropriately to the environment that is encountered. The company officer should remain approximately 3 to 5 feet (0.9 to 1.5 m) away from the vehicle in a defensive posture throughout the survey, while being cognizant of any sudden forward or rearward lurching of the vehicle. The best position and approach to the vehicle is from the front driver-

side corner of the hood, establishing immediate verbal contact with the patient **Figure 7-13 ▼**. The goal is to gain the attention of the patient and prevent the patient from moving his or her head or body; patients may immediately track a rescuer's voice, so it is important to tell them to keep still and not to move their head. From a psychological position, if the patient is conscious, he or she will require the reassurance that you are present and you are going to get him or her out. Compassion should never be omitted in our role as a public servant.

Figure 7-13 The company officer conducting the inner survey must make immediate verbal contact with the patient, identifying himself or herself, and directing the patient to keep his or her head and body still.

The entire inner survey is a quick yet thorough survey that should last no longer than 45 seconds with two or less vehicles. The following information should be collected during the inner survey:

- IDLH hazards
- Type of vehicle (conventional, hybrid, fuel cell, etc.)
- Status of the vehicle (Is it running? Is it in drive or park?)
- Number of patients
- Entrapment
- Entrapment classification type (minor, moderate, heavy)
- Obvious trauma to the patient(s)
- Position and stability of the vehicle
- Activated SRS air bag system—determined by visibly deployed air bags, with the possibility of live nondeployed air bags still present
- **Primary and secondary vehicle access points** to identify the main point of entry into the vehicle and an alternate means of entry if the main access fails

Rescue Tips

For clear, uninterrupted radio transmissions, ask for a tactical or operational channel, if available, and implement NIMS by establishing command. This is a good step in establishing scene control and leadership.

If the company officer, upon his or her initial exam of the incident, determines that the victim requires an immediate and rapid extrication for the best chance of survival, the company officer may quickly complete the inner and outer surveys with his or her crew, conduct a brief and basic vehicle stabilization, and then take the appropriate action for the safe and rapid extrication of the patient. It is imperative to remember that safety must never be compromised in any situation. If the vehicle is positioned where stability threatens the safety of the crew, the company officer may proceed with full stabilization of the vehicle and then move on to extricate the patient. The company officer will not put the crew's life in jeopardy by performing a haphazard procedure that can potentially cause harm or serious injury. In an emergency situation, there is no textbook or template that determines whether a company officer should call for a rapid extrication. The tactical decision for a rapid extrication is a judgment call. This can be one of the hardest decisions to make because of the grave consequences that may occur. One such issue is a potential for paralysis of the patient due to the lack of proper vehicle stabilization and patient immobilization and packaging. This decision has to weigh heavily in favor of benefiting the patient's life. The saying "life over limb" is often used by rescue personnel to justify their actions.

As part of the inner survey, it is important to accurately identify the total number of patients in order to determine the need for additional resources. While the company officer is conducting the inner survey, another rescuer should perform the outer survey, moving in the opposite direction of the company officer around the perimeter of the inner survey.

The Outer Survey

As stated previously, the **outer survey** is to be conducted simultaneously with the inner survey, but the rescuer performing the outer survey should move in the opposite direction of the rescuer performing the inner survey. It is important to note that the rescuer's distance from the vehicle will vary with each incident. Generally, a distance of 25 to 50 feet (7.6 to 15.2 m) starting from the perimeter of the inner survey position outward is adequate for most incidents, but adding in factors such as speed of travel and type of roadway (highway vs. residential roadway) could cause the survey to extend as far as 100 feet (30.5 m) or greater. The reasoning behind maintaining these survey areas is simple. The higher the speed of travel, the greater the distances the objects, vehicles, and victims could be thrown from the site of impact. To better illustrate this point, if the vehicle is traveling at 100 miles per hour (161 kilometers per hour [kph]), the occupant is traveling at the same speed as the vehicle; when the vehicle abruptly stops from hitting something, the occupant, unless held in by a seat belt, will continue to travel at that same original speed. This is why ejections occur. The company officer will have to use his or her best judgment when determining the proper distance from the vehicle. The best position and approach to the vehicle is from the front side, directly opposite the company officer conducting the inner survey, scanning for any of the following:

- IDLH hazards
- Patients who have been ejected
- Walking wounded (patients who were involved in the accident in some way and could have sustained injuries, whether minor or major)
- Additional vehicles
- Infant or adolescent car seats

Again, any IDLH hazards must be called out immediately and all members ordered to "freeze" until the hazard has been mitigated, or made safe to continue.

The inner and outer surveys should be completed at about the same time, and the rescuers will meet to discuss their findings, compile the information, and formulate an incident action plan, which will include the primary and secondary access points.

Once the inner and outer surveys have been completed and the scene is deemed safe to operate, a rapid triage of the victims can be implemented. Remember that vehicles must be properly stabilized before entering them. A quick assessment or rapid triage of a patient can be accomplished without entering a vehicle, possibly through an open window. **Triage** is the process of sorting patients based on the severity of each patient's condition. Once all patients have been triaged, rescuers will establish treatment and transport priorities. This process will help allocate personnel, equipment, and resources to provide the most effective care to everyone. Chapter 9, *Victim Access and Management*, will discuss this process in more detail.

■ Incident Action Plan

A clear, concise incident action plan (IAP) is essential in guiding the initial incident management decision process and continuing collective planning activities of incident management teams. An IAP can be developed formally for large-scale or major inci-

Near Miss REPORT

Report Number: 07-936
Report Date: 05/31/2007

Synopsis: Apparatus placement, placed so that it protected the responders from oncoming traffic, was contributory in also acting as a shield from the falling tree.

Event Description: On 05/31/07 @ 1545 we were operating at a one-vehicle motor vehicle collision (MVC). Initial response was an engine and a rescue, but since extrication was required, a second engine had been requested.

While doing a roof removal to free two victims, weather conditions changed from heavy rain to horizontal rain with lightning, then to golf ball–sized hail. Lightning struck a tree directly across the roadway from the vehicle being worked on. A 24-inch (610-mm) diameter section of tree fell from more than 40 feet (12.2 m) in the air, striking the parked apparatus, breaking into two large pieces, then striking a fire fighter involved in the roof removal in the hands and arms, a police sergeant in the head, and a police vehicle.

The police sergeant was pinned under the tree and required extrication and removal. The fire fighter was immediately ambulatory. Both the police vehicle and the engine received substantial damage. The engine's windshields were broken out, the roof partially collapsed, holes were punched in the roof, and the extended front bumper was bent and twisted. Further investigation of damage is underway. The police vehicle had a partial roof collapse and the back window was broken out. It should be noted that a police officer was sitting in the driver's seat of the police vehicle at the time of impact, and he was uninjured.

The IC requested an additional engine, an additional heavy rescue and two additional ambulances, but because of apparent partial failure of the radio system (believed, at this time, to be due to the severe weather) the request for the engine and the rescue was never heard by the dispatcher. The additional engine that was due heard the call, as they were in quarters having just returned from their own call, so they responded. The extrication continued, and eventually the MVC victims, the fire fighter, and the police sergeant were transported to local hospitals. The fire fighter was released several hours later, and it appears that the police sergeant will be kept overnight with a possible concussion.

Apparatus placement, placed so that it protected the responders from oncoming traffic, was contributory in also acting as a shield from the falling tree. Training that stresses the importance of everyone wearing their full PPE at MVCs was also contributory in limiting injuries.

Lessons Learned:

- Always stay alert to changing weather conditions and the effects on the environmental conditions surrounding you!
- PPE works!

City of Centreville
INCIDENT ACTION PLAN

July 10, 2006
Operational Period-0700 to 1900

Tornado Impact
Centreville North Incident

Action Plan Contents

1) Incident Objectives   5) Communications Plan

2) Organizational Chart   6) Transportation Plan

3) Division Assignments   7) Medical Plan

4) Safety Message

Figure 7-14 A sample incident action plan.

Table 7-2	IAP Components	
Component of IAP	**Corresponding ICS Form (if applicable)**
Incident objectives	ICS 202
Organization list or chart	ICS 203
Assignment list	ICS 204
Communications plan	ICS 205
Logistics plan	
Responder medical plan	ICS 206
Incident map	
Health and safety plan	

Other possible components of a formal IAP may include the following:

- Air operations summary
- Traffic plan
- Decontamination plan
- Waste management or disposal plan
- Demobilization plan
- Operational medical plan
- Evacuation plan
- Site security plan
- Investigative plan
- Evidence recovery plan

Vehicle extrication in itself is a technical process that requires a plan consisting of structured successive steps to produce favorable results. Almost everything technical in life requires planning with some form of procedures or successive steps to follow in order to reach a successful outcome. For example, when attempting to build something, you will need to follow an outline, plan, or blueprint, whether using your imagination or drawing it on paper. The first step is to create a base or foundation and then continue building on that until completion. If steps are skipped or the process is done haphazardly, then a poor product will be the end result. The same holds true for managing an extrication incident; the plan needs to start with the basic fundamentals or foundation and then continue building in successive steps. By following a detailed plan of action, the technical rescuer will be better prepared to mitigate any unforeseen situations when they occur.

dents requiring incident management teams, or informally, through a quick mental reference process for smaller incidents **Figure 7-14 ▲**. For both the formal and informal process, the development of an IAP should proceed through five primary phases:

1. Understand the situation.
2. Establish the incident objectives and strategy.
3. Develop the plan.
4. Prepare and disseminate the plan.
5. Evaluate and revise the plan.

Whether dealing with a small- or large-scale incident, the IAP will encompass a complete risk–benefit analysis and information about hazard mitigation, resource activation and staging, and confirmation on the incident strategy and tactics, including establishing and assigning operational and support tasks. Specific components of a complete formal IAP and corresponding Incident Command System (ICS) forms are found in **Table 7-2 ▶**.

Establishing Scene Safety Zones

Once the IAP is established and the entire area has been surveyed, a **scene safety zone** or **operational zone** needs to be divided into hot, warm, and cold zones. These zones are strictly enforced by a designated incident safety officer. The boundaries of the three safety or operational zones should be established in a manner that ensures the safety of the crews operating within the zones and that limits the exposure of personnel outside the zones to any potential hazards or debris **Figure 7-15 ▶**:

- **Hot zone:** This area, also known as an action zone, is for entry teams and rescue teams only. It immediately surrounds the dangers of the incident, and entry into

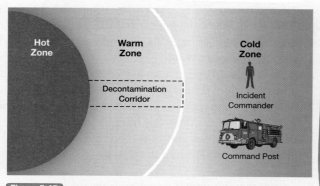

Figure 7-15 The area of operation is divided into three safety/operational zones: the hot zone, warm zone, and cold zone.

this zone is restricted to protect personnel outside the zone.

- **Warm zone:** This area is for properly trained and equipped personnel only. The warm zone is where personnel and equipment decontamination and hot zone support take place, including a debris area for material that is removed from the vehicle(s).
- **Cold zone:** This area is for staging vehicles and equipment, and contains the command post.

The IC will assign personnel or use law enforcement to establish another perimeter outside of the cold zone to keep the public and media out of the operational zones.

The most common method of establishing the safety/operational zones for an emergency incident site is to use law enforcement, fire line tape, or barriers. Once the controlled

Figure 7-16 Lockout/tagout systems are methods of ensuring that systems and equipment have been shut down and that switches and valves are locked and cannot be turned on at the incident scene.

zones have been marked off, the incident safety officer and personnel ensure that the restrictions associated with the various zones of the emergency scene are strictly enforced.

The size of each zone will vary depending on the complexity and size of the incident. Anyone who enters the warm or hot zone must be wearing full PPE, without exception. This is strictly enforced to maintain control of the incident and continuity of safety. Some company officers may, at times, enter the scene without wearing any PPE in order to take a closer look or to give an order; this is jokingly known as the "white wave." If this occurs, treat them with respect and politely ask them to put on some protective gear so they will not be injured by flying debris.

Lockout/tagout systems should be used to secure a safe environment **Figure 7-16 ◄**. These are methods of ensuring that systems and equipment have been shut down and that switches and valves are locked and cannot be turned on at the incident scene. Lockout/tagout control systems are most often used at machinery entrapment situations or confined space incidents to eliminate the movement of equipment, the flow of fluids, or the release of stored energy, but they can also be utilized at vehicle extrication incidents. For example, during an extrication incident, suppose a hydraulic ram is used to lift the roof of the vehicle off of the victim and needs to be held in place to prevent it from coming back down. The ram can be locked out in the extended position by disconnecting the hydraulic lines attached to the tool. This action will prevent the possibility of the hydraulic ram being accidentally closed or opened by another rescuer.

Specific Hazards

Fire Hazards

Because there is a significant risk of spilled fuel in many motor vehicle crashes, a minimum 1¾-inch (44-mm) diameter hose line should be deployed and charged, ready to protect personnel and victims by suppressing any potential fires or hazards that can occur. Other forms of protection include an ABC dry chemical portable fire extinguisher; however, the level of protection is greater with a charged hose line in place. Crashes that pose large fire hazards or actual fires may require additional fire suppression resources, which should be requested as soon as possible. Small fuel spills can be handled by using an absorbent or adsorbent material to isolate a fuel spill from the area around the damaged vehicle **Figure 7-17 ►**. The absorbent material can then be removed from the area by a licensed hazardous material agency. Most tow agencies are licensed to remove this type of hazardous substance once the incident has been terminated.

A post-crash fire can occur for many reasons, including a short in the electrical system or sparks created during the crash igniting spilled fuel. These fires may trap the occupants of the vehicle and require fire suppression. This is why a minimum 1¾-inch (44-mm) diameter charged hose line should be deployed on all vehicle extrication incidents. There are times when fire suppression may be necessary, and according to NFPA 1006, a requisite skill is to supply fire protection and to manage a fire using various extinguishing devices should a fire occur. Personnel should possess the necessary skills, training, and

Figure 7-17 Decide which absorbent/adsorbent material is best suited for use with the spilled product, and use it to isolate a fuel spill from the area around the damaged vehicle.

qualifications before attempting to extinguish a fire. Those with the necessary skills and training will approach a vehicle fire from an uphill, upwind position, moving in from the side at a 45-degree angle. Skill Drill 7-2 offers a basic overview of how to extinguish a vehicle fire. SOPs for vehicle fires may vary

between agencies and jurisdictions. To extinguish a vehicle fire under the hood, follow the steps in **Skill Drill 7-2 ▾** :

1. Don full PPE including SCBA, enter the accountability system, and work as a team.
2. Perform size-up and give an arrival report. Call for additional resources if needed.
3. Ensure that apparatus is positioned uphill and upwind, and that it protects the scene from traffic.
4. Perform the inner and outer surveys, which entail a 360-degree walk-around of the vehicle(s) and the entire scene.
5. Ensure that the crew is protected from any hazards.
6. Identify the type of fuel used in the vehicle and look for fuel leaks; this may require the use of a gas atmospheric monitoring device prior to taking action (recommended).
7. Advance a fire attack line of at least 1¾-inch (44-mm) diameter using water or foam (depending on the fuel type and whether a magnesium block engine exists, which will react with water application).
8. Beware of the hazards associated with the potential for exploding piston struts, bumper struts, SRS air bag cylinders, and tires with split-rim configurations; these are all potential killers.
9. Suppress the fire. Overhaul all areas of the vehicle, passenger compartment, engine compartment, and trunk. (**Step 1**)
10. Use effective water or foam application techniques. Avoid water hammers.
11. Maintain good body mechanics during the fire attack.
12. Notify command when the fire is under control.
13. Investigate the origin and cause of the fire. Preserve any evidence of arson.
14. Return the equipment and crew to service.

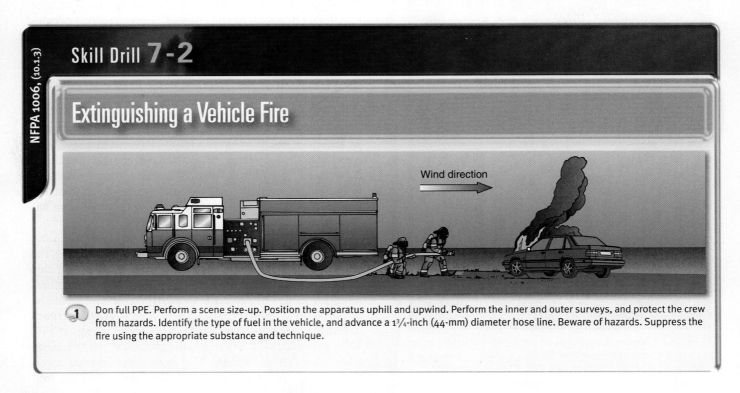

NFPA 1006, (10.1.3)

Skill Drill 7-2

Extinguishing a Vehicle Fire

Wind direction

1 Don full PPE. Perform a scene size-up. Position the apparatus uphill and upwind. Perform the inner and outer surveys, and protect the crew from hazards. Identify the type of fuel in the vehicle, and advance a 1¾-inch (44-mm) diameter hose line. Beware of hazards. Suppress the fire using the appropriate substance and technique.

Electrical Hazards

Down electrical lines present a serious hazard to fire fighters and other rescue personnel. Many vehicle collisions occur during the night, which creates problems for emergency response personnel because of the lack of visibility to identify down electrical lines. In this situation, attention to detail is very important. The technical rescuers must recognize signs dealing with power outages caused by the vehicle. Street lights that are normally lit may not be functional or residents or homes in the area may be without lighting. These signs should alert the emergency responders to proceed very slowly into the suspect location.

Once the electrical line has been identified, contact the appropriate utility provider and isolate the hazard from contact with any personnel or pedestrians Figure 7-18 ▼. Several methods can be used in isolating electrical hazards, such as placing traffic cones around a designated safe-distance perimeter to the electrical line, placing fluorescent snap lights around the safe-distance perimeter, or using barrier tape to tape off the area. Utilize law enforcement for street or access closures if needed. At no time should a rescuer assume that a down power line is dead; utility companies are programmed to energize a line several times at various intervals to determine where there is a break in the line. Power lines are known to jump several feet or meters when energized, so keep a safe distance, and never attempt to throw anything over a down line to contain or move it.

Some locations do not have suspended electrical supply lines or electrical transformers; instead, electrical supply lines are buried and electrical transformers are positioned at ground level. In such a case, the hazards are the same as with suspended lines, but a little closer to the responder and surrounding properties. Like overhead electrical lines, those positioned underground are vulnerable to damage during construction activities and during vehicular crashes. Similarly, electrical transformers positioned at ground or suspended levels may create a number of problems for rescuers, such as open high voltage, toxic smoke and gas, intense heat, potential for explosion from oil-filled equipment, explosion and flying debris from glass insulators

Figure 7-18 Safety is of primary importance when dealing with electricity.

and porcelain insulators, and, with some systems, release of pressurized gas.

Rescue Tips

Apparatus should be placed a minimum of two poles away from down wires or wires with items hanging on them.

Fuel Sources

The most common fuel sources today are gasoline and diesel fuel, although the list of alternative fuels is growing. Liquefied petroleum gas (LPG), or propane, continues to be one of the most widely used fuel alternatives to gasoline and diesel on a worldwide basis, primarily because of the environmental improvements that use of LPG brings. More than 500,000 vehicles in the United States use propane gas, most of which have spark-ignition engines that can operate on either propane or gasoline.

Gas explosions and leaks normally involve natural gas and propane gas. Natural gas incidents usually involve a ruptured supply line or a line that has failed due to corrosion, surface shift, or human intervention. Rescuers should evaluate any suspected releases with gas and air monitoring devices to determine the actual release point. Once the release point has been located, all buildings in the immediate area should be monitored before the area is considered safe.

In its natural state, propane is odorless, and when it is sent through a distribution system, an odorant is added. If the propane is being transferred by pipeline, however, the odorant may not have been added, creating a major concern for rescuers when incidents involve pipelines. Propane is heavier than air, so it will lie close to the ground surface, seeking an ignition source.

As mentioned, a majority of propane leaks or releases occur from leaking tanks, or damaged tanks due to impacts or overpressurization. One of the concerns with dealing with incidents that involve propane is the potential for a **boiling liquid, expanding vapor explosion (BLEVE)** created when fire impinges on the tank, resulting in temperature and pressure increases within the cylinder. As both temperature and pressure increase, the relief valve activates (opens), allowing the pressure to be released. If the relief valve fails to activate or is overcome, the tank will fail, resulting in a violent explosion.

Fuel Run-off

Fuel run-off can be very dangerous if not controlled by fire fighters or rescue personnel. Several concerns arise when dealing with fuel run-off, including the presence of ignition sources, environmental concerns, and reactions of fuels when they mix with each other and other products involved in the incident.

Damming, diverting, diluting, and absorbing are all methods used to control fuel run-off. Fire departments most commonly use floor/oil dry (oil absorbent), soil, and foam agents to dilute and/or suppress ignitable fumes, or slow run-off; this technique makes clean-up proceed much more quickly, especially following a vehicle incident.

Ignition Sources

With active fuel leaks, rescuers must be aware of all potential ignition sources when working a vehicle or machinery incident. Atmospheric monitoring should be established. Depending on the type of fuel, readings, and proximity to potential ignition sources, all ignition sources should be eliminated, if possible, without additional exposure or risk. Additionally, it may be necessary for all apparatus and other vehicles to remain in a staging area upwind beyond the established isolation zone. All individuals not associated with the incident must be removed, and no apparatus or portable equipment is to be started until the hazards are identified and/or mitigated.

Hazardous Materials

Responders, no matter what their level of training, must constantly be aware of the complexity, impact, and potential harm that various types of hazardous materials can present. They must also know how to avoid exposure, whether by inhalation, absorption, ingestion, or injection. The most intelligent course of action is to contact appropriately trained and equipped personnel to handle any incident involving hazardous materials. Responders must be aware of the threats that such materials pose to health, property, and the environment. In today's world, as remote as the danger may seem, responders must also be trained to recognize indicators of a potential terrorist incident as a part of their size-up.

Hazardous materials can come in a variety of forms: solids, liquids, and gas. Such materials can have radioactive, flammable, explosive, toxic, corrosive, biohazardous, oxidizer, asphyxiant, pathogenic, allergenic, or other characteristics that make them hazardous in specific circumstances. Among the hazardous materials that rescuers encounter almost routinely at incidents are gasoline, diesel fuel, kerosene, nitric acid, toluene, and acetone.

Vehicles that carry flammable and nonflammable pressurized gases (e.g., nitrogen, hydrogen, and oxygen), as well as transporters that carry flammable and nonflammable cryogenic liquids, including liquid nitrogen, liquid hydrogen, liquid oxygen (LOX), and liquefied natural gas (LNG) transporters, will also occasionally be involved in incidents Figure 7-19 . Solid materials, including explosives and flammable solids; oxidizers and organic peroxides; and poisons and corrosives are also sometimes involved in incidents. Common examples are fertilizers, pesticides, caustic powders, water-treatment chemicals,

and class 9 materials, which are especially prevalent around mining or construction activities.

There are two primary NFPA standards and one OSHA regulation that apply to hazardous materials and training for emergency responders:

- NFPA 472, *Standard for Competence of Responders to Hazardous Materials/Weapons of Mass Destruction Incidents*
- NPFA 473, *Standard for Competencies for EMS Personnel Responding to Hazardous Materials/Weapons of Mass Destruction Incidents*
- 29 CFR 1910.120, Hazardous Waste Operations and Emergency Response

The DOT defines hazardous materials as "any substance or material in any form or quantity that poses an unreasonable risk to safety and health and to property when transported in commerce." The Environmental Protection Agency's (EPA) definition of *hazardous material* is "any chemical that, if released into the environment, could be potentially harmful to the public's health or welfare." Rescuers should be aware of these threats and contact the appropriate resources to handle these hazards.

Other Hazards

Environmental conditions can lead to unique hazards at a crash scene. Crashes that occur in rain, sleet, or snow, for example, present an added hazard for rescuers and the victims of the crash. Crashes that occur on hills are harder to handle than those that occur on level ground.

Be especially alert for the presence of infectious bodily substances. Be prepared for the presence of blood, and follow standard (body substance isolation) precautions. Specifically, do not let blood or other bodily fluids come in contact with skin, and wear gloves that will protect from both contaminated fluids and sharp objects that are present at a crash site. If you or your clothes become contaminated, report the contamination, document it, and then clean and wash the affected clothes and equipment.

Some crash scenes may present threats of violence. Intoxicated people or those who are upset with other motorists may pose a threat to you or to other people present at the scene. Be alert for weapons that are carried in civilian vehicles. Law enforcement should always be dispatched on all motor vehicle collision incidents.

Occasionally, animals become a hazard at crash scenes. Dogs and other family pets may be protective of their owners and threaten rescuers. Farm animals or horses that have been involved in a crash may need care. You may need to call in specialized resources such as a large animal rescue unit or team to respond and assist with this type of incident.

Air Medical Operations

Air ambulances are used to evacuate medical and trauma patients. They land at or near the scene and transport patients to trauma facilities every day in many areas. There are two basic types of air medical units: fixed-wing and rotary-wing, otherwise known as helicopters Figure 7-20 . Fixed-wing aircraft generally are used for interhospital patient transfers over distances greater than 100 to 150 miles (161 to 241 kilometers

Figure 7-19   The MC-331 pressure cargo tanker carries materials such as propane, ammonia, Freon, and butane.

A.

B.

Figure 7-20 Air ambulances. **A.** Fixed-wing aircraft are generally used to transfer patients from one hospital to another over distances greater than 100 to 150 miles (161 to 241 km). **B.** A rotary-wing aircraft, or helicopter, is used to help provide emergency medical care to patients who need to be transported quickly over shorter distances.

[km]). For shorter distances, ground transport or rotary-wing aircraft are more efficient.

Rotary-wing aircraft have become an important tool in providing emergency medical care. Trauma patient survival is directly related to the time that elapses between injury and definitive treatment. Most air ambulances fly well in excess of 100 mph (161 kph) in a straight line, without road or traffic hazards, directly to a hospital helipad. The crew may include EMTs, paramedics, flight nurses, or physicians.

You should be familiar with the capabilities, protocols, and methods for accessing air ambulances in your area. Every agency has specific criteria for the type of patient who may receive air evacuation, and how and when to call for an air ambulance.

Helicopter Medical Evacuation Operations

A medical evacuation is commonly known as a medivac and is generally performed by helicopters. Most rural and suburban EMS jurisdictions and many urban systems have the capability to perform helicopter medivacs or have a mutual aid agreement with another agency such as police or hospital-based medivac to provide such service. You should become familiar with the helicopter medivac capabilities, protocols, and procedures of your

particular EMS because it varies from service to service. The following are some general guidelines you should be familiar with when considering whether to initiate a helicopter medivac operation.

Establishing a Landing Zone

Establishing a landing zone is the responsibility of the ground crew through a coordinated effort with the flight crew. The landing zone must be agreed upon by both parties. Determining the location involves more than simply looking for a clear space. The ground crew must be prepared to take action to make certain that the flight crew is able to land and take off safely. To prevent any miscommunication, when the request is made for a helicopter medivac response, the request should include a ground contact radio channel (typically a preestablished mutual aid channel), as well as a call sign of the unit that the helicopter medivac should make contact with. Some agencies preestablish landing zones throughout their jurisdiction and mark them on specialized district maps that are updated quarterly. This type of incident preplanning can be coordinated with air rescue and is a good idea to avoid confusion or conflict that may occur during an incident.

Things to do when selecting and establishing a landing zone include the following:

- Make sure the area is a hard or grassy level surface that measures 100 by 100 feet (30 by 30 m) (recommended) and no less than 60 by 60 feet (18 by 18 m) **Figure 7-21 ▼**. If the site is not level, the flight crew must be notified of the steepness and direction of the slope **Figure 7-22 ▶**. Always check with your local medical helicopter for specific landing zone requirements.
- Clear the area of any loose debris that could become airborne and strike either the helicopter or the patient and crew. Such objects include branches, trash bins, flares, accident tape, and medical equipment and supplies.
- Survey the immediate area for any overhead or tall hazards such as power lines or telephone cables, antennas, and tall or leaning trees. The presence of these hazards must be relayed immediately to the flight crew because an alternate landing site may be required. The flight crew may request that the hazard be marked or illuminated by

Figure 7-21 A landing area should be a level surface measuring 100 by 100 feet (30 by 30 m).

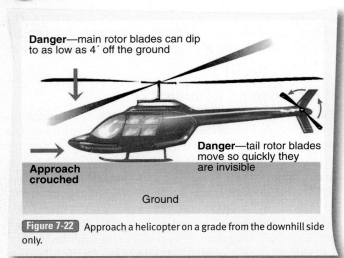

Danger—main rotor blades can dip to as low as 4′ off the ground

Danger—tail rotor blades move so quickly they are invisible

Approach crouched

Ground

Figure 7-22 Approach a helicopter on a grade from the downhill side only.

Landing Zone Safety

Helicopter safety is a combination of good sense and a constant awareness of the need for personal safety. You should be sure to do nothing near the helicopter and go only where the pilot or crew member directs you. The most important rule is to keep a safe distance from the aircraft whenever it is on the ground and "hot," which means the helicopter blades are spinning; most of the time the rotor blades will remain running because the flight crew does not expect to remain on the ground for a long time. This means that all rescuers should stay outside the landing zone perimeter unless directed to come to the aircraft by the pilot or a member of the flight crew. Usually, the flight crew will have their own equipment and will not require any assistance inside the landing zone. If you are asked to enter the landing zone, stay away from the tail rotor; the tips of its blades move so rapidly that they are invisible. Never approach the helicopter from the rear, even if it is not running. If you must move from one side of the helicopter to another, go around the front. Never duck under the body, the tail boom, or the rear section of the helicopter. The pilot cannot see you in these areas.

Another area of concern is the height of the main rotor blade. On many aircraft, it is flexible and may dip as low as 4 feet (1.2 m) from the ground **Figure 7-23 ▼**. When you approach the aircraft, walk in a crouched position. Wind gusts can alter the blade height without warning, so be sure to protect equipment as you carry it under the blades. Air turbulence created by the rotor blades can blow off hats and loose equipment. These objects, in turn, can become a danger to the aircraft and personnel in the area.

SOPs/SOGs must be established for all air operations and landing zone procedures.

weighted cones or by positioning an emergency vehicle with its lights turned on next to or under the potential hazard. Treat all down wires as if they are energized.

- To mark the landing site, use weighted cones, or place lights under these cones for illumination. This procedure is essential during night landings. Never use accident tape or people to mark the site. The use of flares is also not recommended; not only can they become airborne, but they also can potentially start a fire or cause an explosion. There are products on the market for lighting or marking a landing zone, but always refer to an agency representative on the proper procedures, devices, and/or regulations for marking a landing zone.
- Make sure all nonessential persons and vehicles are moved to a safe distance outside of the landing zone.
- If the wind is strong, radio the direction of the wind to the flight crew. They may request that you improvise some form of wind directional device to aid their approach. A bed sheet tightly secured to a tree or pole may be used to help the crew determine wind direction and strength. Never use tape.

Rescue Tips

To prevent any miscommunication, when the request is made for a helicopter medivac response, the request should include a ground contact radio channel (typically a preestablished mutual aid channel), as well as a call sign of the unit that the helicopter medivac should make contact with.

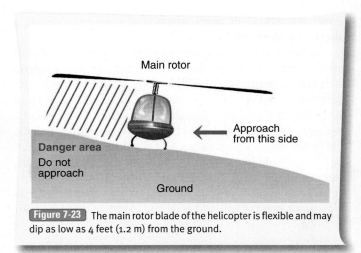

Main rotor

Approach from this side

Danger area
Do not approach

Ground

Figure 7-23 The main rotor blade of the helicopter is flexible and may dip as low as 4 feet (1.2 m) from the ground.

Wrap-Up

Ready for Review

- Site operations, vehicle stabilization, and victim management are parts of a successive three-phase process in vehicle extrication.
- Ensuring that proper safety procedures are followed in any operation is paramount for any organization.
- Because many injured patients require definitive care in less than an hour, the "Golden Hour" is now commonly referred to as the "Golden Period."
- Before the technical rescuer can start the operational process, he or she must be operationally ready, not only in the area of mental and physical readiness, but particularly in the area of personal protection.
- The protective ensemble includes the body piece, helmet, primary eye protection, gloves, footwear, and oftentimes reflective safety vests and respiratory protection.
- Per the federally mandated National Incident Management System (NIMS), common language is the goal for greater continuity and interoperability between agencies.
- The main goal of defensive apparatus placement is to block and protect the scene from the flow of traffic.
- Crowd control is an absolute necessity in protecting responders from individuals who might attempt to enter into the emergency site.
- As an incident grows, the need for additional resources and personnel can grow as well.
- A properly equipped rehabilitation unit providing shelter and thermal control options should be established early during prolonged extrication incidents.
- When responding to vehicle and machinery incidents, it is vital to know what additional specialized equipment is available to assist in managing an incident.
- Proper documentation always pays dividends for an agency when legal issues such as lawsuits occur.
- Size-up begins at the time the incident is dispatched, not at the time the unit arrives at the scene.
- A size-up report gives an update to the units responding so they can organize their response.
- There are generally three categories of entrapment: minor, moderate, and heavy.
- The inner and outer surveys are 360-degree inspections of the scene that are completed by two personnel.
- An incident action plan (IAP) may be formal or informal.
- Scene safety zones or operational zones are divided into hot, warm, and cold zones. These zones are strictly enforced by a designated incident safety officer.
- Hazards at vehicle extrication incidents may include fire hazards, electrical hazards, fuel hazards, ignition sources, and hazardous materials. Personnel should possess the necessary skills, training, and qualifications before attempting to mitigate any of these hazards.
- You should be familiar with the capabilities, protocols, and methods for accessing and landing helicopters in your area.

Hot Terms

Boiling liquid, expanding vapor explosion (BLEVE) An explosion that occurs when a tank containing a volatile liquid is heated.

Cold zone A safe area at an incident for those agencies involved in the operations. The incident commander, command post, EMS providers, and other support functions necessary to control the incident should be located in the cold zone.

Defensive apparatus placement The positioning of apparatus to block and protect the scene from the flow of traffic.

Golden Period The time during which treatment of shock and traumatic injuries is most critical and the potential for survival is best accomplished through rapid medical intervention.

Hot zone The area, also known as an action zone, accessible to entry teams and rescue teams only. It immediately surrounds the dangers of the incident, and entry into this zone is restricted to protect personnel outside the zone.

Immediate danger to life and health (IDLH) Any condition that would do one or more of the following: pose an immediate or delayed threat to life, cause irreversible adverse health effects, or interfere with an individual's ability to escape unaided from a hazardous environment.

Incident clock A procedure where dispatch will automatically notify the incident commander at 10-minute intervals until the incident becomes static.

Inner survey A four-point inspection of the vehicle's front, driver's side, rear, and passenger's side, including the undercarriage on all sides of the vehicle. This survey is conducted approximately 3 to 5 feet (0.9 to 1.5 m) from the vehicle and is performed by the first arriving company officer or experienced personnel.

Lockout/tagout systems Methods of ensuring that systems and equipment have been shut down and that switches and valves are locked and cannot be turned on at the incident scene.

Outer survey A survey conducted simultaneously with the inner survey; the rescuer performing the outer survey moves in the opposite direction as the rescuer performing the inner survey. Distance from the vehicle will vary with each incident, but it is generally a distance of 25 to 50 feet (7.6 to 15.2 m) starting from the perimeter of the inner survey position outward.

Primary and secondary vehicle access points The main point of entry and alternate point of entry that provides a pathway to the trapped and/or injured victim(s).

Scene safety zone/operational zone Zones that are divided into hot, warm, and cold zones. These zones are strictly enforced by a designated incident safety officer.

Scene size-up The systematic and continual evaluation of information presented in either visual or audible form.

Triage The process of establishing treatment and transportation priorities according to severity of injury and medical need.

Warm zone The area located between the hot zone and cold zone at an incident. The decontamination corridor is located in this zone.

Technical Rescuer *in Action*

Just after dinner, you receive a call for the report of a two-car collision on a busy street at an intersection during rush hour. It is dark outside. You quickly go to your engine and don your PPE. As the engine heads toward the scene, you review site operation procedures in your head as dispatch shares important information.

1. Stabilizing an incident requires the technical rescuer to manage site operations, stabilize the vehicle, and:
   A. stabilize the situation.
   B. stabilize the braking system.
   C. stabilize the patient.
   D. stabilize his or her crew.

2. In compliance with the National Incident Management System (NIMS), all emergency response personnel are required to utilize _____ for emergency radio communication.
   A. 10-codes
   B. native language
   C. Q-codes
   D. plain language

3. The incident size-up begins at the:
   A. time of dispatch.
   B. arrival of units on scene.
   C. moment all units respond.
   D. time the company officer arrives.

4. What is the "Golden Period"?
   A. The time it takes for the vehicular rescue team to arrive
   B. The time during which treatment of shock and traumatic injuries is most critical and the potential for survival is best accomplished through rapid medical intervention
   C. The time beginning at dispatch until the victim is packaged
   D. A 20-minute period from the time of dispatch to the victim's removal

5. An incident clock is a procedure where dispatch will automatically notify the incident commander at _____ intervals until the incident becomes static.
   A. 10-minute
   B. 15-minute
   C. 20-minute
   D. 30-minute

6. Describe the protective ensemble required during this particular incident, which involves a hazardous area (a vehicle accident scene on a roadway) and low visibility.
   A. Helmet, primary eye protection, footwear
   B. Hard hat, ear protection, respiratory protection
   C. Reflective vest, helmet, boots
   D. Body protection, helmet, eye protection, gloves, footwear, and reflective safety vests

7. Once your engine arrives on scene at the busy intersection, what is the best method for protecting the rescuers from traffic flow?
   A. Position apparatus around the corner from the incident.
   B. Position apparatus where the side-mounted pump panel is facing the passing traffic.
   C. Use law enforcement to protect the rescuers.
   D. Position apparatus at a 30- to 45-degree angle to the wreckage.

8. What information does the DOT's *Emergency Response Guidebook* (ERG) provide?
   A. Guidance regarding traffic control
   B. Guidance regarding traffic control devices
   C. Guidance to rescuers who may potentially operate at a hazardous material incident
   D. Guidance regarding the positioning of apparatus

9. The inner survey is a _____ point inspection of the vehicle.
   A. one-point
   B. two-point
   C. three-point
   D. four-point

10. When conducting the inner survey, the proper safe distance from the vehicle is approximately:
    A. 6 feet (1.8 m)
    B. 2½ feet (0.7 m)
    C. 3 to 5 feet (0.9 to 1.5 m)
    D. 30 feet (9.1 m)

Vehicle Stabilization

NFPA 1006 Standard

Vehicle and Machinery Rescue

10.1.4 Stabilize a common passenger vehicle or small machine, given a vehicle and machinery tool kit and personal protective equipment, so that the vehicle or machinery is prevented from moving during the rescue operations; entry, exit, and tool placement points are not compromised; anticipated rescue activities will not compromise vehicle or machinery stability; selected stabilization points are structurally sound; stabilization equipment can be monitored; and the risk to rescuers is minimized. (pages 174–190)

(A) Requisite Knowledge. Types of stabilization devices, mechanism of common passenger vehicle and small machinery movement, types of stabilization points, types of stabilization surfaces, AHJ policies and procedures, and types of vehicle and machinery construction components as they apply to stabilization. (pages 174–190)

(B) Requisite Skills. The ability to apply and operate stabilization devices. (pages 174–190)

10.1.5 Isolate potentially harmful energy sources, given vehicle and machinery tool kit and personal protective equipment, so that all hazards are identified, systems are managed, beneficial system use is evaluated, and hazards to rescue personnel and victims are minimized. (pages 190–191)

(A) Requisite Knowledge. Types and uses of personal protective equipment, types of energy sources, system isolation methods, specialized system features, tools for disabling hazards, and policies and procedures of the AHJ. (pages 190–191)

(B) Requisite Skills. The ability to select and use task- and incident-specific personal protective equipment, identify hazards, operate beneficial systems in support of tactical objectives, and operate tools and devices for securing and disabling hazards. (pages 190–191)

Knowledge Objectives

After studying this chapter, you will be able to:

- Explain the basis for stabilizing a vehicle. (pages 174–175)
- Explain the five types of wood box cribbing configurations. (pages 175–176)
- Explain the five directional movements of a vehicle. (page 176)
- Describe how to stabilize a vehicle in its normal position. (pages 176–180)
- Describe the proper procedure for deflating tires. (pages 178–179)
- Describe how to stabilize a vehicle resting on its side. (pages 180–184)
- Describe how to stabilize a vehicle upside down or resting on its roof. (pages 184–186)
- Describe how to marry two vehicles or objects together. (pages 186–190)
- Describe ways to mitigate vehicle electrical hazards. (pages 190–191)

Skills Objectives

After completing this chapter, you will be able to perform the following skills:

- Demonstrate the proper procedure for stabilizing a vehicle in its normal position. (pages 179–180, Skill Drill 8-1)
- Demonstrate the proper procedure for stabilizing a vehicle resting on its side. (pages 182–183, Skill Drill 8-2)
- Demonstrate the proper procedure for stabilizing a vehicle resting on its roof. (pages 185–186, Skill Drill 8-3)
- Demonstrate the proper procedure for marrying two vehicles. (pages 188–189, Skill Drill 8-4)
- Demonstrate the procedure for mitigating vehicle electrical hazards at a motor vehicle collision. (page 191, Skill Drill 8-5)

*y*ou are on the scene of an extrication incident with a car on its side and one victim trapped. The officer asks you and your partner to set up a triangle configuration utilizing a tension buttress system with a chain package. You and your partner have attempted this setup only one time on a training drill and do not feel confident performing this technique.

1. Do you advise the officer that you are not familiar with this technique and suggest another, simpler technique that is equally effective?
2. Which section of this vehicle will you crib first?

Introduction

"Few things are impossible to diligence and skill. Great works are performed not by strength, but perseverance."
—Samuel Johnson

Chapter 7, *Site Operations*, Chapter 8, *Vehicle Stabilization*, and Chapter 9, *Victim Access and Management*, outline a successive three-phase process that the technical rescuer should follow at every extrication incident **Figure 8-1 ▼**. This chapter will discuss the second step of this process, vehicle stabilization.

If not controlled, unstable vehicles are serious threats to rescuers and to those injured in a motor vehicle collision (MVC). The shape, size, and resting positions of vehicles after a collision can create many challenges for rescuers. Proper vehicle stabilization provides a solid foundation from which to work, ensuring safety for the emergency personnel as well as the victims **Figure 8-2 ▶**.

There are numerous methods for cribbing and stabilizing vehicles, such as box cribbing, struts, step chocks, wedges, shims, ratchet lever jacks, stabilizer jacks, rope, chain, cable, winches, ratchet straps, and tow trucks. This chapter will dis-

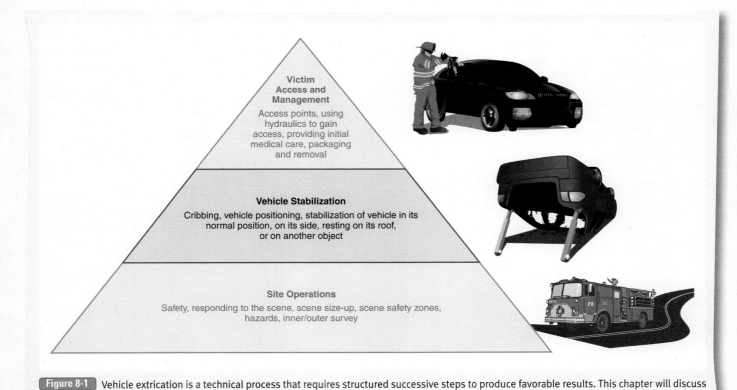

Victim Access and Management
Access points, using hydraulics to gain access, providing initial medical care, packaging and removal

Vehicle Stabilization
Cribbing, vehicle positioning, stabilization of vehicle in its normal position, on its side, resting on its roof, or on another object

Site Operations
Safety, responding to the scene, scene size-up, scene safety zones, hazards, inner/outer survey

Figure 8-1 Vehicle extrication is a technical process that requires structured successive steps to produce favorable results. This chapter will discuss the second step of this process, vehicle stabilization.

Figure 8-2 Ensure vehicles are stabilized before performing any operational techniques on the vehicles.

cuss how to stabilize common passenger vehicles in multiple resting positions following a crash incident.

To facilitate the learning comprehension of this chapter, review Chapter 6, *Tools and Equipment*, for the various stabilization tools and cribbing configurations.

> ### Rescue Tips
>
> During stabilization, responders should always be aware of the potential for vehicles to shift.

Cribbing

As discussed in Chapter 6, *Tools and Equipment*, cribbing is the most basic physical tool used for vehicle stabilization. Cribbing is commonly available as wood or composite materials, with some products being made of steel. Several cribbing designs are utilized for extrication incidents, such as step chocks, wedges, shims, and the basic 4- by 4-inch (102- by-102 millimeter [mm]) sections of timber (commonly referred to simply as "four-by-fours") that are cut at various lengths, most commonly 18 to 20 inches (457 to 508 mm) and 3 to 6 feet (1.2 to 2.4 meter [m]) or longer.

Wood Characteristics

Understanding the basic characteristics of wood used for cribbing is essential to the technical rescuer, whether you are stabilizing a vehicle, shoring a structure, or supporting a load.

Wood is heterogeneous in nature, meaning that it is composed of a mixture of different materials or has a lack of uniformity. Wood is also considered anisotropic in that the properties of each wood species are different according to its growth ring placement and direction of the grain. These two facts are important because not all wood types are suitable for cribbing and/or shoring. As was stated in Chapter 6, *Tools and*

Equipment, soft woods such as Southern Yellow Pine or Douglas Fir are commonly used for cribbing because they are well suited for compression-type loads. Hard wood, such as an oak species, is very strong but may split easily under certain stresses.

When considering wood species for cribbing, the primary concern is the measurement of applied stress (which is a unit of force), without failure, of that particular species of wood. The applied stress factors may be compression, tension, or shear. All of these stress factors, including the proportional strain of the wood, occur when a force is applied and a section of wood bends, which is considered the elastic performance of the wood.

Stress and strain are proportional, which means that any incremental increase in stress is proportional to an incremental increase of strain. The maximum stress and proportional strain on an object beyond its proportional limit will result in the failure of the material. Wood is considered elastic up to its proportional limit; beyond that limit, failure occurs.

The American Society for Testing and Materials (ASTM) has adopted standardized testing guidelines for measuring the relative stress resistance or strength values of particular species of wood. The maximum stress a board can be subjected to without exceeding the elastic range or proportional limit is known as its fiber stress at proportional limit (FSPL) rating.

When the downward force of an object rests or applies pressure on the surface of a section of wood and it is perpendicular to the grain, such as when a section of cribbing is set up in a crosstie or box crib configuration, the cribbing strength is determined using a formula. The dimension of the surface area at the **contact point**, or weight-bearing section of cribbing, is multiplied by the FSPL rating of that particular species of wood. For example, the FSPL rating of the Southern Yellow Pine or Douglas Fir is 500 pounds per square inch (psi) (3447 kilopascals [kPa]), meaning 500 psi (3447 kPa) is the load capability in pounds for each contact point. Now apply this formula to a 4- by 4-inch (102- by 102-mm) box crib. Multiply the surface area of contact of a 4- by 4-inch (102- by 102-mm) section, which is generally configured to be 3.5 by 3.5 inches (89 by 89 mm), or 12.25 square inches (7921 mm²):

> 500 psi (3447 kPa) (FSPL rating) × 12.25 square inches (7921 mm²) (surface area of contact) = 6125 pounds (3.0625 short tons) per area contact point

The load capacity is 6125 pounds (3.0625 short tons) per area contact point. If you have a simple box crib configuration with only four points of contact, then this setup will support a uniform load capacity of 24,500 pounds (12.25 short tons):

> 6125 pounds (3.0625 short tons) × 4 points of contact = 24,500 pounds (12.25 short tons)

Wood Box Cribbing

NFPA 1006, *Standard for Technical Rescuer Professional Qualifications*, discusses five types of wood box cribbing configurations that the technical rescuer needs to be familiar with:

- Two-piece layer crosstie **Figure 8-3A ▶**
- Three-piece layer crosstie **Figure 8-3B ▶**

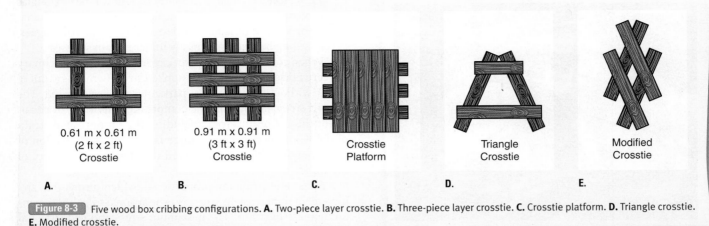

0.61 m × 0.61 m
(2 ft × 2 ft)
Crosstie

A.

0.91 m × 0.91 m
(3 ft × 3 ft)
Crosstie

B.

Crosstie
Platform

C.

Triangle
Crosstie

D.

Modified
Crosstie

E.

Figure 8-3 Five wood box cribbing configurations. **A.** Two-piece layer crosstie. **B.** Three-piece layer crosstie. **C.** Crosstie platform. **D.** Triangle crosstie. **E.** Modified crosstie.

- Crosstie platform **Figure 8-3C** ▲
- Triangle crosstie **Figure 8-3D** ▲
- Modified crosstie **Figure 8-3E** ▲

The two-piece layer crosstie, the three-piece layer crosstie, and the crosstie platform are the most commonly utilized wood box crib configurations; these will be used to demonstrate a majority of the various cribbing scenarios discussed in this chapter. The triangle crosstie and modified crosstie are unique types of wood box crib configurations that are specific to the type of stabilization incident presented; they are generally utilized for tight or odd-shaped spaces.

When you are setting up a basic two- or three-piece crosstie crib configuration, make sure that all the individual sections are uniform, with one on top of the other, providing a 1- to 1½-inch (25- to 38-mm) gap from the ends. Most ends on a cut section of a four-by-four are cracked or splintered, creating a potential for failure. Always avoid placing the contact point, or the weight-bearing section of cribbing, at the ends; doing so will provide you with a safety margin if the load shifts and the crosstie moves **Figure 8-4** ▼. Remember, when using a four-by-four, either Southern Yellow Pine or Douglas Fir, each contact point has an estimated weight or load-bearing capacity of 6000 pounds (3 short tons).

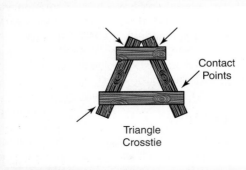

Contact
Points

Triangle
Crosstie

Figure 8-4 Each contact point has an estimated weight-bearing capacity of 6000 pounds (3 short tons).

Vehicle Positioning

There are five directional movements that the officer or technical rescuer in charge must consider during the process of vehicle stabilization:

1. **Horizontal movement.** Vehicle moves forward or rearward on its longitudinal axis or moves horizontally along its lateral axis **Figure 8-5A** ▶.
2. **Vertical movement.** Vehicle moves up and down in relation to the ground while moving along its vertical axis **Figure 8-5B** ▶.
3. **Roll movement.** Vehicle rocks side to side while rotating about on its longitudinal axis and remaining horizontal in orientation **Figure 8-5C** ▶.
4. **Pitch movement.** Vehicle moves up and down about its lateral axis, causing the vehicle's front and rear portions to move left or right in relation to their original position **Figure 8-5D** ▶.
5. **Yaw movement.** Vehicle twists or turns about its vertical axis, causing the vehicle's front and rear portions to move left or right in relation to their original position **Figure 8-5E** ▶.

There are four common postcollision vehicle positions that can be encountered at an accident scene: The vehicle may be in a regular or normal upright position resting on all four tires, it may be resting on its side, it may be resting on its roof, or it may be on top of another vehicle or another object.

The Vehicle in Its Normal Position

Why do you stabilize a vehicle that is upright and resting on all four tires **Figure 8-6** ▶? The first thought that probably comes

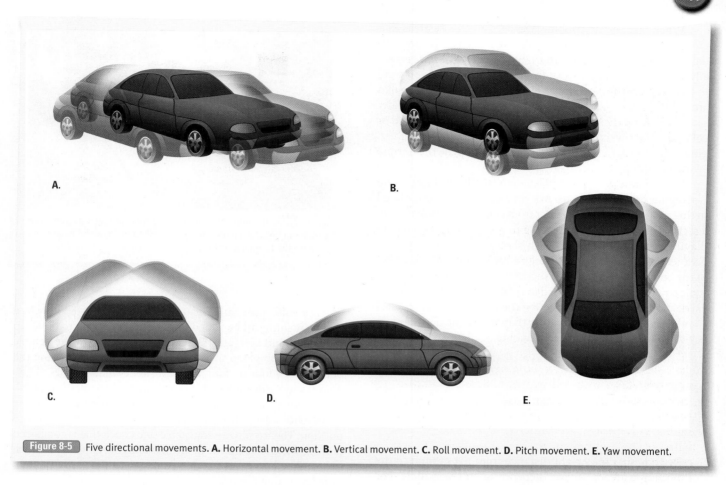

Figure 8-5 Five directional movements. **A.** Horizontal movement. **B.** Vertical movement. **C.** Roll movement. **D.** Pitch movement. **E.** Yaw movement.

Figure 8-6 A vehicle in its normal upright position.

to mind is that you do not want the vehicle to roll away. This is absolutely true, but the main reason we stabilize the upright vehicle is to gain control of all vehicle movement by minimizing the vehicle's suspension system and creating a solid and safe base to work from. A vehicle's suspension system can cause the body of the vehicle to move up and down, potentially causing further injury to a victim. A victim with a suspected spinal injury needs to be properly immobilized immediately; any vehicle movement can exacerbate a spinal injury, potentially causing paralysis of the victim. The goal is to create a balanced platform to work from and minimize the vehicle's suspension system.

To better illustrate how to create this balanced platform and minimize the vehicle suspension system with cribbing, look at the shape of the underside of a vehicle. If the vehicle's frame,

undercarriage, underside, or platform is rectangular or square in design, and if all of the vehicle's upper body components are removed, including side panels, parts, and wheels, you will be left with a rectangular or square frame or platform. In order to balance out this object that is shaped like a rectangle or square, the best practice is to access four or more solid points or areas under the object and insert cribbing equally at these points to establish a balance, whether you build a wood box crib configuration or insert a step chock. In a perfect world you would always have access to all four sides of a vehicle, but the reality is that you may only have access to one or two sides of the vehicle. In such scenarios, use your best judgment and crib the sides that you have access to. Properly cribbing just one side of the vehicle will help to minimize movement of the vehicle. The overall objective for crib placement is to position cribbing in four or more solid areas spread out equally to create that balanced platform.

Also, one must consider placing cribbing at the front and rear tires to eliminate the potential for forward or backward movement of the vehicle. This is particularly a factor when the tires remain inflated. If the tires are deflated and the vehicle is resting firmly on cribbing, then the need for placing cribbing at the front and rear is not a high priority unless the vehicle is positioned on an elevation or decline. This decision to add additional cribbing is up to the technical rescuer in charge of the operation.

When placing the cribbing, the need to choose areas that are solid cannot be stressed enough. Areas such as the firewall/dash section or the area just in front of the rear tires are generally very solid points to work from. For example, if you were to change a flat tire, you would place the car jack under a solid area of the vehicle and avoid weaker areas such as the fender sections behind the rear tire; these weaker sections can fold or collapse under weight or pressure. Also, avoid areas that can potentially block the extrication process or impede the normal swing of a door.

To eliminate potential problems, always think ahead before placing any cribbing sections. Ask yourself, "If cribbing is placed in this area, will it block my door from coming off? Will it catch a section of metal that I am trying to remove?" When you are playing a game of billiards, each shot you take sets up your next series of actions; you strategize each shot and placement of the cue ball to set up each successive step in advance. This same strategy should be applied to cribbing placement—and to extrication as a whole. Each action the technical rescuer takes should set up the next step rather than impede it. Therefore, it is vital to know where to place cribbing.

Rescue Tips

The technical rescuer should always think several steps ahead.

Determining the height distance from the ground to the bottom frame area will vary, depending on the vehicle. For example, the amount of cribbing needed to stabilize a large sport utility vehicle (SUV) will be significantly more than that needed to stabilize a small sports car. Step chocks will remove a lot of the guesswork because of the increased height adjustment on each successive step. When using cribbing, whether it is a basic four-by-four or a step chock, the goal is to make the contact area from the ground to the undercarriage tight, filling up any void spaces Figure 8-7 ▼ .

Figure 8-7   When using cribbing, whether it is a basic four-by-four or a step chock, the goal is to make the contact area from the ground to the undercarriage tight, filling up any void spaces.

Figure 8-8   If a void space still exists after inserting a step chock, a wedge can be added under the step chock to build up the height and increase the contact area between the vehicle and the cribbing.

If a void space still exists after inserting a step chock, a wedge or shim can be added under the step chock to build up the height and increase the contact area between the vehicle and the cribbing; tap the wedge section in position using the butt end of a four-by-four, or use a rubber mallet Figure 8-8 ▲ . Also, if you are dealing with a vehicle that is high off the ground, such as an SUV, a crosstie platform crib configuration can be set up with a step chock placed on top of it and then set into position. There is also an adjustable step chock that is composed of steel, which has a built-in mechanism that allows the chock to be manually adjusted and locked to meet the various height differences that the rescuer will encounter.

One question that is continually asked is whether the vehicle's suspension can be lifted manually, just enough to insert the cribbing, and then let back down to rest on the cribbing that is now properly adjusted to the required height. This is a loaded question. If done correctly, then yes, this method can be attempted and is very effective. The proper technique includes positioning your back against the body of the vehicle near the front or rear wheel well, lifting with your legs and not your back, and lifting the suspension only and not the vehicle itself. The problem comes from a poor lifting posture, or from overexertion by an adrenalin-fueled rescuer who tries to lift the vehicle rather than just move the suspension; injuries will absolutely occur when the latter happens. The decision to use or recommend this technique rests solely upon the officer in charge of the operation, or should come from a directive outlined in a departmental policy or standard operating guideline (SOG), which should also include the approval of an agency's risk manager. Also to be considered when determining to attempt this technique are the position of the vehicle, approximate estimated weight of the vehicle, and, obviously, the physical condition of the rescuer or rescuers who will be performing the lift.

Tire Deflation

An often debated topic is whether the tires of the vehicle should be deflated after the cribbing has been inserted. One benefit for deflating the tires on a passenger vehicle is that it will force the vehicle to rest firmly on the cribbing, creating a solid base to work from. As sections of the vehicle are removed, such as the doors or roof, the vehicle becomes lighter. With a vehicle's tires still inflated and the vehicle becoming lighter, the suspension system will cause the vehicle to rise and the cribbing to come loose. When the tires are deflated, the vehicle settles down onto the cribbing with the suspension system virtually eliminated.

One drawback of tire deflation is that the stability of the vehicle may shift. Or, if there is an object or another vehicle positioned on top of the vehicle, then that object or vehicle can also shift. Determining whether to deflate the tires is purely a judgment call by the officer in charge and can only be determined at the time of the incident and by the type of situation being presented.

Some agencies do not advocate deflating tires because doing so can interfere with law enforcement's investigation by eliminating a means of measuring tire pressure. If your agency does not support tire deflation, then an alternative would be to insert a wedge section of cribbing under the crib configuration and strike the end of it with a mallet or the butt end of a four-by-four until the desired height or stability is achieved.

There are several tools that can be used to deflate a tire. One of the easiest and safest ways is to remove the tire stem. The best overall tool to use for this job is a stem puller or a channel lock wrench **Figure 8-9 ▾**. When using a channel lock wrench, grab hold of the tire stem and rotate the tool so that the head of the wrench rests on the tire rim. Using the rim as a leverage point, move the tool downward, causing the stem to dislodge from its housing.

Rescue Tips

An initial surge of air pressure rapidly releases following tire deflation.

Another option for tire deflation is to use the forked end of a Halligan bar. This technique can only be applied to a rimmed

Figure 8-9 When pulling a tire stem, always use leverage to dislodge the stem from its housing. The best overall tool is a stem puller or a channel lock wrench.

Figure 8-10 Sliding the flat section of the fork of a Halligan bar forward over the area of the rubber tire stem will cut off the end of the tire stem at the base where it protrudes from the rim.

wheel with a protruding stem valve, in addition to the hubcap being removed. Slide forward one side of the fork inside the tire rim just over the area where the stem protrudes from the wheel; the key is positioning the blade of the tool at the correct angle, which can take some practice to achieve. Sliding this flat section of the fork forward over that area will cut off the end of the rubber tire stem at the base where it protrudes from the rim **Figure 8-10 ▴**. Remember that this technique will only work on rimmed wheels; any hubcaps will need to be removed prior to attempting this technique. Also, never use the spiked end of a Halligan bar to puncture the tire. This is a very dangerous practice that can potentially cause injury when the tool rebounds violently off the tire. This approach is also very unprofessional; as technical rescuers we strive to demonstrate skill through control, not through force.

Lastly, in addition to inserting cribbing for stabilization, the technical rescuer must always consider the basic or simple internal forms of stabilizing a vehicle. These steps include placing the vehicle in park, turning off the engine, and applying the parking brake. These are steps that can easily be overlooked because the focus is to quickly put cribbing or step chocks into place; yet these basic steps require very little effort to accomplish.

To stabilize a common passenger vehicle in its normal upright position, follow the steps in **Skill Drill 8-1 ▸**:

1. Don PPE, including SCBA if needed.
2. Enter the secure work area safely.
3. Assess the scene for hazards and complete the inner and outer scene surveys.
4. Lay out a tarp at the edge of the secure work area for staging tools and equipment, if indicated.
5. Apply basic or simple internal forms of stabilization by placing the vehicle in park, turning off the engine, and/or applying the parking brake. (**Step 1**)
6. Insert a step chock or build a wood box crib/crosstie configuration under four or more solid points of the vehicle.
7. If your agency supports tire deflation, deflate the tires to force the vehicle to rest firmly on the cribbing by pulling

NFPA 1006, (10.1.4)

Skill Drill 8-1

Stabilizing a Common Passenger Vehicle in Its Normal Position

1   Don PPE. Enter the secure work area safely. Assess the scene for hazards and complete the inner and outer scene surveys. Lay out a tarp at the edge of the secure work area for staging tools and equipment, if indicated. Apply basic or simple internal forms of stabilization by placing the vehicle in park, turning off the engine, and/or applying the parking brake.

2   Insert a step chock or build a wood box crib/crosstie configuration under four solid points of the vehicle. If your agency supports tire deflation, deflate the tires by pulling or cutting the tire stem. If your agency does not advocate tire deflation, use wedges to force the vehicle to rest firmly on the cribbing and consider placing additional cribbing at the front and rear of the tires to prevent any unexpected forward or backward movement. Reassess to confirm stabilization. Perform all of these tasks in a safe manner. Notify command that the vehicle has been stabilized.

or cutting the tire stem. If your agency does not advocate tire deflation, use wedges to force the vehicle to rest firmly on the cribbing and consider placing additional cribbing at the front and rear of the tires to prevent any unexpected forward or backward movement.

8. Reassess all of the cribbing to confirm position and stabilization.

9. Perform all of these tasks in a safe manner.

10. Notify command that the vehicle has been stabilized. **(Step 2)**

■ The Vehicle Resting on Its Side

A vehicle resting on its side is a very dangerous scenario **Figure 8-11 ▶**. As with other scenarios, the vehicle needs to be properly stabilized before any operations can be conducted. This section will demonstrate a technique using cribbing and tensioned buttress struts. Struts are structural supports used ·as a "buttress" to stabilize and reinforce an object. Because a technical rescuer must always be cognizant of time, this basic technique is designed for a very rapid and simplistic setup. When done correctly, this technique, including the inner and

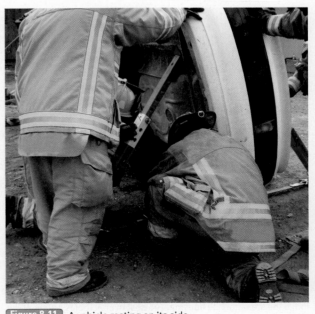

Figure 8-11   A vehicle resting on its side.

outer surveys, should take no longer than 5 minutes to complete. Keep the techniques basic unless you and your crew are well versed in multiple advanced stabilization scenarios; the time to attempt a new technique is not on the emergency scene. There are many ways to stabilize a vehicle resting on its side, and it is in the technical rescuer's best interest to research the various manufacturers of stabilization equipment in order to become familiar with the wide array of tools, to learn more about how to utilize the various stabilization equipment, and to determine which tools are best suited to meet the needs of a particular organization.

One of the greatest advances in the rescue industry, with the exception of hydraulic tools, was the introduction of buttress stabilization struts with a tensioning attachment Figure 8-12 ▾ . These tools have simplified the stabilization process tremendously and make it much safer to conduct emergency operations on a vehicle.

Rescue Tips

Make sure your crew fully understands the stabilization technique before attempting it. Stabilization is a team effort and every member needs to be on the same page.

A vehicle resting on its side has a center of gravity that is very high and a comparatively narrow track or base, which will cause it to topple over very easily. The center of gravity is the area of the object upon which all of the weight is centered; in this case, the center of gravity is high because the vehicle is on its side. The goal is to lower the vehicle's center of gravity by

Figure 8-12 Buttress stabilization struts are structural supports that can be made of steel, aluminum, composite/Kevlar® wrap, and wood. They are used as a "buttress" to stabilize and reinforce an object, expanding a vehicle's base or footprint while also lowering its center of gravity.

expanding the vehicle's **footprint**, or the area of the vehicle in contact with the ground. This can be accomplished with strategically placed struts, cribbing, and ratchet strapping. Ratchet straps are designed for tie-down purposes only and not for lifting. The main objective is to position the struts to form an A-frame configuration. The A-frame configuration will stabilize the vehicle (just as the outriggers on an aerial apparatus are designed to work).

Rescue Tips

All vehicles have different body designs to some degree; some may incorporate more rounded features, and some will incorporate square features. It is important to consider these differences when you are trying to locate an area to position the tips of the struts.

The first step that the technical rescuer will take is to examine whether the vehicle is leaning in a particular direction. If the vehicle is on its side on a level plane with all of its tires intact, then the tendency is for the vehicle to roll toward the roof side; the roof side is considered the most unstable and will need to be the first side that is stabilized. To accomplish the scenario of stabilizing a vehicle resting on its side, the crew members should have a full complement of cribbing sections and struts to work with, such as wedges, shims, four-by-fours, and step chocks.

After the inner and outer surveys are completed and all hazards are clear, the officer or technical rescuer in charge of the operation will place his or her hands on the front or rear section of the vehicle to feel for any shifting or movement of the vehicle as the other crew members begin cribbing. This safety technique allows the officer or rescuer in charge to warn crew members if the vehicle is going to roll. Crew members who are inserting the cribbing are generally unable to determine vehicle movement because their focus is at ground level where the cribbing is being placed. This safety technique gives the officer full control of the operation by providing visibility to both sides of the vehicle Figure 8-13 ▸ . Also, when operating at ground level around an unstable vehicle, such as when inserting or positioning cribbing, the technical rescuer must always work from one knee, in a semi-kneeling stance. This provides better mobility for moving quickly to avert any unexpected events, as opposed to being planted with both knees on the ground. The scene safety officer or the technical rescuer in charge of the operation must keep a keen eye open for any improper techniques and remind personnel to always work safely.

Rescue Tips

When operating at ground level around an unstable vehicle, such as when inserting or positioning cribbing, the technical rescuer must always work from one knee in a semi-kneeling stance.

Figure 8-13 The position of the technical rescuer in charge of the operation should be at the front or rear of the vehicle; this provides full control of the operation by providing visibility to both sides of the vehicle.

To stabilize a vehicle resting on its side using buttress stabilization struts, follow the steps in **Skill Drill 8-2 ▶**:

1. Don PPE, including SCBA if needed.
2. Enter the secure work area safely.
3. Assess the scene for hazards and complete the inner and outer scene surveys.
4. Lay out a tarp at the edge of the secure work area for staging tools and equipment, if indicated.
5. Position an officer at the front or rear section of the vehicle. The officer should place a free hand on the vehicle to feel for any shifting or movement of the vehicle as the other crew members begin cribbing on either side of the vehicle.
6. Build up cribbing under the hood section of the vehicle by placing a four-by-four parallel to the vehicle, close to the hood line.
7. Insert two wedges on top of the four-by-four and lightly strike them in place with the butt end of a four-by-four or using a rubber mallet. Do not strike them too hard because you will have to reset all of the cribbing once the struts are in place.
8. Move to the rear of the vehicle. Utilize a step chock or the four-by-four wedge combination to stabilize the rear of the vehicle. The step chock can be positioned either right side up or upside down to match the anatomical position of the vehicle. In this situation, step chocks generally work best upside down to match the anatomical position of the vehicle. A wedge or two may need to be inserted on top of the step chock to fill any void spaces and create a tight fit. (**Step 1**)

9. As the initial cribbing is conducted, another crew member should set up the tensioned buttress system on both sides of the front quarter of the vehicle. The first strut should be placed at a solid section of the undercarriage so that the tip of the strut can push off from it (push point). Adjust the strut height to maintain an angle of not less than 45 degrees to the vehicle, and then properly lock the height position of the strut in place. (**Step 2**)
10. Move to the opposite (hood) side of the vehicle. Mark a purchase point location in the hood by first extending the strut, measuring the height needed, and marking off the proper location where the strut will be placed. A **purchase point** is the location where access can best be gained. A purchase point is often needed to hold a strut in place.
11. Create the purchase point by placing the spike end of a Halligan bar on the designated mark and striking the back of the tool with a flat-head axe until the hood has been penetrated. If the hood of the vehicle is made of fiberglass or polycarbonate material, then make the purchase point at a higher location where the hood seam and the upper rail join together; the strut tip will grab hold and lock in place when pressure is applied. This purchase point can also be made by utilizing an air chisel or a cordless drill with a large #9 step bit for penetrating the sheet metal. Setup time must be considered when using one of these tools. Remember, time is always working against you.
12. With the spike of the Halligan bar buried in the hood, spin the tool 180 degrees clockwise or counterclockwise, positioning the fork end of the Halligan bar straight up and prying or pulling straight down on the bar, creating an outward lip on the top of the purchase point. (**Step 3**)
13. Place the tip of the strut into the purchase point and adjust the strut height to maintain an angle of not less than 45 degrees to the vehicle. Lock the height position of the strut into place.
14. Lay out a ratchet strap that comes with the strut kit. Connect both struts together by attaching the ratchet strap to the base of each strut. Make certain that the base plates are facing one another and are not turned outward. This would cause the struts to be pulled out of position and fall when the ratchet strap is tightened.
15. Once both struts are attached to one another, the officer or technical rescuer in charge of the operation (the person who has a hand placed on the front section of the vehicle) will have the two crew members double-check the placement of the struts, making sure they are in the proper position before they are tensioned and locked into place. One crew member will then take up all the webbing slack on the ratchet strap before ratcheting; this will ensure that there is no overspooling of the drum on the ratchet strap mechanism. With one hand on the webbing slack and the other hand on the ratchet lever, he or she will pull tight on the webbing slack, guiding it in the spool as the lever is ratcheted. This action will cause the tips of the struts to drive deep into a locking position, slightly lifting the vehicle and causing its footprint to expand outward onto the strut's ground pads.

NFPA 1006, (10.1.4)

Skill Drill 8-2

Stabilizing a Vehicle Resting On Its Side

1 Don PPE. Enter the secure work area safely. Assess the scene for hazards and complete the inner and outer scene surveys. Lay out a tarp at the edge of the secure work area for staging tools and equipment, if indicated. Position an officer at the front or rear of the vehicle. The officer should position a free hand on the vehicle to look and feel for movement or shifting of the vehicle. Build up cribbing under the hood and rear section of the vehicle using step chocks, wood cribbing, and wedges.

2 Place a tensioned buttress strut at a solid section of the undercarriage at the front of the vehicle. Adjust the strut height to maintain an angle of not less than 45 degrees to the vehicle and lock it into place.

3 Move to the opposite (hood) side of the vehicle. Measure and then mark a purchase point location in the hood. Create a purchase point in the hood by using the spike end of a Halligan bar and by rotating the tool 180 degrees and prying or pulling down on the bar to create a lip on the top of the purchase point.

4 Place the tip of the strut into the purchase point, adjust the strut height, and lock it into place. Attach the hooks of the ratchet strap to the base of each strut. Double-check the placement of the struts before ratcheting. Tighten the ratchet strap, locking the struts into place. Reseat all cribbing to be sure the vehicle is stabilized.

16. Once these steps have been completed, reseat all cribbing by striking each section firmly with the butt end of a four-by-four or rubber mallet. (**Step 4**)

Initial crib placement will focus on the most unstable area, which, in this particular scenario will be the front roof side of the vehicle. The objective here is to set up an A-frame configuration using a tension buttress system, thus building up cribbing

under the hood and rear sections of the vehicle and leaving the roof area unobstructed and open to work on.

As an additional safety factor, another set of struts can be applied in the same manner to the rear section of the vehicle for extra stability, but generally one set of struts in the front section of the vehicle with the cribbing configurations at the rear is sufficient to accomplish the task. Also, if the grade level of

Near Miss REPORT

Report Number: 10-452
Report Date: 03/11/2010

Synopsis: The vehicle had been chocked, but no one had checked to see if it was in park or if the emergency brake was on.

Event Description: We responded to a motor vehicle accident with injury. The victim in the driver's seat was restrained. A small amount of extrication was required. The fire crew was already on the scene, stating they needed a medic in the vehicle. The vehicle was off the road on top of a hill with a pond on both sides of the hill. I arrived on the scene and got in the backseat to hold c-spine and do patient care during extrication.

The vehicle had been chocked, but no one had checked to see if it was in park or if the emergency brake was on. As the extrication started, the vehicle started moving down the hill toward the pond. The patient and I were in the vehicle. It stopped prior to making it down the hill and we finished the extrication and transported the patient to the hospital.

Lessons Learned:

- Scene safety—Always be aware of surroundings.
- Teamwork—Work together and remind each other of possible safety issues.
- Command—Just because scene is established and units are on scene does not mean all safety aspects have been taken care of.
- Training—Always chock vehicles and make sure they are in park with the emergency brake set.

---

the surface is sloping, depending on the position of the vehicle, additional cribbing or struts can be added to the front or rear to prevent any potential shifting. A basic operation should take the technical rescuer and crew no longer then 3 to 5 minutes to deploy and complete the stabilization of a vehicle resting on its side. The main advantage of utilizing an A-frame technique is that the roof area is free of any cribbing and an uncomplicated roof removal can be accomplished if called for by the officer in charge. Other techniques require cribbing to be inserted under the roof line in the area of the A, B, C, or greater posts and this can impede a roof removal operation.

The Vehicle Upside Down or Vehicle Resting on Its Roof

When a vehicle is involved in a roll-over, the roof posts can be compromised by the crash impact and subsequent weight of the vehicle now on the posts; this makes the vehicle unstable **Figure 8-14 ▶**. Roof posts that have been compromised from a roll-over impact are not guaranteed to support the weight of the vehicle; therefore, the roof needs a solid artificial support system before any operation can be conducted. Federal Motor Vehicle Safety Standard (FMVSS) 216, *Roof Crush Resistance*, establishes the strength requirements for the passenger compartment roof. Currently, the roof strength requirements for vehicles 6000 pounds (3 short tons) or less is equal to 1½ times the unloaded vehicle weight of the vehicle. This tells us that the

strength of the roof under normal conditions will support 1½ times the vehicle's weight. New standards going into effect in 2012 increase the requirement for roof strength to 3 times the weight of a vehicle weighing 6000 pounds (3 short tons) or less. This is an incredible safety feature for the consumer, but rescuers must not get lulled into a false sense of security by thinking that compromised roof posts are secure. The FMVSS test was

Figure 8-14 A vehicle upside down or resting on its roof.

not designed for postcrash roof supports. Always take the extra safety precaution and properly support the roof structure with struts and cribbing.

Stabilizing a vehicle on its roof involves using struts and applying cribbing, at a minimum, in a four-point configuration. Looking at the vehicle's position, the weight of the engine will normally drive the hood or front area of the vehicle lower to the ground, with the trunk area presenting much higher. This scenario is based on the standard American automobile with a front-end engine compartment, where the roof area has not been completely flattened. With unobstructed access, there are usually three points of entry: the driver's side, the passenger's side, and the trunk area. Stabilization should always be set up to keep these potential entry points open and unobstructed.

Initial crib placement should focus on the most unstable area. In this particular scenario, the unstable area is the trunk side of the vehicle. The objective is to set up an A-frame configuration at the rear of the vehicle using struts, building up cribbing under the rear roof section and hood/dash areas of the vehicle. It is also possible to use crosstie box cribbing configurations that are stacked on top of one another and placed under the trunk area on both sides. The rule of thumb is to never stack box cribbing any higher than two times its width. For example, a 20-inch (508-mm) wide box crib formation should not exceed 40 inches (1016 mm) in height (roughly 3.5 feet [1.1

m]). Understand that by placing box cribbing under the trunk, you eliminate any possible trunk entry utilizing a tunneling option. **Tunneling** is the process of gaining entry through the rear trunk area of a vehicle; this technique is more commonly used for a vehicle resting on its roof. Remember that it is always important to keep all options open to quickly compensate and change directions for any unexpected events.

To stabilize a vehicle resting on its roof, the crew members should have a full complement of cribbing sections and struts to work with such as, wedges, shims, four-by-fours, and step chocks.

To stabilize a vehicle resting on its roof, follow the steps in **Skill Drill 8-3 ▼** :

1. Don PPE, including SCBA if needed.
2. Enter the secure work area safely.
3. Assess the scene for hazards and complete the inner and outer scene surveys.
4. Lay out a tarp at the edge of the secure work area for staging tools and equipment, if indicated.
5. Place step chocks in the two void spaces between the ground and rear section of the roof just before the trunk area on the roof rail. Always place the step chock on the roof rail where it is stronger and avoid the center of the roof, which is weaker and has a tendency to fold or crease. You will need to determine the position in which the step chock

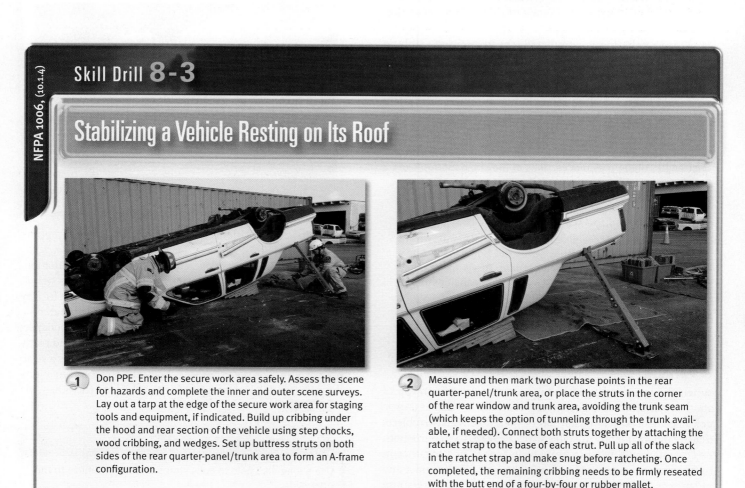

NFPA 1006, (10.1.4)

Skill Drill 8-3

Stabilizing a Vehicle Resting on Its Roof

1. Don PPE. Enter the secure work area safely. Assess the scene for hazards and complete the inner and outer scene surveys. Lay out a tarp at the edge of the secure work area for staging tools and equipment, if indicated. Build up cribbing under the hood and rear section of the vehicle using step chocks, wood cribbing, and wedges. Set up buttress struts on both sides of the rear quarter-panel/trunk area to form an A-frame configuration.

2. Measure and then mark two purchase points in the rear quarter-panel/trunk area, or place the struts in the corner of the rear window and trunk area, avoiding the trunk seam (which keeps the option of tunneling through the trunk available, if needed). Connect both struts together by attaching the ratchet strap to the base of each strut. Pull up all of the slack in the ratchet strap and make snug before ratcheting. Once completed, the remaining cribbing needs to be firmly reseated with the butt end of a four-by-four or rubber mallet.

fits best, either right side up or upside down, to best match the anatomical position of the vehicle. In addition, you may have to insert several wedges on top or bottom of the step chock to fill any void spaces and create a tight fit. The same step chock configuration can be inserted directly across on the opposite roof rail to balance it out. Placing cribbing in this area is not done to hold up the vehicle, even though it will add some support; the main purpose is to eliminate the potential for any rocking.

6. Place cribbing around the front sides of the vehicle under the dash/hood area and the front bumper area.

7. Set up buttress struts on both sides of the trunk area to form an A-frame configuration. (**Step 1**)

8. Option 1: Mark and create a purchase point on both sides of the rear quarter-panel/trunk area with the Halligan tool. The purchase point will hold the strut in place. Option 2: Place the struts in the corner of the rear window and trunk area, avoiding the trunk seam and thereby keeping the option of tunneling through the trunk available, if needed.

9. Connect both struts together by attaching the ratchet strap to the base of each strut.

10. Pull up all of the slack in the ratchet strap and make snug before ratcheting. Watch the strut base on both sides to ensure that the struts are not twisting out of position when the ratchet straps are tightened (if this occurs, reposition the struts).

11. Once ratcheting is completed, the remaining cribbing needs to be firmly reseated with the butt end of a four-by-four or rubber mallet. (**Step 2**)

Rescue Tips

Federal Motor Vehicle Safety Standards (FMVSS) 216, *Roof Crush Resistance*, establishes a minimum requirement for roof strength. Under previous versions of the standard, roofs of vehicles weighing 6000 pounds (3 short tons) or less were required to withstand 1½ times the weight of the vehicle when it is sitting on its roof. The new standard going into effect in 2012 states that roofs of vehicles weighing 6000 pounds (3 short tons) or less are required to withstand 3 times the weight of the vehicle when it is sitting on its roof. Some manufacturers may impose an even higher standard. This is one example of how changes to the standard will affect our abilities during vehicle extrication.

Vehicle on Vehicle or an Object on Top of a Vehicle

When the technical rescuer encounters a vehicle on top of another vehicle or an object on top of a vehicle, such as a large pole or cement post, he or she may be presented with two objects that are independently unstable (Figure 8-15 ▶). To stabilize both objects, they will need to be **married**, or joined together in their current position. Marrying vehicles will eliminate any independent movement of the two objects. For the purpose of this discussion, we will use the example of a vehicle on top of another vehicle. Joining the two vehicles together is best

Figure 8-15   A vehicle on another vehicle.

Figure 8-16   Industrial-grade ratchet strapping.

accomplished by utilizing industrial-grade ratchet strapping Figure 8-16 ▲ .

Because of the high degree of instability of this situation, it is extremely important to marry the two vehicles before operations are conducted. Stabilize the bottom vehicle first by inserting cribbing where there is access. Never crawl under the top vehicle, as you may become trapped by a sudden collapse; always work around the vehicle, remaining aware of and ready for any potential failure or collapse. If you need to pass a strap under a vehicle to the other side, hook the strap end to a long pike pole and safely pass it under the vehicle to the other side Figure 8-17 ▶ .

Rescue Tips

Use a long pike pole to pass straps under the vehicle to the other side.

Voices of Experience

A sedan carrying an older couple left a divided highway one afternoon. The area next to the highway was a deep depression bordered by rock outcroppings about 20 to 30 feet (6 to 9 m) high. The car left the roadway at a high rate of speed—about 60 miles per hour (mph) (97 kilometers per hour [kph])—and nosed into the rock. The rear of the car came to rest on the upslope of the depression, causing the car to turn into a U-shape, with the dashboard firmly entrapping both occupants.

Because the ground was relatively soft, the front of the car was actually embedded in the ground, with the bumper resting on the ground. Stabilization looked like it would be a breeze. After all, the front of the car was rock stable, and the back of the car was on the ground as well. We placed elevated step chocks under the sides and felt pretty confident we had a stable work platform.

Due to the nature of the entrapment, we had to perform a dash-lift technique. With the front end on the ground, we had to do a modified raise, cutting the structural beam in the driver's side front quarter panel. We first addressed popping open the front doors … and found our first problem. We'd placed our cribbing on the sides of the car, effectively blocking the door from opening. The entire rescue came to a grinding halt while our stabilization was reevaluated, removed, replanned, and replaced. The outcome was 50-50: one dead, one critical but alive. We'll never know if the time lost in restabilizing the vehicle was a factor in our patient's death. By the time we were able to put a medic into the car, he was already dead.

When you stabilize a vehicle for a rescue, keep your strategic and tactical goals in mind. Don't let your stabilization block access or disentanglement efforts. Time is life.

W. Buck Heath
Overland Park Fire Department
Overland Park, Kansas

> **"We'd placed our cribbing on the sides of the car, effectively blocking the door from opening."**

Figure 8-17 If you must pass a strap under a vehicle to the other side, hook the strap end to a long pike pole and safely pass it under the vehicle to the other side.

Other guidelines to remember about marrying vehicles utilizing ratchet straps are:

- Always look at the top vehicle and determine where it wants to roll, shift, or move. Strap it in the opposite direction, pulling it in that opposite direction and locking it into place.
- Always ratchet toward yourself and not away, and try to wrap the ratchet strap around the object and hook it back into itself. If you must use the hook into the object, make certain that the object is strong material that will not tear and that the hook itself is the double-wire type.

Various factors will determine how the operation will be conducted. What are the positions of both vehicles? How is the top vehicle resting on the bottom vehicle? Is any section of the top vehicle touching the ground? Where are the victims in relation to the top vehicle? Are there any victims inside of either vehicle? Where are the entry or access points to both vehicles? Will strapping or marrying the vehicles together block any access points? These are just a few questions that need to be addressed before any operation can be conducted, but additional questions will probably come up. This section will review only one of the many possible vehicle-on-vehicle scenarios.

The objective is to stabilize the bottom vehicle and then marry the two vehicles together, eliminating any independent movement. To stabilize or marry a vehicle on top of another vehicle, follow the steps in **Skill Drill 8-4 ▶**:

1. Don PPE, including SCBA if needed.
2. Enter the secure work area safely.
3. Assess the scene for hazards and complete the inner and outer scene surveys.
4. Lay out a tarp at the edge of the secure work area for staging tools and equipment, if indicated.
5. Position an officer at the front or rear section of the vehicle. The officer should place a free hand on the vehicle to feel for any shifting or movement of the vehicle as the other crew members begin cribbing on either side of the vehicle. (**Step 1**)

6. Stabilize the bottom vehicle. Follow the procedures discussed in Skill Drill 8-1, *Stabilizing a Common Passenger Vehicle in Its Normal Position*. Do not deflate the tires of the bottom vehicle; doing so can cause a shifting of the top vehicle, which can cause a potential collapse. Once the top vehicle has been successfully married to the bottom vehicle, the decision to deflate the bottom vehicle's tires can be addressed, if needed.
7. Cover the victim in the bottom vehicle with a blanket and then remove glass from the window frames. Glass removal procedures are discussed in Chapter 9, *Victim Access and Management*.
8. Place cribbing between the top and bottom vehicles to fill all void spaces. (**Step 2**)
9. Anchor the ratcheting section of the ratchet strap to the lowest area of the passenger-side A-post of the bottom vehicle, and then loop the loose end of the ratchet strap through the B-post of the top vehicle and back through and secure it to itself. Make certain that the hook is facing up and outward, not downward, to avoid the potential of the hook dislodging when tension is applied. If the top vehicle is too high to reach, a pike pole can be used to pass the strap through. If possible, avoid the use of ratchet straps within close proximity of the victim.
10. Fill the void spaces where the rocker panel of the top vehicle meets the hood section of the bottom vehicle.
11. Tighten the ratchet strap, and lock or marry the top vehicle to the bottom vehicle. (**Step 3**)
12. Once the top vehicle is secured to the bottom vehicle, the remaining cribbing needs to be firmly reseated with the butt end of a four-by-four or rubber mallet.
13. Once the top vehicle is secured to the bottom vehicle, a rescuer can be placed inside the bottom vehicle to treat and package the victim.

This marrying configuration gives you access to the victim through the entire door and roof area **Figure 8-18 ▼**. Keep in mind that there are several additional cribbing options that can

Figure 8-18 Marrying two vehicles utilizing the technique in Skill Drill 8-4 gives you access to the victim through the entire door and roof area. Even if the A-post were cut for a roof removal, the lower area of that post is strong enough to remain a solid anchor point.

NFPA 1006, (10.1.4)

Skill Drill **8-4**

Stabilizing/Marrying a Vehicle on Top of Another Vehicle

1 Don PPE. Enter the secure work area safely. Assess the scene for hazards. Complete the inner and outer scene surveys. Lay out a tarp at the edge of the secure work area for staging tools and equipment, if indicated. Position an officer at the front or rear of the vehicle to look and feel for movement or shifting of the vehicle.

2 Stabilize the bottom vehicle. Cover the victim in the bottom vehicle with a blanket and then remove glass from the window frames. Place cribbing between the top and bottom vehicle.

3 Anchor the ratcheting section of the ratchet strap to the lowest area of the passenger-side A-post of the bottom vehicle. Loop the loose end of the ratchet strap through the B-post of the top vehicle, and then loop the strap back through and secure it to itself. Fill the void spaces where the rocker panel of the top vehicle meets the hood section of the bottom vehicle. Tighten the ratchet strap, and lock or marry the top vehicle to the bottom vehicle.

4 Reseat all cribbing. Once the top vehicle is secured to the bottom vehicle, a rescuer can be placed inside the bottom vehicle to treat and package the victim.

Figure 8-19 There are several additional cribbing options that can be added to the side of the vehicle on top to prevent any potential sliding as the bottom vehicle is being stabilized.

Figure 8-20 Hidden dangers such as this portable propane tank may be found in the trunk of a vehicle.

be added to the side of the vehicle on top to prevent any potential sliding as the bottom vehicle is being stabilized **Figure 8-19 ▲**.

Hidden Dangers and Energy Sources

While vehicle fires and down power lines are visibly prominent and require immediate action and mitigation before the vehicle is stabilized, there are other potential hazards that may be hidden and cause havoc or injury to personnel on scene. For example, suppose you find a portable propane cylinder that someone was taking to get refilled in the vehicle's trunk or back passenger compartment, or suppose a short from a damaged electrical system starts a postcrash engine fire **Figure 8-20 ▶**.

Once the vehicle has been stabilized, the proactive technical rescuer can mitigate hidden potential hazards by, for example, removing the portable propane tank and disabling the energy system of the vehicle. Again, the best practice is to stabilize the vehicle and establish a solid base or platform for the rescuer to work from, in addition to minimizing vehicle movement, which can potentially exacerbate patient injuries. Unless there is an immediately dangerous to life and health

(IDLH) hazard that will affect the safety of the operation, stabilizing the vehicle should precede opening a hood of a vehicle to eliminate power. This practice is a guideline only that should be applied for conventional vehicles; alternative fueled, hybrid, and fuel cell vehicles require special procedures where disabling the electrical system is incorporated into stabilizing the vehicle and may have to be accomplished before cribbing is applied. The sequence of these actions is a judgment call that the officer in charge will have to make based on the type of incident presented.

Isolating or eliminating a vehicle's electrical system may be as simple as disabling the vehicle's 12-volt DC battery, removing fuses from the fuse box, and/or removing any smart keys out of the range area **Figure 8-21 ▼**. A smart key is an electronic key that allows the driver to remotely start the vehicle from a general range of 15 to 20 feet (4.6 to 6.1 m) away. Before the electri-

Figure 8-21 Isolating or eliminating a vehicle's electrical system may be as simple as disabling the vehicle's 12-volt DC battery, removing fuses from the fuse box, and/or removing any smart keys out of the range area.

cal system is disabled, make sure any electrically controlled devices defined by NFPA 1006, *Standard for Technical Rescuer Professional Qualifications*, as a beneficial system, such as vehicle seats, automatic steering wheel adjustments, or power windows, do not have to be utilized to gain access or create space for the victims. This needs to be coordinated between members of the rescue team and the incident commander.

Another issue that can be encountered in some conventional vehicles is the possibility of multiple batteries or batteries that may be located in places other than the front hood area, such as in the rear trunk, under the front or rear seat, or under the right or left wheel well. Vehicle manufacturers such as Buick, Mercedes, and BMW install 12-volt DC batteries in the trunk or under the rear seats. You must be aware of this possibility. Also be aware that some manufacturers have only provided access to the negative battery cable for purposes of disconnecting the electrical system. Another method to disconnect the energy system is to locate the fuse box and remove the fuses, which will isolate all of the electrical components and disable them. Although the 12-volt DC battery will remain live, this is the second best option. Supplemental restraint system (SRS) air bag control units come equipped with an energy capacitor that can store power, keeping the air bag system active and live even when the power has been disconnected for a varied amount of time, which differs among manufacturers. Supple-

mental restraint systems are discussed in detail in Chapter 5, *Supplemental Restraint Systems*.

To disable a conventional vehicle's electrical system and mitigate the potential electrical hazards at an MVC, follow the steps in **Skill Drill 8-5 ▾**:

1. Don PPE, including SCBA if needed.
2. Stabilize the scene by conducting the inner and outer scene surveys.
3. Stabilize the vehicle with the appropriate stabilization technique(s).
4. Check the vehicle for a smart key and move it out of range, approximately 15 to 20 feet (4.6 to 6.1 m) from the vehicle.
5. Open the vehicle's hood.
6. Locate the 12-volt DC battery system in order to eliminate the source of ignition.
7. Remove the negative terminal connection first and then the positive terminal. This can be accomplished by cutting a section out of the negative battery cable and a section out of the positive battery cable.
8. Fold back and tape the cut sections of cable to prevent the cables from touching the terminals. Always remember to remove the negative before the positive to avoid possible electrical shock or arching. (**Step 1**)

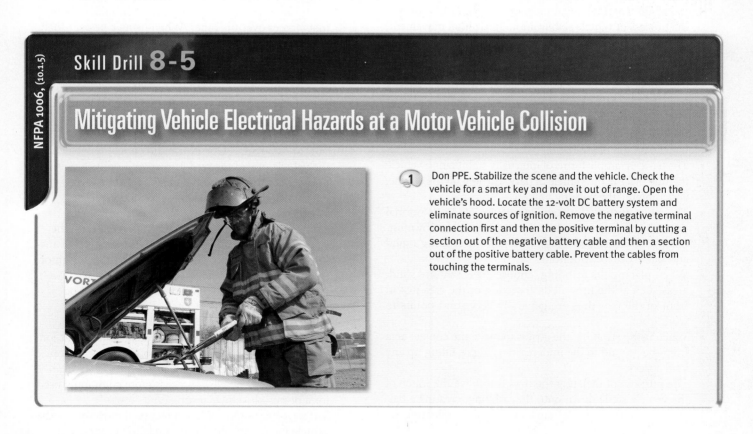

NFPA 1006, (10.1.5)

Skill Drill 8-5

Mitigating Vehicle Electrical Hazards at a Motor Vehicle Collision

1. Don PPE. Stabilize the scene and the vehicle. Check the vehicle for a smart key and move it out of range. Open the vehicle's hood. Locate the 12-volt DC battery system and eliminate sources of ignition. Remove the negative terminal connection first and then the positive terminal by cutting a section out of the negative battery cable and then a section out of the positive battery cable. Prevent the cables from touching the terminals.

Wrap-Up

Ready for Review

- Vehicle stabilization is a critical component of the extrication process.
- Proper vehicle stabilization provides a solid foundation to work from, which ensures safety for the emergency personnel as well as the victim and bystanders.
- Cribbing is the most basic physical tool used in vehicle stabilization.
- Soft woods are commonly used for cribbing because they are well suited for compression-type loads. Hard wood is very strong but may split easily under certain stresses.
- NFPA 1006 discusses five types of wood box cribbing configurations: two-piece layer crosstie, three-piece layer crosstie, platform crosstie, triangle crosstie, and the modified crosstie.
- There are five directional movements to consider during the process of vehicle stabilization: horizontal movement, vertical movement, roll movement, pitch movement, and yaw movement.
- There are four common postcollision vehicle positions that can be encountered at a collision scene: The vehicle may be in a regular or normal upright position resting on all four tires, it may be resting on its side, it may be resting on its roof, or it may be on top of another vehicle or an object may be on top of a vehicle.
- The basic or simple forms of internally stabilizing a vehicle include placing the vehicle in park, turning off the engine, and applying the parking brake.
- The main purpose for stabilizing a vehicle in its normal position is to gain control of all vehicle movement by minimizing the vehicle's suspension system and creating a solid and safer base to work from.
- When placing the cribbing, choose areas that are solid; areas such as the firewall/dash section or the area just in front of the rear tires are generally very solid points to work from.
- When using cribbing, the goal is to make the contact area from the ground to the undercarriage tight, filling up any void spaces.
- The purpose of deflating the tires is to have the frame of the vehicle settle down onto the cribbing, creating a balanced platform to work from and virtually eliminating the suspension system.
- The goal of stabilizing a vehicle on its side is to lower its center of gravity by expanding the vehicle's footprint.
- When a vehicle is involved in a roll-over, the roof posts will be compromised by the impact and weight of the vehicle, making the vehicle unstable. The objective is to set up an A-frame configuration at the rear of the vehicle using struts, building up cribbing under the rear roof section and hood/dash areas to maintain balance.
- When the technical rescuer encounters a vehicle on top of another vehicle or an object on top of a vehicle, he or she is presented with two objects that are independently unstable. These objects need to be joined together, or married, to eliminate any independent movement.
- Once the vehicle is stabilized, the technical rescuer should mitigate any potential postcrash vehicle electrical hazards that can occur, which may require disabling the vehicle's electrical system.

Hot Terms

<u>Contact point</u> When sections of cribbing are set on top of one another, the weight-bearing section of cribbing that crosses over the other. When using a 4- by 4-inch (102- by 102-mm) piece of timber, each contact point has an estimated weight-bearing capacity of 6000 pounds (3 short tons).

<u>Footprint</u> A generic term used to describe an object's balance in relation to its center of gravity, as determined by how much of the object's base touches the surface and how much of the object spans the surface.

<u>Horizontal movement</u> One of five directional movements; the vehicle moves forward or rearward on its longitudinal axis or moves horizontally along its lateral axis.

<u>Marrying (vehicles)</u> The process of joining vehicles together to eliminate any independent movement.

<u>Pitch movement</u> One of five directional movements; the vehicle moves up and down about its lateral axis, causing the vehicle's front and rear portions to move left or right in relation to their original position.

<u>Purchase point</u> The location where access can best be gained.

<u>Roll movement</u> One of five directional movements; the vehicle rocks side to side while rotating about on its longitudinal axis and remaining horizontal in orientation.

<u>Tunneling</u> The process of gaining entry through the rear trunk area of a vehicle, a process more commonly used for a postcrash vehicle resting on its roof.

<u>Vertical movement</u> One of five directional movements; the vehicle moves up and down in relation to the ground while moving along its vertical axis.

<u>Yaw movement</u> One of five directional movements; the vehicle twists or turns about its vertical axis, causing the vehicle's front and rear portions to move left or right in relation to their original position.

Technical Rescuer *in Action*

Your unit responds to a report of a MVC. Upon your arrival, you discover that multiple cars are involved. The incident commander assigns you the task of performing stabilization on vehicle number two.

As you perform your size-up of the vehicle, you note a late-model SUV that has come to rest on its roof. The vehicle has suffered major damage from an apparent roll-over accident. There are parties trapped inside the vehicle who will require emergency medical care. You have access from all sides of the vehicle, and you do not see any secondary hazards such as power lines or leaking fuel as you and your crew complete the inner and outer surveys.

1. Stabilizing the vehicle will provide:
   A. tasks to keep the fire fighters occupied.
   B. time for other resources to arrive.
   C. a stable foundation to work from.
   D. an alternative to rolling a vehicle back onto its wheels.

2. The most basic physical tool in vehicle stabilization is cribbing.
   A. True
   B. False

3. A vehicle resting on its side on level ground with all four wheels intact tends to roll towards its:
   A. trunk side.
   B. roof side.
   C. undercarriage side.
   D. The vehicle will not roll.

4. A vehicle that has been in a roll-over and comes to rest on its roof is compromised because of the:
   A. engine block altering the center of gravity.
   B. potential for leaking fluids.
   C. potential for victims being underneath the vehicle.
   D. potential crash impact that compromised the integrity of the roof posts.

5. In this situation, struts and cribbing should be placed at a minimum of:
   A. one point.
   B. two points.
   C. three points.
   D. four points.

6. A vehicle that has come to rest on its roof provides a number of unique complications.
   A. True
   B. False

7. If no other routes of entry into the vehicle are available, such as if the sides are blocked, the victim can be reached through the rear via a method called:
   A. burrowing.
   B. tunneling.
   C. worm-holing.
   D. belly crawling.

8. To stabilize the vehicle resting on its roof, the technical rescuer would likely use in addition to cribbing:
   A. electric winches.
   B. struts.
   C. chain hoists.
   D. come alongs.

9. The goal of the rescuer should be to create a strut configuration that is a(n):
   A. A-frame.
   B. lean-to.
   C. V-point.
   D. pivotless point.

10. Stabilization of a vehicle resting on its side should be focused on lowering the vehicle's:
    A. position.
    B. height.
    C. weight.
    D. center of gravity.

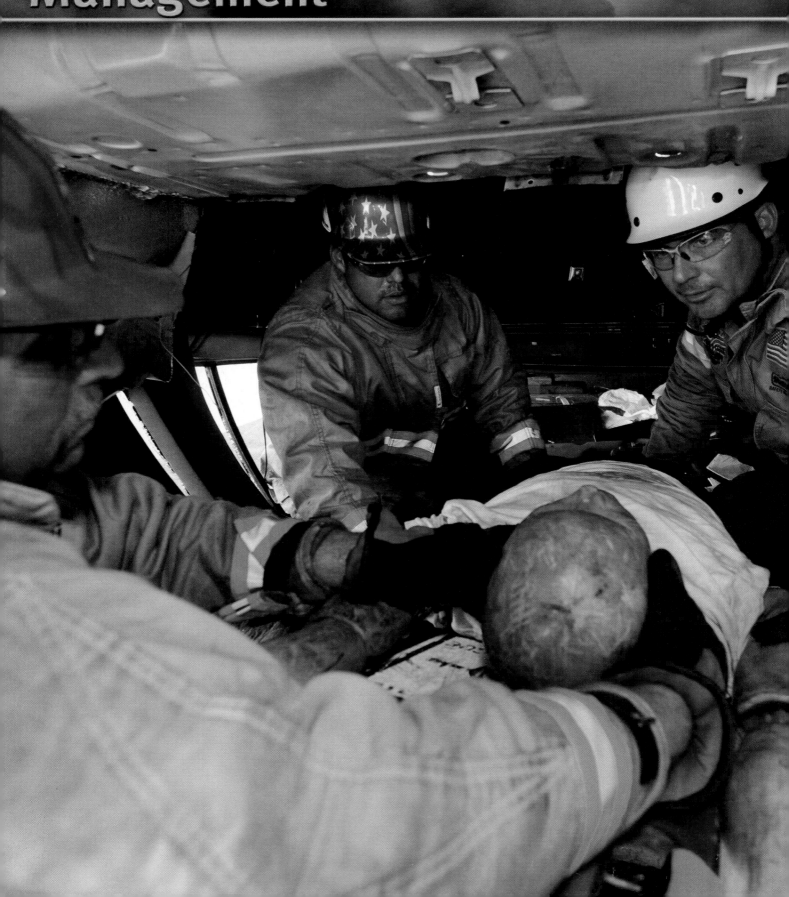

NFPA 1006 Standard

Chapter 5, Job Performance Requirements

5.3 **Victim Management.**

5.3.1 Triage victims, given triage tags and local protocol, so that rescue versus recovery factors are assessed, triage decisions reflect resource capabilities, severity of injuries is determined, and victim care and rescue priorities are established in accordance with local protocol. (pages 235–236)

(A) Requisite Knowledge. Types and systems of triage according to local protocol, resource availability, methods to determine injury severity, ways to manage resources, and prioritization requirements. (pages 235–236)

(B) Requisite Skills. The ability to use triage materials, techniques, and resources and to categorize victims correctly. (pages 235–236)

5.3.2 Move a victim in a low-angle environment, given victim transport equipment, litters, other specialized equipment, and victim removal systems specific to the rescue environment, so that the victim is moved without undue further injuries, risks to rescuers are minimized, the integrity of the victim's securement within the transfer device is established and maintained, the means of attachment to the rope rescue system is maintained, and the victim is removed from the hazard. (pages 236–237)

(A) Requisite Knowledge. Types of transport equipment and removal systems, selection factors with regard to specific rescue environments, methods to reduce and prevent further injuries, types of risks to rescuers, ways to establish and maintain victim securement, transport techniques, rope rigging applications and methods, and types of specialized equipment and their uses. (pages 236–237)

(B) Requisite Skills. The ability to secure a victim to transport equipment, assemble and operate environment-specific victim removal systems, and choose an incident-specific transport device. (pages 236–237)

5.3.3 Transfer a victim to emergency medical services (EMS), given local medical protocols, so that all pertinent information is passed from rescuer to EMS provider, and the victim can be transported to a medical care facility. (pages 236–237)

(A) Requisite Knowledge. Medical protocols for victim transfer; uses for checklists, triage tags, or report forms utilized for this purpose by the AHJ; risks, laws, and liabilities related to victim transfer; and information needed by the EMS provider. (pages 236–237)

(B) Requisite Skills. The ability to report victim condition and history to the EMS provider and to complete reports and checklists, and verbal communications skills. (pages 236–237)

Chapter 10, Vehicle and Machinery Rescue

10.1.6 Determine the common passenger vehicle or small machinery access and egress points, given the structural and damage characteristics and potential victim location(s), so that victim location(s) is identified; entry and exit points for victims, rescuers, and equipment are designated; flow of personnel, victim, and equipment is identified; existing entry points are used; time constraints are factored; selected entry and egress points do not compromise vehicle stability; chosen points can be protected; equipment and victim stabilization are initiated; and AHJ safety and emergency procedures are enforced. (pages 198–208)

(A) Requisite Knowledge. Common passenger vehicle or small machinery construction/features, entry and exit points, routes and hazards operating systems, AHJ standard operating procedure, and emergency evacuation and safety signals. (pages 198–208)

(B) Requisite Skills. The ability to identify entry and exit points and probable victim locations, and to assess and evaluate impact of vehicle stability on the victim. (pages 198–208)

10.1.7 Create access and egress openings for rescue from a common passenger vehicle or small machinery, given a vehicle and machinery tool kit, specialized tools and equipment, personal protective equipment, and an assignment, so that the movement of rescuers and equipment complements victim care and removal, an emergency escape route is provided, the technique chosen is expedient, victim and rescuer protection is afforded, and vehicle stability is maintained. (pages 208–217)

(A) Requisite Knowledge. Common passenger vehicle or small machinery construction and features; electrical, mechanical, hydraulic, pneumatic, and alternative entry and exit equipment; points and routes of ingress and egress; techniques and hazards; agency policies and procedures; and emergency evacuation and safety signals. (pages 208–217)

(B) Requisite Skills. The ability to identify common passenger vehicle or small machinery construction features, select and operate tools and equipment, apply tactics and strategy based on assignment, apply victim care and stabilization devices, perform hazard control based on techniques selected, and demonstrate safety procedures and emergency evacuation signals. (pages 208–217)

10.1.8 Disentangle victim(s), given an operations level extrication incident, a vehicle and machinery tool kit, personal protective equipment, and specialized equipment, so that undue victim injury is prevented, victim protection is provided, and stabilization is maintained. (pages 217–232)

(A) Requisite Knowledge. Tool selection and application, stabilization systems, protection methods, disentanglement points and techniques, and dynamics of disentanglement. (pages 217–232)

(B) Requisite Skills. The ability to operate disentanglement tools, initiate protective measures, identify and eliminate points of entrapment, and maintain incident stability and scene safety. (pages 217–232)

10.1.9   As a member of a team, remove a packaged victim to a designated safe area, given a victim transfer device, designated egress route, and personal protective equipment, so that the team effort is coordinated, the designated egress route is used, the victim is removed without compromising victim packaging, undue injury is prevented, and stabilization is maintained. (pages 236–237)

(A) Requisite Knowledge. Patient handling techniques; incident management system; types of immobilization, packaging, and transfer devices; types of immobilization techniques; and uses of immobilization devices. (pages 236–237)

(B) Requisite Skills. Use of immobilization, packaging, and transfer devices for specific situations; immobilization techniques; application of medical protocols and safety features to immobilize, package, and transfer; and all techniques for lifting the patient. (pages 236–237)

Knowledge Objectives

After studying this chapter, you will be able to:

- Explain the process of gaining access through doors. (page 198)
- Explain the process of gaining access through windows. (pages 198–199)
- Explain when the backboard slide technique would be used. (page 199)
- Explain the differences between tempered and laminated glass. (pages 199–206)
- Describe polycarbonate windows and ballistic-rated glass. (pagess 206–207)
- Describe some of the methods to remove glass safely. (pages 199–207)
- Describe how hydraulic rescue tools can be used to gain door or roof access. (pages 208–217)
- Describe methods of removing the vehicle from around the victim. (pages 217–219)
- Describe when and how to relocate a dashboard, steering wheel, and steering column. (pages 219–232)
- Describe the tasks involved when providing initial medical care to a victim. (pages 233–235)
- Describe triage. (pages 235–236)
- Explain the important aspects of victim packaging, removal, and transport. (pages 236–237)

Skills Objectives

After studying this chapter, you will be able to perform the following skills:

- Assist with or perform the backboard slide technique. (pages 199–200, Skill Drill 9-1)
- Break tempered glass using a spring-loaded center punch. (pages 201–202, Skill Drill 9-2)
- Break tempered glass using a glass handsaw. (pages 202–203, Skill Drill 9-3)
- Remove a windshield using a glass handsaw. (pages 203–205, Skill Drill 9-4)
- Remove a windshield using a reciprocating saw. (pages 206–207, Skill Drill 9-5)
- Remove a windshield from a partially ejected victim. (pages 208–209, Skill Drill 9-6)
- Release a door from its frame or perform the vertical spread. (pages 210–212, Skill Drill 9-7)
- Crush a wheel well using the wheel well crush technique. (page 213, Skill Drill 9-8)
- Gain access using a complete side-removal technique or side-out technique. (pages 214–216, Skill Drill 9-9)
- Remove the roof of an upright vehicle. (pages 219–221, Skill Drill 9-10)
- Roll the dash away from the victim using the dash roll technique. (pages 222–234, Skill Drill 9-11)
- Lift the dash away from the victim using the dash lift technique with hydraulics. (pages 226–228, Skill Drill 9-12)
- Lift the dash away from the victim using the dash lift technique without hydraulics. (pages 228–229, Skill Drill 9-13)
- Relocate a steering wheel and steering column to free a victim. (pages 230–232, Skill Drill 9-14)
- Extricate a victim from a passenger car. (pages 236–237, Skill Drill 9-15)

you are on an extrication incident where a vehicle has impacted a cement light pole, which has collapsed on the front end and hood of the vehicle, trapping the victim under the dash. The company officer orders your crew to perform a dash roll technique utilizing the hydraulic rams, but you know that this is not the proper technique for this type of entrapment.

1. As the technical rescuer on scene, what do you do?
2. Would you explain to the company officer that a dash roll technique will not work in this situation and that a dash lift is the correct technique?
3. Would you perform a dash lift technique to release the dash off of the victim?

Introduction

"You have to learn the rules of the game. And then you have to play better than anyone else." —Albert Einstein

Chapter 7, *Site Operations*, Chapter 8, *Vehicle Stabilization*, and Chapter 9, *Victim Access and Management*, outline a successive three-phase process that the technical rescuer should follow at every extrication incident Figure 9-1 ▾. This chapter will discuss the third step of this process, victim access and management. With the scene and vehicle stabilized, it is time to access, manage, and transfer the victim. Managing the victim involves victim access, care, packaging, and removal. The main objective is not to remove the victim from the vehicle but to remove

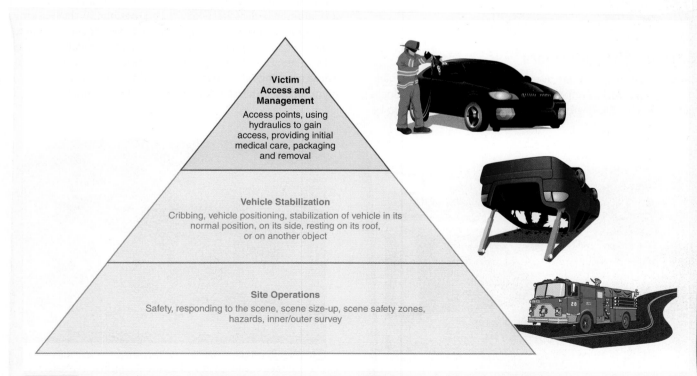

Figure 9-1 Vehicle extrication is a technical process that requires structured successive steps to produce favorable results. This chapter will discuss the third step of this process, victim access and management.

the vehicle from the victim by creating a large opening with systematic and precise techniques.

Access Points

After stabilizing a vehicle, a technical rescuer must gain access into the passenger compartment in order to stabilize, protect, and disentangle the victim. With the rescuer carrying basic medical gear, access into the vehicle may be obtained in a number of ways. For example, access may be obtained via a door adjacent to the victim or by sliding in through a rear or side window utilizing an assisted backboard slide technique. Options such as these may be referred to as primary and secondary vehicle access points. The objective is to gain access to render immediate care.

Primary access refers to the existing openings of doors and/or windows that provide a pathway to the trapped and/or injured victim. Secondary access refers to openings created by rescuers that provide a pathway to trapped and/or injured victims. These two types of access constitute a process of having a plan A and plan B in place; if the team cannot gain entry through existing openings (the established plan A), then plan B is implemented to create an access point. With additional resources and personnel on scene, at times plan A and plan B can be incorporated and conducted simultaneously.

Rescue Tips

The main objective is not to remove the victim from the vehicle but to remove the vehicle from the victim by creating a large opening with systematic and precise techniques.

Access Through Doors

One of the simplest ways to access a victim is to open a vehicle door [Figure 9-2 ▼]. It is important to manually try all of the doors before other methods are used, even if the doors appear to be

badly damaged. The first rule of forcible entry is "Try before you pry." It is an embarrassing waste of time and energy to open a jammed door with heavy rescue equipment only to find that a door could have been opened easily without special equipment. Attempt to unlock and open the least damaged door first. Make sure the locking mechanism is released. Then try the outside and inside handles at the same time if possible. If the doors are locked, you might consider breaking a window utilizing the techniques shown in this text, and then attempt to manually release the locking mechanism to unlock the doors.

Rescue Tips

A fundamental rule of forcible entry is "Try before you pry."

Access Through Windows

If a victim's medical status is serious enough to require immediate care and you cannot enter through a door, consider breaking a window and attempting to manually release the door's locking mechanism or to conduct an assisted backboard slide technique (presented later in this chapter) to render immediate aid. The side and rear windows are commonly made of tempered glass, which will break easily into small pieces when a tool such as a spring-loaded center punch is used. Less commonly, side and rear windows are made of a laminate or polycarbonate material. The rescuer must be prepared to take the appropriate action, which is explained in this chapter.

There are two basic ways to tell what type of glass you are dealing with. First, all glass contains small etched or embossed markings stating that it is "safety" or tempered glass, and/or that it is laminated glass [Figure 9-3 ▼]. Internationally, the markings may appear in another language such as German. *Verbund-Sicherheitsglas* (VSG) is translated in English to mean laminated safety glass; *EV-Verglasung* (ESG) means tempered glass glazing. These markings are generally very difficult to see, especially at night, because the manufacturers attempt to maintain the clear

Figure 9-2 Manually attempt access into the vehicle through the doors first, if possible.

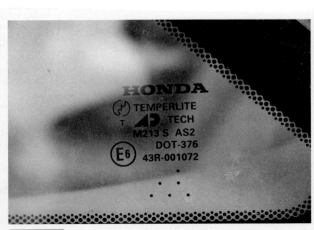

Figure 9-3 All glass contains small markings in one of the corners that is etched or embossed on the glass stating that it is "safety" or tempered glass, and/or that it is laminated glass. The glass shown in this photo is tempered glass.

unobstructed view and natural aesthetics of the glass. The second way to determine what type of glass you are dealing with requires the use of the center punch technique. If the glass is laminated, then the center punch will not be able to penetrate the glass and will only partially fracture the outer glass section. This will be evident by a small spider-web ring around the point of impact. If the glass is comprised of a polycarbonate material, then again the center punch will not be able to penetrate it; more than likely the tool will spring back.

Entering through the windshield, although it may not be a common procedure, will require more aggressive steps utilizing hand and/or power tools such as a reciprocating saw to gain entry. This method of entry may be required for a vehicle resting on its side. Remember that the front windshield on a common passenger vehicle is made of a laminated glass, and it is generally sealed and held in place with a mastic-type adhesive that requires a cutting action of the glass itself to be properly removed.

Glass Removal Safety

When performing glass removal, it is always safer to remove all of the glass rather than leaving some segments intact. The process of removing a roof or a major structural component of the vehicle requires all of the glass to be removed from the vehicle. This is a safety measure that must be performed to avoid glass unexpectedly shattering or falling on the victim when the roof is cut and removed, or when a door is spread with a hydraulic tool; such rescue efforts will compress and twist the entire body of the vehicle, potentially causing a side or rear tempered glass section that has not been removed to shatter.

When breaking glass, make certain that all personnel operating around the vehicle and the victim are aware that the vehicle's glass is going to be broken out. The statement "Breaking glass!" must be made before the action of breaking the glass begins. This safety practice ensures there are no surprises, as can occur when someone haphazardly takes out glass without warning anyone. Ensure that the victim is fully aware of the action taking place and that the victim and the rescuer are covered before any glass is broken, unless all access to the victim is blocked. Attempt to break glass beginning at the farthest point from the victim and remember to clean out the tempered glass fragments from all the window frames using a hand tool or a piece of 4- by 4-inch (102- by 102-millimeter [mm]) cribbing (or "four-by-four"). Be extremely cautious when removing broken sections of laminated glass from the casing of the window with a gloved hand; even gloves are susceptible to a penetrating glass shard. It is safer to use a glass saw to cut sections of laminated glass out. All glass that has been removed, whether it is laminated windshield glass or tinted tempered glass, should be placed in a debris pile in a safe area in order to avoid potential injuries.

Rescue Tips

Attempt to break glass beginning at the farthest point from the victim. Remember to clean out the tempered glass fragments from all window frames using a hand tool or a piece of 4- by 4-inch (102- by 102-mm) cribbing.

■ The Backboard Slide Technique

The overall objective for using the backboard slide technique is to gain access to the patient as rapidly as possible to render care. If the vehicle doors are locked, blocked, or not operable, initial access into the passenger compartment may be accomplished through a rear or side window utilizing an assisted __backboard slide technique__.

To perform the assisted backboard slide technique through a rear window, follow the steps in **Skill Drill 9-1 ▶**:

1. Don appropriate PPE, including mask, eye, and respiratory protection.
2. Assess the scene for hazards and complete the inner and outer surveys.
3. Stabilize the vehicle.
4. Manually try all doors from the inside and the outside, if readily accessible, to confirm they are locked or inoperable. (**Step 1**)
5. If there is access, ensure that the victim inside the vehicle is properly protected using a blanket as cover from flying glass particles.
6. Remove the rear window glass, if needed, utilizing the appropriate techniques discussed in this text. (**Step 2**)
7. Place a tarp or blanket in the window frame to protect against any glass fragments that may still be in the window frame.
8. Place a long backboard up on the trunk area of the vehicle with the front end just resting on the inside of the rear window frame. (**Step 3**)
9. A technical rescuer will position himself or herself on the backboard, either headfirst or feetfirst depending on the type of vehicle, maneuverability, and the size of the opening. This decision is up to the individual entering the vehicle.
10. Once in position, two additional technical rescuers will grab hold of the board on opposite sides and raise the board to slide the technical rescuer positioned on the board safely into the vehicle. (**Step 4**)
11. Once the rescuer is inside of the vehicle, all appropriate medical gear is passed into the vehicle, and patient care is rendered. (**Step 5**)

■ Tempered Safety Glass

There are several tools designed for glass removal; some of these were discussed in Chapter 6, *Tools and Equipment*. The spring-loaded center punch is the most basic and common of all glass removal tools **Figure 9-4 ▶**. It is used for tempered glass only.

When using a center punch to break tempered glass, there are several safety rules that the technical rescuer must follow. First, wearing full PPE, including eye protection, position a gloved hand palm down against the lower corner of the window and frame area that is going to be broken (if this area is accessible). Your hand should be positioned with the thumb pointing upward, flush against the window. With your free hand, take the center punch and rest the body of the tool on the outer ridge of the palm, in the V of the hand, between the thumb and index finger. Place the tip of the tool at a 90-degree angle to and

Skill Drill 9-1

The Assisted Backboard Slide Technique

1 Don appropriate PPE. Assess the scene for hazards and complete the inner and outer surveys. Stabilize the vehicle. Manually try to open all doors before beginning the technique.

2 Ensure that the victim inside the vehicle is properly protected from flying glass particles. Remove the rear window glass, if needed, utilizing the appropriate techniques.

3 Place a tarp or blanket in the window frame to protect against any glass fragments. Place a long backboard up on the trunk area with the front end just resting on the inside of the rear window frame.

4 A technical rescuer will position himself or herself on the backboard, either headfirst or feetfirst depending on the type or vehicle, maneuverability, and the size of the opening. Two additional technical rescuers will grab hold of the board on opposite sides and raise the board to slide the technical rescuer on the board safely into the vehicle.

5 All appropriate medical gear is passed into the vehicle, and patient care is rendered.

against the window. The positioning of your hand against the corner of the window and frame area will act as a safety stop, preventing you from accidentally putting your hand through the window when it breaks **Figure 9-5 ▶**. Once ready, give the warning "Breaking glass!" and then commence plunging the tool into the window.

If the tempered glass is clear with no tinting film, use a hand tool or a four-by-four to clean out the window frame, removing the remaining fragments of glass. If the window is covered with tinting film, the glass will usually hold together even after it is broken. Use the back section of the center punch,

not your hand, and chip out a small section of glass in the corner of the window. Make the opening large enough to fit a gloved hand inside. Now, insert your gloved hand and grab hold of the glass. Pull the glass up and out, taking the entire section of glass out of the frame **Figure 9-6 ▶**. If there is a large amount of accumulated glass fragments on the ground, use a broom or shovel to sweep it under the vehicle or place the debris outside of the hot zone in a designated debris pile.

Tempered safety glass goes through a process where the glass is heated and then quickly cooled; this process gives the glass its strength and resistance to impacts. When tempered

Figure 9-4 The spring-loaded center punch is the most basic and common of all glass removal tools.

Figure 9-5 The technical rescuer must follow safety rules when utilizing a spring-loaded center punch to break tempered glass. This technique prevents the rescuer from accidentally putting a hand through the window as the glass is broken.

Figure 9-6 Tinting film makes it very easy to remove the entire section of glass when it is broken.

glass is fractured, it is designed to break into small pieces, with no long shards. Tinting on tempered glass can be an added benefit for the technical rescuer because when the glass is broken, the tinting normally holds all of these small fragments together,

making the glass easy to remove and dispose of. Adding tape strips or large adhesive sheets to simulate the effect of window tinting may be just as effective, but remember that time is a critical factor in patient care and such actions normally take more time. There are products that come predesigned in sections of adhesive film that are placed on the outside of the glass prior to breaking it. This film is designed to hold the glass fragments together similar to tinting film. Once again, the time expended for the application process must be considered.

Rescue Tips

Always use a hand tool such as a Halligan bar or a four-by-four to clean out the remaining tempered glass fragments in a window frame.

To break tempered glass using a spring-loaded center punch, follow the steps in **Skill Drill 9-2 ▶**:

1. Don appropriate PPE, including mask and eye protection.
2. Assess the scene for hazards and complete the inner and outer surveys.
3. Stabilize the vehicle.
4. Ensure that the victim is properly protected from flying glass particles.
5. Warn personnel with the verbal command "Breaking glass!"
6. Using the window farthest from the victim, place the palm of your hand facedown against the lower corner of the window and frame, with your index finger and thumb facing upward.
7. Position and rest the body of the tool in the ridge section of the palm (V-section of your hand) between your index finger and thumb. (**Step 1**)
8. With the point of the center punch directly on the glass, apply forward pressure on the center punch until the spring is activated and the glass breaks. (**Step 2**)
9. Once the glass breaks, remove remaining glass segments around the window frame using a tool, such as a Halligan bar or a short section of four-by-four. Follow this procedure until all glass has been removed from the frame. (**Step 3**)

The glass handsaw is a manually operated glass removal tool that has several unique features built into it, including a hand guard, a hollow slot for a center punch, and a notched section that fits over the top lip of the glass, which, when turned, causes the glass to fracture. The glass handsaw is capable of breaking tempered or cutting laminated glass.

When the glass saw is utilized to break tempered glass, a center punch is set inside the hollow slot located in the handle of the tool. The hand guard is then used as a bracing mechanism by placing it against the outer steel section of the window frame with the point of the center punch placed against the corner of the window. The tool is rolled forward, plunging the center punch into the glass. The glass handsaw can then be used to clean the remaining window frame of glass fragments.

Skill Drill 9-2

Breaking Tempered Glass Using a Spring-Loaded Center Punch

1 Don appropriate PPE. Assess the scene for hazards and complete the inner and outer surveys. Stabilize the vehicle. Ensure that the victim is properly protected from flying glass particles. Warn all personnel and victims that you will be breaking the glass. Using the window farthest from the victim, place the palm of your hand facedown against the lower corner of the window and frame, with your index finger and thumb facing upward. Position and rest the body of the tool in the ridge section of the palm (V-section of your hand) between your index finger and thumb.

2 With the point of the center punch directly on the glass, apply forward pressure until the spring is activated and the glass breaks.

3 Remove all loose tempered glass fragments from around the window frame using a hand tool or a short section of four-by-four.

To break tempered glass using a glass handsaw, follow the steps in **Skill Drill 9-3 ▶** :

1. Don appropriate PPE, including mask and eye protection.
2. Assess the scene for hazards and complete the inner and outer surveys.
3. Stabilize the vehicle.
4. Ensure that the victim is properly protected from flying glass particles.
5. Place a center punch inside the hollow slot located in the handle of the glass handsaw. Utilizing the hand guard as a bracing mechanism, place the guard against the outer steel section of the window frame using the window farthest from the victim. (**Step 1**)
6. Warn all personnel and victims that you will be breaking the glass by announcing, "Breaking glass!"
7. With the point of the center punch placed against the corner of the window and the guard safety braced against the frame, roll the tool forward, plunging the center punch into the glass.

8. Remove all loose tempered glass fragments from around the window frame using the glass handsaw. (**Step 2**)

■ Laminated Safety Glass

Laminated safety glass is created by heating a layer of clear plastic-type film between two layers of plate glass. This process holds the two pieces of glass together. This type of glass is used for windshields and can sometimes be used in rear and side windows as well. When the window is broken, the plastic-type film between the two layers of glass prevents big shards of glass from flying in on the occupant.

As discussed, laminated glass found in a common passenger vehicle is generally sealed and held in place with a mastic-type adhesive that requires a cutting action of the glass itself to be properly removed. Attempting to scrape this adhesive seal out is difficult at best and extremely time consuming; doing so should be avoided. Procedures for removing this type of gasket-set windshield will be discussed in Chapter 11, *Commercial/ Heavy Vehicles*.

NFPA 1006, (10.1.7)

Skill Drill 9-3

Breaking Tempered Glass Using a Glass Handsaw

1 Don appropriate PPE. Assess the scene for hazards and complete the inner and outer surveys. Stabilize the vehicle. Ensure that the victim is properly protected from flying glass particles. Place a center punch inside the hollow slot located in the handle of the glass handsaw. Utilizing the hand guard as a bracing mechanism, place the guard against the outer steel section of the window frame using the window farthest from the victim.

2 Warn all personnel and victims that you will be breaking the glass. With the point of the center punch placed against the corner of the window and the guard safety braced against the frame, roll the tool forward, plunging the center punch into the glass. Remove all loose tempered glass fragments from around the window frame using the glass handsaw.

Accessing a victim through a windshield is not a common procedure, but this approach could be performed with a vehicle resting on its side. The main purpose for removing the windshield glass is for safety. Utilizing a technique such as scoring the bottom of the windshield with an axe and removing part of the windshield is not a recommended practice. This technique is not as effective or safe as completely removing the entire section of glass. It is highly recommended for safety purposes and consistency to remove the entire windshield by cutting it out with a glass handsaw or reciprocating saw; it takes only seconds to complete the procedure.

The technique to remove the front laminated windshield is best accomplished using two technical rescuers positioned on opposite sides of the vehicle. Sawing through laminated glass requires the use of respiratory protection, such as an N-95 filtered nose and mouth filter/respirator, eye protection, and protection of exposed areas around the face and neck Figure 9-7 ▶. The technical rescuer will use the saw blade of the tool to cut out the glass. The teeth of the blade are set at an inward angle, which will throw a large amount of glass particles back at the technical rescuer on each upward and dragging stroke. The cutting action of the blade causes microparticles of glass to float through the air, potentially damaging the respiratory system

of an unprotected rescuer who is in the immediate proximity or downwind. A filtered nose and mouth mask rated N-95 or higher is specially designed to block fine particles such as these. Also, prior to cutting the windshield, attempt to cover the patient with a blanket that sufficiently protects against glass fragments and particles.

The proper technique for removing a windshield using a glass handsaw can be accomplished by following the steps in Skill Drill 9-4 ▶:

1. Don appropriate PPE, including mask, eye, and respiratory protection.
2. Assess the scene for hazards and complete the inner and outer surveys.
3. Stabilize the vehicle. (Step 1)
4. Ensure that the victim and the other rescuer inside the vehicle are properly protected using a blanket to cover them from flying glass particles.
5. Before the glass removal procedure is started, look inside the vehicle to see if the rearview mirror is attached to the back of the windshield glass. If it is attached, it will need to be removed. If the mirror remains in place, the striking action of the spiked end of the glass tool (required to make

Figure 9-7 The technical rescuer must be wearing full PPE including eye and respiratory protection to protect against flying glass particles and dust.

an access point for the blade) onto the glass will generally dislodge the rearview mirror at a high velocity, potentially hitting the victim or the technical rescuer inside.

6. Instead, using the spiked end of the glass handsaw, make a hole at the center top of the windshield. (**Step 2**)

7. Using the spiked end of the glass handsaw, make a hole at the center bottom of the windshield. (**Step 3**)

8. The holes should be large enough to fit the blade of the tool into them.

9. Insert the blade into the top hole and begin a steady pulling/dragging of the blade rather than short cuts. With one continuous cut and without removing the blade, cut a line all the way around the top of the roof line to just inside the A-post, working the tool to the bottom of the windshield. Try not to saw the glass or make short strokes because the continual downward insertion of the blade pushes glass fragments into the passenger compartment and onto the victim and rescuer.

10. Pull the blade out and insert it into the center bottom hole to finish the cut on the bottom section of the windshield glass. Because the windshield and dashboard meet at a unique angle, the blade of the glass handsaw must be turned sideways to prevent the blade from hitting the dash when cutting.

11. With the cut completed, hand the tool to the technical rescuer on the opposite side of the vehicle so he or she can complete the cut on that side.

12. Remember to support the windshield section that you just cut to prevent it from falling in on the victim once the other side is cut. When all the glass has been completely cut, push the loose windshield from inside the vehicle outward, toward the hood. Fold the glass in half on top of itself and place it in a debris pile outside of the hot zone. (**Step 4**)

Removing the Windshield Using the Reciprocating Saw

With the reciprocating saw, the high-speed reciprocating action of the blade makes the process of cutting out laminated windshield glass very fast, leaving a straight edge on the glass. This occurs because the heat of the blade as it passes through the glass melts the lamination that is between the two layers of glass. The cutting process is very fast and different than using a glass handsaw; be prepared for the difference in speed, and ensure that everyone, including the patient and the rescuer inside the vehicle, are fully protected and aware of your action.

A reciprocating saw can be very effective for cutting through laminated glass when used correctly and following a plan of action. Always plan a few steps ahead. If the plan is to incorporate the removal of the roof and the windshield, then a reciprocating saw would be highly effective. The ideal situation would be to incorporate cutting through both A-posts with cutting out the windshield as opposed to using the reciprocating saw to cut out only the windshield. The technical rescuer must take into account the increased setup time to hook up an electric reciprocating saw versus using another glass tool such as a glass handsaw, which requires no setup. Another option would be to use a battery-operated reciprocating saw, which minimizes the time of deployment to operation. The complexity of the incident and the number of skilled technicians on scene should determine the choice of action.

When using a reciprocating saw to cut into laminated glass, there are two schools of thought when choosing the TPI rating for the saw blade. (TPI measures the number of teeth on the blade per inch.) The first option is to use a bimetal blade with a high TPI, such as 14. This will produce a fine cut and minimize the chance of large particles of glass flying out. However, the high TPI ratio will produce and throw a large amount of fine glass particles, mostly in a powderlike cloud. The second option is to use a bimetal blade with a variable TPI of approximately 6 and 9. However, utilizing the larger size teeth at a 6–9 TPI ratio will produce and throw large particles of glass everywhere. When utilizing a reciprocating saw to cut out laminated glass, regardless of the blade used, it will produce large amounts of glass fragments, particles, and dust comprised of glass particulate, which can potentially damage the respiratory system and eyes when unprotected. Proper respiratory and full eye protection is a must.

Another issue of concern is the potential for an air bag cylinder to be located in the A-post **Figure 9-8 ▶**. An air bag cylinder can be visualized by pulling back the plastic or fabric molding around the post with your gloved hand or prying it back with a large flat-head screwdriver or other type of small prying tool. If an air bag cylinder is located, cut in an area that avoids that cylinder. If an air bag cylinder is located, cut in an area that avoids the cylinder (whether it is high or low). The actual nylon air bag that extends out of the cylinder and up around the roof

NFPA 1006, (10.1.7)

Skill Drill 9-4

Removing the Windshield Using a Glass Handsaw

1 Don appropriate PPE. Assess the scene for hazards and complete the inner and outer surveys. Stabilize the vehicle.

2 Ensure that the victim and the other rescuer inside the vehicle are properly protected from flying glass particles. Remove the rearview mirror if it is still in place. Using the spiked end of the glass handsaw, make a hole at the center top of the windshield.

3 Using the spiked end of the glass handsaw, make a hole at the center bottom of the windshield.

4 Insert the blade into the top hole and begin a steady pulling/dragging of the blade all the way around the top of the roof line to just inside the A-post, working the tool to the bottom of the windshield. Pull the blade out and insert it into the center bottom hole to finish the cut on the bottom section of the windshield glass. Hand the tool to the technical rescuer on the opposite side of the vehicle so he or she can complete the cut on that side. Support the windshield section that you just cut to prevent it from falling in on the victim once the other side is cut. Push the loose windshield from inside the vehicle outward, toward the hood. Fold the glass in half on top of itself and place it in a debris pile outside of the hot zone.

rail can be cut. Avoid the cylinder and attached electrical components. Remember the saying, "If an air bag cylinder is found, then cut around!" All personnel on scene must be made aware of the air bag location. Once you are through the post, the rest of the windshield can be quickly cut, removed, and placed in a debris area outside the hot zone.

Remember that before the windshield is cut, another technical rescuer with full PPE, including respiratory protection, must be placed in position to support the windshield from falling in on the victim when the cut is made. With the windshield removed, the technical rescuer will proceed to the other A-post and follow the same steps just described. Also remember that

Figure 9-8 If an air bag cylinder is located, cut in an area that avoids that cylinder.

when you cut through any vehicle posts, there must be enough personnel to assist with supporting the roof to prevent it from collapsing on the victim. Although the B-posts are still intact, always yield to safety first. This may seem like an obvious statement, but this is a common occurrence because all of the focus is often on the cut and removal of the roof and not on supporting the roof once it is cut. One factor you can count on to be present 100% of the time is gravity. Proper procedures for performing a roof removal utilizing a reciprocating saw will be discussed in Chapter 10, *Alternative Extrication Techniques*.

The proper technique for removing a windshield using a reciprocating saw can be accomplished by following the steps in **Skill Drill 9-5 ▸**:

1. Don appropriate PPE, including mask, eye, and respiratory protection.
2. Assess the scene for hazards and complete the inner and outer surveys.
3. Stabilize the vehicle.
4. Ensure that the victim and the other rescuer inside the vehicle are properly protected using a blanket to cover them from flying glass particles.
5. If the A-posts will be incorporated into the cutting process, then examine the posts for any air bag cylinders by removing or pulling back the moldings. If air bag cylinders are found, then cut around. (**Step 1**)
6. Utilizing a reciprocating saw (battery or electric), create a purchase point to start the cut. This can be accomplished by starting the cut at the top of the A-post or by creating a hole in the glass with the spiked end of a Halligan tool or the spiked end of a glass handsaw. (**Step 2**)
7. Starting at the top of one A-post, make one continuous cut. Cut all the way through the post and continue the cut into the top of the windshield and around the entire windshield frame, staying as close to the edge as possible. (**Step 3**)
8. Ensure that another technical rescuer wearing full PPE including respiratory protection is supporting the cut windshield from falling in on the patient.

9. When all the glass has been completely cut, push the loose windshield from inside the vehicle outward, toward the hood. Fold the glass in half on top of itself and place the section in a debris pile outside of the hot zone.
10. Once the glass has been removed, the other A-post can be cut if a roof removal operation has been called for by the officer in charge. (**Step 4**)

Polycarbonate Windows and Ballistic-Rated Glass

Polycarbonate window material is a thermoplastic material used in vehicle window applications in lieu of traditional vehicle glass, whether tempered or laminated. As discussed in Chapter 3, *Mechanical Energy and Vehicle Anatomy*, polycarbonate is a lighter, durable plastic that is up to 250 times stronger than glass; it is naturally designed to resist direct impacts by any striking tool carried on the apparatus.

When polycarbonate windows are encountered on a vehicle, the best technique to address this type of material is to treat it as a part of the vehicle body, removing the entire section as one piece, whether it is an entire roof or a door structure with the window. Another option is to utilize a prying tool or the hydraulic spreader to pry the polycarbonate window section out of its frame casing, which is held in place by a mastic or similar type of adhesive. If there happens to be a purchase point section caused by a vehicle's crash deformity, you may be able to place the tips of a hydraulic spreader into the opening and release the section containing the polycarbonate material from its casing. Polycarbonate material that has a bend or some type of deformity caused by an impact may be loaded and can release from its casing, either on its own or from the force of a tool. It is designed to conform back to its original shape.

If removing a window section of polycarbonate is a must, then a purchase point can be made by crushing a section of the roof rail or section of accessible metal in which the seam of the window to be removed is seated. This action should cause an opening or purchase point in the area just large enough to insert the tips of the hydraulic spreader and force the window out of its casing. The use of saws, such as a reciprocating saw, is not effective because the heat produced by the friction of the blade will melt the thermoplastic and reseal itself in some areas as the cut is being made. Utilizing a powered rotary saw (K-12 saw) and reversing the carbide tip blade is effective for cutting this type of material, but this saw is impractical to use and time consuming to set up. One tool that shows promise is the smaller hand-held electric 5-inch (127-mm) dual-action circular saw with counter-rotating blades **Figure 9-9 ▸**. This tool, with the action of the counter-rotating blades, throws out the thermoplastic material as it cuts and can cut out a line large enough to prevent the material from resealing itself, as occurs when a reciprocating saw is used.

Ballistic or bullet-resistant glass utilizes multiple layers of tempered glass, laminated material, and polycarbonate thermoplastics, all sandwiched together to the desired thickness. The weight and thickness of the glass will increase depending on each increased level of protection, which can be as high as 3

Skill Drill 9-5

Removing the Windshield Using a Reciprocating Saw

1 Don appropriate PPE. Assess the scene for hazards and complete the inner and outer surveys. Stabilize the vehicle. Ensure that the victim and the other rescuer inside the vehicle are properly protected from flying glass particles. If the A-posts will be incorporated into the cutting process, then examine the posts for any air bag cylinders.

2 Utilize a reciprocating saw (battery or electric) to create a purchase point to start the cut.

3 Begin at the top of one A-post and make one continuous cut all the way through the post; continue into the top of the windshield and around the entire windshield frame, staying as close to the edge as possible.

4 Another technical rescuer should be supporting the weight of the windshield. When all of the glass has been cut, push the loose windshield from inside the vehicle outward, toward the hood. Once the glass has been removed, the other A-post can be cut if needed.

or more inches (76 or more mm), depending on the consumer's design request and the customization of the vehicle to fit and hold the weight of the glass. When ballistic glass is encountered on a vehicle, the best technique is to approach it just as you would polycarbonate material, by treating it as a part of the vehicle body and removing the entire section as one piece, whether it is an entire roof or a door structure.

Rescue Tips

When polycarbonate windows are encountered on a vehicle, the best technique is to treat this material as a part of the vehicle body, removing the entire section as one piece, whether it is an entire roof or a door structure with the window.

Figure 9-9 One tool useful for cutting polycarbonate windows is the hand-held electric 5-inch (127-mm) dual-action circular saw with counter-rotating blades.

Removing the Windshield from a Partially Ejected Victim

Occupants who are not properly restrained by a seat belt system can easily be ejected when a collision occurs. Remember the points brought up in Chapter 3, *Mechanical Energy and Vehicle Anatomy*, regarding the kinetics of energy and the law of motion. A body in motion will remain in motion until acted upon by anther force or object. If the vehicle is traveling at 50 miles per hour (mph) (80 kilometers per hour [kph]) and abruptly stops due to a collision, the unrestrained occupant(s) will continue to travel at 50 mph (80 kph) until stopped by an object or force. In some incidents, the occupant is partially ejected with his or her head or torso breaching the front windshield, trapped in a constricting web of shattered glass, held in place by the laminate. The technical rescuers will be challenged in many ways to render immediate care and release the victim from the glass entrapment. When encountering an occupant who has been partially ejected with his or her head protruding from the windshield, follow the steps in **Skill Drill 9-6 ▶**:

1. Don appropriate PPE, including mask and eye protection.
2. Assess the scene for hazards and complete the inner and outer surveys.
3. Immediately support the victim's head from outside of the vehicle.
4. Stabilize the vehicle.
5. Place a technical rescuer inside the vehicle to give additional support to the head and upper torso of the victim. (**Step 1**)
6. Carefully insert towels around the head and neck of the victim from the direction the windshield was impacted. (**Step 2**)
7. With the head fully supported from both inside and outside the vehicle, a technical rescuer positioned outside of the

vehicle will use trauma shears to slowly cut away sections of laminated glass that is entrapping the victim. (**Step 3**)
8. When enough space has been created to safely remove the victim's head, maintain cervical support and properly immobilize and package the victim for removal from the vehicle. (**Step 4**)

Rescue Tips

When treating a victim who has been partially ejected through the windshield, slowly insert towels around the head and neck of the victim from the direction the windshield was impacted.

Using Hydraulic Rescue Tools to Gain Door and Roof Access

If technical rescuers cannot gain access by the previously mentioned techniques, they must use heavy extrication tools to gain access to the victim. Hydraulic rescue tools for extrication have been around for several decades, originating from the auto racing industry and quickly becoming the staple for vehicle extrication across the world. As advances in technology continue to grow, new tools emerge that are faster and more powerful, designed to make vehicle extrication for the technical rescuer less complicated and cumbersome. Our knowledge in the use of these tools needs to also evolve and grow with new technology. Today's technical rescuer must shed the old style of spreading and tearing apart vehicles and look at the extrication process through the eyes of a surgeon, dissecting sections and fully understanding the dynamics of moving metal. This section is about simplifying things, "working smarter," and accomplishing our goal of removing trapped victims in the safest, fastest, and most efficient way.

Making a purchase point is the process of gaining an access area to insert and better position a tool for operation. For example, a purchase point may be needed to expose the locking/latching mechanism or hinges of a door enough to insert a hydraulic cutter; this technique is known as **expose and cut**. A hydraulic spreader is the best hydraulic tool for making a purchase point **Figure 9-10 ▼**.

Figure 9-10 A hydraulic spreader is the best hydraulic tool for making a purchase point.

NFPA 1006, (10.1.7)

Skill Drill 9-6

Removing the Windshield from a Partially Ejected Victim

1. Don appropriate PPE. Assess the scene for hazards and complete the inner and outer surveys. Immediately support the victim's head from outside of the vehicle. Stabilize the vehicle. Place a technical rescuer inside the vehicle to give additional support to the head and upper torso of the victim.

2. Carefully insert towels around the head and neck of the victim from the direction the windshield was impacted.

3. A technical rescuer positioned outside of the vehicle will use trauma shears to slowly cut away sections of laminated glass that is entrapping the victim.

4. When enough space has been created to safely remove the victim's head, maintain cervical support and properly immobilize and package the victim for removal from the vehicle.

One traditional technique that is not effective is to use a Halligan bar and flat-head axe to create a purchase point to gain access to the door in order to insert a hydraulic tool. This method requires two personnel and multiple tools to accomplish what one technical rescuer can accomplish using the hydraulic spreader to create an even larger, more effective opening Figure 9-11 ▶ . The goal again is to "work smarter, not harder."

Most hydraulic cutters today are rated to cut through hinges and locking/latching mechanisms located in vehicles; however, it is a good idea to check with the manufacturer to see if the tool that your organization uses is rated to do so. If the manufacturer does not recommend this procedure for the tool or your organization is absolutely against cutting into this type of material, then some alternative steps can be utilized, which are described in the next section.

Figure 9-11 A purchase point can be created anywhere in a vehicle where a seam exists and the metal surrounding this area can be crushed or spread, such as the rocker panel section around the door seam.

Figure 9-12 When performing the vertical spread, keep the tips of the spreader in the general vicinity of the steel backer or reinforcement plates directly behind the latch.

Door Access from the Latch Side: The Vertical Spread

Trying to release a door from its frame can sometimes be very difficult. Depending on the level of intrusion and the integrity and type of the material that is being spread, this process can challenge the best technical rescuer.

An older traditional method mentioned in the previous section would have two technical rescuers create a purchase point using a striking and prying tool to bend the sheet metal at the edge or seam of the door near the latching/locking mechanism, just below the door handle. Today, hydraulic tools are likely to be used. The problem occurs when a hydraulic spreader is inserted into the purchase point. When the spreader is opened, the action causes the metal above and below the area of spreading to collapse around the tool, limiting the opening and eliminating the possibility of inserting a hydraulic cutter to cut the latching mechanism. This is an example of why a full understanding of how metal reacts and moves by the force of a hydraulic spreader is so important. This technique would cause the technical rescuer to continue spreading the door until it tears off the latch, which is exactly what you want to avoid doing. Forcing a door off the latch or hinge with the shear power of the tool is a very dangerous and obsolete technique. Once the integrity and strength of the door frame are lost because of poor spreading technique, there is no longer a vantage point to push from and the metal in this area becomes weakened and starts to tear and shred, much like an aluminum can would, with the technical rescuer fighting the door. This wastes valuable time and causes additional stress and physical exhaustion for the technical rescuer utilizing the tool. There are some techniques that require the use of tearing metal through force, but this is not one of them. It is critical that you fully understand how metal reacts and moves by the force of a hydraulic spreader. Expose and cut!

Expose the latch with a **vertical spread**, insert the cutter, and cut the latching mechanism. If the cutter is not rated for cut-

ting the latching mechanism, then, as an alternative step, continue to work the spread vertically down and outward, until the door can be rolled off of the latch with controlled movements by working the tips of the spreader around the latch. There are heavier gauge steel backer or reinforcement plates directly behind the latch, as well as the Nader pin or U-bolt, which will prevent the metal from shredding; keep the tips of the spreader in this general vicinity **Figure 9-12 ▲**. This technique is a faster and smarter way of gaining access, requiring the technical rescuer to expend less physical energy and limiting the possibility of the door being violently released. Remember that you control the technique; the technique does not control you! Consider the basic design of the door frame; the best method of exposing the latch and creating enough room for the cutter blades is to start the spread at the top of the door and work the window frame, also known as a **D-ring**, down and out, away from the latching mechanism. The D-ring is a generic term used to describe the window frame of the door. The term came from the idea that the frame of the window resembles the letter "D" on its side.

Rescue Tips

Remember that you control the technique, the technique doesn't control you!

To release a door from its frame or perform the vertical spread, follow the steps in **Skill Drill 9-7 ▶**:

1. Don appropriate PPE, including mask and eye protection.
2. Assess the scene for hazards and complete the inner and outer surveys.
3. Stabilize the vehicle.
4. Ensure that the victim and the other rescuer inside the vehicle are properly protected using a blanket to cover them from flying glass particles.
5. Remove all vehicle glass utilizing the appropriate technique demonstrated in this chapter. **(Step 1)**

NFPA 1006, (10.1.7)

Skill Drill 9-7

Releasing a Door from Its Frame or Performing the Vertical Spread

1. Don appropriate PPE. Assess the scene for hazards and complete the inner and outer surveys. Stabilize the vehicle. Ensure that the victim and the other rescuer inside the vehicle are properly protected from flying glass particles. Remove all vehicle glass utilizing the appropriate technique.

2. Place the hydraulic spreader vertically in the door's window, close to where the door handle is normally located. Position the tips of the spreader with the bottom arm resting on the door's window sill and the top arm in position to catch the underside of the roof rail when fully opened. Push off of the roof rail.

3. As the tool starts to open, adjust the positioning of the arms appropriately by lifting up on the back end of the spreader to maximize the spreading capacity of the tool. Do not fail to make this adjustment.

4. With the tool in a semivertical position, push the window frame (D-ring) out and downward out of the way.

5. When the top corner of the door opens, reposition the spreader by dropping the tips of the tool down into this opening with the tool positioned almost vertically. The position of the tool at this point is critical for a successful outcome. Spread the metal outward, making the door start to fold out and down. At this point, the latching mechanism should be visible, or it should be fully accessible to position the hydraulic cutter to cut. If the hydraulic cutter is not rated to cut this type of material, an alternative step is to continue to work the spread vertically down and outward with controlled movements by working the tips of the spreader around the latch until the door can be rolled off of the latch.

6. Place the hydraulic spreader vertically in the door's window, close to where the door handle is normally located. Position the tips of the spreader with the bottom arm resting on the door's window sill and the top arm in position to catch the underside of the roof rail when fully opened. Push off of the roof rail; do not push off of the door's window frame (D-ring). This action will only cause the window frame to tear off and compromise the technique. The tool operator should be positioned on the side of the tool opposite the door swing; this is a defensive position of safety just in case the door is unexpectedly jarred open by the tool's force. **(Step 2)**

7. With the hydraulic spreader in position and as the tool starts to open, adjust the positioning of the arms appropriately by lifting up on the back end of the spreader to maximize the spreading capacity of the tool. Do not fail to make this adjustment; it could cause the tool to move forcefully inside the vehicle, negating any spreading ability and potentially causing further injury to a victim. Proper positioning of the hydraulic spreader will cause the window frame to buckle out, giving the tool operator a purchase point to start the second part of the technique. **(Step 3)**

8. Position the tool in a semivertical position where the tips of the spreader are placed in the purchase point area. Note that the initial positioning/angle of the spreader will vary depending on the location of the purchase point. The angle and spreading action will cause the top corner of the door (where the window frame meets the door) to open. Push the window frame (D-ring) out and downward out of the way. **(Step 4)**

9. When the top corner of the door opens, reposition the spreader by dropping the tips of the tool down into this opening with the tool positioned almost vertically. The position of the tool at this point is critical for a successful outcome. Spread the metal outward, making the door start to fold out and down. At this point, the latching mechanism should be visible, or it should be fully accessible to position the hydraulic cutter to cut. As discussed, if the hydraulic cutter is not rated to cut this type of material, an alternative step is to continue to work the spread vertically down and outward with controlled movements by working the tips of the spreader around the latch until the door can be rolled off of the latch. There are heavier gauge steel backer or reinforcement plates that are directly behind the latch, as well as the Nader pin or U-bolt, which will prevent the metal from shredding; keep the tips of the spreader in this general vicinity.

10. A second technical rescuer should stand with a hydraulic cutter in hand, powered up with the blades fully opened and ready to cut the latching bolt, which encompasses the latching mechanism. Once the latching mechanism has been cut and the door released/opened, the technical rescue team can transition into the next operation, which will be determined by the company officer or incident commander. **(Step 5)**

The importance of crew members working in tandem, understanding the technique that is being performed, and being prepared with the appropriate tools in hand and ready for action, cannot be stressed enough. As trained technical rescuers, all of these techniques need to flow and transition seamlessly without needless interruptions such as trying to locate a tool or waiting for the blades of a hydraulic cutter to be opened.

Door Access from the Hinge Side: Front Wheel Well Crush

Gaining entry into the interior of a vehicle by removing a door from the hinge side is not a common procedure because removing the door in this fashion goes against the natural swing of the door, making it very difficult to remove once the procedure progresses to the latch side. However, this technique is needed for certain crash scenarios.

One such hypothetical crash scenario involves the front of a vehicle impacting a wall. The impact in this scenario crushes the front end of the vehicle, compressing both doors, which will now require forcible entry using hydraulic tools. The vehicle is fully equipped with supplemental restraint devices and two front and side-impact air bags located in both driver and passenger doors. The two front air bags deploy as designed, but the two side air bags remain live because there is no direct impact to any of the side-impact sensors. Any attempt at a door removal by the latch side could potentially trip the door sensors and activate the side air bag, deploying it on the occupants. One possible solution in this scenario is to enter the door from the hinge side **Figure 9-13 ▼**, cutting the hinges and then pulling the door back and away from the occupant and releasing it from the latch side. This is not a perfect science and an accidental air bag deployment may occur, but it provides the technical rescuer a viable option, including disconnecting the vehicle's 12-volt DC battery. The best option would be to enter the vehicle by removing the roof, but if the occupants are trapped under the dash, then the doors will have to come off anyway if a dash displacement technique is to be applied.

One of the more favorable techniques for gaining access to door hinges from the outside is a **wheel well crush technique**

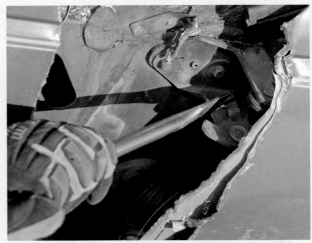

Figure 9-13   When door removal is impossible from the latch side due to air bag sensors, one possible solution is to enter the door from the hinge side.

using the hydraulic spreader and cutter. Using the hydraulic spreader to crush the wheel well creates a purchase point at the door's seam, allowing the hydraulic spreader space to get in and expose the hinges. The hydraulic cutter can then be inserted so the hinge can be cut. To perform the wheel well crush technique, follow the steps in **Skill Drill 9-8**:

1. Don appropriate PPE, including mask and eye protection.
2. Assess the scene for hazards and complete the inner and outer surveys.
3. Stabilize the vehicle.
4. Ensure that the victim and the other rescuer inside the vehicle are properly protected using a blanket to cover them from flying glass particles.
5. Remove all vehicle glass utilizing the appropriate technique. (**Step 1**)
6. Disconnect the vehicle's 12-volt battery if the engine compartment is accessible.
7. Prepare to crush the wheel well by locating an area just between the strut tower and the dash/firewall section. Begin by opening the arms of the spreader and placing the tip of the top arm on the hood or top section of the wheel

well of the vehicle, making certain that the tip of the top arm is flush with the hood. It should not be positioned at an angle.

8. Ensure that the bottom arm, as it rises upward, clears the tire and strut coil, falling into position under the wheel well. When done correctly, this will seem like an optical illusion where the bottom arm of the spreader appears to be the only arm moving with the top arm level and stationary. (**Step 2**)
9. As the arms of the spreader lock onto the wheel well, the tool will want to slide off the angle and try to conform with the slope of the wheel well. Hold the tool in position, as described in the preceding steps, to prevent this from occurring.
10. The arms of the spreader will form a crease in the wheel well and upper rail area, causing the panel to buckle outward at the door seam where the panel and door meet, which in turn exposes the door's hinges. This creates a purchase point for the hydraulic spreader to create enough space around the door's hinges to insert a hydraulic cutter so the hinges can be cut. (**Step 3**)

Skill Drill 9-8

NFPA 1006, (10.1.7)

The Wheel Well Crush Technique

1. Don appropriate PPE. Assess the scene for hazards and complete the inner and outer surveys. Stabilize the vehicle. Ensure that the victim and the other rescuer inside the vehicle are properly protected from flying glass particles. Remove all vehicle glass utilizing the appropriate technique.

2. Disconnect the vehicle's 12-volt battery if the engine compartment is accessible. Locate an area just between the strut tower and the dash/firewall section. Open the arms of the spreader and place the tip of the top arm on the hood or top section of the wheel well of the vehicle, making certain that the tip of the top arm is flush with the hood. Ensure that the bottom arm, as it rises upward, clears the tire and strut coil, falling into position under the wheel well.

3. As the arms of the spreader lock onto the wheel well, prevent the tool from sliding off the angle and trying to conform with the slope of the wheel well. Hold the tool in position. The arms of the spreader will form a crease in the wheel well and upper rail area, causing the panel to buckle outward at the door seam where the panel and door meet, exposing the door's hinges. This creates a purchase point for the hydraulic spreader to create enough space around the door's hinges to insert a hydraulic cutter so the hinges can be cut.

The Complete Side Removal Technique: The Side-Out

There are four basic types of impacts that a vehicle can sustain during a collision—a frontal impact, a side impact, a rear impact, or a roll-over/roof impact. Side impacts have a higher rate of occurrence according to yearly statistics compiled by the Department of Transportation (DOT). The **side-out technique** is designed specifically for four-door vehicles involved in a side-impact collision. The technique allows technical rescuers to remove the front and rear door as one unit on the same side of a four-door vehicle. This technique was first referenced in *Fire Engineering* magazine in November 1999 and has made a tremendous impact for fire rescue agencies worldwide by dramatically reducing the time it takes to gain access through the doors of a four-door vehicle involved in a side-impact collision.

Understanding what occurs to the body structure of a vehicle after it has been involved in a side-impact collision is vital to comprehending the effectiveness of the side-out technique. The intrusion that occurs from a side-impact collision causes the entire door frame to fracture or partially fracture, which also causes the directional force of both doors at the B-post to move inward toward occupants **Figure 9-14 ▼**. If the technical rescuer attempts to spread the driver's side door at the latching mechanism utilizing the hydraulic spreader, he or she will only cause the B-post, including both front and rear doors, to continue to move inward and collapse on the victim. This occurs because the directional force of the fracture caused by the impact is pushing inward and wants to continue in that direction. Due to the force applied to the metal, the metal will move or seek to find the least path of resistance—in this case inward.

The correct action is to push the doors and B-post out and away from the victim. This technique utilizes the natural swing or directional movement of the doors and pushes or forces the doors and B-post outward, away from the occupant. This technique is best accomplished when two technical rescuers are working in tandem; one technical rescuer should carry out the hydraulic spreader assignment, and the other should carry out the hydraulic cutter assignment. The technique begins at the rear door and progresses forward.

Figure 9-14 A side-impact collision causes the entire door frame to fracture or partially fracture, which also causes the directional force of both doors at the B-post to move inward toward occupants.

To perform the side-removal/side-out technique, follow the steps in **Skill Drill 9-9 ▶**:

1. Don appropriate PPE, including mask and eye protection.
2. Assess the scene for hazards and complete the inner and outer surveys.
3. Stabilize the vehicle.
4. Ensure that the victim and the other rescuer inside the vehicle are properly protected using a blanket to cover them from flying glass particles.
5. Remove all vehicle glass utilizing the appropriate technique. **(Step 1)**
6. Disconnect the vehicle's 12-volt battery if the engine compartment is accessible.
7. Cut the seat belt straps. Inspect all posts and roof rail areas for the possibility of air bag cylinders or seat belt pretensioning systems. If any are found, cut around them.
8. Release the rear door from the latching mechanism utilizing the vertical spread technique described in this chapter.
9. Open the door and position the cutter at the bottom of the B-post, just above the area where the B-post meets the rocker panel. Make a small relief cut into the bottom of the post. Do not make a sectional or pie cut to get the cutter blades in deeper; it is not necessary, does not enhance the procedure, and wastes valuable time. Also, do not make the mistake of positioning the blades incorrectly and accidentally cutting into the rocker panel; if the integrity of the area is compromised, the rocker panel and floor area will tear away instead of the B-post, posing a critical failure of the technique. **(Step 2)**
10. When the relief cut has been completed in the B-post, move the cutter directly up toward the top of the B-post and roof rail area. Make an upward-angled cross-cut on both sides of the top section of the post and roof rail section. This cross-cut removes the jagged stump that would be left by just making one cut across the post. **(Step 3)**
11. As the technical rescuer on the cutter is completing the cross-cut section on the opposite side of the B-post, the technical rescuer who is operating the hydraulic spreader should start to position the tool to spread the B-post off the rocker panel. **(Step 4)**
12. The initial position of the hydraulic spreader is at the area where the relief cut was made at the bottom of the B-post. The objective is to angle the hydraulic spreader in a general 40- to 45-degree range, where the tip of the bottom arm is placed on the rocker panel and the tip of the top arm is angled near the bottom section of the rear door. **(Step 5)**
13. Once the spreader is in position, use cribbing to thoroughly shore up under the rocker panel in the area where the bottom arm/tip of the hydraulic spreader rests. The cribbing must be inserted after the spreader has been positioned in place because the resting area of the bottom arm of the spreader will vary each time. The placement of the cribbing needs to be precise. It is very important that this area is fully shored because once the hydraulic spreader is engaged, the tip of the tool can easily penetrate through the hollow rocker panel, tearing out the floor section. This will cause a critical failure of the technique. **(Step 6)**

14. Before the spreader is engaged, as a safety measure, attach strapping, rope, or webbing to the rear door in order to apply a constant slight outward and upward pull, assisting with the movement of the door from a safe distance. Never butt up or brace against a door that is being spread; the force of the hydraulic spreader can cause the door to violently release, driving it into the rescuer. As the spreader is opened and the B-post starts to tear away from the panel, the angle of the spreader may need to be readjusted to gain better leverage. If the bottom of the B-post is spot-welded to the rocker panel, it should tear off fairly easily once force is applied by the opening of the spreader. If the B-post is molded as part of the rocker panel and force is applied, tearing can occur, separating the rocker panel in two sections. If this occurs, simply cut through the remaining small section of metal using the hydraulic cutter. (**Step 7**)

NFPA 1006, (10.1.7)

Skill Drill 9-9

The Complete Side-Removal/Side-Out Technique

1 Don appropriate PPE. Assess the scene for hazards and complete the inner and outer surveys. Stabilize the vehicle. Ensure that the victim and the other rescuer inside the vehicle are properly protected from flying glass particles. Remove all vehicle glass utilizing the appropriate technique.

2 Disconnect the vehicle's 12-volt battery if the engine compartment is accessible. Cut the seat belt straps. Cut around any air bag cylinders or seat belt pretensioning systems. Release the rear door from the latching mechanism utilizing the vertical spread technique. Open the door and position the cutter at the bottom of the B-post, just above the area where the B-post meets the rocker panel. Make a small relief cut into the bottom of the post.

3 When the relief cut has been completed in the B-post, move the cutter directly up toward the top of the B-post and roof rail area. Make an upward-angled cross-cut on both sides of the top section of the post and roof rail section.

4 As the technical rescuer on the cutter is completing the cross-cut section on the opposite side of the B-post, the technical rescuer who is operating the hydraulic spreader should start to position the tool to spread the B-post off the rocker panel.

5 The initial position of the hydraulic spreader is at the area where the relief cut was made at the bottom of the B-post.

6 Once the spreader is in position, use cribbing to thoroughly shore up under the rocker panel in the area where the bottom arm/tip of the hydraulic spreader rests.

(Continues)

NFPA 1006, (10.1.7)

Skill Drill 9-9

The Complete Side-Removal/Side-Out Technique *(Continued)*

7 Before the spreader is engaged, as a safety measure, attach strapping, rope, or webbing to the rear door in order to apply a constant slight outward and upward pull, assisting with the movement of the door from a safe distance. If the bottom of the B-post is spot-welded to the rocker panel, it should tear off fairly easily once force is applied by the opening of the spreader. If the B-post is molded as part of the rocker panel and force is applied, tearing can occur, separating the rocker panel in two sections. If this occurs, simply cut through the remaining small section of metal using the hydraulic cutter.

8 Once the doors and B-post release, widen or cut the door section off from the hinges. To widen the door, position the spreader in the front door's jamb around the midpoint area between the hinges. (If there is a swing bar, then position the tool just above that bar.) The technical rescuer will place his or her backside against the inside of the door.

9 As the tool is engaged to open, the rescuer pulls back on the tool, using it as leverage, and slowly pushes backward against the door, bending the door back toward the front wheel and widening the door to create access.

10 Cover any jagged metal that the procedure may have exposed. This procedure provides a large access point to safely remove the victim.

---

15. Once the doors and B-post release, widen the door opening or cut the door section off from the hinges. The fastest and most efficient technique is to widen the door opening, eliminating the time spent on spreading or cutting hinges. To widen the door, position the spreader in the front door's jamb around the midpoint area between the hinges. (If there is a swing bar, then position the tool just above that bar.) Ensure that the second technical rescuer is still assisting with the swing of the door using strapping, rope, or webbing. With the spreader in position, the technical rescuer places his or her backside against the inside of the door. **(Step 8)**

16. As the tool is engaged to open, the technical rescuer pulls back on the tool, using it as leverage as he or she slowly pushes backward against the door. This action bends the door back toward the front wheel, widening the door enough to provide equally sufficient access in half the time

it would take to completely spread or cut the door off from the hinges. **(Step 9)**

17. Use a floor mat or tarp to cover any jagged metal that the procedure may have exposed. This procedure provides plenty of room for the technical rescuers to safely remove the victim. **(Step 10)**

Rescue Tips

Never lean against a door that is being spread. The force of the hydraulic spreader can cause the door to violently release, driving it into you. Tie the door off with webbing and stand at a safe distance.

The side-out technique can be a very fast access technique when performed correctly. This technique, performed by skilled technical rescuers, has been accomplished in many scenarios in less than 5 minutes. The key is that the two technical rescuers operating the hydraulic spreader and cutter thoroughly know the technique, working in tandem by transitioning seamlessly between spreading and cutting. The perfect side-out technique will flow from the rear of the vehicle to the front of the vehicle with each step performed one after the other without interruption.

Removing the Vehicle from the Victim

At times the technical rescuer may encounter a situation where the intrusion of the B-post and doors is so severe that they have almost encapsulated the victim. Any spreading of the doors in this situation, regardless of the location of the tool, will cause the metal to crush down more on the victim. Creating enough room to place tools in the most advantageous position for a technique to be applied may be necessary in extreme entrapments. With the multitude of possible scenarios that can be presented, it is difficult to provide one technique that will be most effective. The technical rescuer will have to use his or her best judgment on removing the vehicle from the victim.

In an entrapment situation, one possible solution is to push the B-post or door section off of the victim from the inside using a hydraulic ram. There are many factors that will influence the correct positioning of the ram; the main factor is accessibility, which again cannot be predicted. The key is to locate an effective base for the hydraulic ram, preferably a telescoping ram, to be positioned and operated from **Figure 9-15 ▼**.

The area inside the vehicle where the transmission hump is located seems to be an effective base to work from, if it can be accessed. If you have gained access to this area, the next step will be to position the base of the ram using the hump of the transmission to push from. As the hydraulic ram is engaged, maneuver the tip of the tool to meet the area that will best force the metal off of the victim; the tip may have to be maneuvered

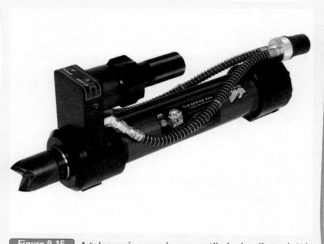

Figure 9-15 A telescoping ram is a versatile hydraulic tool. It is compact when closed and can be used in tight spaces to create a large opening when fully extended.

in several different locations to accomplish this task. Once enough room has been established, the side-out technique can be initiated.

Roof Removal

One of the fastest ways to gain access and extricate a victim is by removing the roof. Victims are often needlessly manipulated by rescuers attempting to remove them through a door, when removing the roof would provide better access and keep the victims in-line as they are packaged and removed without excessive manipulation. Properly packaging patients by placing an immobilization device on them, along with keeping them in-line as they are moved onto a backboard, provides the best patient care. Other benefits of removing the roof include the ability to have multiple rescuers in the vehicle attending to and packaging the patient, the ability to see the entrapment more clearly, and increased maneuverability with less obstruction to operate the tools.

The process of removing a roof can involve multiple tools, such as hydraulic tools, power tools, pneumatic tools, and hand tools. This section will describe the process of removing a roof using hydraulic tools exclusively; the following chapter will discuss the use of additional tools.

When removing a roof, the technical rescuer must expose the interior of each post and the roof rail prior to cutting; this will reveal any problems that can potentially cause injury or halt or delay the process. Some of these problems may include air bag cylinders, boron rods or advanced high-strength steel plates or blanks, seat belt harnesses, seat belt adjustment bars, or seat belt pretension systems. If any of these obstacles are encountered, the easiest solution is to avoid them by cutting above or below the object or area of concern. When cutting a roof post, the blade of the hydraulic cutter should be perpendicular to the object being cut. If the hydraulic cutter blades are not perpendicular to the object being cut, then the blades of the tool will start to bend sideways, potentially causing blade separation, failure of the blades, or multiple cutting attempts. It is recommended that the cuts be made as low as possible on the posts to keep the jagged post ends out of the way. Another option may be to cut where the least amount of metal is showing. In some instances, it may be better to make one single cut high on a post rather than having to make several cuts at a lower section of the post because of the width or thickness of the post. Having to make multiple cuts will take valuable time. If the post ends are a concern, cover them with a precut hose sleeve or a heavy blanket.

Another situation that commonly occurs when cutting a post is tool movement; when the tool is in the beginning stage of the cut, it can start to move forcefully inward toward the vehicle, or outward. This movement is caused by the blades of the tool trying to fracture and cut the metal where it finds the path of least resistance; the entire tool will move and the blades will begin to make their own groove to cut into. To combat this problem and gain full control of the tool, be prepared to respond at the first instance of tool movement. As the tool closes around the post and movement is detected by the tool, the tool will want to forcefully shift away from you. Push or pull forcefully a few

Voices of Experience

One December afternoon, our department was dispatched to a traffic accident on one of our main highways. This particular accident involved a semi-truck that was overturned. We had a full complement of response, including engines and rescue units, with a medical helicopter on standby. This accident occurred on a four-lane highway with tremendous traffic flow. The lane that the accident occurred in was closed for extrication operations and personnel safety.

Upon our arrival, the semi-truck was off the side of the highway, up on an embankment. The cab was severely damaged, with only one occupant—the truck driver. EMS arrived on the scene with ALS components and paramedics on board. After an initial scene size-up and stabilization, a human chain was formed to allow the distribution of medical equipment and extrication tools. The victim was viable, but critical. We notified dispatch to launch the medical helicopter due to the victim's status and the extrication time window. Because of the mechanism of injury and damage to the semi cab, we knew this would be a true challenge.

Knowing that the extrication would be extensive, paramedics established full ALS and IV lines for patient stabilization. We continued to talk to and support the victim. Knowing the predicament, he was capable of communicating his chief complaints. Full immobilization and packaging were conducted, followed by extrication. Once the victim was extricated, we quickly loaded him onto the helicopter.

> **"The points to consider, among many, are that site operations, vehicle stabilization, and victim management, in conjunction with the extrication process, are imperative to the successful outcome of an extrication operation."**

There were many challenges during this incident. We needed to determine who was capable of performing intermediate and advanced life-saving skills. Personnel accountability and equipment considerations were addressed, as were medical training and certification concerns. Due to the mechanism of injury, head, neck, and spinal precautions were a priority, along with the basic ABCs (airway, breathing, and circulation). Bloodborne pathogens were also a primary concern; all personnel wore PPE. Getting the victim to the edge of the highway was a challenge because the semi-truck was positioned on an embankment. The rescuers had to move the victim down the embankment to the waiting helicopter on the roadway. It is in these circumstances that IV lines, 12 leads, and other adjunct equipment can become lost, disconnected, or damaged in the removal process. The human chain that was formed allowed us to move the patient slowly down the embankment to the awaiting helicopter crew. The victim, medical equipment, and prior patient care were reassessed, and the victim was airlifted to a local trauma center. He survived the ordeal.

The points to consider, among many, are that site operations, vehicle stabilization, and victim management, in conjunction with the extrication process, are imperative to the successful outcome of an extrication operation.

Rob Hitt
Glassy Mountain Fire Department
Landrum, South Carolina

times in the direction opposite of the tool's movement and at the same time continue to engage the throttle, applying the cutting action of the blades. These two actions will force the blades of the hydraulic cutter to make a different groove in the metal, giving you full control of the cut and position of the tool.

When addressing the C-posts of a common passenger vehicle, there are several cutting options that the technical rescuer can take. Rear C-posts come in a variety of sizes and shapes. Wide C-posts may require several cuts using a hydraulic cutting tool because of the limited size of the opening created by the blades; a reciprocating saw would normally be the tool of choice for this situation. An option to minimize the number of cuts required on a wide C-post using a hydraulic cutter is to make cuts on both sides of the post and then position the tips of the hydraulic spreader in the cut sections with the tool set perpendicular to the post. With the hydraulic spreader in place, close the tool on the cut; the tips of the spreader will tear through and crush the remaining section of metal. This will open up enough room to make one final cut in the center of the post with the hydraulic cutter. Because the C-post is hollow, make certain that the tips of the spreader carry past the inside wall of the post; doing so will prevent the tips from crushing only the outside wall and not the inside wall, thus avoiding a common error. This technique, when carried out correctly, can eliminate multiple cuts that would otherwise have to be made utilizing a hydraulic cutter.

Rescue Tips

The C-post is hollow.

To remove the roof of an upright vehicle, follow the steps in **Skill Drill 9-10 ▶**:

1. Don appropriate PPE, including mask and eye protection.
2. Assess the scene for hazards and complete the inner and outer surveys.
3. Stabilize the vehicle.
4. Ensure that the victim and the other rescuer inside the vehicle are properly protected using a blanket to cover them from flying glass particles.
5. Remove all vehicle glass utilizing the appropriate technique. (**Step 1**)
6. Disconnect the vehicle's 12-volt battery if the engine compartment is accessible.
7. Expose the interior of each post and the roof rail prior to cutting to determine if there are air bag cylinders, seat belt harnesses, seat belt adjustment bars, or seat belt pretension systems.
8. The proper order of cutting posts on a typical A-B-C-post roof structure will depend on the location of the patient. The last cut should be the post closest to the patient if this is an option. In this scenario we will begin at the A-post. Position several rescuers on either side of the vehicle to assist in supporting the roof. With sufficient personnel on scene to assist in supporting the roof, the order of operation

should take the least possible steps to accomplish the goal at hand.

9. Start the cut on the A-post. The proper cutting angle of the hydraulic cutter should be perpendicular to the object being cut. (**Step 2**)
10. Work toward the B-post. Check again for any seat belt slide bars or reinforcement plates and avoid cutting in this area, if possible. Before the cut is made, another rescuer must be positioned to support the roof when it is cut. (**Step 3**)
11. Once the B-post is cut, begin to cut the rear C-post. If the C-post is wide, make cuts on both sides of the post. (**Step 4**)
12. Position the tips of the hydraulic spreader in the cut sections of the C-post with the tool set perpendicular to the post. With the hydraulic spreader in place, close the tool on the cut; the tips of the spreader will tear through and crush the remaining section of metal, leaving one final cut to be made. Because the C-post is hollow, make certain that the tips of the spreader carry past the inside wall of the post; doing so will prevent the tips from crushing only the outside wall and not the inside wall, thus avoiding a common error. (**Step 5**)
13. Cut the remaining section of metal on the C-post. (**Step 6**)
14. With crew members supporting the roof, move to the opposite side and perform the same steps cutting all of the remaining posts. Before the last cut is made, ensure that the roof is fully supported by personnel on both sides, preferably at all four posts, to prevent any accidental collapse. (**Step 7**)
15. Walk the roof off of the front or back of the vehicle, depending on where the victim is located. To avert any miscommunication and the accidental dropping of the roof on the victim, this step must be a coordinated effort between the crew members supporting the roof; it is best if one person takes the lead and directs the entire movement.
16. Place the roof in an area outside of the hot zone in a designated debris pile. (**Step 8**)

Rescue Tips

Expose the interior of each post and the roof rail prior to cutting to determine if there are air bag cylinders, seat belt harnesses, seat belt adjustment bars, or seat belt pretension systems.

Relocating the Dashboard, Steering Wheel, and Steering Column

Occupants can become trapped under the dash area of the vehicle following a myriad of crash scenarios. There are several techniques that are designed to remove or lift the dash section that is entrapping the victim. Each of these techniques is designed to resolve a specific type of entrapment scenario that the technical rescuer may encounter. Utilizing the wrong

Skill Drill 9-10

Removing the Roof of an Upright Vehicle

1 Don appropriate PPE. Assess the scene for hazards and complete the inner and outer surveys. Stabilize the vehicle. Ensure that the victim and the other rescuer inside the vehicle are properly protected from flying glass particles. Remove all vehicle glass utilizing the appropriate technique.

2 Disconnect the vehicle's 12-volt battery if the engine compartment is accessible. Expose the interior of each post and the roof rail prior to cutting to determine if there are air bag cylinders, seat belt harnesses, seat belt adjustment bars, or seat belt pretension systems. The proper order of cutting posts on a typical A-B-C-post roof structure will depend on the location of the patient. The last cut should be the post closest to the patient if this is an option. In this scenario, we will begin at the A-post. Position several rescuers on either side of the vehicle to assist in supporting the roof. With the hydraulic cutter perpendicular to the object being cut, start the cut on the A-post.

3 Work toward the B-post. Check again for any seat belt slide bars or reinforcement plates, and avoid cutting in this area, if possible. Before the cut is made, another rescuer must be positioned to support the roof when it is cut.

4 Once the B-post is cut, begin to cut the rear C-post. If the C-post is wide, make cuts on both sides of the post.

technique for an entrapment that requires a specific application can potentially complicate the overall operation. Knowing when to use one technique over the other can greatly reduce the time it takes to safely release and remove the victim without causing further harm. No one technique is always 100% effective; every incident will present differently, and in some cases, combining techniques may improve the overall outcome. Problems can always occur, even after a technique has been initiated. A skilled technician will think outside of the box, make adjustments, and overcome any issues that may arise.

Skill Drill 9-10

Removing the Roof of an Upright Vehicle *(Continued)*

5 Position the tips of the hydraulic spreader in the cut sections of the C-post with the tool set perpendicular to the post. With the hydraulic spreader in place, close the tool on the cut.

6 Cut the remaining section of metal on the C-post.

7 With crew members supporting the roof, move to the opposite side and perform the same steps, cutting all of the remaining posts. Ensure that the roof is fully supported by personnel on both sides.

8 Walk the roof off of the front or back of the vehicle, depending on where the victim is located. Place the roof in an area outside of the hot zone.=

■ The Dash Roll Technique

The traditional **dash roll technique** has been the standard technique for displacing the dashboard for many years; it is one of the least technical dash displacement applications to apply in the field. The process involves pushing or rolling the entire front end of the vehicle, which encompasses the dashboard, steering wheel, and steering column, off of the entrapped occupant utilizing hydraulic rams. The technique begins by removing the roof and gaining access to the front doors of both sides of the vehicle. Better maneuverability can be achieved if the doors are removed, but the technique can also be performed with the doors intact in the opened position. This technique can be performed with one ram positioned on the entrapment side or with the combination of two hydraulic rams positioned on both sides of the vehicle for a more symmetrical push. The telescoping ram (about 20 to 60 inches [508 to 1524 mm]) is the most effective type of hydraulic ram for this application because it eliminates the need to premeasure the opening.

When the dash roll technique is performed correctly, the vehicle's entire front end, including the dash, will lift up and forward, hinging from the relief cuts made on both sides. These relief cuts should tear slightly, giving the extra room needed to

Near Miss REPORT

Report Number: 09-394
Report Date: 04/09/2009

Synopsis: We regularly work in concert with EMS workers who are not familiar with the hazards associated with vehicle extrication.

Event Description: Our fire department regularly supports EMT refresher training of local EMS by providing vehicle extrication re-familiarization training and realistic vehicle extrication evolutions. We use donated damaged vehicles and rescue mannequins.

During one such evolution, a rookie fire fighter, working in conjunction with two veteran rescue fire fighters and two EMS responders, was attempting to remove a door from a vehicle using a hydraulic rescue tool. The door had already been forced open and the rookie fire fighter was attempting to separate the door from the vehicle at the hinges. When the second (top) hinge broke free from the car, the door actually launched approximately 3 feet (1 meter) from the vehicle, hitting the arm of an EMS worker. Just a few seconds before, the EMS worker was leaning against the outer edge of the door. A senior rescue fire fighter had moved the EMS worker away from the edge of the door while the rookie was working with the hydraulic rescue tool to separate the door from the vehicle.

Had the senior rescue fire fighter not been vigilant in his overview of the entire evolution, the EMS worker would have been struck by the car door as it released from the vehicle, possibly causing minor to severe injuries.

Lessons Learned:

- We regularly work in concert with EMS workers, some of which may not be familiar with the hazards associated with vehicle extrication.
- While they understand the concepts of extrication, they do not regularly perform extrication.
- Responders must be extremely vigilant of all personnel on the scene.

remove the victim. Another option is to insert cribbing wedges into the relief cuts once the push has been made with the dash lifted; this will assist in keeping the dash up and in place, or if one of the hydraulic rams is accidentally released or inadvertently moved, the cribbing wedges will prevent the dash from recollapsing onto the victim. Once the dash area is displaced, the victim should be properly packaged and removed toward the rear of the vehicle; the opened position of the hydraulic rams will prevent removal from the side.

To perform the dash roll technique, follow the steps in **Skill Drill 9-11 ▶**:

1. Don appropriate PPE, including mask and eye protection.
2. Assess the scene for hazards and complete the inner and outer surveys.
3. Stabilize the vehicle.
4. Ensure that the victim and the other rescuer inside the vehicle are properly protected using a blanket to cover them from flying glass particles.
5. Remove all vehicle glass utilizing the appropriate technique.

6. Disconnect the vehicle's 12-volt battery if the engine compartment is accessible.
7. Scan the vehicle for all/any SRS components (air bags), including exposing all the roof posts and roof liner.
8. Release and open both front doors utilizing the appropriate technique; the doors should be removed but can remain if so chosen.
9. Remove the roof of the vehicle utilizing the appropriate technique.
10. Position the hydraulic cutter to make an angled relief cut just under the bottom hinge of both front doors, where the firewall meets the rocker panel. (**Step 1**)
11. A relief cut will need to be completed on each side of the vehicle to be effective. (**Step 2**)
12. Premeasure for the appropriately sized hydraulic ram, or utilize a small to large telescopic model. Position the base of the ram at the bottom corner of the B-post and rocker panel with the tip of the tool angled upward, extending out to reach the bottom corner of the A-post where the dash and A-post join together. (**Step 3**)

NFPA 1006, (10.1.7)

Skill Drill 9-11

The Dash Roll Technique

1 Don appropriate PPE. Assess the scene for hazards and complete the inner and outer surveys. Stabilize the vehicle. Ensure that the victim and the other rescuer inside the vehicle are properly protected from flying glass particles. Remove all vehicle glass utilizing the appropriate technique. Disconnect the vehicle's 12-volt battery if the engine compartment is accessible. Scan the vehicle for all/any SRS components (air bags). Release and open both front doors utilizing the appropriate technique; the doors should be removed, but can remain in place if so chosen. Remove the roof of the vehicle utilizing the appropriate technique. With both front doors in the opened position, use the hydraulic cutter to make an angled relief cut on each side of the vehicle, just under the bottom hinge where the firewall meets the rocker panel.

2 A relief cut will need to be completed on each side of the vehicle to be effective.

3 Premeasure for the appropriately sized hydraulic ram, or utilize a small to large telescopic model. Position the base of the ram at the bottom corner of the B-post and rocker panel, with the tip of the tool angled upward, extending out to reach the bottom corner of the A-post.

4 If the B-post has been removed, measure the proper distance of the ram as described in Step 12 and then crimp the rocker panel in the area where the base of the ram rests using the hydraulic spreader. This creates a gap for the base of the ram to sit in and to push from.

5 Insert and stagger the height of cribbing under the rocker panel beginning at the area where the gap was created so there is an incremental increase in steps and space.

6 Position hard protection such as a backboard between the patient and the hydraulic ram if needed. Position and open the ram. Once the tip is in the correct position, push down on the base of the hydraulic ram as it is opened to resist the force of the tool kicking out. Operate the tool until enough room has been created to access and remove the victim.

13. If the B-post has been removed, measure the proper distance of the ram as described in Step 12 and then crimp the rocker panel in the area where the base of the ram rests using the hydraulic spreader. This creates a gap for the base of the ram to sit in and to push from. (**Step 4**)

14. Insert and stagger the height of cribbing under the rocker panel beginning at the area where the gap was created so there is an incremental increase in steps and space. This formation will go from lower to higher, with the highest section of cribbing shored tightly under the rocker panel preventing it from fully collapsing. (**Step 5**)

15. Position hard protection such as a backboard between the patient and the hydraulic ram if needed.

16. Position and open the ram. Once the tip is in the correct position, push down on the base of the hydraulic ram as it is opened to resist the force of the tool kicking out. Applying pressure as the tool is opened will force the rocker panel downwards, causing it to conform to the gap, filling the space created by the incremental cribbing formation which in turn also creates an artificial wall to push off from.

17. Operate the tool until enough room has been created to access and remove the victim. (**Step 6**)

18. Cribbing can be placed in the relief cut opening to maintain the position of the dash if needed.

19. If more than one hydraulic ram is utilized, ensure that the procedure is coordinated and symmetrical.

As discussed, there are a few options to consider if the B-post has been removed and there is not an area for the base of the hydraulic ram to push off from. If the vehicle is a two-door model, there is a product known as an L-bracket, which is made of steel and fits over the rocker panel and preferably up against the back of the door frame (the section that has the Nader pin or U-bolt attached to it). The welded steps on the L-bracket give the rescuer several sizing options for the ram to push from.

Another option to consider if the B-post has been removed is to stagger the height of the cribbing under the rocker panel so there is an incremental increase in steps and space. This formation will go from lower to higher, with the highest section of cribbing shored tightly under the rocker panel. Place the base of the hydraulic ram just over the lower section of cribbing where a gap has been purposely created, holding it in place until the ram is opened and the tip of the tool is placed into position. The tip of the hydraulic ram should be positioned at the bottom corner of the A-post where the dash and A-post join together **Figure 9-16 ▶**.

Once the tip is in the correct position, push down on the base of the hydraulic ram as it is opened to resist the force of the tool kicking out. Applying pressure on the base of the hydraulic ram as the tool is opened will crush the hollow rocker panel down, causing it to conform to the gap created by the incremental cribbing formation, which in turn also creates an artificial wall to push off from. This technique takes a lot of practice to perfect, but it is very effective when performed correctly.

A less desirable option is to drive the spiked end of a Halligan bar into the rocker panel, creating an artificial push point. The problem with this method is that it breaches the

Figure 9-16   The second option if the B-post has been removed and there is not an area for the base of the hydraulic ram to push off from is to stagger the height of the cribbing under the rocker panel so there is an incremental increase in steps and space.

rocker panel, which weakens the area and can cause the Halligan bar to push back and tear open the hollow wall of the rocker panel when force is applied. If this occurs, there is a possibility of the Halligan bar dislodging forcefully, potentially causing injury.

Rescue Tips

Do not use the hydraulic spreader to crush the rocker panel, leaving the spreader in the clamped position as a push point for the hydraulic ram. The hydraulic spreader was never designed for this type of maneuver or pressure; this can damage the arms of the hydraulic spreader. There are many techniques that are far better and are designed to use the tools in the right application.

Rescue Tips

No one technique will always be 100% effective; every incident will present differently, and in some cases, combining techniques may improve the overall outcome.

■ The Dash Lift Technique

Let's say a technical rescuer arrives on the scene of a motor vehicle collision where the vehicle involved has impacted a cement light post, forcing the vehicle's front end and dash onto the driver, or the vehicle has collided with the rear end of a semi-truck causing an underride-type entrapment **Figure 9-17 ▶**. A dash roll technique would not be effective in either of these situations. The weight and position of the cement pole is locking and holding down the dash area, and the rear end of the semi-truck is also locking and holding down the dash area. If the dash roll technique were used, the entire floorboard and rocker panel area, where the relief cuts would be made, would push up and create what is called a **tepee effect**. The dash area would not be able to effectively release off of the victim.

Figure 9-17 The dash lift technique releases and lifts the dash area independently from the front end of the vehicle. It is the best technique to use in this situation.

Rescue Tips

Use the right technique for the specific type of entrapment. Knowing when to use one technique over the other can greatly reduce the time it takes to safely release and remove the patient without causing further harm.

Because the dash area needs to be released and lifted independently from the front end of the vehicle, a **dash lift technique** is the best option. Releasing a section of the dash from the front end of the vehicle requires a hinge point to be created at the upper rail area of the hood between the strut tower and the dash **Figure 9-18 ▾**. The strut tower, which is normally located above the center point area of the wheel, is attached to the vehicle's upper rail frame, making it one of the strongest parts of that

Figure 9-18 Releasing a section of the dash from the front end of the vehicle requires a hinge point to be created at the upper rail area of the hood between the strut tower and the dash.

entire area. The goal is to release the strut tower from the dash/firewall section and create a hinge point for the dash to rise up and away from the trapped occupant.

To create this hinge effect, relief cuts are made in two areas. The first relief cut is made through the upper rail section as just described, and the second relief cut is made through the firewall area between the two hinges where the door was attached. When the technique is performed correctly, this section of the dash, including the steering wheel and steering column, will lift straight up and off of the occupant, leaving the front end of the vehicle, including the front wheels, stationary. This is very important because the front section of the vehicle with the cement pole or the other vehicle resting on top of it will not move when the dash section is lifted. This technique is designed to lift the dash on one side of the vehicle only; the steps will have to be repeated on the opposite side if that side of the dash needs to be lifted as well.

Once all of the initial steps have been completed, the hood area closest to the dash must be exposed and the hinge cut so the area can be examined for any possible hydraulic or gas-filled piston struts that are installed to assist in lifting the vehicle's hood **Figure 9-19 ▾**. These piston struts are generally installed in the area of operation and must be removed or disabled for safety purposes. To verify whether there is a hydraulic or gas-filled piston, insert the tips of the hydraulic spreader under the top

Figure 9-19 Piston struts assist in lifting the vehicle's hood.

Figure 9-20 To disable a hydraulic or gas-filled piston strut, with the spreader still in place holding open the hood and the hood's hinge cut, insert a hydraulic cutter and cut the section of the piston where it attaches to the vehicle.

corner area of the hood where the dash and hood meet. Slowly open the spreader to lift up the corner of the hood just enough to get the hydraulic cutter in to cut the hinge attachment for the hood, releasing that side of the hood from the dash. If the hood is not cut from the dash at that particular hinge attachment, then the hood could impede the lifting of the dash. At the same time, presence of a hydraulic or gas-filled piston strut can be confirmed. If a piston is found, verify if it is in the area of operation where the upper rail will be crushed or cut. If the piston strut is in the way, then it will need to be disabled or relocated. To disable a hydraulic or gas-filled piston strut, with the spreader still in place holding open the hood and the hood's hinge cut, insert a hydraulic cutter and cut the section of the piston where it attaches to the vehicle **Figure 9-20 ▲**. Do not cut into the cylinder body because there will be a rapid release of hydraulic fluid and gas under pressure. Addressing these items can quickly clear the way to proceed with the dash lift technique.

The dash lift technique is an advanced technique that first requires preparatory work to be completed and has multiple steps that must be carried out with strict discipline. Leaving out a step, which can happen in a high-stress environment, will almost always result in a critical failure; take the time to practice being proficient. To perform the dash lift technique using hydraulics, follow the steps in **Skill Drill 9-12 ▶**:

1. Don appropriate PPE, including mask and eye protection.
2. Assess the scene for hazards and complete the inner and outer surveys.
3. Stabilize the vehicle.
4. Ensure that the victim and the other rescuer inside the vehicle are properly protected using a blanket to cover them from flying glass particles.
5. Remove all vehicle glass utilizing the appropriate technique. (**Step 1**)
6. Disconnect the vehicle's 12-volt DC battery if readily accessible.
7. Remove the vehicle's roof. This technique can still be performed with the roof on; the only requirement is that a

section of the A-post connecting the dash area to be lifted must be removed. For best access to the victim, the roof should be removed. Remember to check the posts and roof rails for SRS components.

8. Remove the vehicle's door on the side to be lifted.
9. Lift and cut the vehicle's hood at the hinge attachment closest to the dash on the side to be lifted and verify if a hydraulic or gas piston strut exists. If a piston is found, then disable or relocate the item as a safety measure. (**Step 2**)
10. Remove all or part of the wheel well using the wheel well crush technique. This will expose the door's hinges and the upper rail frame section of the engine compartment. Once a purchase point has been created, insert the tips of the spreader in the space between the firewall and the wheel well panel. Angle the spreader and force the panel off the front attachments, which are generally two or three small connection points. Slowly work the spreader across the inside of the wheel well panel, which will tear away if done correctly. If the bottom section of the wheel well panel releases, which is normally spot-welded in, and tool leverage is lost, then reposition the spreader, clamp down on the panel where it is connected to the upper frame rail, and peel it back off of the area; this is thin sheet metal and will tear easily. (If it is a polycarbonate panel, as found in Saturn's vehicles, then it will either break off or come off as one section; the same goes for fiberglass panels.) (**Step 3**)
11. Once the wheel well panel is removed and the upper frame rail is fully exposed, use the hydraulic cutter to make a complete cut through the upper rail frame between the strut tower and dash to ensure the proper release of the dash from the front end of the vehicle. As an option, the hydraulic spreader can be applied before the cut is made to precrush the upper rail section, which will also assist in the cutting action of the blades. (**Step 4**)
12. Using the hydraulic cutter, make a complete cut through the entire firewall area directly between the top and bottom hinge. The amount of cuts may vary depending on the type of vehicle construction or width of the area. If a large firewall area is encountered or there is not enough room to get the cutter in, use the hydraulic spreader to crush the firewall. This is accomplished by making two smaller cuts on both sides of the firewall and inserting both tips of the hydraulic spreader into the cuts and compressing the metal enough to provide room for one final cut. As the final cut is made, there may be a slight visual release of the dash where the dash and steering wheel lift an inch (25 mm) or so. This is an indication that the cut is made all the way through. (**Step 5**)
13. Insert cribbing under the rocker panel firewall area for support, and make certain that there are sufficient pieces of cribbing readily available to be utilized as the evolution progresses.
14. Position the hydraulic spreader vertically with one arm over the other, and insert the tips in the area of the firewall that was cut. Slowly open the tool, using the top and bottom hinge, which are the strongest parts of that area, as guides and as push points. As the arms open, carefully watch the

NFPA 1006, (10.1.7)

Skill Drill 9-12

The Dash Lift Technique (Hydraulic)

1 Don appropriate PPE. Assess the scene for hazards and complete the inner and outer surveys. Stabilize the vehicle. Ensure that the victim and the other rescuer inside the vehicle are properly protected from flying glass particles. Remove all vehicle glass utilizing the appropriate technique.

2 Disconnect the vehicle's 12-volt DC battery if readily accessible. For best access to the victim, remove the roof. Remove the vehicle's door on the side to be lifted. Lift and cut the vehicle's hood at the hinge attachment closest to the dash on the side to be lifted and verify if a hydraulic or gas piston strut exists. If a piston is found, disable or relocate the item as a safety measure.

3 Remove all or part of the wheel well by performing the wheel well crush technique. This will expose the door's hinges and the upper rail frame section of the engine compartment. Once a purchase point has been created, insert the tips of the spreader in the space between the firewall and the wheel well panel. Angle the spreader and force the panel off the front attachments. Slowly work the spreader across the inside of the wheel well panel until it tears away.

4 Once the wheel well panel is removed and the upper frame rail is fully exposed, use the hydraulic cutter to make a complete cut through the upper rail frame between the strut tower and dash to ensure the proper release of the dash from the front end of the vehicle.

5 Using the hydraulic cutter, make a complete cut through the entire firewall area directly between the top and bottom hinge. If a large firewall area is encountered or there is not enough room to get the cutter in, use the hydraulic spreader to crush the firewall.

6 Insert cribbing under the rocker panel firewall area for support. Position the hydraulic spreader vertically with one arm over the other, and insert the tips in the area of the firewall that was cut. Slowly open the tool, using the top and bottom hinge, which are the strongest parts of that area, as guides and as push points. The dash will lift up and hinge at the section that was cut in the upper frame rail, taking the vehicle's front end completely out of the lift and supplying ample room for victim removal.

tool for any signs of slippage or twisting. If this occurs, insert cribbing to support the dash from coming back down on the victim and quickly readjust the tool.

15. With the technique applied correctly, the dash will lift up and hinge at the section that was cut in the upper frame rail, taking the vehicle's front end completely out of the lift and supplying ample room for victim removal. **(Step 6)**

Alternatively, to perform the dash lift technique without using hydraulics, follow the steps in **Skill Drill 9-13 ▾** :

1. Don appropriate PPE, including mask and eye protection.
2. Assess the scene for hazards and complete the inner and outer surveys.
3. Stabilize the vehicle.
4. Ensure that the victim and the other rescuer inside the vehicle are properly protected using a blanket to cover them from flying glass particles.
5. Remove all vehicle glass utilizing the appropriate technique **(Step 1)**.
6. Disconnect the vehicle's 12-volt DC battery.
7. Scan the vehicle for all/any SRS components (air bags), including exposing all the roof posts and roof liner.
8. Release the vehicle's door on the latch-side using a combination of a pneumatic air chisel and a Hi-Lift jack. **(Step 2)**
9. Remove the vehicle's roof. This technique can still be performed with the roof on; the only requirement is that a section of the A-post connecting the dash area to be lifted must be removed. Perform this step using a reciprocating saw or pneumatic air chisel. **(Step 3)**

10. Remove the door at the hinge-side by cutting the hinges with a pneumatic air chisel.
11. Remove all or part of the wheel well panel using a pneumatic air chisel. This will expose the upper rail frame section of the engine compartment.
12. Expose the vehicle's engine compartment hood at the hinge attachment closest to the dash on the side to be lifted and verify if a hydraulic or gas piston strut exists. If a piston is found, then disable or relocate the item as a safety measure. Cut the hinge attachment connecting the hood to the dash.
13. With the wheel well panel removed and the upper frame rail fully exposed, use the pneumatic air chisel or reciprocating saw to make a complete cut through the upper rail frame between the strut tower and dash to ensure the proper release of the dash from the front end of the vehicle. **(Step 4)**
14. Using the pneumatic air chisel or reciprocating saw, make a complete cut through the entire firewall area directly between the top and bottom hinge. A larger section may have to be cut out to make room for the lip of the Hi-Lift jack. **(Step 5)**
15. Insert cribbing under the rocker panel firewall area for support, and make certain that there are sufficient pieces of cribbing readily available to be utilized as the evolution progresses.
16. Position the Hi-Lift jack on the rocker panel and insert the top lip of the tool in the area of the firewall that was cut. Slowly open the tool, using the top and bottom hinge,

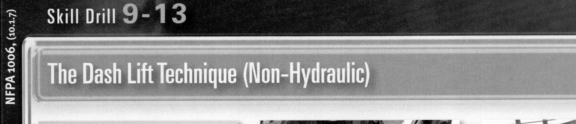

NFPA 1006, (10.1.7)

Skill Drill **9-13**

The Dash Lift Technique (Non-Hydraulic)

1   Don appropriate PPE. Assess the scene for hazards and complete the inner and outer surveys. Stabilize the vehicle. Ensure that the victim and the other rescuer inside the vehicle are properly protected. Remove all vehicle glass.

2   Disconnect the vehicle's 12-volt DC battery. Scan the vehicle for all/any SRS components (air bags). Release the vehicle's door on the latch-side using a combination of a pneumatic air chisel and a Hi-Lift jack.

3   Remove the vehicle's roof. This technique can still be performed with the roof on; the only requirement is that a section of the A-post connecting the dash area to be lifted must be removed. Perform this step using a reciprocating saw or pneumatic air chisel.

NFPA 1006, (10.1.7)

Skill Drill 9-13

The Dash Lift Technique (Non-Hydraulic) *(Continued)*

4 Remove the door at the hinge-side. Remove all or part of the wheel well panel using a pneumatic air chisel. Expose the vehicle's engine compartment hood at the hinge attachment closest to the dash on the side to be lifted and verify if a hydraulic or gas piston strut exists. If a piston is found, then disable or relocate the item as a safety measure. Cut the hinge attachment connecting the hood to the dash. With the wheel well panel removed and the upper frame rail fully exposed, use the pneumatic air chisel or reciprocating saw to make a complete cut through the upper rail frame between the strut tower and dash to ensure the proper release of the dash from the front end of the vehicle.

5 Using the pneumatic air chisel or reciprocating saw, make a complete cut through the entire firewall area directly between the top and bottom hinge. A larger section may have to be cut out to make room for the lip of the Hi-Lift jack.

6 Insert cribbing under the rocker panel firewall area for support. Position the Hi-Lift jack on the rocker panel and insert the top lip of the tool in the area of the firewall that was cut. Slowly open the tool, using the top and bottom hinge, which are the strongest parts of that area, as guides and as push points. As the tool opens, carefully watch the tool for any signs of slippage or twisting. If this occurs, insert cribbing to support the dash from coming back down on the victim and re-adjust the tool.

7 With the technique applied correctly, the dash will lift up and hinge at the section that was cut in the upper frame rail, taking the vehicle's front end completely out of the lift and supplying ample room for victim removal.

which are the strongest parts of that area, as guides and as push points. As the tool opens, carefully watch the tool for any signs of slippage or twisting. If this occurs, insert cribbing to support the dash from coming back down on the victim and re-adjust the tool. (**Step 6**)

17. With the technique applied correctly, the dash will lift up and hinge at the section that was cut in the upper frame rail, taking the vehicle's front end completely out of the lift and supplying ample room for victim removal. (**Step 7**)

One possible complication of the dash lift technique is the presence of dash brackets. As discussed in Chapter 3, *Mechanical Energy and Vehicle Anatomy*, dash brackets are located in the center console area where the radio, air conditioning unit, and other various components are located. The metal brackets are bolted or welded into the floorboard of the vehicle and are designed to lock the dash in place to minimize any movement resulting from an impact.

Dash brackets, because of their inherent design, can at times resist the upward movement of the dash lift technique, halting any progress. If dash brackets are in place, the stopping of the dash's upward movement will occur almost immediately at the beginning of the spread. It is extremely important for the technical rescuer to recognize the signs of this occurring. If the dash does not release and rise as the lift is being performed, and/or the firewall area tears away and folds outward toward the technical rescuer, there is the possibility that dash brackets are in place, keeping the dash from releasing. To get the proper lift of the dash in order to create enough room to remove the entrapped victim, the two dash brackets will need to be cut before the dash will release. To accomplish this objective, the technical rescuer must gain access to the front passenger compartment opposite the driver and start to expose the brackets by removing the plastic molding surrounding the brackets. This plastic molding should break away and release with little effort **Figure 9-21 ▾** .

Once the passenger side dash bracket is exposed, use the hydraulic cutter to cut through the entire section and then verify if there is enough access to cut through the driver's side dash bracket. If this is not possible, use the hydraulic spreader to push up and expose the other dash bracket. This should provide adequate room to insert the hydraulic cutter to sever the driver's side dash bracket. Once the dash brackets are cut, resume the dash lift technique and complete the lift.

If access is not possible to the passenger side because of an additional trapped occupant or object, attempt another approach or angle, if possible, or consider another technique such as a dash roll or steering wheel pull to gain access. Remember to always have a plan B in place for unexpected occurrences.

Steering Wheel and Steering Column Relocation

Relocating a steering column utilizing a rated come along may be a reliable option when other dash displacement techniques, such as a dash lift or dash roll technique, are not possible or a hydraulic tool fails. This technique has come under a lot of scrutiny in the past based on an unsubstantiated belief that under the extreme force applied from the come along, the steering column can come apart and break off, driving the entire column into the victim. One technique that is not recommended is using hydraulic tools along with a rated chain package to relocate the steering column. This technique is very dangerous because the technical rescuer has no control over the amount of force that is applied; nor can the technical rescuer feel the

Figure 9-21 Remove the plastic molding to expose the dash brackets.

force that is applied, as he or she can when using a rated come along.

When the come along and chain package has been properly applied, the technical rescuer has complete control over and feel of the entire movement. In addition, there is a safety feature for the handle component of the come along that is rated to fail by bending when a predetermined force is exerted. The properly rated come along for this technique is generally in the 2000- to 4000-pound (1- to 2-short tons) category; it has a rated handle and a complete chain package, as described in Chapter 6, *Tools and Equipment*.

There may also be a misconception about the amount of pull that is needed to release the steering column and free a trapped victim. The come along does not have to be overexerted where the steering wheel/column is pulled through the dash to be effective; just a few inches of upward movement will create enough space to free the victim. If needed, more room can be created by cutting the section of the steering wheel ring that was impinging on the victim, but always make sure before attempting to cut a section of the steering wheel ring that it is not still impinging on the victim. The force and cutting action of the hydraulic tool as the ring is cut can drive a section of the steering wheel deeper into the victim before it releases. There is also a ratcheting cutting tool similar to bolt cutters that is very effective in cutting steering wheel rings and is less aggressive than a hydraulic cutter. Again, it is best to relieve the pressure of the impinging steering wheel before attempting to cut the ring.

Before starting the technique, ensure that the following tools and hardware are available: several four-by-fours, one **ladder crib** section, one rated come along with operation handle, a complete chain package consisting of two chains, and several bungee cords or fastening devices.

Two technical rescuers will be needed to complete the evolution, with both initial assignments being performed simultaneously. To relocate the steering wheel and steering column, follow the steps in **Skill Drill 9-14 ▸** :

1. Don appropriate PPE, including mask and eye protection.
2. Assess the scene for hazards and complete the inner and outer surveys.
3. Stabilize the vehicle.
4. Ensure that the victim and the other rescuer inside the vehicle are properly protected using a blanket to cover them from flying glass particles.
5. Remove all vehicle glass utilizing the appropriate technique. (**Step 1**)
6. Technical Rescuer 1: While positioned at the front of the vehicle, take the section of ladder cribbing and lay it over the front hood/bumper area of the vehicle. Using a bungee cord with hooks or another type of fastener, quickly attach the ladder cribbing to the hood to temporarily hold the cribbing in place while the chain section is set up.
7. Technical Rescuer 1: Rest the support ring located at one side of the chain's end section on top of the ladder crib at the lower part of the hood (the come along will be attached to this).
8. Technical Rescuer 1: Using the other end of the chain, locate an area under the vehicle to secure to; this is known

as an anchor point. Make certain that this anchor point is secured to an area of the vehicle that will support the force applied by the come along, such as a frame section of the undercarriage. Loop the chain through and take up all of the slack utilizing the chain-shortener hook, which will lock in that loose section, securing the anchor point. To correctly position the come along, the flywheel will always be in the right hand of the operator. Place the come along on top of the ladder cribbing and secure it to the support ring of the chain. (**Step 2**)

9. Technical Rescuer 2: Use three (3) four-by-fours to build a <u>slide crib</u> configuration on top of the dash directly in line with the steering wheel.

10. Technical Rescuer 2: Utilizing the second section of chain in the chain package, weave the support ring of the chain through the bottom section of the steering wheel ring starting from behind. Continue through the opening and go around the front main base of the steering wheel, ending

back through the top section of the steering wheel ring. Pull the support ring through and lay it on the cross-section of the slide crib. Fasten one end of a bungee cord with hooks or another fastening device to the front hood and the other end to the support ring to hold it in place. (**Step 3**)

11. Technical Rescuer 2: Wrap the other end of the chain clockwise around the steering wheel column twice, just behind the steering wheel ring. Pull it snug. To help maneuver the chain with minimal difficulty, use both hands, with one hand on top of the steering wheel column and the other hand underneath the steering wheel column, to pull the end of the chain over, through, and around the steering wheel column.

12. Technical Rescuer 2: Once the chain has been tightly secured around the steering wheel column, take the end section of the chain, and mirroring the front section of chain, weave it through the bottom section of the steering wheel ring starting from behind. Continue through the opening and

NFPA 1006, (10.1.7)

Skill Drill 9-14

Relocating the Steering Wheel and Steering Column

1 Don appropriate PPE. Assess the scene for hazards and complete the inner and outer surveys. Stabilize the vehicle. Ensure that the victim and the other rescuer inside the vehicle are properly protected from flying glass particles. Remove all vehicle glass utilizing the appropriate technique.

2 Technical Rescuer 1 (the technical rescuer positioned at the front of the vehicle): Take the section of ladder cribbing and lay it over the front hood/bumper area of the vehicle. Use a fastener to attach the ladder cribbing to the hood to temporarily hold the cribbing in place. Rest the support ring located at one side of the chain's end section on top of the ladder crib at the lower part of the hood (the come along will be attached to this). Using the other end of the chain, locate an area under the vehicle to secure to. Loop the chain through and take up all of the slack utilizing the chain-shortener hook, which will lock in that loose section, securing the anchor point. Place the come along on top of the ladder crib-bing and secure it to the support ring of the chain.

3 Technical Rescuer 2: Use three (3) four-by-fours to build a slide crib configuration on top of the dash directly in line with the steering wheel. Utilizing the second section of chain in the chain package, weave the support ring of the chain through the bottom section of the steering wheel ring starting from behind. Continue through the opening and go around the front main base of the steering wheel, ending back through the top section of the steering wheel ring. Pull the support ring through and lay it on the cross-section of the slide crib. Fasten one end of a bungee cord with hooks or another fastening device to the front hood and the other end to the support ring to hold it in place.

(Continues)

Skill Drill 9-14

Relocating the Steering Wheel and Steering Column *(Continued)*

4 Technical Rescuer 2: Wrap the other end of the chain clockwise around the steering wheel column twice, just behind the steering wheel ring. Pull it snug. Pull the end of the chain over, through, and around the steering wheel column. Once the chain has been tightly secured around the steering wheel column, take the end section of the chain, and mirroring the front section of chain, weave it through the bottom section of the steering wheel ring starting from behind. Continue through the opening and go around the front main base of the steering wheel, ending back through the top section of the steering wheel ring. Pull up all of the slack in the chain, and secure the chain to the support ring utilizing the chain shortener.

5 With both chain sections in place, technical rescuer 1 takes hold of the come along with the right hand on the flywheel and the left hand on the free spool lever located on the top left section of the device. Technical rescuer 2 will take hold of the hook-and-wire cable attachment of the come along and pull out enough cable to attach it to the support ring located on top of the slide cribbing. Technical rescuer 1 will release the free spool lever and slowly let out the wire cable by rotating the flywheel counterclockwise.

6 Once the wire cable has been attached and all of the slack has been taken up by turning the flywheel clockwise, complete a safety check of all connections, ensuring that all of the hooks are secure and facing upward and that the wire cable has not crossed on itself. Technical rescuer 1 will now insert the control handle into the come along and slowly push the handle forward and back, engaging the device's gear mechanism. When enough room has been created to free the victim, technical rescuer 1 will remove the control handle of the come along and relocate it to the front of the device. This will ensure that the device is not accidentally engaged and released.

go around the front main base of the steering wheel, ending back through the top section of the steering wheel ring. Pull up all of the slack in the chain, and secure the chain to the support ring utilizing the chain shortener. **(Step 4)**

13. With both chain sections in place, technical rescuer 1 will take hold of the come along with the right hand on the flywheel and the left hand on the free spool lever located on the top left section of the device. Technical rescuer 2 will take hold of the hook-and-wire cable attachment of the come along and pull out enough cable to attach it to the support ring located on top of the slide cribbing. Technical rescuer 1 will release the free spool lever and slowly let out the wire cable by rotating the flywheel counterclockwise. Make certain that the release of the cable is controlled to prevent the cable from backlashing; this would be a critical error ending any possibility of continuing the technique. **(Step 5)**

14. Once the wire cable has been attached and all of the slack has been taken up by turning the flywheel clockwise, complete a safety check of all connections, ensuring that all of the hooks are secure and facing upward and that the wire cable has not crossed on itself.

15. Technical rescuer 1 will now insert the control handle into the come along and slowly push the handle forward and back, engaging the device's gear mechanism. As the tool is operated, technical rescuer 2 will be positioned with the victim and will verbally signal to technical rescuer 1 when enough clearance of the steering wheel has been established.

16. When enough room has been created to free the victim, technical rescuer 1 will remove the control handle of the come along and relocate it to the front of the device. This will ensure that the device is not accidentally engaged and released. **(Step 6)**

Providing Initial Medical Care

Remember to always follow departmental protocols on rendering medical care; life support/medical protocols are designated by the authority having juristiction (AHJ). Once inside the vehicle, the technical rescuer must render basic medical attention to the victim, such as managing airway, breathing, and circulation. In addition to providing medical care, there are several other steps the rescuer inside the vehicle must carry out. Some of these tasks may include cutting all of the seat belts to assist in roof removal, identifying any undetected air bag or SRS components and relaying this information to the crew conducting the operation, attempting to see if any of the seat adjustment mechanisms are operational, and providing soft and hard protection for the victim and himself or herself if tool operation will be in close proximity.

Initial care of the victim also involves maintaining control of all life-threatening problems, including immobilization of the head, neck, and body to avoid any potential spinal injuries and controlling external bleeding [Figure 9-22 ▾]. A rescuer positioned outside the vehicle can maintain manual temporary spinal immobilization of the victim. The rescuer inside the vehicle should have all of the necessary medical gear inside the vehicle to render aid based on the level of medical service the jurisdiction provides. If the victim has a patent airway, the medical director may advise the administration of oxygen. The procedure for administering oxygen is a medical directive authorized only by medical protocol for the jurisdiction.

Airway

After you complete the inner and outer scene surveys, if you observe that the patient has airway problems and is lying on the seat or floor, attempt to access the vehicle safely in order to apply the standard jaw-thrust maneuver [Figure 9-23 ▸]. Use the jaw-thrust maneuver if there is any possibility that the collision could have caused a head or spinal injury. Remember that the safety of yourself and your crew always precludes entering a vehicle that is unsafe, regardless of the incident that is presented.

If the patient is in a sitting or a semi-reclining position, approach him or her from the side by leaning in through the

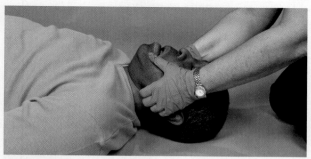

Figure 9-23   Use the jaw-thrust maneuver if the patient is lying on the seat or floor and there is any possibility that the collision could have caused a head or spinal injury.

window and across the front seat. Grasp the patient's head with both hands and put one hand under the patient's chin and the other hand on the back of the patient's head just above the neck. Raise the head to a neutral position to open the airway [Figure 9-24 ▾].

A.

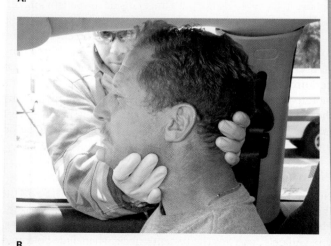

B.

Figure 9-24   A. To open the airway of a patient in a sitting or a semi-reclining position, place one hand under the patient's chin and the other hand on the back of the patient's head just above the neck. B. Raise the head to a neutral position to open the airway.

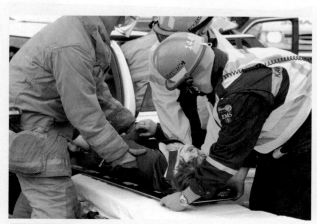

Figure 9-22   Victim packaging and removal.

Table 9-1   Normal Rates for Respirations

Age	Range (breaths/min)
Adults and adolescents	12 to 20
Children (1 to 12 years)	15 to 30
Infants	25 to 50

Note: Ranges presented in other courses may vary.

Table 9-2   Normal Ranges for Pulse Rates

Age	Range (beats/min)
Infant: 1 month to 1 year	100 to 160
Toddler: 1 to 3 years	90 to 150
Preschool age: 3 to 6 years	80 to 140
School age: 6 to 12 years	70 to 120
Adolescent: 12 to 18 years	60 to 100
Adult	60 to 100

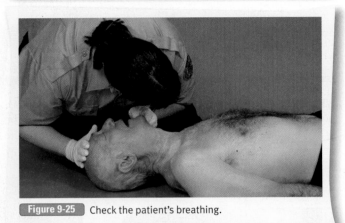

Figure 9-25 Check the patient's breathing.

Figure 9-26 Check an unconscious patient's circulation by checking the carotid pulse.

Breathing

If the patient is conscious, assess the rate **Table 9-1 ▲** and quality of the patient's breathing. Does the chest rise and fall with each breath or does the patient appear to be short of breath? If the patient is unconscious, check for breathing by placing the side of your face next to the patient's nose and mouth. You should be able to hear the sounds of breathing, see the chest rise and fall, and even feel the movement of air on your cheek **Figure 9-25 ▲**. If the patient is having difficulty breathing or if you hear unusual sounds, check for an object in the patient's mouth, such as food, vomitus, dentures, gum, chewing tobacco, or broken teeth, and remove it; however, do not put a finger inside the patient's mouth without the use of a bite block for protection.

If you cannot detect any movement of the chest and no sounds of air are coming from the nose and mouth, breathing is absent. Take immediate steps to open the patient's airway and perform rescue breathing. Because trauma is suspected after a motor vehicle collision, protect the cervical spine by keeping the patient's head in a neutral position and using the jaw-thrust maneuver to open the airway. Maintain cervical stabilization until the head and neck are immobilized.

Circulation

Next, check the patient's circulation (heartbeat) **Table 9-2 ▶**. If the patient is unconscious, check the carotid pulse **Figure 9-26 ▶**. Place your index and middle fingers together and touch the larynx (Adam's apple) in the patient's neck. Then slide your two fingers off the larynx toward the patient's ear until you feel a slight notch. Practice this maneuver until you are able to find a carotid pulse within 5 seconds of touching the patient's larynx. If you cannot feel a pulse with your fingers in 5 to 10 seconds, begin CPR.

If the patient is conscious, assess the radial pulse rather than the carotid pulse **Figure 9-27 ▼**. Place your index and middle fingers on the patient's wrist at the thumb side. You should practice taking the radial pulse often to develop this skill.

Next, quickly check the patient for any severe external bleeding. If you discover severe bleeding, you must take immediate action to control it by applying direct pressure over the wound.

Quickly assess the patient's skin color and temperature. It is important to check the color of the patient's skin when you arrive

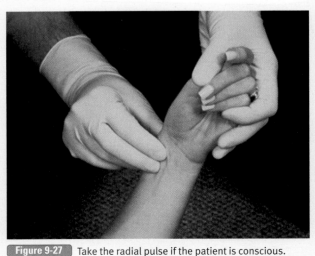

Figure 9-27 Take the radial pulse if the patient is conscious.

at the scene so that you can monitor the patient's skin for color changes as time goes on. Skin color is described as follows:

- **Pale.** Whitish, indicating decreased circulation to that part of the body or to all of the body. This could be caused by blood loss, poor blood flow, or low body temperature.
- **Flushed.** Reddish, indicating excess circulation to that part of the body
- **Blue.** Also called cyanosis, indicating lack of oxygen and possible airway problems
- **Yellow.** Indicating liver problems
- **Normal.**

Patients with deeply pigmented skin may show color changes in the fingernail beds, in the whites of the eyes, on the palm of the hand, or inside the mouth.

Rescue Tips

Remember to wear gloves to avoid contact with body fluids that may contain blood.

Rescue Tips

Rescue operations are those activities directed at locating endangered persons at an emergency incident, removing those persons from danger, and treating injured victims and providing for transport to an appropriate healthcare facility. *Recovery operations* are nonemergency activities undertaken by responders to retrieve property or remains of victims.

Table 9-3 Triage Priorities

Triage Category	Typical Injuries
Red tag: first priority (immediate) Patients who need immediate care and transport Treat these patients first, and transport as soon as possible	Airway and breathing difficulties Uncontrolled or severe bleeding Severe medical problems Signs of shock (hypoperfusion) Severe burns Open chest or abdominal injuries
Yellow tag: second priority (delayed) Patients whose treatment and transport can be temporarily delayed	Burns without airway problems Major or multiple bone or joint injuries Back injuries with or without spinal cord damage
Green tag: third priority, minimal (walking wounded) Patients who require minimal or no treatment and transport can be delayed until last	Minor fractures Minor soft-tissue injuries
Black tag: fourth priority (expectant) Patients who are already dead or have little chance for survival; treat salvageable patients before treating these patients	Obvious death Obviously nonsurvivable injury, such as major open brain trauma Respiratory arrest (if limited resources) Cardiac arrest

Triage

Some incidents may involve multiple victims. Rapid and accurate triage will help bring order to the chaos of a scene with multiple victims and allow the most critical patients to be transported first. Triage simply means to sort patients based on the severity of their injuries.

There are four common triage categories. They can be remembered using the mnemonic IDME, which stands for Immediate (red), Delayed (yellow), Minor or Minimal (green; hold), and Expectant (black; likely to die or dead) Table 9-3.

It is important that each patient involved in the incident has a tag or label indicating his or her condition. Triage tags should be weatherproof and easy to read Figure 9-28. They should be color coded and should clearly show the category of the patient. Triage tags will become part of the patient's medical record. Most have a tear-off receipt with a number corresponding to the number on the tag. When torn off by the transportation officer, it will assist him or her in tracking a patient.

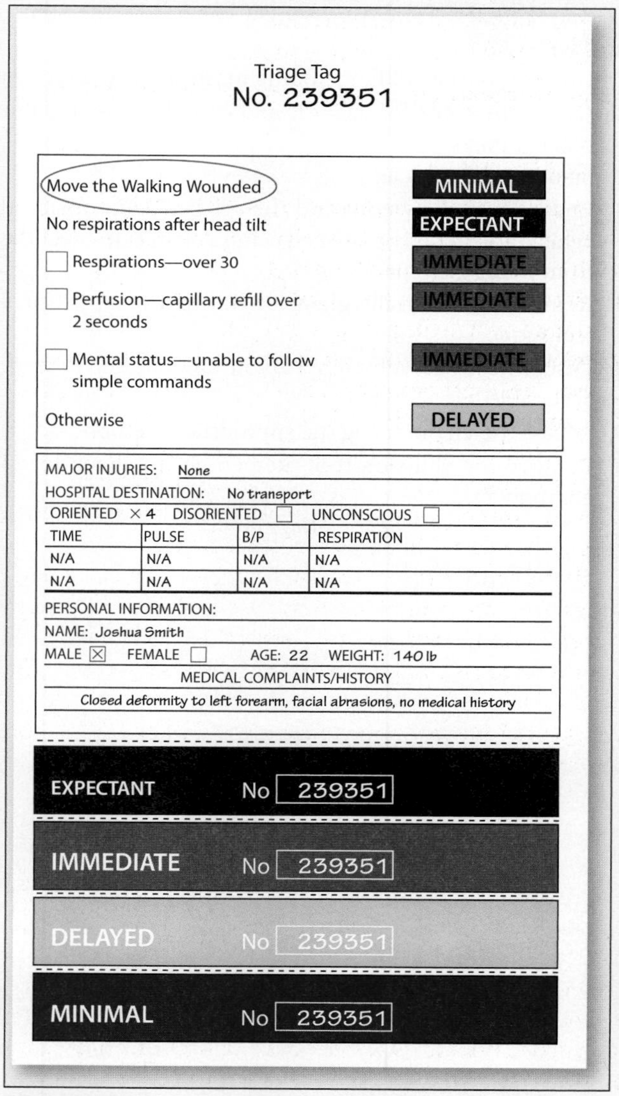

Triage Tag
No. 239351

Move the Walking Wounded	MINIMAL
No respirations after head tilt	EXPECTANT
☐ Respirations—over 30	IMMEDIATE
☐ Perfusion—capillary refill over 2 seconds	IMMEDIATE
☐ Mental status—unable to follow simple commands	IMMEDIATE
Otherwise	DELAYED

MAJOR INJURIES: None
HOSPITAL DESTINATION: No transport
ORIENTED × 4 DISORIENTED ☐ UNCONSCIOUS ☐

TIME	PULSE	B/P	RESPIRATION
N/A	N/A	N/A	N/A
N/A	N/A	N/A	N/A

PERSONAL INFORMATION:
NAME: Joshua Smith
MALE ☒ FEMALE ☐ AGE: 22 WEIGHT: 140 lb
MEDICAL COMPLAINTS/HISTORY
Closed deformity to left forearm, facial abrasions, no medical history

EXPECTANT No 239351
IMMEDIATE No 239351
DELAYED No 239351
MINIMAL No 239351

Figure 9-28 A sample triage tag.

START triage is one of the easiest methods of triage. START stands for Simple Triage And Rapid Treatment. The staff members at Hoag Memorial Hospital in Newport Beach, California, developed this method of triage. It is easily mastered with practice and will enable you to rapidly categorize patients. START triage uses a limited assessment of the patient's ability to walk, respiratory status, hemodynamic status, and neurologic status.

Lou Romig, MD, recognized that the START triage system does not take into account the physiologic and developmental differences of pediatric patients. She therefore developed the JumpSTART triage system for pediatric patients. JumpSTART is intended for use in children younger than 8 years or who appear to weigh less than 100 lb (45 kg).

Victim Packaging and Removal

Maintaining alignment is one of the main objectives when removing a victim from a vehicle. The following technique is one procedure that can be used to meet this objective. To extricate a victim from a passenger car follow the steps in **Skill Drill 9-15 ▶**:

1. Don appropriate PPE, including mask and eye protection.
2. Assess the scene for hazards and complete the inner and outer surveys.
3. Stabilize the vehicle.
4. Ensure that the victim and the other rescuer inside the vehicle are properly protected using a blanket to cover them from flying glass particles.
5. Remove all vehicle glass utilizing the appropriate technique.
6. Disconnect the vehicle's 12-volt DC battery if readily accessible.
7. Access the victim using the appropriate method.
8. Remove the vehicle's roof. Remember to check for SRS components.

9. Place an immobilization device on the patient. (**Step 1**)
10. Insert a backboard behind the patient. If the seat adjustment is still operational and can be reached, support the patient and adjust the back of the seat in the down position, if possible. (**Step 2**)
11. Position rescue personnel in a three-point position around the victim with a rescuer on each side of the victim and one at the head of the victim. The rescuer at the head of the victim will give the command to move the patient as a unit, lifting and sliding the patient up onto the backboard.
12. Lift the backboard and the victim as a unit onto the trunk/rear dash area. Properly secure the patient to the backboard. (**Step 3**)
13. Once the patient has been properly packaged, use a four-point carry to transfer the patient to an awaiting stretcher. When performed correctly, this technique will maintain patient alignment with minimal manipulation required. (**Step 4**)

Transport

Once the victim has been removed from the vehicle, transfer all pertinent medical information to the EMS personnel who will transport the patient to the appropriate medical facility. Skilled verbal communication and written documentation will make it possible for you to effectively coordinate the transfer of care. Depending on the methods used in your jurisdiction, checklists, triage tags, or forms may be used to transfer the necessary information, including patient condition and history, to EMS. Documentation serves as an excellent record that the care delivered was appropriate, guarantees the proper transfer of responsibility, and ensures the continuity of patient care. The type of transport used to deliver the patient to the emergency department will vary depending on the severity of the victim's injuries and the distance to the medical facility.

NFPA 1006, (10.1.9)

Skill Drill 9-15

Extricating a Victim from a Passenger Car

1 Don appropriate PPE. Assess the scene for hazards and complete the inner and outer surveys. Stabilize the vehicle. Ensure that the victim and the other rescuer inside the vehicle are properly protected from flying glass particles. Remove all vehicle glass utilizing the appropriate technique. Disconnect the vehicle's 12-volt DC battery if readily accessible. Access the victim using the appropriate method. Remove the vehicle's roof. Place an immobilization device on the patient.

2 Insert a backboard behind the patient. Adjust the back of the seat in the down position, if possible.

3 Position rescue personnel in a three-point position around the victim. The rescuer at the head of the victim will give the command to move the patient as a unit. Lift the backboard and the victim as a unit onto the trunk/rear dash area. Properly secure the patient to the backboard.

4 Use a four-point carry to transfer the patient to an awaiting stretcher.

Wrap-Up

Ready for Review

- Victim management involves vehicle entry, patient packaging, and patient removal.
- The main objective is not to remove the victim from the vehicle but to remove the vehicle from the victim by creating a large opening with systematic and precise techniques.
- Primary access points are the existing openings into the vehicle, and secondary access points are openings created by rescuers.
- One of the simplest ways to access a victim is to open a vehicle door. It is important to manually try all of the doors before other methods are used, even if the doors appear to be badly damaged.
- When conducting extrication operations on a vehicle, it is safer to remove all of the vehicle's glass as opposed to leaving some segments intact.
- Polycarbonate glass is a lighter, durable plastic that is up to 250 times stronger than glass; it is naturally designed to resist direct impacts by any striking tool carried on the apparatus.
- Making a purchase point is the process of gaining an access area to insert and better position a tool for operation. Once a purchase point has been established, the goal is to create a wide enough opening with the hydraulic spreader to expose the locking/latching mechanism or hinges and to insert a hydraulic cutter; this technique is known as expose and cut.
- For gaining door access, a vertical spread technique gives the technical rescuer the best vantage point to expose the latch and create enough room for the cutter blades to get in and cut the latching mechanism.
- One of the more favorable techniques utilized to gain access to door hinges from the outside is a wheel well crush technique using the hydraulic spreader.
- The complete side removal technique (side-out) is a highly effective technique for four-door side-impact collisions; it pushes the door frame and B-post outward and away from the occupant, utilizing the door's natural directional movement.
- When performing a roof removal, position rescuers on both sides of the vehicle to support the roof as the posts are cut.
- The dash roll technique involves pushing or rolling the entire front end of the vehicle, which encompasses the dashboard, steering wheel, and steering column, off of the entrapped occupant utilizing hydraulic rams.
- The dash lift technique involves lifting the dash upward with the hydraulic spreader by making precise relief cuts in the hood's upper rail and between the hinges of the firewall area separating the dash section from the front end of the vehicle.
- Dash brackets can at times resist the upward movement of the dash lift technique, halting any progress.
- Relocating a steering column utilizing a rated come along may be a reliable option when other dash displacement techniques, such as a dash lift or dash roll technique, are not an option or a hydraulic tool fails.
- Once inside the vehicle, the technical rescuer must render basic medical attention to the victim, such as managing airway, breathing, and circulation.
- Because trauma is suspected after a motor vehicle collision, protect the cervical spine by keeping the patient's head in a neutral position and using the jaw-thrust maneuver to open the airway. Maintain cervical stabilization until the head and neck are immobilized.
- Some incidents may involve multiple victims. Triage simply means to sort patients based on the severity of their injuries.
- There are four common triage categories. They can be remembered using the mnemonic IDME, which stands for Immediate (red), Delayed (yellow), Minor or Minimal (green; hold), and Expectant (black; likely to die or dead).
- Once the victim has been removed from the vehicle, transfer all pertinent medical information to the EMS personnel who will transport the patient to the appropriate medical facility.
- Skilled verbal communication and written documentation will enable you to effectively coordinate the transfer of care.

■ Hot Terms

Backboard slide technique An initial access technique that is used if the vehicle doors are locked, blocked, or inoperable. May be accomplished through a rear or side window.

Dash lift technique A technique used to lift and release a section of the dash from the front end of the vehicle using the hydraulic spreader. It is performed by making precise relief cuts in the hood's upper rail and between the hinges of the firewall area, separating the dash section from the front end of the vehicle.

Dash roll technique The standard technique for many years for displacing the dashboard. It is one of the least technical applications to apply in the field. The process involves using hydraulic rams to push or roll the entire front end of the vehicle upward and forward, including the dashboard, steering wheel, and steering column, off of the entrapped occupant.

D-ring A generic term used to describe the window frame of the door.

Expose and cut The process of creating a wide enough opening with the hydraulic spreader to expose the locking/latching mechanism or hinges and to insert a hydraulic cutter to cut.

Ladder crib Several 2- by 4-inch (51 by 102 mm) sections of wood attached together by a strip of webbing running along the sides.

Primary access The existing openings of doors and/or windows that provide a pathway to the trapped and/or injured victim.

Secondary access Openings created by rescuers that provide a pathway to trapped and/or injured victims.

Side-out technique A technique used to gain access to a four-door vehicle involved in a side-impact collision.

Slide crib Two sections of 4- by 4-inch cribbing positioned parallel to each other with a third section of cribbing on top traversing the two bottom sections.

Tepee effect The negative effect of a dash roll technique where the dash cannot release or push forward and the entire floorboard/rocker panel area, where the relief cuts were made, tepees upward and impinges on the victim.

Vertical spread A door access procedure utilizing a hydraulic spreader; the tool is placed vertically in the window frame of the door and pushes off of the roof rail and window frame to create an access point to the door's latching mechanism.

Wheel well crush technique A favorable technique utilized to gain access to door hinges from the outside.

Technical Rescuer *in Action*

Your unit is the first on scene at a multi-vehicle accident. It appears that all three vehicles have suffered major damage. You notice that there are patients who have not self-extricated and appear to be trapped. In one vehicle, a patient has been partially ejected through the windshield. You take command, complete the inner and outer surveys, check for further hazards, stabilize the vehicles, and complete a preliminary triage of the patients. Your crew is now ready to begin managing the victims.

1. The most desirable way for the rescuer to access the patient who has been partially ejected is to support the ejected head or torso from outside the vehicle and gain vehicle access:
   A. by removing the roof.
   B. through the rear window.
   C. through an open door.
   D. through the windshield.

2. Glass that will easily break into small fragments is also know as:
   A. laminated glass.
   B. tempered glass.
   C. safety glass.
   D. polycarbonate glass.

3. The tool of choice to remove the windshield away from the partially ejected victim is a:
   A. reciprocating saw.
   B. glass saw.
   C. pair of trauma shears.
   D. pry axe.

4. To gain access to an area of metal on the vehicle to insert a tool, the rescuer should create a:
   A. position point.
   B. pivot point.
   C. pry point.
   D. purchase point.

5. The technique that refers to the removal of the front and rear door as one unit on the same side of a four-door vehicle is known as the:
   A. total-out technique.
   B. cut-out technique.
   C. blow-out technique.
   D. side-out technique.

6. The rescuer should lean against the door that is being spread to provide counter pressure.
   A. True
   B. False

7. When cutting posts with the hydraulic cutter, the blades should be positioned _____ to the posts to prevent the blades from twisting or separating.
   A. vertically
   B. horizontally
   C. perpendicular
   D. tangentially

8. The weight of the dashboard could be pushed or rolled off from the patient using a:
   A. block and tackle.
   B. hydraulic ram.
   C. electric winch.
   D. bottle jack.

9. Exposing the interior of the post and the roof rail will reveal:
   A. air bag cylinders.
   B. cutting points.
   C. glass channels.
   D. crumple zones.

10. The rescuer's main objective should be to remove the patient from the vehicle.
    A. True
    B. False

Alternative Extrication Techniques

NFPA 1006 Standard

There are no objectives for this chapter.

Knowledge Objectives

After studying this chapter, you will be able to:

- Describe tunneling and when to use it. (pages 244–247)
- Describe how to remove a front seat and provide an example of when this should be performed. (pages 247–248)
- Describe the cross ram technique and provide an example of when this should be performed. (pages 248–251)
- Describe how to stabilize an impaled object. (pages 252–253)
- Describe how to remove a roof using an air chisel. (pages 252–256)
- Describe how to remove a roof using a reciprocating saw. (pages 256–258)
- Describe how to remove a roof from a vehicle resting on its side. (pages 258–259)
- Describe how to remove a door on the hinge side using an air chisel. (pages 260–262)
- Describe how to perform a side removal on a vehicle that is upside down or resting on its roof. (pages 262–264)
- Describe how to relocate a pedal. (pages 264–265)

Skills Objectives

After studying this chapter, you will be able to perform the following skills:

- Tunnel through the trunk. (pages 245–247, Skill Drill 10-1)
- Remove a front seat. (pages 247–248, Skill Drill 10-2)
- Perform a cross ram operation utilizing a hydraulic ram. (pages 250–251, Skill Drill 10-3)
- Stabilize an impaled object. (pages 252–253, Skill Drill 10-4)
- Remove a roof using an air chisel. (pages 254–256, Skill Drill 10-5)
- Remove a roof using a reciprocating saw. (pages 256–258, Skill Drill 10-6)
- Remove a roof from a vehicle resting on its side. (pages 258–259, Skill Drill 10-7)
- Remove a door on the hinge side using an air chisel. (pages 260–262, Skill Drill 10-8)
- Perform a side removal on a vehicle that is upside down or resting on its roof. (pages 262–264, Skill Drill 10-9)
- Relocate a pedal. (pages 264–265, Skill Drill 10-10)

ou arrive on scene of a motor vehicle collision where a common passenger vehicle is resting on its crushed roof. The inner and outer surveys reveal that one occupant is trapped in the rear passenger area and both doors are blocked and inaccessible because of positioning. There are no hazards, and the vehicle is not running.

1. What are some of the possible solutions to this scenario?
2. What steps are needed to stabilize the vehicle?
3. Will you perform a trunk tunneling technique to free the victim?

Introduction

"Before anything else, preparation is the key to success."
—Albert Einstein

Well-rounded technical rescuers have a diverse repertoire of alternative techniques that they are familiar with and they can apply at any incident. This chapter will discuss several alternative methods that can be utilized on unique incidents that may be encountered. Although some of these techniques may never be used, it is best to be prepared with as many options as possible to resolve an incident the moment a unique scenario is presented.

Tunneling

As discussed briefly in Chapter 8, *Vehicle Stabilization*, tunneling is the process of gaining entry through the rear trunk area of a vehicle. The technique is more commonly performed for a vehicle that is resting on its roof, but it can be performed for a vehicle in any resting position following a multitude of crash scenarios, such as a vehicle underride where the impact has caused access to the doors and roof to be blocked off. At first glance, a tunneling scenario can be intimidating and can appear time consuming; however, with practice this technique can be accomplished fairly quickly and can enable in-line removal of the victim, thus minimizing the need for victim manipulation. A word of caution about attempting any tunneling operation involving hybrid, fuel cell, and alternative fuel vehicles: These vehicles may have high-pressure storage fuel tanks and/or battery packs in the trunk area, so an alternative method of entry is recommended.

Let's consider a sample scenario. The vehicle on the right in Figure 10-1 ▶ suffered a side-impact collision that flipped the vehicle over, blocking entry on both sides. The best option for this scenario is to tunnel through the back of the vehicle to gain access and extricate the victim. Using this scenario, start the technique by stabilizing the scene and stabilizing the vehicle.

There are several tools that need to be set up and ready to utilize, such as an air chisel with several spare air bottles, an electric-powered reciprocating saw, large bolt cutters, and the hydraulic spreader and cutter. With the vehicle properly stabilized, the focus will be on removing the trunk in order to create a very large opening. The objective is to expose the latching mechanism using the hydraulic spreader and then cut the latching mechanism to release the trunk cover. In this scenario, gaining access to the latching mechanism will be demonstrated using a combination of the hydraulic spreader/cutter. This tech-

Figure 10-1 This vehicle on the right suffered a side-impact collision that flipped it over, blocking entry on both sides.

nique can also be accomplished using a reciprocating saw and/or air chisel. The decision for what tool or combination of tools to be used will depend on the type of vehicle, the body design, and the presentation of the vehicle.

Follow the steps in **Skill Drill 10-1 ▾** to tunnel through the trunk of a common passenger vehicle:

1. Don appropriate personal protective equipment (PPE), including mask and eye protection.
2. Assess the scene for hazards and complete the inner and outer surveys.
3. Stabilize the vehicle. (**Step 1**)
4. If accessible, ensure that the victim is covered with a blanket for protection from flying debris.
5. Remove all attainable vehicle glass using the appropriate technique.
6. Locate the area where the trunk lip meets the rear bumper; this area will have to be forced open by creating a purchase point opening. Use the hydraulic spreader to squeeze/crush the area at the downward bend of the trunk cover. This will cause the lip section of the trunk to crumple inward and raise just enough to create a purchase point. Gaining access to the latching mechanism on the trunk can also be accomplished with the reciprocating saw and/or an air chisel.

7. Once a purchase point is established, utilize the hydraulic spreader and work the area around the latching mechanism in order to expose the latching mechanism. Do not try to force the trunk open by spreading it off its latch; this will only cause the entire vehicle to shift and jeopardize the integrity of the stabilization struts and cribbing that were just established. Determine if there are any hazardous or heavy storage items in the trunk. (**Step 2**)
8. As a safety measure, place a temporary section of cribbing/strut under the trunk to support the trunk from falling open when the latching mechanism is cut.
9. Once the area has been cleared of any storage hazards, insert the hydraulic cutter and cut the latch area to release the trunk cover. (**Step 3**)

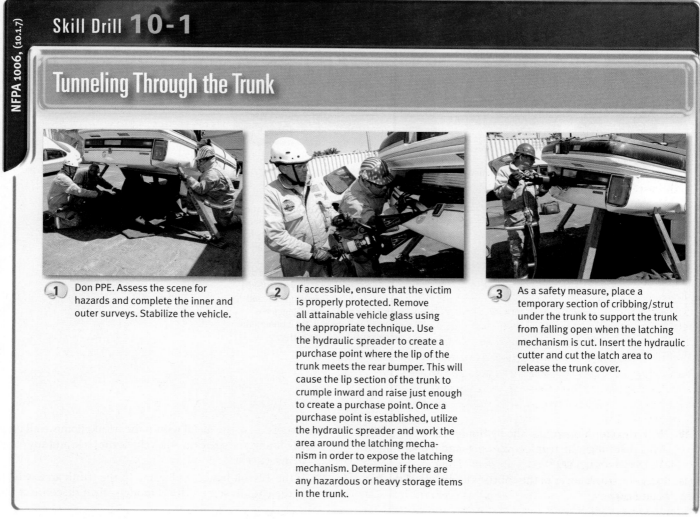

Skill Drill 10-1

NFPA 1006, (10.1.7)

Tunneling Through the Trunk

1 Don PPE. Assess the scene for hazards and complete the inner and outer surveys. Stabilize the vehicle.

2 If accessible, ensure that the victim is properly protected. Remove all attainable vehicle glass using the appropriate technique. Use the hydraulic spreader to create a purchase point where the lip of the trunk meets the rear bumper. This will cause the lip section of the trunk to crumple inward and raise just enough to create a purchase point. Once a purchase point is established, utilize the hydraulic spreader and work the area around the latching mechanism in order to expose the latching mechanism. Determine if there are any hazardous or heavy storage items in the trunk.

3 As a safety measure, place a temporary section of cribbing/strut under the trunk to support the trunk from falling open when the latching mechanism is cut. Insert the hydraulic cutter and cut the latch area to release the trunk cover.

(Continues)

Skill Drill 10-1

Tunneling Through the Trunk *(Continued)*

4 With the trunk open, use the hydraulic cutter to cut the arms that hinge the trunk cover open and totally remove the trunk cover.

5 Place the trunk cover in the debris pile. Expose all of the metal within the trunk. Remove all interior lining, the spare tire, the vehicle tire jack, and any other items.

6 If the 12-volt battery is located in the trunk area, disconnect the battery. If the trunk uses tension rods, which assist in lifting the trunk cover, maintain a safe position and cut and remove using bolt cutters or a hydraulic cutter.

7 Remove the rear dash using a reciprocating saw, an air chisel, or a combination of both.

8 Remove the rear seats using the appropriate technique. Remove the front seat backs. This is best accomplished by exposing the hinges on the seatback by cutting away the seat material with a knife or trauma shears. Also, if the 12-volt battery has not been disconnected and/or cannot be accessed, scan the seats and roof structure for any indications of supplemental restraint system (SRS) components (air bags) and proceed with caution. Once the hinges are exposed, cut them using a pneumatic air chisel or hydraulic cutter.

---

10. With the trunk open, use the hydraulic cutter to cut the arms that hinge the trunk cover open and totally remove the trunk cover. (**Step 4**)

11. Place the trunk cover in the debris pile outside of the hot/action zone.

12. Expose all of the metal within the trunk. Remove all interior lining, the spare tire, the vehicle tire jack, and any other items. (**Step 5**)

13. If the 12-volt battery is located in the trunk area, which is the case in some vehicle models, then disconnect the battery utilizing the techniques described in Chapter 8, *Vehicle Stabilization*, Skill Drill 8-5.

14. If the trunk uses tension rods, which assist in lifting the trunk cover, they will have to be cut and removed. The rods are connected to the trunk cover's hinge arms and are normally composed of spring steel, which uses the backseat or rear dash area to push against to gain tension. The rods are under tension and may release forcefully when cut. Maintain a safe position and temporarily remove personnel from the area until after the rods have been removed. Utilizing industrial bolt cutters, carefully remove the rods by cutting them closest to the ends where they attach to the trunk hinge arms; this tool provides better maneuverability and will be much lighter to work with than the hydraulic cutter. The hydraulic cutter will have to be utilized if the rods cannot be cut by bolt cutters, but again, maneuverability and weight of the tool will be a factor. An air chisel or reciprocating saw will not work effectively in this situation because there is too much free movement of the rods for either tool to penetrate the metal. (**Step 6**)

15. Remove the rear dash; this is the area that separates the trunk from the rear passenger compartment. The vehicle's stereo speakers may be mounted to this section of metal. The best tool for this application will either be a reciprocating saw or an air chisel, or a combination of both; the choice will be based on the amount of room or access that is presented to insert either tool. (**Step 7**)

16. Remove the rear seats using the appropriate technique. Rear seats can be attached in multiple ways, so it is difficult to explain one single way to remove them which encompasses all of the techniques. It is best to access the attached areas and utilize the appropriate tool for removing the seat. In some instances, the rear seats may be free floating, which makes removal a simple process.

17. Remove the front seat backs. This is best accomplished by exposing the hinges on the seatback by cutting away the seat material with a knife or trauma shears. Also, if the 12-volt battery has not been disconnected and/or cannot be accessed, scan the seats and roof structure for any indications of SRS components (air bags) and proceed with caution. Once the hinges are exposed, cut them using a pneumatic air chisel or hydraulic cutter (**Step 8**).

18. Immobilize and package the patient for removal on a backboard.

Rescue Tips

Before opening the trunk of a vehicle on its roof, determine if there are any hazardous or heavy storage items that may be in the trunk that can potentially come crashing down.

Seat Removal

Once a rescuer has tunneled inside the vehicle through the trunk, he or she may encounter a variety of patient entrap-

ment scenarios. If the opening is large enough, two rescuers should be positioned inside the passenger compartment. One rescuer will perform the patient disentanglement by gaining any additional access and removing sections of the vehicle that are entrapping the patient; the other rescuer will perform patient care as described in Chapter 9, *Victim Access and Management*. If the entrapment is heavy, evaluate what is impinging the victim and utilize the correct tool or tools to release the victim; it may take an air chisel or hydraulic cutter to remove the seats or a hydraulic ram to create more room.

Front Seat Removal

Removal of a front seat can be difficult at times because of the various types of seat frames and advanced designs that may be encountered. Seats are attached to the floorboards of the vehicle in four or more places, with positional adjustments on slide tracks that are either fully automatic/motorized or manually operated. The floor attachments can be accessible in the simplest designs, or they may be fully enclosed and motorized with advanced designs. Seats will also contain hinges that allow the seat backs to adjust forward or rearward either automatically/motorized or manually. The seat back hinges are normally the best area of access for removing the seat back from the seat frame. The material covering the seats may have to be removed using a knife or trauma shears to expose the area of attachment (the steel hinges) where the seat back is attached to the bottom seat section.

To remove a front seat from a vehicle, follow the steps in **Skill Drill 10-2 ▶**:

1. Don appropriate PPE, including mask and eye protection.
2. Assess the scene for hazards and complete the inner and outer surveys.
3. Stabilize the vehicle.
4. Ensure that the victim is covered with a blanket for protection from flying debris.
5. Remove all attainable vehicle glass using the appropriate technique. (**Step 1**)
6. Manually check all of the seat adjustment mechanisms for operability if needed.
7. Disconnect the 12-volt battery utilizing the techniques described in Chapter 8, *Vehicle Stabilization*, Skill Drill 8-5.
8. Carefully use a hand knife or trauma shears to cut away the upholstery of the front seat in order to expose the metal frame of the seat and the areas of attachment; this technique will allow you to avoid having to perform any blind cuts. (**Step 2**)
9. Assess the seat for any side air bag inflators that may be located in the seat backs. Fully expose the inflator and move any wires out of cutting areas.
10. Remove the front seat by cutting the seat back hinges or cutting the frame attachment at the floorboard using an air chisel or hydraulic cutter; this procedure may vary with each vehicle because of the differences in patient location, interior design, and frame structure of the seat. (**Step 3**)

NFPA 1006, (10.1.7)

Skill Drill 10-2

Removing the Front Seat

1 Don PPE. Assess the scene for hazards and complete the inner and outer surveys. Stabilize the vehicle. Ensure that the victim is properly protected. Remove all attainable vehicle glass using the appropriate technique.

2 Check all seat adjustment mechanisms if needed. Disconnect the 12-volt battery. Cut away the upholstery of the front seat in order to expose the metal frame of the seat.

3 Assess the seat for any side air bag inflators that may be located in the seat backs. Fully expose the inflator and move any wires out of cutting areas. Remove the front seat by cutting the seat back hinges or cutting the frame attachment at the floorboard using an air chisel or hydraulic cutter. This procedure may vary with each vehicle because of the differences in patient location, interior design, and frame structure of the seat.

Impingement and Penetrating Objects

Impingement

The <u>cross ram technique</u> is used when the impingement of metal on the victim requires a unique mechanism of movement that cannot be completed with a hydraulic spreader. For example, suppose an extremity gets trapped between the interior section of the B-post and the side of the seat back. This scenario requires very precise movements to avoid further impingement of metal; any spreading or cutting in the area before the pressure of the entrapment is relieved can potentially drive the impingement further onto the victim. The objective for this scenario is to push the B-post back to its original position or just slightly beyond. A hydraulic ram, particularly a telescoping ram, can accomplish this task by providing the ability to maneuver the tool's arm to push out multiple areas, thus relieving the pressure of the impinging metal on the victim **Figure 10-2 ▶**. In the cross ram application, the hydraulic ram has the advantage over a hydraulic spreader because the tool can be used in tight spaces; for example, the ram can be placed to push off the center transmission hump or the opposite B-post. The hydraulic ram can also be used to push up a crushed roof section, creating additional room and eliminating the potential for the roof to

push down on the victim while it is being cut for removal. A crushed roof with impingement on the victim needs to be raised off the victim using a hydraulic tool before any posts or sections are cut, because the metal will draw down on the victim as it is cut.

When performing a cross ram technique, have several 4- by 4-inch (102- by 102-millimeter [mm]) sections of cribbing (or "four-by-fours"), wedges, and shims to insert between the B-post and seat back, above and below the extremity, once the impingement is relieved by the hydraulic ram **Figure 10-3 ▶**. The cribbing sections will retain the space created by the ram, preventing any potential collapse of the metal back onto the extremity if the door will be removed.

Be aware of the potential for tool slippage. This can occur during a cross ram operation when the hydraulic ram is extended under pressure and is forcing the B-post and door frame to spread beyond their vertical platform. The surface contact of the hydraulic ram tip is very small and can easily slip off, firing up or down under extreme force when a sloping angle is created by the B-post/door frame pushing outward. It is imperative that all personnel be made aware of this potential hazard. Slippage can be prevented by recognizing the signs and repositioning the tool. Adding a four-by-four between the B-post/door frame and the tip of the hydraulic ram may work in

Voices of Experience

I currently serve as a captain on a technical rescue team for the Broward Sheriff's Office Fire Rescue Department. Our team responded via mutual aid to a city department early one morning to assist with a two-vehicle accident with people trapped in both vehicles. Upon arrival my unit was assigned to extricate the driver of a two-door Lexus. The damage to the vehicle was so extensive that the front end had been pushed into the interior, pinning the driver's lower extremities and chest. The rear of the vehicle also sustained major damage from hitting a tree when it spun around from the initial impact of the head-on collision.

On our first attempt we had difficultly gaining access with the hydraulic tools. I then backed my crew out to reorganize, and I made the decision to do something we had trained on years earlier. I used two tow trucks to open the vehicle, which looked like a clamshell that had closed around the driver. One tow truck was positioned in front of the vehicle with the tow cable secured around the A-posts. The second tow truck was positioned to the rear of the vehicle with the tow cable secured to the C-posts. We made a relief cut across the roof of the vehicle; the vehicle had a sun roof, so we didn't have much roof material to cut. Once this was accomplished, both wreckers slowly started simultaneously to take up slack on the cables. This caused the vehicle to open up like a clam. This winch process was done very slowly and precisely. The operation took only a moment once the relief cut was performed, and we had complete access to the vehicle. We were able to open up the vehicle so much that we could actually walk through it to render care atop the patient.

> **"The operation took only a moment once the relief cut was performed, and we had complete access to the vehicle."**

We were successful on this call because we have a working relationship with the wrecker/tow companies in our area. We were able to receive training from these companies, share ideas, and show them our capabilities and equipment.

Looking at the call in retrospect, a few items come to mind. Fire/rescue personnel must perform as they are trained. Working on an actual call is not the time to start winging it. Rescue operations that are successful are due to the training and structured application the crews have received and practiced. Do not attempt to do something that you just saw or have not performed successfully several times in training. Also, have a plan B. It's okay to change your approach if plan A is not working.

If you properly and consistently train with your crew, then have faith in them. As a company officer, you need to lead by example. Your crew is an extension of you.

Mike Nugent
Broward Sheriff's Office Department of Fire Rescue
Broward County, Florida

Figure 10-2 A hydraulic ram, particularly a telescoping ram, can push the B-post back by providing the ability to maneuver the tool's arm to push out multiple areas, thus relieving the pressure of the impinging metal on the victim.

theory by increasing the footprint of the ram, but in reality the four-by-four will normally roll, break, or kick out under extreme force; therefore, this is not a recommended practice. If the seat belt harness is positioned low enough on the B-post, then this would be a good place to position the tip of the hydraulic ram to push from. Remember from the vehicle anatomy chapter that there is a steel reinforcement plate located inside the B-post directly behind the seat belt harness, giving this area added strength and support.

The following Skill Drill outlines a cross ram technique. In this scenario, an occupant traveling in the rear passenger area has an extremity trapped at the B-post area following a side-impact collision. Tools needed to perform a cross ram technique include a large hydraulic ram, preferably one that is telescopic and extends to 50 inches or greater, assorted cribbing sections, and a backboard.

Figure 10-3 When performing a cross ram technique, have several four-by-four sections of cribbing, wedges, and shims to insert between the B-post and seat back, above and below the extremity, once the impingement is relieved by the hydraulic ram.

To perform a cross ram operation utilizing a hydraulic ram, follow the steps in **Skill Drill 10-3 ▶**:

1. Don appropriate PPE, including mask and eye protection.
2. Assess the scene for hazards and complete the inner and outer surveys.
3. Stabilize the vehicle.
4. Position a rescuer inside the vehicle to ensure that the victim is properly protected from flying debris.
5. Remove all vehicle glass using the appropriate technique.
6. If the 12-volt battery can be accessed, then disconnect it utilizing the techniques described in Chapter 8, *Vehicle Stabilization*, Skill Drill 8-5.
7. Scan the vehicle for all/any SRS components (air bags), including exposing all the roof posts and roof liner. (**Step 1**)
8. Remove the roof of the vehicle utilizing the appropriate technique described in this chapter. The roof may need to be lifted first if it is impinging on the patient. (**Step 2**)
9. Position several rescuers inside and around the vehicle for support.
10. The patient will require certain medical intervention depending on the injuries presented. Utilize department medical protocols.
11. Properly immobilize the patient, and be prepared to insert cribbing sections around the crushed extremity once the B-post is pushed back into position.
12. Place the base of the hydraulic ram at the B-post farthest from the patient, and position a backboard to protect the patient. The backboard is for hard protection only; the hydraulic ram does not rest on the backboard. (**Step 3**)
13. Slowly open the hydraulic ram and position the tip at the B-post that is crushing the patient's extremity (look for the seat belt harness for support). (**Step 4**)
14. Open the hydraulic ram and push the B-post back to its normal vertical position just enough to remove the extremity. Be careful not to overextend the post beyond the verti-

NFPA 1006, (10.1.7)

Skill Drill **10-3**

Performing a Cross Ram Operation Utilizing a Hydraulic Ram

1 Don PPE. Assess the scene for hazards and complete the inner and outer surveys. Stabilize the vehicle. Ensure that the victim is properly protected. Remove all vehicle glass. If the 12-volt battery can be accessed, then disconnect it. Assess for SRS components.

2 Remove or lift the roof of the vehicle utilizing the appropriate technique.

3 Position rescuers inside and around the vehicle for support. Properly immobilize the patient, and be prepared to insert cribbing sections around the crushed extremity once the B-post is pushed back into position. Place the base of the hydraulic ram at the B-post farthest from the patient, and position a backboard to protect the patient.

4 Slowly open the hydraulic ram, and position the tip at the B-post that is crushing the patient's extremity.

5 Open the hydraulic ram and push the B-post back to its normal vertical position just enough to remove the extremity. Place cribbing underneath the extremity to support the post from potentially collapsing back in.

6 Support the extremity from both sides and perform a backboard slide technique of the patient.

cal position because this can cause the hydraulic ram to slip from its position.

15. Place cribbing underneath the extremity to support the post from potentially collapsing back in if the hydraulic ram were to slip or be removed. (**Step 5**)

16. Support the extremity from both sides and perform a backboard slide technique of the patient utilizing the technique described in Chapter 9, *Victim Access and Management*, Skill Drill 9-1. (**Step 6**)

Rescue Tips

Purchase an inexpensive tape measure, and using a permanent marker, mark on the tape measure the closed and open lengths (in inches) of each ram carried on the rescue unit. Use this tape measure to quickly determine which ram(s) will or will not fit in a given space before trying to get the ram into the damaged vehicle.

Penetrating Objects

Penetrating or impaled objects in patients can occur from sources outside the vehicle, such as road debris, vehicle components, or vehicle cargo, or from objects inside the vehicle the passenger is traveling in. When an object impales an occupant, the main objective is to stabilize the penetrating object at the entry point and the exit point, if an exit point exists. Normally the penetrating object will need to be cut at the entry and exit areas so the patient can be properly extricated from the vehicle.

There are multiple impalement scenarios that can occur as well as multiple types of penetrating objects. This section will cover one such scenario involving cargo that came loose from a truck traveling at approximately 50 miles per hour (mph) (80 kilometers per hour [kph]) and entered the vehicle traveling behind it. The cargo consisted of sections of ½-inch (13-mm) steel rebar and steel galvanized fencing post. One of these objects penetrated the vehicle from the front windshield and impaled the occupant in the front passenger seat at the area of the left shoulder, passing approximately 2 feet (1 meter) through the seat back. Upon your arrival, the patient is in stable condition. The vehicle has not sustained any major damage, and no air bags have been deployed.

This is a very specific type of scenario that will encompass a majority of essential steps required for similar types of incidents that you may encounter. Obviously, some incidents will require additional steps or other interventions, but these are situational and can only be determined at the time the incident occurs. The cutting tools required for this scenario will be a metal-cutting circular saw with a carbide-tip blade, or a pneumatic cutoff tool (whizzer tool) with a metal-cutting carbide blade, and several spare air cylinders. In addition, there should be a basic apparatus toolbox containing several types of hand tools, various cribbing sections, a water extinguisher, webbing material, towels, blankets, and pillows from a transport rescue. Utilizing a hydraulic cutter, a reciprocating saw, or a handsaw for this scenario will cause too much movement or torquing of the impaled object, possibly causing further injury and pain to the patient.

To stabilize an impaled object, follow the steps in **Skill Drill 10-4 ▶** :

1. Don appropriate PPE, including mask and eye protection.
2. Assess the scene for hazards and complete the inner and outer surveys.
3. Stabilize the vehicle.
4. If the 12-volt battery can be accessed, then disconnect it utilizing the techniques described in Chapter 8, *Vehicle Stabilization*, Skill Drill 8-5.
5. Ensure that the victim is properly protected from flying debris.
6. Remove all vehicle glass using the appropriate technique.
7. Scan the vehicle for all/any SRS components (air bags), including exposing all the roof posts and roof liner.
8. Consider removal of the passenger side door that provides direct access to the patient as well as the roof. This may seem extreme, but this type of incident requires a large amount of room in order to access the victim, providing a minimal amount of patient and object manipulation.
9. Manually immobilize the patient and stabilize the impaled object. Determine medical interventions necessary depending on the injuries presented. Utilize department medical protocols. (**Step 1**)
10. Ensure that the victim is still covered with a blanket for protection from flying debris when the impaled object is cut.
11. As a precaution, position a rescuer inside the vehicle with a water extinguisher to be applied, if needed. The impaled metal object may need to be cooled as it is cut or the rescuer may need to suppress any sparks that may occur. This is rare with a metal-cutting circular saw but can be expected when using a cutoff saw. Using the cutoff saw will require more cutting time to penetrate and cut through the material. To suppress sparks, fan the tip of the water extinguisher with a gloved finger to mist the blade. The blanket or towels used to cover the patient may be wet for additional protection.
12. Safely position the metal-cutting circular saw over the metal impaled object, whether it is the steel rebar or steel galvanized fencepost. Pull back the safety arm to release the blade; with two hands, operate the tool and cut off a section of the impaled object as close to the patient as possible. (**Step 2**)
13. Once the impaled object is cut, remove it and place it in a debris pile outside of the hot zone.
14. Cut the center out of two or more pillows and slip the pillows over the front end of the impaled object and secure with tape. (**Step 3**)
15. Cut the seat back hinges utilizing the steps outlined in this chapter. Using a knife or trauma shears, expose the seat back hinges. If the seat back is light enough, it can be utilized as support of the object and held in place with tape. If the seat back can be removed with minimal manipulation, then do so (this is a situational procedure that can only be determined at the time of the incident). The impaled object may require support from pillows if the seat back is removed. (**Step 4**)
16. Remove the patient in a seated position and transfer immediately to a waiting stretcher. (**Step 5**)

Roof Removal

When cutting a roof with an air chisel or reciprocating saw, examine the roof damage for any tensioning or torsion deformities caused by the impact of the crash. Understanding this is very important because it will determine the proper angle to start the cut. Remember to remove all of the plastic molding on the posts and roof rail, and always keep half of the tip of the chisel blade showing through the entire cut, which will prevent the blade from getting stuck.

When a post is under tension or has a torsional bend to it, it is under a lot of pressure. Cutting such a post in the wrong location can cause the chisel blade or the blade of a recipro-

NFPA 1006, (10.1.8)

Skill Drill 10-4

Stabilizing an Impaled Object

1 Don PPE. Assess the scene for hazards and complete the inner and outer surveys. Stabilize the vehicle. If the 12-volt battery can be accessed, disconnect it. Ensure that the victim is properly protected from flying debris, and remove all vehicle glass. Scan the vehicle for SRS components. Consider removal of the passenger side door that provides direct access to the patient as well as the roof. Manually immobilize the patient and stabilize the impaled object. Determine medical interventions necessary.

2 Ensure that the patient is still covered with a blanket for protection from flying debris. Position a rescuer inside the vehicle with a water extinguisher to be applied, if needed. Safely position the metal-cutting circular saw or cutoff saw over the metal impaled object, and pull back the safety arm to release the blade. With two hands, cut off a section of the impaled object as close to the patient as possible.

3 Place the impaled object in a debris pile. Cut the centers out of two or more pillows and slip both pillows over the front end of the impaled object and secure with tape. If the seat back is light enough, it can be utilized as support of the object and held in place with tape.

4 If the seatback can be removed with minimal manipulation, then do so (this is a situational procedure that can only be determined at the time of the incident).

5 Remove the patient in a seated position and transfer to a stretcher.

cating saw to pinch off and become stuck between the metal, stopping or jamming the tool. Always cut in an area away from the bend or torsion to minimize the chance of the chisel blade or reciprocating saw blade getting trapped by the metal collapsing around it from the pressure release. Take a moment and visualize how the post will react and where the metal will move when it is cut. The concept is similar to cutting off a hanging tree limb using a hand or chain saw. A tree limb is under natural tension from supporting itself; if a straight cut is made through

the center of the branch, chances are that midway through the cut the blade will become pinched off and stuck because both sides of the limb are bending by the increasing pressure to separate. It would be wiser to cut the branch at an area closer to the trunk and use the support and weight of the entire tree. The same theory can be applied to cutting a roof post under tension or torsional pressure; cut at the base of the post using the weight of the vehicle, or cut at the roof line area if that area is not compromised.

It is important to watch the reaction of the post throughout the entire cut; there may be a time when only half of the post can be cut at one angle, and then another angle may be needed to complete the cut to avoid the tension releasing and trapping the chisel or reciprocating saw blade. An option that can assist the cutting action for either tool is using a steel wedging tool such as a large flat-head screwdriver or another flat chisel blade, which can be inserted in the cut behind the tool and used as a block to prevent the tension of the cut roof metal from collapsing on the blade as it cuts through the rest of the roof area.

Some additional tips to remember when using an air chisel or reciprocating saw include the following:

- When cutting through a large or wide post, a pie cut or inspection cut will have to be made to peel back the outer metal panel to reveal the inside panel so the cut can be completed. An air chisel blade cannot cut through a wide two-piece roof post without completing this step first.
- When using a reciprocating saw, try to cut at a semi-downward angle to take advantage of the vehicle's center of gravity and thereby limit tool vibration; to further limit vibrations, remember to keep the shoe of the tool firmly against the object being cut.
- As a post is being cut, if there seems to be an area that is very difficult to cut through where the penetration seems to halt, the probability of an impact or reinforcement bar located in the post is very high. Attempt the cut at a lower section of the post, preferably at the base.
- Remember that regardless of the tool utilized, hard protection should always be used to protect the victim and the rescuer inside the vehicle; if it is possible, try to angle or position all cuts away from the victim to eliminate any potential accidents.

■ Roof Removal Using the Air Chisel

As discussed in Chapter 6, *Tools and Equipment*, a high-pressure air chisel is an excellent tool for extrication purposes when it is in the hands of a well-trained technical rescuer. An air chisel requires continual repetitious training to keep skills at the optimum level needed to be effective on extrication incidents.

Multiple blades come with most air chisel kits; the two most widely used are the panel cutter blade, or T-blade, and the flat/curved blade **Figure 10-4 ▶**. The two blades are different in use and appearance. The panel cutter blade is used for cutting shallow, straighter cuts on small-gauge sheet metal, and the flat/curved blade is used for cutting through medium- to heavy-gauge steel.

Success with an air chisel is based on maintaining control throughout the entire operation. With this high-powered pneumatic system, it is very easy for an inexperienced operator to lose control of the blade, while expending a bottle of compressed air within minutes and having very little progress to show from it.

To remove a roof using an air chisel, follow the steps in **Skill Drill 10-5 ▶**:

1. Don appropriate PPE, including mask and eye protection.
2. Assess the scene for hazards and complete the inner and outer surveys.

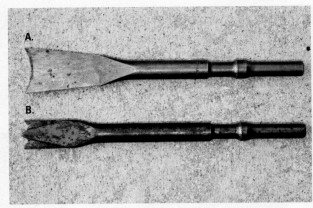

Figure 10-4　The two most widely used air chisel blades. **A.** The standard/flat curved blade (top). **B.** The panel cutter blade (bottom).

3. Stabilize the vehicle. (**Step 1**)
4. If the 12-volt battery can be accessed, disconnect it utilizing the techniques described in Chapter 8, *Vehicle Stabilization*, Skill Drill 8-5. (**Step 2**)
5. Ensure that the victim is covered with a blanket for protection from flying debris.
6. Remove all vehicle glass utilizing the appropriate technique.
7. If possible, position a rescuer inside the vehicle to provide medical support and immobilization.
8. Insert the long flat/curved blade into the air chisel. Hook up the regulator to the air cylinder and attach the hose to the regulator and the air chisel. Turn the air cylinder valve on with the blade and air chisel facing the ground to avoid an accidental discharge of the blade. Never free-fire the air chisel; doing so can damage the tool. Always check the pressure gauge by firing the tool against a solid surface and then dial up to the appropriate pressure in the general range of 150 to 200 pounds per square inch (psi) (1034 to 1379 kilopascals [kPa]) (refer to the manufacturer air pressure specifications for the model of tool that your organization utilizes).
9. Remove all of the plastic and molding around the posts and roof rail to identify any potential air bag cylinders. This can be completed quickly with a large flat-head screwdriver, other small hand tool, or, with practice, the end of the flat/curved chisel blade. Try to avoid cutting into any plastic or rubber material because the air chisel blade will only bounce off the material as opposed to cutting through it.
10. Check the roof for areas of damage causing metal tension and torsion. Cut around this area to avoid getting a blade pinched off by the metal when the pressure is suddenly relieved by the cut.
11. Ensure that all the seat belts have been cut.
12. Starting at the A-post, create a purchase point with the air chisel. Grip the blade in one hand, midshaft between the tip and collar; this will provide better control to maneuver

NFPA 1006, (10.1.7)

Skill Drill 10-5

Removing a Roof Using an Air Chisel

1 Don PPE. Assess the scene for hazards and complete the inner and outer surveys. Stabilize the vehicle.

2 If the 12-volt battery can be accessed, then disconnect it.

3 Ensure that the victim is properly protected from flying debris. Remove all vehicle glass utilizing the appropriate technique. If possible, position a rescuer inside the vehicle to stabilize the victim. Insert the long flat/curved blade into the air chisel. Hook up the regulator to the air cylinder and attach the hose to the regulator and the air chisel. Turn the air cylinder valve on. Remove all of the plastic and molding around the posts and roof rail to identify any potential air bag cylinders. Check the roof for areas of damage causing metal tension and torsion. Cut around this area. Ensure that all seat belts have been cut. Starting at the A-post, create a purchase point with the air chisel. Once the blade has penetrated the metal, position the tip so at least half of the blade is always showing.

4 If there is a cut strip of glass attached to the A-post that was left from cutting out the windshield, then ensure that the majority of the glass in proximity to the area to be cut is removed prior to cutting the post, and that all personnel are wearing the appropriate eye and respiratory protection. If glass is present, apply duct tape around the post in the area to be cut; the tape will hold a lot of the glass in place as the blade passes through it. If a boron rod or other advanced-strength steel-reinforced section is encountered, avoid the material. Work the tool around the A-post, mimicking the curvature of the post. Once a section of metal is loosened, pull or bend it out of the way to get to the inside of the post. Once the A-post is cut, move to the B-post. Choose an area at least 2 inches (51 mm) above or below the seat belt bracket. Another rescuer should be assigned to support the roof.

5 The technique used to cut the C-post will depend on the width of the post. A wide C-post is composed of two pieces of steel with a hollow center. It will require a pie cut or sectional cut on the outside panel to expose the inside panel so the cut can be completed through both sections. Once the pie cut has been made, the blade of the tool can be inserted to assist in peeling back the section cut.

6 Move to the posts on the other side of the vehicle and repeat the steps. Two rescuers must be assigned to support both sides of the roof as it is cut. Place the roof in a designated debris pile outside of the hot zone.

the blade. Place one edge of the blade on the metal and depress the trigger of the tool to make a purchase point. Once the blade has penetrated the metal, position the tip so at least half of the blade is always showing. It is extremely important to remember to keep half of the blade showing throughout the entire movement to avoid burying/trapping the blade; this is a common critical error made by inexperienced operators. (**Step 3**)

13. If there is a cut strip of glass attached to the A-post that is left from cutting out the windshield, then ensure that the majority of the glass in proximity to the area to be cut is removed prior to cutting the post, and that all personnel are wearing the appropriate eye and respiratory protection. To minimize glass fragmentation, apply duct tape around the post in the area to be cut; the tape will hold a lot of the glass in place as the blade passes through it. If a boron rod or other advanced-strength steel-reinforced section is encountered, then reposition the cut higher or lower on the post to avoid the material. Work the tool around the A-post, mimicking the curvature of the post; do not keep the tool in one place. Once a section of metal is loosened, pull or bend it out of the way to get to the inside of the post. A-posts are generally composed of rolled sheets of metal for reinforcement purposes; once one layer is removed, another layer may be revealed underneath. If the technical rescuer fails to remove the layers as he or she works, then there is a chance of burying the blade because the blade is not wide enough to maintain keeping half the blade out. If uncertain whether the post has been cut all the way through, a hand light can be shone through the back side of the cut.

14. Once the A-post has been completely cut, move to the B-post. Avoid cutting into the seat belt bracket, which has a reinforced steel backing plate. Choose an area at least 2 inches (51 mm) above or below the seat belt bracket. At this point, another rescuer should be assigned to support the roof. (**Step 4**)

15. The technique used to cut the C-post will depend on the width of the post. A wide C-post is composed of two pieces of steel with a hollow center. It will require a pie cut or sectional cut on the outside panel to expose the inside panel so the cut can be completed through both sections. Once the pie cut has been made, the blade of the tool can be inserted to assist in peeling back the section cut. The inside panel may have holes in it that are used to pass wires or other materials through. These holes are beneficial because they mean less metal for the operator to cut. (**Step 5**)

16. Move to the posts on the other side of the vehicle and repeat the steps performed on the first side. Two rescuers must be assigned to support both sides of the roof as it is cut.

17. Before the roof is removed, ensure that all attachments such as plastic interior liners or wires have been cut. Removal of the severed roof must be a coordinated and controlled movement, with one technical rescuer giving the order and direction of travel. Place the roof in a designated debris pile outside of the hot zone. (**Step 6**)

Roof Removal Using the Reciprocating Saw

A reciprocating saw is an excellent tool for removing a vehicle's roof. It is fast and efficient, especially on vehicles with large C-posts. An electric reciprocating saw with a power output of around 11 to 15 amps will enable the operator to effectively cut through a wide range of grades and types of steel. Use a 9-inch (229-mm) bi-metal cutting blade with a TPI (teeth per inch) rating between 9 and 14. A lower rating, such as 5 TPI, will cause the tool to vibrate heavily, and the teeth will break off because of the inability of the blade to penetrate the metal effectively. A blade with a higher TPI rating, such as 18, is not aggressive enough and will dull quickly. Some reciprocating saws come equipped with a speed setting dial or switch; this should be set at the highest setting. It is recommended by some manufacturers to start the cut at a lower speed setting to establish the cut and then gradually build up to a higher setting as the cut progresses. To limit the possibility of tool failure or voiding the manufacturer's warranty, always check with the tool manufacturer for the proper tool operation and safety specifications.

To remove a roof using a reciprocating saw, follow the steps in **Skill Drill 10-6 ▸** :

1. Don appropriate PPE, including mask and eye protection.
2. Assess the scene for hazards and complete the inner and outer surveys.
3. Stabilize the vehicle.
4. If the 12-volt battery can be accessed, then disconnect it utilizing the techniques described in Chapter 8, *Vehicle Stabilization*, Skill Drill 8-5.
5. If possible, position a rescuer inside the vehicle to provide medical support and immobilization.
6. Ensure that the victim is covered with a blanket for protection from flying debris.
7. Remove all glass utilizing the appropriate technique. (**Step 1**)
8. Remove all of the plastic and molding around the posts and roof rail to identify any potential air bag cylinders. This can be completed quickly with a large flat-head screwdriver or other small hand tool. If the material cannot be removed, wrap the post with duct tape in the area to be cut; doing so will help keep everything together as the cut is made.
9. Ensure that all the seat belts have been cut.
10. Check the roof for areas of damage causing metal tension and torsion. Cut around these areas to avoid getting a blade pinched off by the metal under pressure when the pressure is suddenly relieved by the cut. (**Step 2**)
11. Begin at the A-post. If there is a cut strip of glass attached to the A-post that is left from cutting out the windshield, then ensure that the majority of the glass in proximity to the area to be cut is removed prior to cutting the post, and that all personnel are wearing the appropriate eye and respiratory protection. To minimize glass fragmentation, apply duct tape around the post in the area to be cut; the tape will hold a lot of the glass in place as the blade passes

NFPA 1006, (10.1.7)

Skill Drill 10-6

Removing a Roof Using a Reciprocating Saw

1 Don PPE. Assess the scene for hazards and complete the inner and outer surveys. Stabilize the vehicle. If the 12-volt battery can be accessed, then disconnect it. If possible, position a rescuer inside the vehicle to provide medical support and immobilization. Ensure that the victim is properly protected from flying debris. Remove all vehicle glass utilizing the appropriate technique.

2 Remove all of the plastic and molding around the posts and roof rail to identify any potential air bag cylinders. Ensure that all seat belts have been cut. Check the roof for areas of damage causing metal tension and torsion. Cut around this area.

3 Begin at the A-post. Ensure that the majority of the glass in proximity to the area to be cut is removed prior to cutting the post, and that all personnel are wearing the appropriate eye and respiratory protection. To minimize glass fragmentation, apply duct tape around the post in the area to be cut. If a boron rod or other advanced-strength steel-reinforced section is encountered, then reposition the cut higher or lower on the post to avoid the material.

4 Once the A-post is cut, move to the B-post. Choose an area at least 2 inches (51 mm) above or below the seat belt bracket. Another rescuer should be assigned to support the roof.

5 Once the B-post is cut, move to the rear C-post. Keep the shoe of the reciprocating saw firmly against the metal and cut on a semidownward angle to take advantage of the vehicle's center of gravity and thereby limit vibrations.

6 Move to the posts on the other side of the vehicle and repeat the steps. Two or more rescuers must be assigned to support both sides of the roof as it is cut. Place the roof in a designated debris pile outside of the hot zone.

through it. If a boron rod or other advanced-strength steel-reinforced section is encountered, then reposition the cut higher or lower on the post to avoid the material. **(Step 3)**

12. Once the A-post is cut, move to the B-post. The technique for cutting the B-post will be exactly the same as the A-post, but avoid cutting into the seat belt bracket, which has a reinforced steel backing plate or adjustment bar. This area is heavy-gauge steel and will require a lot of time to cut through, so choose an area at least 2 inches (51 mm) above or below the seat belt bracket, or cut at the base of the post or roof line to avoid these stronger metals. At this point, another rescuer should be assigned to support the roof. (**Step 4**)

13. Once the B-post is cut, move to the rear C-post. The reciprocating saw should rip through the metal C-post with ease, working best on a wider post. Keep the shoe of the saw firmly against the metal and cut on a semidownward angle to take advantage of the vehicle's center of gravity; this approach will minimize vibration and the potential for pinching off the blade. (**Step 5**)

14. Move to the posts on the other side of the vehicle and repeat the steps performed on the first side. Two rescuers must be assigned to support both sides of the roof as it is cut.

15. Before the roof is removed, ensure that all attachments such as plastic interior liners or wires have been cut. Removing the severed roof must be a coordinated and controlled movement, with one technical rescuer giving the order and direction of travel. Place the roof in a designated debris pile outside of the hot zone. (**Step 6**)

> ### Rescue Tips
> Before the roof is removed, ensure that all the seat belts have been cut.

Rapid Roof Removal: Vehicle on Its Side

There are multiple techniques to remove the roof of a vehicle that is resting on its side. One technique, designed for rapid entry, involves the use of a reciprocating saw and can be accomplished within minutes. Apply the methods described in Chapter 8, *Vehicle Stabilization*, regarding stabilizing a vehicle on its side before initiating this process. To remove the roof of a vehicle on its side, follow the steps in **Skill Drill 10-7 ▶**:

1. Don appropriate PPE, including mask and face protection.

2. Assess the scene for hazards and complete the inner and outer surveys.

3. If the 12-volt battery can be accessed, then disconnect it utilizing the techniques described in Chapter 8, *Vehicle Stabilization*, Skill Drill 8-5.

4. Stabilize the vehicle using buttress stabilization struts as described in Chapter 8, *Vehicle Stabilization*, Skill Drill 8-2. (**Step 1**)

5. If possible, position a rescuer inside the vehicle to provide medical support and immobilization.

6. Ensure that the victim is covered with a blanket for protection from flying debris.

7. Remove all vehicle glass utilizing the appropriate technique. (**Step 2**)

8. Remove all of the plastic and molding around the posts and roof rail to identify any potential air bag cylinders. This can be completed quickly with a large flat-head screwdriver or other small hand tool. (**Step 3**)

9. Ensure that all the seat belts have been cut.

10. Check the roof for areas of damage causing metal tension and torsion. Cut around these areas to avoid getting a blade pinched off by the metal when the pressure is suddenly relieved by the cut.

11. Insert a long backboard lengthwise from the front or rear window of the vehicle; the rescuer inside the vehicle will use this as hard protection as the reciprocation saw blade passes through the roof.

12. Insert a 6-inch (152-mm) bi-metal cutting blade with a TPI rating of around 9 to 14 into a high-powered electric reciprocating saw. Rooftops are normally 2 to 3 inches (51 to 76 mm) thick, so a shorter blade is recommended to minimize the potential for hitting the victim or rescuer. (**Step 4**)

13. Make two lengthwise cuts to remove the inside section of the roof. Starting at the lower section of the roof (closest to the ground), make the initial cut just above the roof rail on the inside corner of the front A-post or at the rear post of the vehicle, which may be a C-post, D-post, or higher post depending on the type of vehicle. It is important to make the cut at the low end of the roof first; if the higher section were to be cut first, the bottom cut would be impossible to make because the roof would collapse onto itself, pinching off the blade of the saw. Follow a straight horizontal line, completing the cut through the roof to the opposite side. As the technical rescuer outside the vehicle makes the cut, the rescuer inside the vehicle will position the backboard to provide hard protection from the blade. A safety officer should be positioned outside of the vehicle, maintaining visual awareness of the blade proximity to the victim or rescuer. (**Step 5**)

14. After the bottom cut has been made, cut the top section. Follow the same cutting pattern used on the bottom section. Position two personnel on either side of the roof to support it and prevent it from falling in on the victim and rescuer as the cut is completed. In a coordinated effort, place the roof section in a designated debris pile outside of the hot zone. (**Step 6**)

15. For added protection, place a tarp or large blanket over the bottom cut section of the roof to prevent any potential cut injuries. (**Step 7**)

To increase the speed of this technique, two reciprocating saws can be used. As the bottom line is cut midway through the roof, another technical rescuer can start cutting the top line with another reciprocating saw. This is a very aggressive technique that should only be performed on an actual extrication incident by skilled technicians who have trained extensively on this type of procedure. Performing the technique in this fashion, through a coordinated and controlled effort, can lead to roof removal within 2 to 3 minutes. Because two operations are being conducted simultaneously, everyone involved in this technique has to be on the same page. The advantage of this technique, aside from the faster results, is that a section of the roof is actually removed; if the roof flap technique is used, large sections of the roof with cut posts remain and may impede rescue efforts.

Skill Drill 10-7

Removing a Roof from a Vehicle Resting on Its Side

1 Don PPE. Assess the scene for hazards and complete the inner and outer surveys. If the battery can be accessed, disconnect it. Stabilize the vehicle using the buttress stabilization strut technique.

2 If possible, position a rescuer inside the vehicle to provide medical support and immobilization. Ensure that the victim inside the vehicle is covered with a blanket for protection from flying debris. Remove all vehicle glass utilizing the appropriate technique.

3 Remove all of the plastic and molding around the posts and roof rail to identify any potential air bag cylinders.

4 Ensure that all seat belts have been cut. Check the roof for areas of damage causing metal tension and torsion. Cut around these areas. Insert a long backboard lengthwise from the front or rear window of the vehicle; the rescuer inside the vehicle will use this as hard protection as the reciprocation saw blade passes through the roof. Insert a 6-inch (152-mm) bi-metal cutting blade with a TPI rating of around 9 to 14 into a high-powered electric reciprocating saw.

5 Starting at the lower section of the roof (closest to the ground), make the initial cut on the inside corner of the front A-post or the last post of the vehicle, which may be a C-post, D-post, or higher post depending on the type of vehicle, just above the roof rail. Follow a straight horizontal line, completing the cut through the roof to the opposite side. As the technical rescuer outside the vehicle makes the cut, the rescuer inside the vehicle will position the backboard to provide hard protection from the blade.

6 After the bottom cut has been made, cut the top section. Follow the same cutting pattern used on the bottom section. Position two personnel on either side of the roof to support it. Place the roof section in a designated debris pile outside of the hot zone.

7 For added protection, place a tarp or large blanket over the bottom cut section of the roof to prevent any potential cut injuries.

Door Removal Hinge Side

In Chapter 9, *Victim Access and Management*, the technique of removing a door on the hinge side is described using a combination of a hydraulic spreader and hydraulic cutter; in this section the same technique will be accomplished utilizing an air chisel. Utilizing an air chisel to cut through door hinges can be a challenging task, but with training, this is a very effective technique that can be quickly mastered.

To remove a door on the hinge side using an air chisel, follow the steps in **Skill Drill 10-8 ▾** :

1. Don appropriate PPE, including mask and eye protection.
2. Assess the scene for hazards and complete the inner and outer surveys.
3. Stabilize the vehicle.
4. If the 12-volt battery can be accessed, then disconnect it utilizing the techniques described in Chapter 8, *Vehicle Stabilization*, Skill Drill 8-5.
5. If possible, position a rescuer inside the vehicle to provide medical support and immobilization.
6. Ensure that the victim is covered with a blanket for protection from flying debris.
7. Remove all vehicle glass utilizing the appropriate technique.
8. Ensure that all the seat belts have been cut.
9. Examine the interior of the vehicle for air bags, including side door air bags; if a roof removal technique will be performed, expose all of the posts and interior roof rail liners.
10. Insert the long flat/curved blade into the air chisel.
11. Turn the air cylinder valve on with the blade and tool facing the ground to avoid an accidental discharge of the blade. Never free-fire the air chisel into the air to check the pressure; doing so can damage the tool. Check the pressure gauge by placing the blade against a solid surface and dialing up to the appropriate pressure in the general range of

NFPA 1006, (10.1.7)

Skill Drill 10-8

Removing a Door on the Hinge Side Using an Air Chisel

1. Don PPE. Assess the scene for hazards and complete the inner and outer surveys. Stabilize the vehicle. If the 12-volt battery can be accessed, then disconnect it. If possible, position a rescuer inside the vehicle to provide medical support and immobilization. Ensure that the victim inside the vehicle is covered with a blanket for protection from flying debris. Remove all vehicle glass utilizing the appropriate technique. Ensure that all the seat belts have been cut. Examine the interior of the vehicle for air bags, including side door air bags. Insert the long flat/curved blade into the air chisel. If a roof removal technique will be performed, expose all of the posts and interior roof rail liners. Turn the air cylinder valve on with the blade and tool facing the ground. Check the pressure gauge.

2. Cut away an opening in the top section of the quarter panel where the hood and quarter panel meet. The opening should be large enough to expose both hinges. Angle the blade downward, cutting toward the bottom of the panel around the wheel well.

3. Bring the tool back to the top of the quarter panel and prepare to make another vertical downward cut to the bottom end of the panel. Position the blade downward to cut on the inside of the panel directly on the angled 90-degree bend. As the blade moves down, bend the blade outward, manipulating the cut strip of metal away to maintain visibility of the cut.

NFPA 1006, (10.1.7)

Skill Drill 10-8

Removing a Door on the Hinge Side Using an Air Chisel *(Continued)*

4 When the bottom is reached, cut the bottom section horizontally and remove the entire section of the panel; the door hinges will now be exposed. Examine the door hinges to determine the design type and the approach that will need to be taken. Cut the bottom hinge first. To cut through a heavy-gauge two-leaf overlay-type hinge with a center pin and spring, kneel on one knee, positioning the tip of the chisel on the area to be cut. The hinge overlaps (two-leaf) itself. Avoid cutting two sections of the hinge at the same time. With one hand on the trigger and one hand on the shaft of the blade, lean into the gun and while depressing the trigger, apply a slight up-and-down rocking motion while pushing into the gun with the chest and shoulder.

5 Once the bottom hinge is cut, move up to the top hinge and follow the same technique. If the door comes equipped with a swing bar, the cut should be made with the blade angled on the bar section, closest to the door or firewall.

6 With the door hinges severed, pull back the door and cut the wires that pass through the center to disconnect any air bags that may be present.

200 psi (1379 kPa). Refer to the manufacturer's air pressure specifications for the model of tool that your organization utilizes. **(Step 1)**

12. Cut away an opening in the quarter panel large enough to expose both hinges. Make the initial cut at the top section of the quarter panel in the area where the hood and quarter panel meet. Angle the blade downward, cutting toward the bottom of the panel around the wheel well. **(Step 2)**

13. Bring the tool back to the top of the quarter panel and prepare to make another vertical downward cut to the bottom end of the panel. The panel at that area bends at a 90-degree angle and attaches perpendicularly to the outside panel. Position the blade downward to cut on the inside of the panel directly on the angled 90-degree bend. As the blade moves down, bend the blade outward, manipulating the cut strip of metal away to maintain visibility of the cut. **(Step 3)**

14. When the bottom is reached, cut the bottom section horizontally and remove the entire section of the panel; the door hinges will now be exposed.

15. Carefully examine the door hinges to determine the design type and the approach that will need to be taken. Cut the bottom hinge first. Two types of hinges are commonly found: heavy-gauge, two-leaf overlay-type hinges with a center pin and spring, which are fairly common on larger vehicles, and lighter gauge standard hinges, which are used on smaller compact vehicles. To successfully cut through a heavy-gauge, two-leaf overlay-type hinge with a center pin and spring, cut only one hinge leaf at a time. The hinge overlaps itself, and you will want to avoid cutting two sections of the hinge at the same time. Visualize the area of the hinge where the chisel blade will pass through only the one section. Kneel on one knee, positioning the tip of the chisel on the area to be cut. The back section of the air chisel should be braced against your chest. With one hand on the trigger and one hand on the shaft of the blade, lean into the gun and while depressing the trigger, apply a slight up-and-down rocking motion while pushing into the gun with the chest and shoulder. Maintain complete balance while applying pressure with the rocking motion.

Once a groove is cut into the hinge, the rest of the cut will flow with less stoppage. If there is a spring attachment on the hinge, beware of it flying off once the hinge is released. (**Step 4**)

16. Move up to the top hinge and follow the same technique. If the door comes equipped with a swing bar, the cut should be made with the blade angled on the bar section, closest to the door or firewall, not at the center; remember that the air chisel works best by cutting through solid sections of metal with some type of backing. Attempting to cut through the center will only cause the swing bar to bend from the pounding of the chisel. (**Step 5**)

Rescue Tips

A technique that is commonly taught but that is not recommended here is dropping the floor pan just below the victim's feet where the brake and accelerator pedals are located. The reason this technique is not recommended is because of major safety concerns; fuel lines and high-powered electrical lines generally run in the area of the rocker panel/channel, and cutting into this area or tearing into this section can potentially rupture and/or sever one of these lines, causing additional problems or severe injury to the patient and crew.

17. With the door hinges severed, pull back the door and cut the wires that pass through the center; this will disconnect the current to any door air bag unit that may exist. (**Step 6**)

Side Removal: Vehicle Upside Down or Resting on Its Roof

Gaining entry through the side of a vehicle that has rolled over and is resting on its roof can require either a very basic procedure such as forcing a door open or a very involved process with multiple steps. The following Skill Drill will take you through the necessary procedures, starting with the basic steps and evolving to more advanced procedures. The scenario that will be utilized to explain the Skill Drill involves a two-door common passenger vehicle that has rolled over several times and comes to rest on its crushed roof, trapping one victim upside down in the front passenger compartment.

To perform side removal on a vehicle upside down or resting on its roof, follow the steps in **Skill Drill 10-9 ▾**:

1. Don appropriate PPE, including mask and eye protection.
2. Assess the scene for hazards and complete the inner and outer surveys.

NFPA 1006, (10.1.7)

Skill Drill **10-9**

Performing a Side Removal on a Vehicle Upside Down or Resting on Its Roof

1 Don PPE. Assess the scene for hazards and complete the inner and outer surveys. Stabilize the vehicle.

2 If accessible, ensure that the victim inside the vehicle is covered with a blanket for protection from flying debris. If accessible, use the hydraulic cutter to cut through the bottom of the window frame (D-ring) on both sides. Remove the section of the window frame (D-ring) and place it in the designated debris pile. This step releases the door from the ground. Using the hydraulic spreader, create a purchase point on the door to gain access to the latching mechanism.

3 Once the purchase point has been created, work the door down and outward exposing the latching mechanism. Cut the latching mechanism with a hydraulic cutter. Remove the door and place it in the designated debris pile.

NFPA 1006, (10.1.7)

Skill Drill 10-9

Performing a Side Removal on a Vehicle Upside Down or Resting on Its Roof *(Continued)*

4 Using the hydraulic cutter, make a complete cut all the way through the firewall between the bottom and top hinges. Make an additional relief cut into the bottom section of the B-post at the corner of the B-post and rocker panel. This action will enlarge the side opening to provide greater access to the patient.

5 Secure cribbing under the dash area beneath the firewall section. Insert the tips of the hydraulic spreader into the opening in the firewall in a vertical position. Simultaneously, with the hydraulic spreader in place, position a small hydraulic ram with the base of the tool on the roof rail and the tip on the rocker panel closest to the B-post. Use hard protection for the patient if there is a possibility of impingement. Open both tools simultaneously, causing the floor of the vehicle to lift and separate at the relief cuts made in the B-post and firewall. The distance (lift) needed will be determined by the officer in charge of the operation.

6 Immobilize and package the patient according to standard operating procedures.

3. Stabilize the vehicle. (**Step 1**)
4. Ensure that the victim is covered with a blanket for protection from flying debris.
5. If accessible, use the hydraulic cutter to cut through the bottom of the window frame (D-ring) on both sides (by the B-post and through the A-post cutting through the entire A-post if accessible). Remove the section of the window frame (D-ring) and place in the designated debris pile outside of the action zone. This step releases the door from the ground.
6. Using the hydraulic spreader, create a purchase point on the door to gain access to the latching mechanism. Pinch and partially collapse the area around the door seam and rocker panel. Be careful not to tear the metal and lose the integrity of the area to spread from. (**Step 2**)
7. Once the purchase point has been created, start to work the door down and outward exposing the latching mechanism.
8. Once the latching mechanism is exposed, cut the mechanism using a hydraulic cutter. With the latch cut and the

window frame (D-ring) removed, the door should be able to swing open using the natural movement of the hinges.
9. Remove the door by cutting the hinges and place it in the designated debris pile outside the action zone. (**Step 3**)
10. Using the hydraulic cutter, make a complete cut all the way through the firewall between the bottom and top hinges. Make an additional relief cut into the bottom section of the B-post at the corner of the B-post and rocker panel. These two relief cuts will enable the vehicle to open up and provide greater access to the patient. (**Step 4**)
11. Secure cribbing under the dash area just under the firewall section that was just cut. Insert the tips of the hydraulic spreader in a vertical position (this action will be the same as a dash-lift technique performed upside down) into the opening in the firewall. This action will be the same as the dash-lift technique performed upside down.
12. Simultaneously, in conjunction with the hydraulic spreader, position a small hydraulic ram with the base of the tool on the roof rail and the tip on the rocker panel closest to the B-post. The base of the ram has to be positioned on the roof

rail and not on the floor or the vehicle will be lifted off the ground.

13. Use hard protection for the patient if there is a possibility of tool impingent.

14. Open both tools simultaneously causing the floor of the vehicle to lift and separate at the relief cuts made in the B-post and firewall. The distance (lift) that is needed to gain sufficient access to the patient will be determined by the officer in charge of the operation. (**Step 5**)

15. Immobilize and package the patient according to standard operating procedures. Both hydraulic tools must be manned throughout the entire operation to avoid any potential slippage of the two tools. (**Step 6**)

Pedal Displacement and Removal

Acceleration and brake pedals are notorious for trapping or entangling occupants' foot extremities. Some instances occur where there is minor damage to the vehicle and the occupant happens to get a foot stuck under the brake pedal. Brake pedals can be designed with the standard hydraulic assistance and position, or they can be automatically height adjusted to a predetermined position that is programmed by the onboard computer system. Some manufacturers have designed a special air bag component that utilizes a piston rod and/or pillow-type bag that pushes the foot out of the way using the piston rod in conjunction with a rapidly inflating small air bag. This prevents the foot from getting trapped or injured from the impact.

Properly supporting a foot that is trapped under a pedal must be accomplished prior to any attempt of moving the pedal. Normally the rescuer will find that the foot has been fractured or dislocated by the impact of the foot striking the floor panel and the repositioning of the pedal around the foot. The pedal arm can be cut and/or relocated using mechanical tools such as a hydraulic cutter, reciprocating saw, air chisel, or pneumatic cutoff tool (whizzer saw); a come along and chain set; or a very simple tool such as a section of webbing or rope **Figure 10-5 ▶** What tool or combination of tools used will be determined at the time of the incident; there are many possible variables and presentations that will dictate which option is best. Some of the variables that may occur include the type of entrapment, the position of the foot, the type of injury that was sustained to the foot as well as the overall condition of the patient, the involvement of air bags or other SRS components in direct proximity to the pedal, other interventions or extrication procedures that have to be performed on the vehicle to release the occupant, and whether the pedals are standard model or automatic height adjustment pedals, which can be utilized prior to disabling the vehicle's 12-volt DC battery. These are only some of the variables that may occur that must be considered and planned for. As discussed, the process may only require a simple relocation of the pedal to free the foot.

The following Skill Drill describes relocating a standard pedal and does not involve any air bag or automatic height adjustment components. The technique requires the swing of the driver-side door to be operational. This technique can be applied to a door that has to be forced open; the only requirement is that the swing action using the hinges must remain operational.

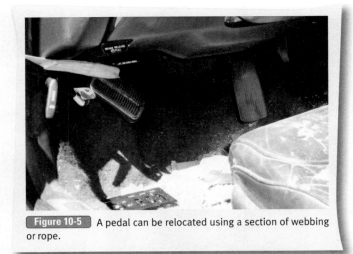

Figure 10-5 A pedal can be relocated using a section of webbing or rope.

To relocate a standard pedal void of any air bag or automatic height adjustment components, follow the steps in **Skill Drill 10-10 ▶**:

1. Don appropriate PPE, including mask and eye protection.

2. Assess the scene for hazards and complete the inner and outer surveys.

3. Stabilize the vehicle.

4. Position a rescuer inside the vehicle to assess the condition of the patient and the entrapment of the extremity. The rescuer must also support the foot throughout the operation.

5. Ensure that the victim is covered with a blanket for protection from flying debris.

6. Remove vehicle glass using the appropriate technique, as needed.

7. If a side-impact air bag is located in the door or seat area, then gain access to the engine compartment and disconnect the 12-volt DC battery.

8. Ensure that the swing of the driver-side door is operational.

9. Firmly attach a long section of webbing measuring no less than 15 feet (5 meters) to the pedal arm, and stretch the webbing to the window frame on the driver-side door. (**Step 1**)

10. With the driver-side door partially opened 1 inch (25 mm) or more, firmly secure the webbing to the window frame (D-ring) at the area closest to the door's latching mechanism. There should be slight tension on the webbing from the door frame to the pedal. (**Step 2**)

11. With the trapped foot extremity firmly supported by a rescuer, slowly open the driver-side door, which will cause the pedal to bend completely sideways, releasing the entrapment. This occurs because of the leveraged pulling force. This step has to be a fully coordinated effort because too much force or an accelerated pulling force can potentially cause further harm to the patient.

12. Once the foot extremity has been released, take the appropriate steps to properly package and remove the patient from the vehicle, and place him or her onto an awaiting stretcher. (**Step 3**)

NFPA 1006, (10.1.8)

Skill Drill 10-10

Relocating a Pedal

  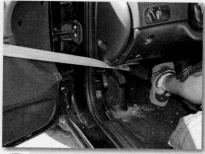

1 Don PPE. Assess the scene for hazards and complete the inner and outer surveys. Stabilize the vehicle. Position a rescuer inside the vehicle to assess the condition of the patient and the entrapment of the extremity as well as support the foot throughout the operation. Ensure that the victim is properly protected from flying debris. Remove vehicle glass using the appropriate technique. If a side-impact air bag is located in the door or seat area, then gain access to the engine compartment and disconnect the 12-volt DC battery. Ensure that the swing of the driver-side door is operational. Firmly attach a long section of webbing to the pedal arm and stretch the webbing to the window frame on the driver-side door.

2 With the driver-side door partially opened 1 inch (25 mm) or more, firmly secure the webbing to the window frame (D-ring) near the door's latching mechanism. There should be slight tension on the webbing.

3 With the trapped foot extremity firmly supported by a rescuer, slowly open the driver-side door to bend the pedal sideways and release the entrapment. Once the foot extremity has been released, take the appropriate steps to properly package and remove the patient from the vehicle, and place him or her onto an awaiting stretcher.

Wrap-Up

Ready for Review

- Tunneling is the process of gaining entry through the rear trunk area of a vehicle. It is most commonly performed for a vehicle that is resting on its roof, but it can be performed for a vehicle in any resting position.
- There are multiple variations of the types of seat frames.
- Seats are attached to the floorboards of the vehicle in four or more places, with positional adjustments on slide tracks that are either fully automatic/motorized or manually operated.
- The material covering the seats may have to be removed using a knife or trauma shears to expose the area of attachment located on the floor or to expose the steel hinges where the seat back is attached to the bottom seat section.
- The cross ram technique is used when the impingement of metal on the victim requires a unique mechanism of movement.
- The hydraulic ram has the advantage over a hydraulic spreader because it can be used in tight spaces.
- When an object impales a vehicle occupant, the main objective is to stabilize the penetrating object at the entry point and the exit point, if an exit point exists.
- The pneumatic air chisel is a powerful tool with multiple blades. The two most widely used blades are the panel cutter, or T-blade, and the flat/curved blade. The panel cutter is used for cutting shallow straight cuts on small-gauge sheet metal, and the flat/curved blade is used for cutting through medium- to heavy-gauge steel.
- A reciprocating saw is an excellent tool to use for removing a vehicle's roof, especially on vehicles with large C-posts.
- The technique of removing a door on the hinge side can be completed using a combination of a hydraulic spreader and hydraulic cutter, or it can be accomplished utilizing an air chisel.
- Gaining entry through the side of a vehicle that has rolled over and is resting on its roof can require either a very basic procedure such as forcing a door open, or it can be a very involved process requiring multiple steps.
- Acceleration and brake pedals are notorious for trapping or entangling occupants' foot extremities. The pedal arm can be cut and/or relocated using mechanical tools such as a hydraulic cutter, reciprocating saw, air chisel, or pneumatic cutoff tool (whizzer saw); a come along and chain set; or a very simple tool such as a section of webbing or rope.

Hot Term

Cross ram technique The use of a hydraulic ram to push off of the opposite door post, B-post, floor transmission hump, or inside rocker panel to move the interior of the vehicle away from the entrapped occupant.

Technical Rescuer *in Action*

You are dispatched for a motor vehicle collision in front of a large shopping center. You arrive to find two vehicles involved in the collision; one of the vehicles is on its roof and wedged in between a parked vehicle on the driver's side and a cement light pole on the passenger's side. There is no access on either side of the vehicle. There is also another passenger vehicle involved in the collision that is positioned upright. Both vehicles have sustained heavy damage. Your engine company is the first to arrive on scene.

1. Tunneling involves the process of making vehicle entry through:
- **A.** performing a side-out.
- **B.** the trunk of the vehicle.
- **C.** the roof of the vehicle.
- **D.** the front windshield of the vehicle.

2. Tunneling is more commonly used for vehicles:
- **A.** on their side.
- **B.** on their roof.
- **C.** down embankments.
- **D.** upright.

3. With the vehicle stabilized, victim protected, and glass removed, the first step of tunneling is to:
- **A.** remove the windshield.
- **B.** remove the dashboard.
- **C.** create an opening in the trunk.
- **D.** create an opening on the side of the vehicle.

4. The second step in tunneling before opening the trunk is to:
- **A.** determine if there are any hazardous or heavy storage items in the trunk.
- **B.** create an opening in the side of the vehicle.
- **C.** expose the victim to fresh air.
- **D.** break all of the vehicle glass.

5. The optimal way to access the trunk would be to:
- **A.** expose and cut the latching mechanism of the trunk.
- **B.** crawl into the trunk area.
- **C.** remove the rear dashboard.
- **D.** remove the rear window frame.

6. The tool used to expose the latching mechanism of the trunk is the:
- **A.** hydraulic spreader.
- **B.** hydraulic cutter.
- **C.** screwdriver.
- **D.** come along.

7. In order to release the trunk cover, you will need to:
- **A.** use a come along to pull the trunk cover off.
- **B.** cut the latching mechanism using a hydraulic cutter.
- **C.** use a Halligan bar to pry the latching mechanism apart.
- **D.** use a circular saw to cut the latching mechanism.

8. The best tool for cutting tension rods located in the trunk of a passenger vehicle are an industrial bolt cutter and:
- **A.** Pneumatic air chisel
- **B.** Hydraulic cutter
- **C.** Wire cutter
- **D.** Reciprocating saw

9. You approach the upright vehicle and determine the roof will need to be removed. A tool that can be used to identify whether air cylinders are present in posts is:
- **A.** a large flat-head screwdriver.
- **B.** a hydraulic cutter.
- **C.** a hydraulic spreader.
- **D.** a pneumatic ram.

10. Seat belts should be cut prior to the roof being removed.
- **A.** True
- **B.** False

11. When utilizing a reciprocating saw to remove the roof of a vehicle on its side:
- **A.** cut the higher section first.
- **B.** cut the lower section first.
- **C.** cut a vertical section.
- **D.** cut a horizontal section.

CHAPTER 11

NFPA 1006 Standard

10.2.1 Plan for a commercial heavy vehicle or large machinery incident, and conduct initial and ongoing size-up, given agency guidelines, planning forms, and operations-level vehicle/machinery incident or simulation, so that a standard approach is used during training and operational scenarios, emergency situation hazards are identified, isolation methods and scene security measures are considered, fire suppression and safety measures are identified, vehicle/machinery stabilization needs are evaluated, and resource needs are identified and documented for future use. (pages 277–278, 311–315)

(A) Requisite Knowledge. Operational protocols, specific planning forms, types of commercial/heavy vehicles and large machinery common to the AHJ boundaries, vehicle/machinery hazards, incident support operations and resources, vehicle/machinery anatomy, and fire suppression and safety measures. (pages 277–278, 311–315)

(B) Requisite Skills. The ability to apply operational protocols, select specific planning forms based on the types of commercial/heavy vehicles and large machinery, identify and evaluate various types of commercial/heavy vehicles and large machinery within the AHJ boundaries, request support and resources, identify commercial/heavy vehicles and large machinery anatomy, and determine the required fire suppression and safety measures. (pages 277–278, 311–315)

10.2.2 Stabilize commercial/heavy vehicles and large machinery, given a vehicle and machinery tool kit and personal protective equipment, so that the vehicle or machinery is prevented from moving during the rescue operations; entry, exit, and tool placement points are not compromised; anticipated rescue activities will not compromise vehicle or machinery stability; selected stabilization points are structurally sound; stabilization equipment can be monitored; and the risk to rescuers is minimized. (pages 278–282)

(A) Requisite Knowledge. Types of stabilization devices, mechanism of heavy vehicle and machinery movement, types of stabilization points, types of stabilization surfaces, AHJ policies and procedures, and types of vehicle and machinery construction components as they apply to stabilization. (pages 278–282)

(B) Requisite Skills. The ability to apply and operate stabilization devices. (pages 278–282)

10.2.3 Determine the heavy vehicle or large machinery access and egress points, given the structural and damage characteristics and potential victim location(s), so that victim location(s) is identified; entry and exit points for victims, rescuers, and equipment are designated; flow of personnel, victim(s), and equipment is identified; existing entry points are used; time constraints are factored; selected entry and egress points do not compromise vehicle or machinery stability; chosen points can be protected; equipment and victim stabilization are initiated; and AHJ safety and emergency procedures are enforced. (pages 282–300, 315–316)

(A) Requisite Knowledge. Heavy vehicle and large machinery construction/features, entry and exit points, routes and hazards, operating systems, AHJ standard operating procedures, and emergency evacuation and safety signals. (pages 282–300, 315–316)

(B) Requisite Skills. The ability to identify entry and exit points and probable victim locations, and assess and evaluate the impact of heavy vehicle or large machinery stability on the victim(s). (pages 282–300, 315–316)

10.2.4 Create access and egress openings for rescue from a heavy vehicle or large machinery, given vehicle and machinery tool kit, specialized tools and equipment, personal protective equipment, and an assignment, so that the movement of rescuers and equipment complements victim care and removal, an emergency escape route is provided, the technique chosen is expedient, victim and rescuer protection is afforded, and vehicle stability is maintained. (pages 282–300, 315–316)

(A) Requisite Knowledge. Heavy vehicle and large machinery construction and features; electrical, mechanical, hydraulic, and pneumatic systems, and alternative entry and exit equipment; points and routes of ingress and egress; techniques and hazards; agency policies and procedures; and emergency evacuation and safety signals. (pages 282–300, 315–316)

(B) Requisite Skills. The ability to identify heavy vehicle and large machinery construction features, select and operate tools and equipment, apply tactics and strategy based on assignment, apply victim care and stabilization devices, perform hazard control based on techniques selected, and demonstrate safety procedures and emergency evacuation signals. (pages 282–300, 315–316)

10.2.5 Disentangle victim(s), given a Level II extrication incident, a vehicle and machinery tool kit, personal protective equipment, and specialized equipment, so that undue victim injury is prevented, victim protection is provided, and stabilization is maintained. (pages 282–300, 315–316)

(A) Requisite Knowledge. Tool selection and application, stabilization systems, protection methods, disentanglement points and techniques, and dynamics of disentanglement. (pages 282–300, 315–316)

(B) Requisite Skills. The ability to operate disentanglement tools, initiate protective measures, identify and eliminate points of entrapment, and maintain incident stability and scene safety. (pages 282–300, 315–316)

▌Knowledge Objectives

After studying this chapter, you will be able to:

- Define a commercial motor vehicle (CMV). (page 271)
- Describe the Federal Motor Carrier Safety Administration (FMCSA) bus categories. (page 272)
- Define a school bus. (page 272)
- Describe the four types of school buses, including the subcategories, that must meet all Federal Motor Vehicle Safety Standards (FMVSS) for school buses. (pages 273–274)
- Describe some of the standards for school bus design. (pages 273–274)
- Describe safety features of school bus design. (pages 274–277)
- Describe SWOT analysis and how it can be helpful for preplanning. (page 277)
- Explain how to stabilize the scene of a school bus extrication. (page 278)
- Explain the four common resting positions of a school bus after a collision. (page 278)
- Describe how to gain or create access into a school bus. (pages 282–301)
- Describe the eight weight categories for CMVs. (page 301)
- Explain the difference between a semi-truck and a tractor-trailer. (pages 301–304)
- Describe the three different kinds of cabs found on commercial trucks. (pages 304–305)
- Explain how the semi-truck is connected to a tractor-trailer. (pages 307–308)
- Describe the braking systems of CMVs. (pages 309–310)
- Explain the nine classifications of hazardous materials. (pages 311–312)
- Describe some of the vehicles used to transport hazardous materials. (pages 312–315)
- Describe the five types or classifications of tow units. (pages 315–316)

▌Skills Objectives

After studying this chapter, you will be able to perform the following skills:

- Stabilize a school bus in its normal position. (pages 278–279, Skill Drill 11-1)
- Stabilize a school bus resting on its side. (pages 278–280, Skill Drill 11-2)
- Stabilize a vehicle resting on its roof. (pages 280–281, Skill Drill 11-3)
- Stabilize/marry a school bus on top of another vehicle. (pages 282–283, Skill Drill 11-4)
- Gain access into a school bus by removing a front windshield. (pages 284–285, Skill Drill 11-5)
- Remove a bench seat from a school bus. (pages 286–287, Skill Drill 11-6)
- Remove a section of the sidewall of a school bus. (pages 287–288, Skill Drill 11-7)
- Gain access through the roof of a school bus on its side. (pages 289–291, Skill Drill 11-8)
- Gain access through the rear door of a school bus in its normal position. (pages 291–292, Skill Drill 11-9)
- Gain access through the rear door of a school bus resting on its side. (pages 291–294, Skill Drill 11-10)
- Gain access through the front door of a school bus in its normal position. (pages 293–296, Skill Drill 11-11)
- Remove a victim from under a school bus resting on its side. (pages 296–299, Skill Drill 11-12)
- Disable the hybrid system on a type C or D school bus. (pages 300–301, Skill Drill 11-13)

As the officer on the engine, you and your crew pull up to an intersection where just moments prior a dump truck side-impacted a type C school bus, flipping the school bus onto its side. There are multiple adolescent students on the bus with injuries.

1. Does your agency have a preestablished emergency plan or mass-casualty incident (MCI) protocols to manage this type of incident?

2. What resources do you have that can be immediately dispatched or utilized?

3. Have you trained for such an incident, and can you initiate a START triage system utilizing the crew on hand?

4. What is your action plan?

Introduction

"Unless you try to do something beyond what you have already mastered, you will never grow." —Ralph Waldo Emerson

This chapter is based on advanced Technician Level II requirements as listed in NFPA 1006, *Standard for Technical Rescuer Professional Qualifications*, and 1670, *Standard on Operations and Training for Technical Search and Rescue Incidents*.

The United States Department of Transportation (DOT) defines a **commercial motor vehicle (CMV)** as a motor vehicle or combination of motor vehicles used in commerce to transport passengers or property if the motor vehicle satisfies at least one of the following criteria:

- Has a **gross vehicle weight rating (GVWR)** of 26,001 pounds (lb) or more (11,794 kilograms [kg] or more) inclusive of a towed unit(s) with a GVWR of more than 10,000 lb (4536 kg)
- Has a GVWR of 26,001 lb or more (11,794 kg or more)
- Is designed to transport 16 or more passengers, including the driver
- Is of any size and is used in the transportation of hazardous materials as defined in this section

Furthermore, as described in Chapter 3, the DOT established a vehicle classification scheme that is separated into categories depending on whether the vehicle carries passengers or commodities. Nonpassenger vehicles are further subdivided by number of axles and number of units, including both power and trailer units.

CMVs travel our roads, streets, and highways every day and are generally associated with the idea of carrying cargo. Many CMVs are utilized or designed for a special purpose, and the word "truck" is often combined with that vehicle's purpose,

such as dump truck, fire truck, tow truck, or garbage truck Figure 11-1 ▾ . CMVs also include box trucks, which utilize a single frame, and semi-tractor trailers, which use two or more separate frames that are equipped to haul machinery, chemicals, and a vast variety of commodities/supplies, as well as livestock. Other specialized vehicles such as concrete mixers, vehicle transporters, buses, and cranes are also examples of CMVs that you may encounter when dealing with extrication. In England and some other European nations, a truck or CMV is referred to as a "lorry."

Rescue Tips

Gross vehicle weight rating (GVWR) consists of ratings established by manufacturers that take into account cargo, people, fuel, and the vehicle itself to determine the maximum total weight of the vehicle according to the manufacturer's specifications.

Figure 11-1 The tower ladder fire apparatus is one example of a commercial motor vehicle designed for a special purpose.

There is a great diversity in the types of buses on the roadways today; accurately describing or classifying a bus can be difficult at best. The Federal Motor Carrier Safety Administration (FMCSA) categorizes buses into carrier types or by function or purpose[*]:

- **School bus**—Any public or private school or district, or contracted carrier operating on behalf of the entity, providing transportation for kindergarteners through grade 12 pupils **Figure 11-2 ▾** .
- **Transit bus**—An entity providing passenger transportation over fixed, scheduled routes, within primarily urban geographic areas **Figure 11-3 ▸** .
- **Intercity bus**—A company providing for-hire, long-distance passenger transportation between cities over fixed routes with regular schedules **Figure 11-4 ▸** .
- **Charter/tour bus**—A company providing transportation on a for-hire basis, usually round-trip service for a tour group or outing. The transportation can be for a specific event or can be part of a regular tour **Figure 11-5 ▸** .
- Other—All bus operations not included in the previous categories. Examples include private companies providing transportation to their own employees, nongovernmental organizations such as churches or nonprofit groups, noneducational units of government such as departments of corrections, and private individuals **Figure 11-6 ▸** .

Figure 11-3   Transit bus.

Figure 11-4   Intercity bus.

Figure 11-5   Charter or tour bus.

Figure 11-2   School bus.

[*]*Source*: Bus Operator Types and Driver Factors in Fatal Bus Crashes: Results From the Buses Involved in Fatal Accidents Survey, Federal Motor Carrier Safety Administration, 2009; http://www.fmcsa.dot.gov/facts-research/research-technology/report/FMCSA-RRA-09-041_BIFA.pdf.

Classifying a particular bus type is further complicated because these buses can be interchanged and used for any of these pur-

Figure 11-6 Other types of buses may include a Department of Corrections bus.

poses, depending on what country or region of the world you are in.

This chapter is divided into two sections. The first section presents bus anatomy and extrication operations utilizing a school bus as the primary example. The second section covers large trucks and semi-trucks.

School Buses

On any given school day there are multiple school buses crossing your path or traveling alongside of you. You may not even acknowledge their presence because they are just as common as any other vehicle on the road. The National Highway Traffic Safety Administration (NHTSA) has reported that there are approximately 474,000 public school buses on the roads traveling approximately 4.3 billion miles annually to transport 23.5 million children and adolescents to and from school and school-related activities.

The NHTSA lists school bus travel as one of the safest forms of transportation, reporting approximately 0.001 accidents per every 100 million miles traveled. As an emergency responder, you may never come across a school bus accident, but it is vital to be prepared and know the makeup, structural components, and different types of school buses that are on the roadways today.

Rescue Tips

The National Highway Traffic Safety Administration (NHTSA) has reported that there are approximately 474,000 public school buses traveling our roadways today.

The DOT, through the NHTSA, issued statute 49 U.S.C. section 30125, which defines a "school bus" as any vehicle that is designed for carrying a driver and more than 10 passengers and that, NHTSA decides, is likely to be "used significantly" to transport "pre-primary, primary, and secondary" students to or from school or related events (including school-sponsored field trips and athletic events). The NHTSA has two classifications

for school buses—small (GVWR of less than 10,000 lb [4536 kg]) and large (GVWR that is equal to or greater than 10,000 lb [4536 kg]). The school bus industry designated four categories or classifications of school buses, with several sub-classifications; be aware that these can differ in various areas of the country or internationally. The occupant capacity for each type will also vary depending upon different manufacturers' specifications. Regardless of the variations, all four types, including the subcategories, must meet all Federal Motor Vehicle Safety Standards (FMVSS) for school buses:

- **Type A school bus:** This type of school bus that is a conversion type or bus constructed utilizing a cutaway front section vehicle with a left side driver's door **Figure 11-7 ▼**. This definition includes two sub-classifications:
  - Type A-1, with a GVWR of 14,500 lb (6577 kg) or less.
  - Type A-2, with a GVWR greater than 14,500 lb (6577 kg) and less than or equal to 21,500 lb (9752 kg).
- **Type B school bus:** This type of school bus is constructed utilizing a stripped chassis. The entrance door is behind the front wheels **Figure 11-8 ▼**. This definition includes two sub-classifications:
  - Type B-1, with a GVWR of 10,000 lb (4536 kg) or less.

Figure 11-7 Type A school bus.

Figure 11-8 Type B school bus.

- Type B-2, with a GVWR greater than 10,000 lb (4536 kg).
- **Type C school bus:** This type of school bus, also known as a *conventional school bus*, is constructed utilizing a chassis with a hood and front fender assembly. The entrance door is behind the front wheels **Figure 11-9 ▾**. These buses have a GVWR greater than 21,500 lb (9752 kg). Eighty-five to ninety percent of all school buses are type C or D.
- **Type D school bus:** This type of school bus, also known as a *"transit-style" rear or front engine school bus*, is constructed utilizing a stripped chassis where the outer body of the bus is mounted to the bare chassis. The entrance door is ahead of the front wheels, and the face or front section of the bus is flat **Figure 11-10 ▾**. Type D buses have a passenger capacity of 80 to 90 people.

■ School Bus Anatomy

A school bus is designed with a body-over-frame construction. The body of the bus is composed of a full skeletal frame system consisting of steel trusses and studs, which are attached and reinforced by steel cross members running the entire length of the bus. This body frame is encapsulated by inner and outer sheet metal panels. These inner and outer panels consist of 22- to 24-gauge steel and generally have fiberglass insulation inside that ranges from 1 to 1.5 inches (25 to 38 millimeters [mm]) thick for noise reduction and thermal protection.

Figure 11-9 Type C school bus.

Figure 11-10 Type D school bus.

The overall design features of the bus, such as metal thickness and spacing of channel beams, may vary among manufacturers, but all must meet the FMVSS for school buses. The standards include:

- Length of the bus shall not be greater than 45 feet (14 meters [m]).
- Width of the bus shall not be greater than 102 inches (2591 mm).
- Inside body height of the bus shall measure 72 inches (1829 mm) or more.
- Width of aisles will be at least 12 inches (305 mm), as measured from seat cushion to seat cushion with the top of the seat backs tapering inward, extending the spacing to a minimum of 15 inches (381 mm) in width.
- Aisle clearance for wheelchair accessibility will be at least 30 inches (762 mm) to the closest emergency door and/or lift area.
- School buses equipped with a power lift or a ramp shall have aisles a minimum of 30 inches (762 mm) wide leading from the wheelchair or other type of mobility device area to the emergency door, power lift, or ramp service entrance.

Chassis Frame

The chassis frame of the school bus consists of two long steel channel beams of 8- to 10-gauge steel that has steel cross members of 14-gauge steel. The cross members can vary in spacing but normally can be spaced up to 12 inches (305 mm) apart. Because of the extremely difficult work involved, the potential time expended, and the large amount of heavy-gauge steel located in this area, attempting to gain entry through the floor should not be attempted unless there are no other areas or alternative techniques to gain entry. Attached to the frame are components such as the engine, suspension system, fuel tank, wheels, and front and rear axles.

> **Rescue Tips**
>
> Because of the extremely difficult work involved, the potential time expended, and the large amount of heavy-gauge steel located in this area, attempting to gain entry through the floor should not be attempted unless there are no other areas or alternative techniques to gain entry.

Floor Deck

The floor deck is comprised of 14-gauge sheet metal panels that are attached to the chassis frame. Plywood of $\frac{1}{2}$-inch (13-mm) or $\frac{5}{8}$-inch (16-mm) thickness is fastened over the steel deck and covered with a thick corrugated rubber or vinyl matting. Because of the thickness of the floor design, attempting any entry into the bus through the floor area is not recommended.

Bow Frame Trusses

The main body frame structure consists of steel sidewall **bow trusses** that run continuously from below the floor level on one side of the bus, vertically raising/bowing over to form the roof structure, which continues extending over and down the other side of the bus past the floor level. These structural mem-

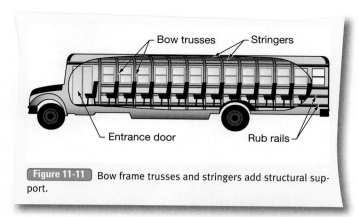

Figure 11-11 Bow frame trusses and stringers add structural support.

Figure 11-12 The rear emergency exit.

bers or ribs make up the support frame of the school bus and are normally comprised of heavier 12-gauge steel. Each roof pillar or side window frame makes up one of these structural members and has the crash and rub rails attached to it along with exterior and interior sheet metal panels **Figure 11-11**. The locations and spacing of these trusses are easy to determine because of the exterior rivets that attach the interior and exterior panels to the trusses. To give these bow frame truss members structural support at the roof level, steel longitudinal **stringers** are added that run continuously from the front of the school bus to the rear. Stringers consist of 16-gauge steel. There can be up to four stringers split with two on each side of the roof emergency hatch.

Rub Rails

Rub rails or guard rails are visible exterior steel attachments that are comprised of 16-gauge corrugated metal. These steel members are 4 inches (102 mm) or more in width and are attached to the bow trusses. They run the entire length of the school bus, wrapping around to the rear of the vehicle. Rub rails are easily identifiable because they are usually painted black against the yellow background of the bus itself; from a distance, they look like black stripes running the length of the vehicle. These rub rails are strategically placed with the bottom rail positioned at the floor line and the mid rail positioned at the area of the seat cushion level. There can also be a top rail that ties in the bottom of the side and rear window frames.

Crash Rail

The **crash rail** is normally composed of 14-gauge steel and extends just above the floor area between the floor and seat rub rail, extending the entire length of the school bus. As the name suggests, the crash rail is designed to protect the students from impact intrusions into the passenger compartment.

Entrance Door

The entrance door of the bus is designed in a two-section, split-type style and opens outward; whereby the door folds to one side or the other. Other opening configurations may be encountered on older school buses, but they are less common than the outward-swinging type. The passenger door is located on the right side of the bus, opposite and in direct view of the driver. The opening can measure 24 inches (610 mm) in width and 68 inches (1727 mm) in height. On a type C bus, the front passenger door is located behind the front wheels, and on a type

D bus, it is located in front of the wheels. The front door can be manually operated with a lever bar that the driver controls from his or her seat. The door may also be opened and closed with an air-actuated mechanism that pressurizes and releases air through a switch. This mechanism is also operated by the driver. In addition, an air-actuated door has a clearly marked emergency release valve that can be located on the upper right side of the entryway, or to the left or the right of the entryway; locations will vary among manufacturers. All entrance door glass is composed of tempered safety or laminate glass and is held in place by a gasket seal that can be manually removed and the glass panels pushed in to release from the frame.

Emergency Exits

Rear doors are designed to open or swing outward from left to right, with the hinges being on the right side **Figure 11-12**. There are multiple types of hinge attachments that the technical rescuer might find on a rear door or side door, such as piano-type hinges, three-bolt hinges, single-strap hinges, or nonexposed hinges, which may be on the inside of the school bus, just to name a few. The rear door on a school bus is a main access or egress point depending on the location of the victims and position of the bus itself.

Transit-style school buses, or type D school buses, do not have a rear door because of the rear-mount engine design. This type of school bus will have an emergency escape window just above the engine access hatch; type D buses also have a side door generally located in the middle section of the bus. There may also be special service entrance doors for repair technicians or wheelchair accessibility doors.

Rescue Tips

The technical rescuer might encounter multiple types of hinge attachments on a rear door or side door, such as piano-type hinges, three-bolt hinges, single-strap hinges, or nonexposed hinges, which may be on the inside of the school bus.

Side Window Exits

Side windows can be composed of either tempered safety or laminate-type glass, depending on the manufacturer. Each side window has an opening of at least 9 inches (229 mm) in height but no more than 13 inches (330 mm), and at least 22 inches (559 mm) in width. The windows are designed to slide down to open. Some buses may also contain one window on each side of the bus that may be less than 22 inches (559 mm) wide. Side windows can be easily removed by cutting the attachment clips or bolts with a pneumatic air-chisel and a prying tool (refer to Skill Drill 11-7, Removing a Section of the Sidewall of a School Bus, for a demonstration of this).

The release mechanism of emergency windows follows FMVSS No. 217, *Bus Emergency Exits and Window Retention and Release*. Emergency windows are designed to release the entire frame, remaining hinged only on one side, thus affording the escapee the full height and width of the opening. Removal of front windshield glass will be discussed later in the chapter, including a skill drill technique for gaining access into and removal of the front windshield.

Emergency Roof Hatches

Emergency roof exits are hinged on the front or forward side and are operable from both inside and outside the vehicle Figure 11-13 ▼. The total opening is 16½ inches (410 mm). The opening is large enough for a person self-escaping or entering the bus, but is too narrow to allow a patient secured to a backboard to pass through.

Bench Seats

Standard bench seats are designed to have a 1-inch (25-mm) tubular steel frame and two outer leg posts attached to the floor by bolts or screws. The inside of the bench seat, which rests against the sidewall of the bus, is screwed or bolted to a lip that extends from the interior "skin" approximately 1 inch (25 mm). The one exception to this design is the seats located in the emergency egress way, whether it is a door or window; in this location, the seat section of the bench lifts up out of the way to make room for escapees.

Driver Seat

The driver/operator seat is typically a single high-back chair with a fully integrated three-point seat belt harness. It is normally designed with an air-actuated suspension-type system (air ride), which is manually operated to the desired height of the driver.

Figure 11-13 Emergency roof exits are operable from both inside and outside the vehicle.

Figure 11-14 School bus batteries are secured on a pull-out sliding tray in the body and are generally located near the front of the vehicle on the driver's side.

Battery Compartment

The school bus batteries are secured on a pull-out sliding tray that is positioned inside the vehicle body and are generally located near the front of the vehicle on the driver's side Figure 11-14 ▲. There are normally two or more 12-volt DC batteries connected in parallel configuration to increase the amperage output. In type A and some type B buses, the batteries are normally located in the engine compartment similar to a conventional vehicle.

Rescue Tips

There are normally two or more 12-volt DC batteries positioned on a pull-out tray that are connected in parallel configuration to increase the amperage output.

Exhaust After-Treatment Device

The Environmental Protection Agency (EPA) set emission control standards for all diesel-fueled engines built after January 2007. The standards called for a 90 percent reduction of particulate matter (soot and ash) and a 55 percent reduction of nitrogen oxide (NOx).

In compliance with the EPA standards, the school bus industry installed an **exhaust after-treatment device** that replaced the standard muffler assembly. This device captures and converts soot to carbon dioxide and water through the combination of a diesel particulate filter (DPF) and a diesel oxidation catalyst (DOC). This conversion process is called regeneration, and there are three stages or processes of regeneration: passive, active, and manual. **Passive regeneration** occurs automatically when the particulate matter (soot) that is caught in the DPF is burned off naturally by the elevated temperatures of the exhaust system. When passive regeneration does not oxidize the soot sufficiently, the onboard computer system operates in an **active regeneration** mode. In this mode, fuel is injected into the system to burn and create higher temperatures, up to 1112° Fahrenheit (F) (600° Celsius [C]). **Manual regeneration** occurs only when the parking brake is set and the engine is running. Manual regeneration is normally completed when the active regeneration fails to clear the system sufficiently. All three regeneration modes will engage automatically by the computer system independent of the driver's actions.

■ Site Operations: School Buses

The greatest concern for the officer in charge at a school bus extrication incident is gaining and maintaining control of the incident through proper scene management. Proper scene management is critical and can mean the difference between a successful, controlled operation and an operation where chaos and freelancing occur. Any extrication with multiple patients (especially children) can be overwhelming to the first arriving officer, but by applying the basic principles of incident management and breaking things down to manageable segments, the officer will gain control of the incident and gain confidence in his or her decision making. When agenices conduct preplanning for school bus accidents, local school authorities should be contacted and included in the planning. Many schools have a transportation emergency plan to ensure the accountability, safety, and security of the children. The first 10 to 15 minutes of the incident usually set the tone for how the operation goes. In these first minutes, everything is coming at the officer in charge at such a fast rate that the process of critical thinking can easily max out. The first rule for gaining control of an incident is self-control. Think about what the priorities and objectives are for the incident:

- Safety
- Scene stabilization
- Resources available and needed (personnel, equipment, and apparatus), including the proper management of those resources
- Vehicle stabilization
- Victim access and management: START triage (patient stabilization)
- Extrication of victims
- Terminating the incident

Planning

Preplanning or strategic planning is the key to success for any organization. The needs assessment study that was discussed in Chapter 2, *Rescue Incident Management*, covers a wide range of topics and basically answers the question of where the organization wants to be in relation to providing the necessary level of emergency services for the type of community it serves. This model incorporates both long- and short-range goals for the organization.

A **SWOT analysis**, which is mainly used in a corporate business-type environment, is a self-examination model that can be adjusted, adapted, and applied to any situation, incident, or project, large or small, that an organization is currently or will be involved in. Not many emergency organizations use this type of self-evaluation model, but it can be an invaluable resource tool that can quickly address deficiencies and elevate an agency to the next level. A SWOT analysis evaluates the **S**trengths, **W**eaknesses, **O**pportunities, and **T**hreats of an organization. If you were to apply a SWOT analysis to operating at a school bus rescue, it may consider questions such as the following:

- **S**trengths: What are the overall capabilities of the agency, including levels and types of training and preestablished response plans, to successfully operate at this type of incident with minimal to no hindrances?
- **W**eaknesses: What are the possible operational deficiencies that could cause the operation to fail? Is there a lack of adequate staffing, equipment, apparatus, or training?
- **O**pportunities: Are there any training opportunities, mass-casualty incident (MCI) drills, equipment needs, mutual aid agreements, or grant funding to help elevate the readiness stage and response capabilities of the organization?
- **T**hreats: Are there any external elements such as inclement weather or time of season that would change the emergency plan or create additional challenges to overcome? Is it 98°F (36.7°C) with full sun exposure or 10°F (−12°C) in snow or ice?

Remember that this SWOT analysis of your organization is fully expandable, but to be effective, the analysis must be complete, honest, and objective.

Progressive agencies have preplanned and trained heavily for such an event and already have preestablished MCI protocols and/or emergency response plans that are in place. Depending on how your agency's MCI plan is set up, a level 3 MCI for your area may involve 50 plus patients, 10 transport units, and 8 fire apparatus automatically dispatched/responding. Even the best plans fail if they are not properly carried out. As a technical rescuer or an officer in charge, you need to be prepared to properly manage this type of incident. If you do not have a preestablished plan, gain control of this situation immediately by assigning a resource/staging officer and add a separate response channel for all incoming units. A lot of air traffic transpires during an incident of this type and scale; you will never be able to get one word of direction in on the radio if a separate response channel is not incorporated. There is absolutely nothing more frustrating to an incident commander (IC), who is fully engaged in the call, than having to stop what he or she is doing to answer a routing request from an incoming unit that is lost. Get all of the incoming or staged units off of the main operations channel and issue an "emergency traffic only" directive for all units operating on the incident. This will eliminate most or all of the unnecessary chatter that can occur at a large incident.

Some agencies also carry MCI kits on their apparatus, which can contain items such as a tactical worksheet or board that has preset benchmarks that the IC can reference and check off.

Such items can help the IC determine what incident priorities need to be addressed/accomplished. Also, some other items may include incident management system (IMS) command vests, which visually establish position assignments for personnel such as command, operations, safety, triage, and staging, and possibly some type of START triage system, which may include patient priority/status tags and flowcharts.

Scene Stabilization

Remember that your size-up of the incident begins the moment you are dispatched. Be proactive and call for additional units and/or specialty units, such as a heavy rescue/technical rescue team, a hazardous materials unit, or a tow agency (preferably a class C rotator unit with an articulating boom), to respond. Put air rescue on standby as a prewarning. These units can always be canceled if the incident is found to be insignificant. In addition, preassignments should be directed to your crew while en route. When approaching a scene, a thorough visual scan of the entire area should be conducted prior to stepping off the apparatus. Easily recognized hazards such as down utility poles, variations in topography such as embankments and slopes, or water hazards such as canals, streams, and lakes must be taken into consideration. Upon your arrival, give a clear, accurate account of what is presented, and conduct an inner and outer survey of the scene to formulate your action plan. Some basic initial considerations that need to be addressed include the following:

- What type of school bus are you dealing with (type C or D)?
- What is the damage level (minor, moderate, heavy)?
- Is the school bus still running?
- What is the resting position (upright, side, roof, override)?
- How many occupants are there (initial estimate)?
- What are additional resources needed (for example, apparatus, tow units, equipment, personnel, command staff)?
- Do you have an established MCI plan that can be implemented?

Other considerations may need to be added here that pertain to your response area or jurisdictional requirements. This list is a start, which you should add to and customize to your agency's needs and capabilities.

Stabilization: School Buses

As with most vehicles, there are basically four positions in which a school bus will present that the technical rescuer will have to stabilize—upright, on its side, on its roof, or on another vehicle. The main goal is to create a solid foundation to work from, which includes lowering the bus's center of gravity and preventing further unnecessary movement. The two more common suspension systems found on the types C and D buses have either a metal leaf spring–type system, or an air ride–type system with air bladders. The air ride suspension system is designed to protect the cargo and the frame system of the vehicle using air bladders. The bladders contain air under pressure ranging up to 120 psi (827 kPa). Stabilizing an upright school bus that utilizes an air bladder is extremely important because in the case of a bag rupture or leak, the bus will list or lean heavily to the side of the rupture. To stabilize a school bus in its normal position, follow the steps in **Skill Drill 11-1 ▶**. Some steps may occur simultaneously.

1. Don appropriate PPE, including mask and eye protection.
2. Assess the scene for hazards and complete the inner and outer surveys.
3. Activate MCI protocols if preestablished and/or if needed based on the number of patients.
4. Call for additional resources, such as an appropriately sized tow truck unit (preferably a class C rotator with an articulating boom), a technical rescue team (TRT) unit, or a hazardous materials unit.
5. Set up the hazard control zones (hot, warm, cold).
6. Set up two 1¾-inch (44-mm) charged hose lines in defensive positions.
7. Ensure that the vehicle is not running. Note: If normal shutdown procedures including disconnecting the battery system cannot be accomplished, then locate the air intake manifold in the engine compartment and discharge a 10-lb (4.5-kg) minimum dry chemical extinguisher into the device, which should suffocate and shut down the engine.
8. Locate and disable the battery system. (**Step 1**)
9. Place two Hi-Lift Jacks at the rear of the school bus, lifting from the bumper. (**Step 2**)
10. Position four tension buttress struts on the school bus with two at the front sides and two at the rear sides in an A-frame setup. Create purchase points for the tips of the struts to be set in place using a Halligan bar and flat-head axe or a pneumatic air chisel. (**Step 3**)
11. Pass the cargo straps under the school bus using a long pike pole to avoid going under the vehicle. (**Step 4**)
12. Additional cross-tie box cribbing can be set in position under the frame at the rear of the school bus and at the front under the bumper area. This will require a lot of cribbing, so ensure that the units on hand can support this task. The cribbing height for a cross-tie box crib should not exceed two times its width.
13. Prevent any forward or backward movement by positioning cribbing or step chocks upside down in a wedge-type setup in front of and behind each tire. (**Step 5**)
14. If access into the school bus can be established safely, then ensure that the air brake has been engaged, if the driver has not already done so.
15. Stabilize the school bus in its normal position. (**Step 6**)

To stabilize a school bus resting on its side, follow the steps in **Skill Drill 11-2 ▶**. Some steps may occur simultaneously.

1. Don appropriate PPE, including mask and eye protection.
2. Assess the scene for hazards and complete the inner and outer surveys. Ensure that there is no fuel leaking. Consider utilizing the appropriate foam for the type of fuel encountered.
3. Activate MCI protocols if preestablished and/or if needed based on the number of patients.
4. Call for additional resources, such as an appropriately sized tow truck unit (preferably a class C rotator with an articulating boom), TRT unit, or hazardous materials unit.

NFPA 1006, (10.2.2)

Skill Drill 11-1

Stabilizing a School Bus in Its Normal Position

1 Don appropriate PPE. Assess the scene for hazards and complete the inner and outer surveys. Activate MCI protocols depending on the number of patients. Call for additional resources. Set up the hazard control zones. Stage two charged hose lines. Ensure that the vehicle is not running. Locate and disable the battery system.

2 Place two Hi-Lift Jacks at the rear of the school bus, lifting from the bumper.

3 Position four tension buttress struts on the school bus with two at the front sides and two at the rear sides in an A-frame setup. Create purchase points for the tips of the struts to be set in place using a Halligan bar and flat-head axe or a pneumatic air chisel.

4 Pass the cargo straps under the school bus using a long pike pole to avoid going under the vehicle.

5 Prevent any forward or backward movement by positioning cribbing or step chocks upside down in a wedge-type setup in front of and behind each tire.

6 If access into the school bus can be established safely, then ensure that the air brake has been engaged, if the driver has not already done so. Stabilize the school bus in its normal position.

5. Set up the hazard control zones (hot, warm, cold).

6. Set up two 1¾-inch (44-mm) charged hose lines in defensive positions.

7. Depending on which side of the school bus is resting on the ground, locate and disable the battery system according to the procedures outlined in this chapter. (**Step 1**)

8. Ensure that the school bus is not running. Note: If normal shutdown procedures including disconnecting the battery system cannot be accomplished, then locate the air intake manifold in the engine compartment and discharge a 10-lb (4.5-kg) minimum dry chemical extinguisher into the device, which should suffocate and shut down the engine.

9. Secure the area around the muffler/regeneration device to ensure that this area is avoided because of burn potential from the heat of the device. (**Step 2**)

10. Because a school bus is relatively stable when positioned on its side on level ground, the stabilization will generally require positioning wedge sections or step chocks upside down in a wedge-type setup under the roof line and the undercarriage or floor line of the school bus. Wedges set on top of a section of 4- by 4-inch (102- by 102-mm) cribbing (or "four-by-four") can be utilized as was described in Skill Drill 8-2 for stabilizing a common passenger vehicle resting on its side. (**Step 3**)

NFPA 1006, (10.2.2)

Skill Drill 11-2

Stabilizing a School Bus Resting on Its Side

1 Don appropriate PPE. Assess the scene for hazards and complete the inner and outer surveys. Ensure that there is no fuel leaking. Consider utilizing the appropriate foam for the type of fuel encountered. Activate MCI protocols depending on the number of patients. Call for additional resources. Set up the hazard control zones. Stage two charged hose lines. Depending on which side of the school bus is resting on the ground, locate and disable the battery system according to the procedures outlined in this chapter.

2 Ensure that the school bus is not running. Note: If normal shutdown procedures including disconnecting the battery system cannot be accomplished, then locate the air intake manifold in the engine compartment and discharge a 10-lb (4.5-kg) minimum dry chemical extinguisher into the device to try to suffocate and shut down the engine. Secure the area around the muffler/regeneration device to ensure that this area is avoided.

3 Stabilize the school bus. Stabilization will generally require positioning wedge sections or step chocks upside down in a wedge-type setup under the roof line and the undercarriage or floor line of the school bus. Wedges set on top of a section of 4- by 4-inch (102- by 102-mm) cribbing (or "four-by-four") can be utilized as was described in Skill Drill 8-2 for stabilizing a common passenger vehicle resting on its side.

A school bus, because of the shape and design of its semi-arched roof, will not normally come to rest on its roof after a collision or roll-over incident. It will tend to remain upright or come to rest on its side. However, as rare as it may be, a school bus can come to rest on its roof. To stabilize a school bus that is resting on its roof, follow the steps in **Skill Drill 11-3 ▶**. Some steps may occur simultaneously.

1. Don appropriate PPE, including mask and eye protection.
2. Assess the scene for hazards and complete the inner and outer surveys. Ensure that there is no fuel leaking. Consider utilizing the appropriate foam for the type of fuel encountered.
3. Activate MCI protocols if preestablished and/or if needed based on the number of patients.
4. Call for additional resources, such as an appropriately sized tow truck unit (preferably a class C rotator with an articulating boom), TRT unit, or hazardous materials unit.
5. Set up the hazard control zones (hot, warm, cold).
6. Set up two 1¾-inch (44-mm) charged hose lines in defensive positions.
7. Locate and disable the battery system if accessible according to the procedures outlined in this chapter.
8. Ensure that the school bus is not running. Note: If normal shutdown procedures including disconnecting the battery

system cannot be accomplished, then locate the air intake manifold in the engine compartment and discharge a 10-lb (4.5-kg) minimum dry chemical extinguisher into the device, which should suffocate and shut down the engine. Access may not be available with a front-mount engine found in type C buses.

9. Secure an area around the muffler/regeneration device to ensure that this area is avoided because of the burn potential from heat retention of the device.
10. Secure wedges and step chocks upside down in a wedge-type setup around the entire roof line to keep the school bus from rocking. (**Step 1**)
11. Position four tension buttress struts on the school bus with two at the front sides and two at the rear sides in an A-frame setup. (**Step 2**)
12. Create purchase points for the tips of the struts to be set in place using a Halligan bar and flat-head axe or a pneumatic air chisel. The cargo straps used for tensioning of the struts can be passed through the windows to connect the struts together. Or, the strap can pass through a window around the truss bow and back through the adjoining window and be secured back onto itself. (**Step 3**)
13. The stabilized school bus resting on its roof. (**Step 4**)

Skill Drill 11-3

Stabilizing a School Bus Resting on Its Roof

1 Don appropriate PPE. Assess the scene for hazards and complete the inner and outer surveys. Ensure that there is no fuel leaking. Consider utilizing the appropriate foam for the type of fuel encountered. Activate MCI protocols depending on the number of patients. Call for additional resources. Set up the hazard control zones. Stage two charged hose lines. Locate and disable the battery system if accessible. Ensure the school bus is not running. Note: If normal shutdown procedures including disconnecting the battery system cannot be accomplished, then locate the air intake manifold in the engine compartment and discharge a 10-lb (4.5-kg) minimum dry chemical extinguisher into the device, which should suffocate and shut down the engine. Access may not be available with a front-mount engine found in type C buses. Secure an area around the muffler/regeneration device to ensure that this area is avoided. Secure wedges and step chocks upside down in a wedge-type setup around the entire roof line to keep the school bus from rocking.

2 Position four tension buttress struts on the school bus with two at the front sides and two at the rear sides in an A-frame setup.

3 Create purchase points for the tips of the struts to be set in place using a Halligan bar and flat-head axe or a pneumatic air chisel. The cargo straps used for tensioning of the struts can be passed through the windows to connect the struts together. Or, the strap can pass through a window around the truss bow and back through the adjoining window and be secured back onto itself.

4 This image shows the stabilized school bus resting on its roof.

Figure 11-15 On August 5, 2010, a type D school bus collided with a pick-up truck and was then struck in the rear by another type D school bus.

The elongated box shape of a school bus is designed to prevent over-rides in which the school bus comes to rest on top of another vehicle after a collision. The overall body design of a school bus is low to the ground, which allows it to bounce off an object rather than projecting upward on top of the object. This design, however, was not effective in an accident that occurred on August 5, 2010. A type D school bus collided with a pick-up truck and was then struck in the rear by another type D school bus. The force projected the first bus over the pick-up truck and onto the cab of a semi-truck **Figure 11-15 ▲** .

Such accidents involving a school bus are extremely rare and would challenge even the best prepared agency. However, once again, if the incident is broken down into manageable segments utilizing the IMS, it can be handled and controlled with a high percentage of success. Multiple stabilization tools and resources are required to properly stabilize a vehicle of this size in this type of resting position, some of which include a large tow unit, minimum class C rotator truck with an articulating boom, large amounts of cribbing, ratchet/cargo straps, come alongs with chain packages, and tension buttress struts. To stabilize or marry a school bus on top of another vehicle, follow the steps in **Skill Drill 11-4 ▶** . Some steps may occur simultaneously.

1. Don appropriate PPE, including mask and eye protection.
2. Assess the scene for hazards and complete the inner and outer surveys. Ensure that there is no fuel leaking. Consider utilizing the appropriate foam for the type of fuel encountered. (**Step 1**)
3. Activate MCI protocols if preestablished and/or if needed based on the number of patients.
4. Call for additional resources, such as an appropriately sized tow truck unit (preferably a class C rotator with an articulating boom), TRT unit, or hazardous materials unit.
5. Set up the hazard control zones (hot, warm, cold).
6. Set up two 1¾-inch (44-mm) charged hose lines in defensive positions.
7. Locate and disable the battery system if accessible according to the procedures outlined in this chapter.

8. Ensure that the school bus is not running. Note: If normal shutdown procedures including disconnecting the battery system cannot be accomplished, then locate the air intake manifold in the engine compartment and discharge a 10-lb (4.5-kg) minimum dry chemical extinguisher into the device, which should suffocate and shut down the engine.
9. Position four tension buttress struts on the school bus with two at the front sides and two at the rear sides in an A-frame setup. Create purchase points for the tips of the struts to be set in place using a Halligan bar and flat-head axe or a pneumatic air chisel. Pass the cargo straps under the school bus using a long pike pole to avoid going under the vehicle. Doing so will lower the center of gravity on the bus and expand its overall footprint, thus preventing vehicle roll. (**Step 2**)
10. Additional cross-tie box cribbing can be set in position under the frame at the rear of the school bus and at the front under the bumper area. This will require a lot of cribbing, so ensure that the units on hand can support this task. The cribbing height for a cross-tie box crib should not exceed two times its width.
11. Prevent any forward or backward movement by positioning cribbing or step chocks upside down in a wedge-type setup in front of and behind each tire that is touching the ground. (**Step 3**)
12. Marry the top vehicle to the bottom vehicle utilizing cargo straps or come alongs with chain packages. (**Step 4**)
13. Attempt to marry the vehicles together in four or more areas for added stability. (**Step 5**)

■ Victim Access: School Buses

Gaining or creating access into a school bus will depend on the vehicle's resting position. The goal is to create a main entry and exit area so the flow of rescuers and patient removal is consistent and controlled. The cagelike structure, with varying heavy-gauge steels throughout, is designed to provide occupant protection and safety. Thus, cutting into a school bus can be a difficult process requiring a lot of work and various specialized techniques. Again, the technique used will depend heavily on the bus's resting position. Because so many different crash scenarios can occur, covering them all in detail would require a separate book. This chapter will focus on procedures that are the most commonly utilized, that offer the highest percentage of success, and that, with practice, can be accomplished in the shortest amount of time. The tools used consist of but are not limited to a large assortment of cribbing sections, struts with tensioning straps, Hi-Lift Jacks, hand tools such as a Halligan bar and a flat-head axe, electric-powered reciprocating saws with 20 to 30 spare blades, air chisels, hydraulic tools, rescue-lift air bags, and an appropriately sized tow truck unit, preferably a class C rotator with an articulating boom **Figure 11-16 ▶** .

Front Window Access

The front windows on a school bus are composed of laminate safety glass that when removed from a bus on its side can provide the rescuer a natural opening that is large enough to pass equipment through and remove patients on backboards with

Skill Drill 11-4

NFPA 1006, (10.2.2)

Stabilizing/Marrying a School Bus on Top of Another Vehicle

1 Don appropriate PPE, including mask and eye protection. Assess the scene for hazards and complete the inner and outer surveys. Ensure that there is no fuel leaking. Consider utilizing the appropriate foam for the type of fuel encountered.

2 Activate MCI protocols depending on the number of patients. Call for additional resources. Set up the hazard control zones. Stage two charged hose lines. Locate and disable the battery system if accessible. Ensure the school bus is not running. Note: If normal shutdown procedures including disconnecting the battery system cannot be accomplished, then locate the air intake manifold in the engine compartment and discharge a 10-lb (4.5-kg) minimum dry chemical extinguisher into the device, which should suffocate and shut down the engine. Position four tension buttress struts on the school bus with two at the front sides and two at the rear sides in an A-frame setup. Create purchase points for the tips of the struts to be set in place using a Halligan bar and flat-head axe or a pneumatic air chisel. Pass the cargo straps under the school bus using a long pike pole.

3 Additional cross-tie box cribbing can be set in position under the frame at the rear of the school bus and at the front under the bumper area. This will require a lot of cribbing. Prevent any forward or backward movement by positioning cribbing or step chocks upside down in a wedge-type setup in front of and behind each tire that is touching the ground.

4 Marry the top vehicle to the bottom vehicle utilizing cargo straps or come alongs with chain packages.

5 Attempt to marry the vehicles together in four or more areas for added stability.

Figure 11-16 A class C rotator with an articulating boom tow truck.

minimal restriction. There are two settings for installing the laminate glass in the vehicle—the gasket push-out type and the beaded mastic type.

The gasket push-out type uses a flat laminate glass and comes in several configurations. For example, the windshield may include one large panel of glass, two large panels of glass separated by a thin metal divider, or four sections of glass where the entire windshield extends outward with two large panels in the front and two small panels that square off the sides. These panels are held in place by a rubber gasket seal, which can be removed and the glass pushed out from the inside of the vehicle, or it can be cut out with an electric reciprocating saw or a hand glass saw. It may be easier to use an electric reciprocating saw, which enables you to cut right through the metal glass divider while cutting the glass without having to stop. Utilizing a reciprocating saw to remove the windshield is the fastest procedure. Keep in mind that using the reciprocating saw will leave a glass edge. The glass edge can easily be pulled out along with the gasket, or a blanket can be placed over that section to protect from the sharp edges.

With the other type of setting on a school bus windshield, the beaded mastic type, adhesive secures the laminate panel around the entire perimeter of the glass; this is similar to the mastic settings used on the front windshield of a standard conventional vehicle. The difference between the gasket setting and the mastic adhesive setting is that the laminate panel that is held in with the mastic adhesive is generally one section or one panel of glass, and it is curved instead of flat. This mastic setting requires that the windshield be cut out using a hand glass saw or an electric reciprocating saw. To gain access into a school bus by removing the front windshield, follow the steps in **Skill Drill 11-5 ▶**. Some steps may occur simultaneously.

1. Don appropriate PPE, including mask and eye protection.
2. Assess the scene for hazards and complete the inner and outer surveys. Depending on which side the school bus is resting on, ensure that there is no fuel leaking. Consider utilizing the appropriate foam for the type of fuel encountered.

3. Activate MCI protocols if preestablished and/or if needed based on the number of patients.
4. Call for additional resources, such as an appropriately sized tow truck unit (preferably a class C rotator with an articulating boom), TRT unit, or hazardous materials unit.
5. Set up the hazard control zones (hot, warm, cold).
6. Set up two 1¾-inch (44-mm) charged hose lines in defensive positions.
7. Depending on which side of the school bus is resting on the ground, locate and disable the battery system according to the procedures outlined in this chapter.
8. Ensure that the school bus is not running. Note: If normal shutdown procedures including disconnecting the battery system cannot be accomplished, then locate the air intake manifold in the engine compartment and discharge a 10-lb (4.5-kg) minimum dry chemical extinguisher into the device, which should suffocate and shut down the engine.
9. If the bus is resting on its side or roof, secure the area around the muffler/regeneration device to ensure that this area is avoided because of burn potential from the heat of the device.
10. If the school bus is resting on its side, it is relatively stable when positioned on level ground. The stabilization will generally require positioning wedge sections or step chocks upside down in a wedge-type setup under the roof line and the undercarriage or floor line of the school bus. Wedges set on top of a four-by-four can be utilized as was described in Skill Drill 8-2 for stabilizing a common passenger vehicle resting on its side. **(Step 1)**
11. Use a large flat-head screwdriver or some type of prying tool to get a section of the gasket off of the glass and then pull it off around the entire perimeter of the window. Use the prying tool to work a corner of the glass away from the frame and then pull the entire glass toward you and out of the casing. This approach does not always work well because the gasket may have dry-rotted and/or become compromised by age, and the glass may break or the gasket may tear apart in sections.
12. Use a Halligan bar or a glass hand saw to make a purchase opening at the top of the window. **(Step 2)**
13. Use a reciprocating saw to cut out the two sections of glass, cutting through the center divider bar and working the blade around the inside perimeter of the windshield. **(Step 3)**
14. Remove the windshield and place it in a designated debris pile outside of the hot zone. **(Step 4)**

Seat Removal

Removing a bench seat can be a fairly easy process that can be accomplished quickly using a combination of a hydraulic spreader and a hydraulic cutter. Other tools such as a pneumatic air chisel or an electric reciprocating saw can also be utilized to accomplish the same task and may need to be used in some instances because of tight spaces, positioning, or precision cutting. For speed of operation, the combination of the hydraulic spreader and cutter is the best tool choice to accomplish the

NFPA 1006, (10.2.4)

Skill Drill 11-5

Gaining Access Into a School Bus by Removing a Front Windshield

1. Don appropriate PPE, including mask and eye protection. Assess the scene for hazards and complete the inner and outer surveys. Ensure that there is no fuel leaking. Consider utilizing the appropriate foam for the type of fuel encountered. Activate MCI protocols depending on the number of patients. Call for additional resources. Set up the hazard control zones. Stage two charged hose lines. Locate and disable the battery system if accessible. Ensure the school bus is not running. Note: If normal shutdown procedures including disconnecting the battery system cannot be accomplished, then locate the air intake manifold in the engine compartment and discharge a 10-lb (4.5-kg) minimum dry chemical extinguisher into the device, which should suffocate and shut down the engine. If the bus is resting on its side or roof, secure the area around the muffler/regeneration device. If the school bus is resting on its side, stabilization will generally require positioning wedge sections or step chocks upside down in a wedge-type setup under the roof line and the undercarriage or floor line of the school bus. Wedges set on top of a four-by-four can be utilized as was described in Skill Drill 8-2 for stabilizing a common passenger vehicle resting on its side.

2. Use a large flat-head screwdriver or some type of prying tool to get a section of the gasket off of the glass and then pull it off around the entire perimeter of the window. Use the prying tool to work a corner of the glass away from the frame and then pull the entire glass toward you and out of the casing. Use a Halligan bar or a glass hand saw to make a purchase opening at the top of the window.

3. Use a reciprocating saw to cut out the two sections of glass, cutting through the center divider bar and working the blade around the inside perimeter of the windshield.

4. Remove the windshield and place it in a designated debris pile outside of the hot zone.

task. To remove a bench seat from a school bus, follow the steps in **Skill Drill 11-6 ▾**. Some steps may occur simultaneously.

1. Don appropriate PPE, including mask and eye protection.
2. Assess the scene for hazards and complete the inner and outer surveys. Depending on which side the school bus is resting on, ensure that there is no fuel leaking. Consider utilizing the appropriate foam for the type of fuel encountered.
3. Activate MCI protocols if preestablished and/or if needed based on the number of patients.
4. Call for additional resources, such as an appropriately sized tow truck unit (preferably a class C rotator with

an articulating boom), TRT unit, or hazardous materials unit.

5. Set up the hazard control zones (hot, warm, cold).
6. Set up two 1¾-inch (44-mm) charged hose lines in defensive positions.
7. Depending on the resting position of the school bus, locate and disable the battery system according to the procedures outlined in this chapter.
8. Ensure that the school bus is not running. Note: If normal shutdown procedures including disconnecting the battery system cannot be accomplished, then locate the air intake manifold in the engine compartment and discharge a

NFPA 1006, (10.2.4)

Skill Drill 11-6

Removing a Bench Seat From a School Bus

1 Don appropriate PPE, including mask and eye protection. Assess the scene for hazards and complete the inner and outer surveys. Ensure that there is no fuel leaking. Consider utilizing the appropriate foam for the type of fuel encountered. Activate MCI protocols depending on the number of patients. Call for additional resources. Set up the hazard control zones. Stage two charged hose lines. Locate and disable the battery system if accessible. Ensure the school bus is not running. Note: If normal shutdown procedures including disconnecting the battery system cannot be accomplished, then locate the air intake manifold in the engine compartment and discharge a 10-lb (4.5-kg) minimum dry chemical extinguisher into the device, which should suffocate and shut down the engine. Stabilize the school bus from movement by utilizing the techniques outlined in this chapter. Position a hydraulic spreader with the arms vertical behind and under the bench frame closest to the lip that extends and attaches the seat to the sidewall. Slowly open the spreader, ensuring that the top arm catches the frame and detaches the seat from the sidewall lip attachment.

2 Use the hydraulic cutter to cut the seat frame where it is bolted to the floor next to the aisle.

3 Remove the entire bench seat and place it in a debris pile outside the hot zone. It is best to hand off debris to someone outside of the bus because of the amount of seats and material that may have to be removed.

10-lb (4.5-kg) minimum dry chemical extinguisher into the device, which should suffocate and shut down the engine.

9. Stabilize the school bus from movement by utilizing the techniques outlined in this chapter.

10. Position a hydraulic spreader with the arms vertical behind and under the bench frame closest to the lip that extends and attaches the seat to the sidewall. Slowly open the spreader, ensuring that the top arm catches the frame and detaches the seat from the sidewall lip attachment. This section of the seat must be removed first or there will not be enough leverage for the spreader to operate effectively. (**Step 1**)

11. Use the hydraulic cutter to cut the seat frame where it is bolted to the floor next to the aisle. Attempting to take the bolts off by utilizing a pneumatic impact wrench will normally just spin the heads and will not remove them. Cutting the frame produces the most success even though a ½-inch (13 mm) section of cut frame protrudes from the floor. (**Step 2**)

12. Remove the entire bench seat and place it in a debris pile outside the hot zone. It is best to hand off debris to someone outside of the bus because of the amount of seats and material that may have to be removed. (**Step 3**)

Side Wall Access

Removing a section of the sidewall can produce a large opening for patient removal. The technique involves removing two windows that are side by side, cross-cutting the roof bow truss that separates those two windows, removing two or more rows of bench seats, cutting down the sides of the sidewall, creating a relief cut on the bottom section of the center roof bow truss, and then pulling the entire section out and down. Multiple tools will be needed to accomplish this task, including, but not limited to, two electric reciprocating saws with multiple blades, a pneumatic air chisel, a hydraulic spreader, and a hydraulic cutter. To remove a section of the sidewall of a school bus, follow the steps in **Skill Drill 11-7 ▶**. Some steps may occur simultaneously.

1. Don appropriate PPE, including mask and eye protection.

2. Assess the scene for hazards and complete the inner and outer surveys. Depending on which side the school bus is resting on, ensure that there is no fuel leaking. Consider utilizing the appropriate foam for the type of fuel encountered.

3. Activate MCI protocols if preestablished and/or if needed based on the number of patients.

4. Call for additional resources, such as an appropriately sized tow truck unit (preferably a class C rotator with an articulating boom), TRT unit, or hazardous materials unit.

5. Set up the hazard control zones (hot, warm, cold).

6. Set up two 1¾-inch (44-mm) charged hose lines in defensive positions.

7. Depending on the resting position of the school bus, locate and disable the battery system according to the procedures outlined in this chapter.

8. Ensure that the school bus is not running. Note: If normal shutdown procedures including disconnecting the battery system cannot be accomplished, then locate the air intake manifold in the engine compartment and discharge a 10-lb (4.5-kg) minimum dry chemical extinguisher into the device, which should suffocate and shut down the engine.

9. Stabilize the school bus from movement by utilizing the techniques outlined in this chapter.

10. Enter the school bus. With a pneumatic air chisel, cut through the screw attachment or cut off the attachments or screw heads that lock the windows to the school bus frame. This task can be accomplished without breaking any of the glass in the windows. (**Step 1**)

11. Once these attachments or screws are removed, the windows will pull right out of the frame casing. (**Step 2**)

12. Remove two or more sets of seat benches utilizing the technique outlined in Skill Drill 11-6.

13. Use the hydraulic cutter to cut the roof bow truss that divides the two windows. Utilize a cross-cut technique to keep the post stub from protruding downward. (**Step 3**)

14. Position two technical rescuers outside of the bus with reciprocating saws. Simultaneously cut downward on two sections of the sidewall. The appropriate location to begin the cutting is just inside the outer section of the window frame. This is where two roof bow trusses will form the outer frame of those two windows that were removed. Hint: Stay to the inside of the rivet heads that outline the position of a roof bow truss. Cut down past the first rub rail, which should be approximately seat level (cutting any farther down will be difficult at best because you will encounter heavier gauge steel with the crash rail causing the reciprocating saw to slow down and the blade to melt or dull quickly). (**Step 4**)

15. Use a Halligan bar and flat-head axe to make two holes on both sides of the lower section of the center roof bow truss in the general area of the seat level rub rail by driving the spiked end of the Halligan bar into the sidewall by striking it with the flat-head axe. The area in which the holes will be made is just below the first rub rail, near seat level; if you go any lower, you will encounter the crash rail area, which is composed of high-gauge steel and is extremely difficult to cut into. (**Step 5**)

16. Insert the blade of a reciprocating saw into one of the holes just created and cut through the roof bow truss; this will now act as a relief cut. (**Step 6**)

17. Grip the top center roof bow truss that was cut and pull out and downward, taking with it the entire sidewall that was just cut out. (**Step 7**)

18. Move all of the debris to the debris pile outside of the hot zone.

Roof Access

Gaining access into the roof of a school bus can be difficult because it requires cutting through multiple structural members composed of heavier gauge steels.

Rescue Tips

Avoid making the purchase point holes in line with any row of rivets because there will be a structural member underneath the sheet metal that is held in place by the rivets.

Skill Drill 11-7

Removing a Section of the Sidewall of a School Bus

1 Don appropriate PPE, including mask and eye protection. Assess the scene for hazards and complete the inner and outer surveys. Ensure that there is no fuel leaking. Consider utilizing the appropriate foam for the type of fuel encountered. Activate MCI protocols depending on the number of patients. Call for additional resources. Set up the hazard control zones. Stage two charged hose lines. Locate and disable the battery system if accessible. Ensure the school bus is not running. Note: If normal shutdown procedures including disconnecting the battery system cannot be accomplished, then locate the air intake manifold in the engine compartment and discharge a 10-lb (4.5-kg) minimum dry chemical extinguisher into the device, which should suffocate and shut down the engine. Stabilize the school bus from movement by utilizing the techniques outlined in this chapter. Enter the school bus. With a pneumatic air chisel, cut through the screw attachment or cut off the attachments or screw heads that lock the windows to the school bus frame.

2 Once these attachments or screws are removed, the windows will pull right out of the frame casing.

3 Remove two or more sets of seat benches utilizing the technique outlined in Skill Drill 11-6. Use the hydraulic cutter to cut the roof bow truss that divides the two windows. Utilize a cross-cut technique to keep the post stub from protruding downward.

4 Position two technical rescuers outside of the bus with reciprocating saws. Simultaneously cut downward on two sections of the sidewall, beginning just inside the outer section of the window frame where two roof bow trusses will form the outer frame of those two windows that were removed. Cut down past the first rub rail.

5 Use a Halligan bar and flat-head axe to make two holes on both sides of the lower section of the center roof bow truss in the general area of the seat level rub rail by driving the spiked end of the Halligan bar into the sidewall by striking it with the flat-head axe. The area in which the holes will be made is just below the first rub rail, near seat level.

6 Insert the blade of a reciprocating saw into one of the holes just created and cut through the roof bow truss; this will now act as a relief cut.

7 Grip the top center roof bow truss that was cut and pull out and downward, taking with it the entire sidewall that was just cut out.

To gain access through the roof of a school bus resting on its side, follow the steps in **Skill Drill 11-8 ▶**. Some steps may occur simultaneously.

1. Don appropriate PPE, including mask and eye protection.
2. Assess the scene for hazards and complete the inner and outer surveys. Depending on which side the school bus is resting on, ensure that there is no fuel leaking. Consider utilizing the appropriate foam for the type of fuel encountered.
3. Activate MCI protocols if preestablished and/or if needed based on the number of patients.
4. Call for additional resources, such as an appropriately sized tow truck unit (preferably a class C rotator with an articulating boom), TRT unit, or hazardous materials unit.
5. Set up the hazard control zones (hot, warm, cold).
6. Set up two 1¾-inch (44-mm) charged hose lines in defensive positions.
7. Depending on which side of the school bus is resting on the ground, locate and disable the battery system according to the procedures outlined in this chapter.
8. Ensure that the school bus is not running. Note: If normal shutdown procedures including disconnecting the battery system cannot be accomplished, then locate the air intake manifold in the engine compartment and discharge a 10-lb (4.5-kg) minimum dry chemical extinguisher into the device, which should suffocate and shut down the engine.
9. Secure the area around the muffler/regeneration device to ensure that this area is avoided because of burn potential from the heat of the device.
10. Because a school bus is relatively stable when positioned on its side on level ground, the stabilization will generally require positioning wedge sections or step chocks upside down in a wedge-type setup under the roof line and the undercarriage or floor line of the school bus. Wedges set on top of a four-by-four can be utilized as was described in Skill Drill 8-2 for stabilizing a common passenger vehicle resting on its side.
11. Locate the area of the roof that will be designated as an exit/egress path. The height of this opening will vary depending on the structure and position of the bus as well as the overall height of the rescuer. An adequate area can measure 6 feet (1.8 m) tall and 5 feet (1.5 m) wide, depending on where the roof bow trusses are located. Be aware that all school buses are not the same, and some may have additional support members such as roof stringers that run perpendicular to roof bow trusses; these can be identified by the presence of rivets.
12. A good rule of thumb is to count three roof bow trusses. Cut the outside dimensions of the opening just *inside* the two outer roof bow trusses; the center roof bow truss should be positioned directly in the middle of your opening. Optional: To maintain an even cut or opening, you can take a spray can of marking paint or some type of marking device and draw your opening to give yourself a cutting guide.
13. Use a Halligan bar and a flat-head axe or appropriate striking tool to create four purchase point holes. The first two holes will be just on the outside of the center roof bow

truss at the top, which will mark the top dimension of your opening. (**Step 1**)

14. The second two holes will be at the bottom of that same center roof bow truss, which will mark the bottom dimension of your opening. (**Step 2**)
15. With the help of another responder, insert the blades of two electric reciprocating saws into the top purchase point holes just created and start the opening cuts; do not cut the top center roof bow truss until last to avoid excessive vibration or the blade becoming pinched and/or stuck in the metal. The two saws should move in opposite directions outward and then down. (Remember to stay inside the rivets, which indicate where a roof bow truss or structural component is.) You may have to cut through a roof stringer, which will slow your saw down because it is an approximately 16-gauge steel beam. (**Step 3**)
16. When you and the other rescuer reach the bottom of the cut, two things can be done: The cut can continue back to the center roof bow truss with the intention of removing the entire section of metal, or the section can be pulled down or flapped. If the plan is to flap, then use one saw to cut the bottom center roof bow truss, inserting the blade into the purchase point opening that was made and cutting completely through the roof bow truss. (**Step 4**)
17. Make the same cut at the top section of the center roof bow truss in the same fashion; remember that this has to be cut last to avoid excessive vibration or the blade becoming pinched and/or stuck in the metal. (**Step 5**)
18. Pull the entire section of metal that was just cut downward. Hint: If you cannot grip the cut section because it has fallen inward, take the adz end of a Halligan bar and position it into a cut opening at the top to gain leverage so the section of metal can be pulled down.
19. Place a tarp or covering over the flapped or cut section to avoid injury. (**Step 6**)

Rescue Tips

Be aware that all school buses are not the same, and some may have additional support members such as roof stringers that run perpendicular to roof bow trusses; these can be identified by the presence of rivets.

Rear Door Access

There are several techniques that can be applied to gain entry into the rear emergency door of a school bus. The resting position of the bus will be a factor in deciding which technique to use. With the vehicle resting in the upright position, gaining access into the rear emergency door of a school bus requires removing the two safety glass panels located in the rear door prior to forcing entry. A vertical spread technique can be performed in one of the window frame openings of the door to create a purchase point opening. The goal is to create a large opening near the latching mechanism, which will need to be cut and released with the hydraulic cutter. A standing platform utilizing two or more cross-tie box cribbing configurations and

Skill Drill 11-8

Roof Access: School Bus on Its Side

1 Don appropriate PPE, including mask and eye protection. Assess the scene for hazards and complete the inner and outer surveys. Ensure that there is no fuel leaking. Consider utilizing the appropriate foam for the type of fuel encountered. Activate MCI protocols depending on the number of patients. Call for additional resources. Set up the hazard control zones. Stage two charged hose lines. Locate and disable the battery system if accessible. Ensure the school bus is not running. Note: If normal shutdown procedures including disconnecting the battery system cannot be accomplished, then locate the air intake manifold in the engine compartment and discharge a 10-lb (4.5-kg) minimum dry chemical extinguisher into the device, which should suffocate and shut down the engine. Secure the area around the muffler/regeneration device to ensure that this area is avoided. Stabilize the school bus from movement by utilizing the techniques outlined in this chapter. Locate the area of the roof that will be designated as an exit/egress path. A good rule of thumb is to count three roof bow trusses. Cut the outside dimensions of the opening just *inside* the two outer roof bow trusses; the center roof bow truss should be positioned directly in the middle of your opening. Use a Halligan bar and a flat-head axe or appropriate striking tool to create four purchase point holes just on the outside of the center roof bow truss at the top, which will mark the top dimension of your opening. The first two holes will be just on the outside of the center roof bow truss at the top, which will mark the top dimension of your opening.

2 The second two holes will be at the bottom of that same center roof bow truss, which will mark the bottom dimension of your opening.

3 With the help of another responder, insert the blades of two electric reciprocating saws into the top purchase point holes just created and start the opening cuts; do not cut the top center roof bow truss until last. The saws should move in opposite directions outward and then down, staying inside the rivets.

4 Once you have reached the bottom of the cut, two things can be done: The cut can continue back to the center roof bow truss with the intention of removing the entire section of metal, or the section can be pulled down or flapped. If the plan is to flap, then use one saw to cut the bottom center roof bow truss, inserting the blade into the purchase point opening that was made and cutting completely through the roof bow truss.

Skill Drill 11-8

NFPA 1006, (10.2.4)

Roof Access: School Bus on Its Side *(Continued)*

5 Make the same cut at the top section of the center roof bow truss in the same fashion; remember that this has to be cut last.

6 Pull the entire section of metal that was just cut downward. Place a tarp or covering over the flapped or cut section to avoid injury.

a backboard may need to be created to gain the proper height to operate the tools. The side emergency door that can be found on a type D bus will have only one glass panel.

Rear Door Access: School Bus in Its Normal Position

To gain access through the rear emergency door of a school bus in its normal position, follow the steps in **Skill Drill 11-9 ▶**. Some steps may occur simultaneously.

1. Don appropriate PPE, including mask and eye protection.
2. Assess the scene for hazards and complete the inner and outer surveys.
3. Activate MCI protocols if preestablished and/or if needed based on the number of patients.
4. Call for additional resources, such as an appropriately sized tow truck unit (preferably a class C rotator with an articulating boom), TRT unit, or hazardous materials unit.
5. Set up the hazard control zones (hot, warm, cold).
6. Set up two 1¾-inch (44-mm) charged hose lines in defensive positions.
7. Depending on the resting position of the school bus, locate and disable the battery system according to the procedures outlined in this chapter.
8. Ensure that the school bus is not running. Note: If normal shutdown procedures including disconnecting the battery system cannot be accomplished, then locate the air intake manifold in the engine compartment and discharge a 10-lb (4.5-kg) minimum dry chemical extinguisher into the device, which should suffocate and shut down the engine.

9. Stabilize the school bus from movement by utilizing the techniques outlined in this chapter.
10. Try to open the door manually. If the door won't open, remove the two safety glass panels located within the rear door frame. This can be accomplished by breaking the glass with a glass tool if it is tempered glass or by using a hand glass saw to cut it out if it is laminate glass. You may attempt to pull the gasket out to release the glass, but it may become a very time-consuming task if the gasket starts to tear apart due to weathering. **(Step 1)**
11. Place the hydraulic spreader with the arms vertical in the lower window frame opening, and perform a vertical spread technique. **(Step 2)**
12. Once a purchase point opening has been established, place the tips of the hydraulic spreader in the door frame and create a larger opening; work the tool upward toward the latching mechanism. **(Step 3)**
13. Once a large enough opening has been created around the latching mechanism, use the hydraulic cutter to cut the latch, which will release the door.
14. With the door released, tie the door back in the open position utilizing rope or webbing. **(Step 4)**

Rear Door Access: School Bus Resting on Its Side

The main goal of gaining access into a school bus is to create an adequate opening to rescue and remove patients as quickly as possible. Cutting out the entire rear section of a school bus is great practice for improving tool-handling skills, but doing so is not necessary and wastes valuable time. Opening the rear door will create a large enough opening to safely remove

NFPA 1006, (10.2.4)

Skill Drill 11-9

Rear Door Access: School Bus in Its Normal Position

1 Don appropriate PPE, including mask and eye protection. Assess the scene for hazards and complete the inner and outer surveys. Activate MCI protocols depending on the number of patients. Call for additional resources. Set up the hazard control zones. Stage two charged hose lines. Locate and disable the battery system if accessible. Ensure the school bus is not running. Note: If normal shutdown procedures including disconnecting the battery system cannot be accomplished, then locate the air intake manifold in the engine compartment and discharge a 10-lb (4.5-kg) minimum dry chemical extinguisher into the device, which should suffocate and shut down the engine. Stabilize the school bus from movement utilizing the techniques outlined in this chapter. Try to open the rear door manually. If the door won't open, remove the two safety glass panels located within the rear door frame. This can be accomplished by breaking the glass with a glass tool if it is tempered glass or by using a hand glass saw to cut it out if it is laminate glass.

2 Place the hydraulic spreader with the arms vertical in the lower window frame opening, and perform a vertical spread technique.

3 Once a purchase point opening has been established, place the tips of the hydraulic spreader in the door frame and create a larger opening; work the tool upward toward the latching mechanism.

4 Once a large enough opening has been created around the latching mechanism, use the hydraulic cutter to cut the latch, releasing the door. Tie the door back in the open position utilizing rope or webbing.

occupants and will save valuable time. To gain access through the rear emergency door of a school bus resting on its side, follow the steps in **Skill Drill 11-10 ▶**. Some steps may occur simultaneously.

1. Don appropriate PPE, including mask and eye protection.
2. Assess the scene for hazards and complete the inner and outer surveys. Depending on which side the school bus is resting on, ensure that there is no fuel leaking. Consider utilizing the appropriate foam for the type of fuel encountered.
3. Activate MCI protocols if preestablished and/or if needed based on the number of patients.
4. Call for additional resources, such as an appropriately sized tow truck unit (preferably a class C rotator with an articulating boom), TRT unit, or hazardous materials unit.
5. Set up the hazard control zones (hot, warm, cold).
6. Set up two 1¾-inch (44-mm) charged hose lines in defensive positions.
7. Depending on which side of the school bus is resting on the ground, locate and disable the battery system according to the procedures outlined in this chapter.
8. Ensure that the school bus is not running. Note: If normal shutdown procedures including disconnecting the battery system cannot be accomplished, then locate the air intake manifold in the engine compartment and discharge a 10-lb (4.5-kg) minimum dry chemical extinguisher into the device, which should suffocate and shut down the engine.
9. Secure the area around the muffler/regeneration device to ensure that this area is avoided because of burn potential from the heat of the device.
10. Because a school bus is relatively stable when positioned on its side on level ground, the stabilization will generally require positioning wedge sections or step chocks upside down in a wedge-type setup under the roof line and the undercarriage or floor line of the school bus. Wedges set on top of a four-by-four can be utilized as was described in Skill Drill 8-2 for stabilizing a common passenger vehicle resting on its side.
11. Build a standing platform to work from so the tools are not elevated and operated over your head. This can be accomplished with two or more cross-tie box cribbing configurations and a large backboard, or by other means.
12. If the door will not open manually, remove all of the safety glass from the rear, which includes the door panels and glass panels adjacent to the rear door, using the appropriate technique described for the type of glass encountered. **(Step 1)**
13. Use the hydraulic cutter to cut through the top section of the door frame; this is the post that connects the door to the adjacent upper window frame. **(Step 2)**
14. Cut diagonally into the bottom corner section of the top window where the door latch and the top window frame meet. The cut line should be directed downward toward the corner of the bottom window frame of the door. **(Step 3)**
15. Cut diagonally upward, toward the upper corner of the bottom door window frame, attempting to meet the first cut that was made. **(Step 4)**

16. If the two cuts do not meet, then take a reciprocating saw and complete the cut or attempt to work the hydraulic cutter upward, cutting away the last remaining section.
17. The goal is to cut around the latching mechanism, which will release the door without having to spread the latch away with the hydraulic spreader. **(Step 5)**

Front Door Access

Entry through the front door of a school bus will depend on the operability of the door itself. Always try to open the door manually. Otherwise, it may only require removing one of the glass panels in the door and inserting a pike pole into the bus to grip the lever bar that releases the door. This is a very fast technique that requires the least amount of forcible entry tactics and tool usage **Figure 11-17 ▼**.

Gaining access through the door can also require cutting it out completely, utilizing hydraulic tools or a combination of hydraulic tools and reciprocating saws. To gain access through the front entry door of a school bus in its normal position, follow the steps in **Skill Drill 11-11 ▶**. Some steps may occur simultaneously.

1. Don appropriate PPE, including mask and eye protection.
2. Assess the scene for hazards and complete the inner and outer surveys.
3. Activate MCI protocols if preestablished and/or if needed based on the number of patients.
4. Call for additional resources, such as an appropriately sized tow truck unit (preferably a class C rotator with an articulating boom), TRT unit, or hazardous materials unit.
5. Set up the hazard control zones (hot, warm, cold).
6. Set up two 1¾-inch (44-mm) charged hose lines in defensive positions.
7. Locate and disable the battery system according to the procedures outlined in this chapter.
8. Ensure that the school bus is not running. Note: If normal shutdown procedures including disconnecting the battery system cannot be accomplished, then locate the air intake manifold in the engine compartment and discharge

Figure 11-17 Entry through the front door of a school bus may only require removing one of the glass panels in the door and inserting a pike pole into the bus to grip the lever bar that releases the door.

Skill Drill 11-10

Rear Door Access: School Bus Resting on Its Side

1. Don appropriate PPE, including mask and eye protection. Assess the scene for hazards and complete the inner and outer surveys. Ensure that there is no fuel leaking. Consider utilizing the appropriate foam for the type of fuel encountered. Activate MCI protocols depending on the number of patients. Call for additional resources. Set up the hazard control zones. Stage two charged hose lines. Locate and disable the battery system if accessible. Ensure the school bus is not running. Note: If normal shutdown procedures including disconnecting the battery system cannot be accomplished, then locate the air intake manifold in the engine compartment and discharge a 10-lb (4.5-kg) minimum dry chemical extinguisher into the device, which should suffocate and shut down the engine. Secure the area around the muffler/regeneration device. Stabilize the school bus from movement utilizing the techniques outlined in this chapter. Build a standing platform to work from so the tools are not elevated and operated over your head. If the door will not open manually, remove all of the safety glass from the rear using the appropriate technique.

2. Use the hydraulic cutter to cut through the top section of the door frame; this is the post that connects the door to the adjacent upper window frame.

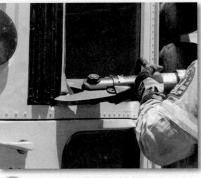

3. Cut diagonally into the bottom corner section of the top window where the door latch and the top window frame meet. The cut line should be directed downward toward the corner of the bottom window frame of the door.

4. Cut diagonally upward, toward the upper corner of the bottom door window frame, attempting to meet the first cut that was made.

5. If the two cuts do not meet, then take a reciprocating saw and complete the cut or attempt to work the hydraulic cutter upward, cutting away the last remaining section. The goal is to cut around the latching mechanism, which will release the door without having to spread the latch away with the hydraulic spreader.

a 10-lb (4.5-kg) minimum dry chemical extinguisher into the device, which should suffocate and shut down the engine.

9. If the door will not open manually, remove all of the safety glass from the front entry door by removing the gasket and

pushing the panels inward or breaking and/or cutting the glass out using the appropriate techniques outlined in this text. (Step 1)

10. After you remove one side of the door and can make entry into the bus, and if the lever arm used to control the open-

ing and closing of the door is inoperable, cut it loose using a hydraulic cutter. (**Step 2**)

11. The door is composed of two sections that release outward. With the glass removed, the appearance of the door frame will resemble a cross, which is the frame section that will be cut out. (**Step 3**)

12. Cutting the frame of both sections should begin at the bottom closest to the hinged section of the frame (do not attempt to remove the piano-type hinge of the door sections), and you will work your way to the top section of the door. Starting at the bottom first minimizes the vibration effect that can occur when using a recip-

NFPA 1006, (10.2.4)

Skill Drill 11-11

Front Door Access: School Bus in Its Normal Position

1 Don appropriate PPE, including mask and eye protection. Assess the scene for hazards and complete the inner and outer surveys. Activate MCI protocols depending on the number of patients. Call for additional resources. Set up the hazard control zones. Stage two charged hose lines. Locate and disable the battery system if accessible. Ensure the school bus is not running. Note: If normal shutdown procedures including disconnecting the battery system cannot be accomplished, then locate the air intake manifold in the engine compartment and discharge a 10-lb (4.5-kg) minimum dry chemical extinguisher into the device, which should suffocate and shut down the engine. If the door will not open manually, remove all of the safety glass from the front entry door by removing the gasket and pushing the panels inward or breaking and/or cutting the glass out using the appropriate techniques outlined in this text.

2 After you remove one side of the door and can make entry into the bus, and if the lever arm used to control the opening and closing of the door is inoperable, cut it loose using a hydraulic cutter.

3 The door is composed of two sections that release outward. With the glass removed, the appearance of the door frame will resemble a cross, which is the frame section that will be cut out.

(continues)

Skill Drill 11-11

Front Door Access: School Bus in Its Normal Position *(Continued)*

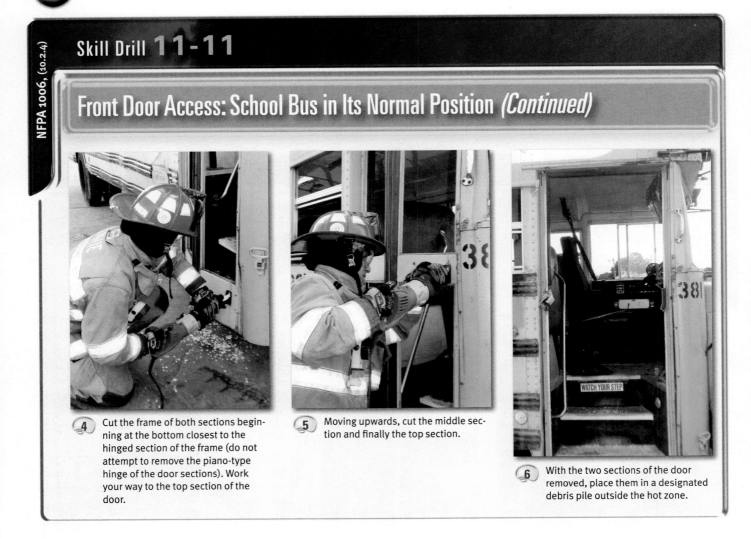

4 Cut the frame of both sections beginning at the bottom closest to the hinged section of the frame (do not attempt to remove the piano-type hinge of the door sections). Work your way to the top section of the door.

5 Moving upwards, cut the middle section and finally the top section.

6 With the two sections of the door removed, place them in a designated debris pile outside the hot zone.

rocating saw (a hydraulic cutter can also be utilized). **(Step 4)**

13. Moving upwards, cut the middle section and finally the top section. **(Step 5)**

14. The A-post and side/front windshield can also be incorporated in the removal process if needed to gain additional access.

15. With the two sections of the door removed, place them in a designated debris pile outside the hot zone. **(Step 6)**

Rescue-lift Air Bag Operation

On any given school bus rescue, an occupant can be ejected or partially ejected and trapped under the bus itself. There are multiple tools that can be used to safely and effectively lift and extricate a victim who has become trapped under a school bus. Tools such as an appropriately sized tow truck unit, preferably a class C rotator with an articulating boom, or rescue-lift air bags, whether they are a high-pressure type or the low-pressure high-lift type, are ideal to accomplish the rescue. Again, preplanning and training for incidents such as this cannot be stressed enough. There is not a tow agency out there that would not be willing to train with fire rescue agencies in sharing resources.

Some fire rescue agencies in Europe, where tow agencies are automatically dispatched on incidents such as this in their jurisdiction, are well ahead of the curve. They are well prepared because many have worked and trained together, and they have response protocols preestablished.

A rescue-lift air bag operation for a person who has been ejected and become trapped under a school bus that is now resting on its side is very similar to an operation for a person who has become trapped under a conventional vehicle. This operation is actually less complicated because there is a much larger surface area to work from on a school bus. The key is to always lift evenly and from areas with structural support such as roof bow trusses. The "lift an inch, crib an inch" expression still applies; there should be ample amounts of cribbing on hand.

One issue that can hamper a rescue-lift air bag operation is unstable terrain such as muck, or loose sand, termed "sugar sand," which can sink a bag when inflated. A solution is to place a large, flat, solid object, such as the steel or aluminum ground pads that are placed under aerial outriggers, under the bottom bag. Doing so expands the bag's footprint and equally distributes the lift point of the bag across the entire dimension of the plate. Remember, the larger the surface area covered, the greater the resistance to sinking. When dealing with unstable terrain, it is best practice to have some digging tools on hand, such as shovels and picks, because you may have to remove sand/dirt to gain a position to insert the bag(s). You may get only one bag in position initially, but once you lift the school bus a few inches and properly secure cribbing on both sides, you will be able to deflate the one bag and gain enough room to insert a second to help perform the full lift.

As a safety precaution, review the rescue-lift air bag information detailed in Chapter 6, *Tools and Equipment*, before performing this skill drill. To perform a rescue-lift air bag operation for a person who has been ejected and is trapped under a school bus resting on its side, follow the steps in **Skill Drill 11-12 ▶**. Some steps may occur simultaneously.

1. Don appropriate PPE, including mask and eye protection.
2. Assess the scene for hazards and complete the inner and outer surveys. Depending on which side the school bus is resting on, ensure that there is no fuel leaking. Consider utilizing the appropriate foam for the type of fuel encountered.
3. Activate MCI protocols if preestablished and/or if needed based on the number of patients.
4. Call for additional resources, such as an appropriately sized tow truck unit (preferably a class C rotator with an articulating boom), TRT unit, or hazardous materials unit.
5. Set up the hazard control zones (hot, warm, cold).
6. Set up two 1¾-inch (44-mm) charged hose lines in defensive positions.
7. Depending on which side of the school bus is resting on the ground, locate and disable the battery system according to the procedures outlined in this chapter.
8. Ensure that the school bus is not running. Note: If normal shutdown procedures including disconnecting the battery system cannot be accomplished, then locate the air intake manifold in the engine compartment and discharge a 10-lb (4.5-kg) minimum dry chemical extinguisher into the device, which should suffocate and shut down the engine.
9. Secure the area around the muffler/regeneration device to ensure that this area is avoided because of the burn potential from the heat of the device.
10. Because a school bus is relatively stable when positioned on its side on level ground, the stabilization will generally require positioning wedge sections or step chocks upside down in a wedge-type setup under the roof line and the undercarriage or floor line of the school bus. Wedges set on top of a four-by-four can be utilized as was described in Skill Drill 8-2 for stabilizing a common passenger vehicle resting on its side. **(Step 1)**
11. Where the patient is trapped under the school bus and how he or she is presenting will determine how many rescue-lift air bags need to be utilized and the best placement of the bags to produce an even lift. Two bags may be sufficient, with the two placed on one side of the patient in conjunction with cribbing, or four bags may be needed, with two bags positioned on both sides of the patient, also in conjunction with cribbing.

12. To properly and safely conduct a rescue-lift air bag operation, a minimum of five personnel are required: one personnel tending to the patient, two personnel on each side of the patient inserting cribbing, one personnel managing the rescue-lift air bag controls, and one personnel as the officer in control/safety to oversee and direct the operation. Medical rescue personnel with ALS-level training will also be needed to immediately treat the patient when he or she is released. Some medical protocols recommend treating crushing injuries immediately, before the patient is released. This will be a decision made by the authority having jurisdiction based on departmental medical protocols and medical direction. **(Step 2)**
13. Position two rescue-lift air bags, one on top of the other with the largest bag on the bottom, under the school bus close to the patient. Remember, bag placement should be under a structural support such as a roof bow truss. Otherwise, the roof section will cave in when the bags are inflated. **(Step 3)**
14. Each technical rescuer on cribbing assignment who is positioned on both sides of the patient must be prepared with a full complement of various sizes and types of cribbing by his or her side. Prebuilding a cross-tie box crib configuration can assist in preventing delays. Always remember to use a longer section of cribbing to push the box crib in position under the school bus; never put your body or extremities in a position where it can become trapped. **(Step 4)**
15. As the order is given to inflate the first (lowest) bag, simultaneously insert the appropriate-sized cribbing on both sides of the patient to support the lift and to maintain an even lift. As the lift is increased by the inflation of the second bag, the cribbing will also need to increase in height. **(Step 5)**
16. Once the proper height is achieved with full support on both sides of the patient, insert a backboard under the patient. Place two or more personnel at each side of the patient and one personnel at the head to maintain spinal immobilization and simultaneously slide the patient in position onto the backboard. Once the patient is clear of the entrapment, secure the patient to the backboard and transfer to a medical team. **(Step 6)**

Rescue Tips

To properly and safely conduct a rescue-lift air bag operation, a minimum of five personnel are needed: one personnel tending to the patient, two personnel on each side of the patient inserting cribbing, one personnel managing the rescue-lift air bag controls, and one personnel as the officer in control/safety to oversee and direct the operation.

Steering Wheel/Column Relocation
Relocating a steering wheel/column off of a driver who has become trapped from a collision can be quickly and safely accomplished using a small tow-truck unit that is equipped

Skill Drill 11-12

Removing a Victim From Under a School Bus Resting on Its Side

1 Don appropriate PPE, including mask and eye protection. Assess the scene for hazards and complete the inner and outer surveys. Ensure that there is no fuel leaking. Consider utilizing the appropriate foam for the type of fuel encountered. Activate MCI protocols depending on the number of patients. Call for additional resources. Set up the hazard control zones. Stage two charged hose lines. Locate and disable the battery system if accessible. Ensure the school bus is not running. Note: If normal shutdown procedures including disconnecting the battery system cannot be accomplished, then locate the air intake manifold in the engine compartment and discharge a 10-lb (4.5-kg) minimum dry chemical extinguisher into the device, which should suffocate and shut down the engine. Secure the area around the muffler/regeneration device. Stabilize the school bus from movement utilizing the techniques outlined in this chapter.

2 Determine the placement and number of rescue-lift air bags needed to produce an even lift. Position five rescuers around the patient; one personnel to tend to the patient, two personnel on each side of the patient inserting cribbing, one personnel managing the rescue-lift air bag controls, and one personnel as the officer in control/safety to oversee and direct the operation. Medical rescue personnel with ALS-level training will also be needed to immediately treat the patient when he or she is released.

3 Position two rescue-lift air bags, one on top of the other with the largest bag on the bottom, under the structural support of the school bus close to the patient.

4 Use a longer section of cribbing to push the box crib in position under the school bus; never put your body or extremities in a position where it can become trapped.

Skill Drill 11-12

NFPA 1006, (10.2.5)

Removing a Victim From Under a School Bus Resting on Its Side *(Continued)*

5 As the order is given to inflate the first (lowest) bag, simultaneously insert the appropriate-sized cribbing on both sides of the patient to support the lift and to maintain an even lift. As the lift is increased by the inflation of the second bag, the cribbing will also need to increase in height.

6 Once the proper height is achieved with full support on both sides of the patient, insert a backboard under the patient. Place two or more personnel at each side of the patient and one personnel at the head. Once the patient is clear of the entrapment, secure the patient to the backboard and transfer to a medical team.

with a stationary boom. Personnel can back the tow unit to the front of the school bus, extend its lower brace bar to brace up against the front of the bus, pull the wire rope from on top of the boom, and wrap the rope securely around the steering wheel ring and column. The bracing bar is normally utilized to extend/slide underneath the undercarriage and tires of a conventional vehicle in order to lift and haul a vehicle away. The stationary boom provides a height leverage advantage that, when operated, can easily relocate a steering column off of an entrapped occupant in minutes. Other techniques can be utilized, such as using a come along and chain package (the same technique that is used for relocating the steering wheel and steering column of a conventional vehicle, demonstrated in Chapter 9, Skill Drill 9-14). This technique can only be applied to a type C bus or a bus with an extended front hood/engine compartment so the equipment can be properly set in place. A transit-style type D school bus will require the use of a tow unit as just described. There are other techniques that may be used, but the tow unit is the safest and fastest way to accomplish the objective.

■ Alternative Fueled Buses

The majority of school buses on the road today utilize conventional diesel as their fuel source, with a smaller proportion of

buses utilizing regular unleaded gasoline. However, just like the auto industry, school bus manufacturers are designing buses to operate on alternative fuels as well as advanced propulsion systems such as hybrid and fuel cell technology. Alternative fuels such as ethanol, methanol, propane, bio-diesel, liquefied natural gas (LNG), and compressed natural gas (CNG) are starting to become more prevalent. Other systems such as bi-fuel or dual-fuel engines are designed to run on two different types of fuels by switching from one tank to another. There are also newer cleaner types of petroleum-based fuels such as clean-diesel, or reformulated gasoline, also known as "oxygenated gasoline" or the E85 flex fuel just to name a few.

Hybrid technology for school buses utilizes the same concept design as conventional vehicles, where parallel (more common) or series propulsion systems are combined with a diesel- or gasoline-powered internal combustion engine (ICE). The industry standard orange-colored high-voltage wire is also used. A parallel system is more commonly used in types C and D school bus designs and is designed to work in tandem with a diesel-operated ICE. The propulsion of the bus can either be operated from the onboard electric generator or through the diesel-operated ICE. In this particular parallel design, the hybrid system is placed behind the transmission, which allows the bus to operate solely on the battery system or the ICE. To engage the hybrid system, the driver must flip an operational switch located on a separate hybrid driver interface panel mounted on the dashboard. A hybrid school bus, regardless of the propulsion system, uses two or more 12-volt DC batteries located on a slide-out tray, normally positioned outside at the front of the bus at the driver's side area. These batteries function in the same fashion as a conventional school bus—that is, start the bus, power the air conditioner, power basic electrical components, and so on.

Rescue Tips

A hybrid school bus, regardless of the propulsion system, uses two or more 12-volt DC batteries located on a slide-out tray, normally positioned outside at the front of the bus at the driver's side area.

Two types of parallel hybrid systems are in use for types C and D school buses; they consist of the **charge-depleting hybrid system**, which is a plug-in hybrid electric vehicle (PHEV), and the **charge-sustaining hybrid system**, which is a standard hybrid that operates on regenerative properties to charge its batteries. The battery pack for both systems is placed in a sealed unit that comes in two designs. One is a single battery pack that weighs up to 1500 lb (680 kg) and is set in the undercarriage on the driver's side. Because of the weight load of this single battery pack, a counterbalance on the opposite side of the vehicle is used that weighs up to 1000 lb (454 kg). The other battery system design splits the battery pack into two battery packs that are set opposite each other on both sides of the undercarriage of the bus, thus eliminating the need for any counterbalance weight. The battery packs also have a service disconnect switch that is designed to cut off voltage from the packs. This switch is easily recognizable and attainable, being mounted on the outside of the pack and in clear view.

Rescue Tips

Two types of parallel hybrid systems are in use for types C and D school buses:
- Charge-depleting hybrid system: Utilizes a sealed lithium-ion (Li-Ion) battery pack comprised of 28 12-volt DC batteries set in series, yielding 336 DC volts and requiring 4 to 8 hours of charging time depending on whether a 120- or 220-volt charge base system is used. The system is designed to utilize the battery system for up to 44 miles (71 kilometers [km]), depending on the terrain, and then must be recharged.
- Charge-sustaining hybrid system: Utilizes nickel metal hydride (NiMh) batteries in 7.2-volt modules operating at a nominal 288 DC volts. Battery recharge is controlled by a continuous supply of power through regenerative braking.

A **series-operated propulsion system**, which is not as commonly found, is a system design that utilizes the electric motor by itself to propel the school bus, and the combustion engine is only used to regenerate the battery pack. These systems may be found on a type A school bus.

Emergency Procedures for the Hybrid Bus

The emergency procedures for disabling the hybrid system on a type C or D bus are directed at the current parallel system described in the preceding section; newer hybrid propulsion drive designs are in the developmental stages and will more than likely have different emergency procedures for disabling these types of systems. It is the responsibility of the technical rescuer to stay current with the latest technology and research development in new types of hybrid propulsion system designs.

The following skill drill assumes that the vehicle has been in a collision and is upright; it will include additional appropriate steps for addressing this type of incident. To properly disable the hybrid system on a type C or D school bus, follow **Skill Drill 11-13 ▼**. Some steps may occur simultaneously. Please note there are no photos for this Skill Drill.

1. Don appropriate PPE, including mask and eye protection.
2. Assess the scene for hazards and complete the inner and outer surveys. Avoid standing at the front or rear of the vehicle to protect against any sudden forward or backward movement of the vehicle.
3. Activate MCI protocols if preestablished and/or if needed based on the number of patients.
4. Call for additional resources, such as an appropriately sized tow truck unit (preferably a class C rotator with an articulating boom), TRT unit, or hazardous materials unit.
5. Set up the control zones (hot, warm, cold).
6. Set up two 1¾-inch (44-mm) charged hose lines in defensive positions.
7. Ensure that the vehicle is not running.
8. Gain entry into the school bus and disable the hybrid system:
   - Turn OFF the operational switch located on the hybrid driver interface panel mounted on the dashboard.
   - Turn OFF and remove the vehicle ignition key.
   - Set the vehicle's parking brake.
   - Exit the bus and turn OFF the battery pack service disconnect switch located on the outside of the battery pack.
9. Locate and disable the conventional 12-volt DC battery system (not the hybrid battery pack) according to the procedures outlined in this chapter.
10. Place two Hi-Lift Jacks at the rear of the school bus, lifting from the bumper.
11. Position four tension buttress struts on the school bus with two at the front sides and two at the rear sides in an A-frame setup. Create purchase points for the tips of the struts to be set in place using a Halligan bar and flat-head axe or a pneumatic air chisel. Pass the cargo straps under the school bus using a long pike pole to avoid going under the vehicle.
12. Additional cross-tie box cribbing can be set in position under the frame at the rear of the school bus and at the front under the bumper area. This will require a lot of cribbing; ensure that the units on hand can support this task. The

Figure 11-18 The technical rescuer will in time encounter a collision involving one or multiple large commercial vehicles.

cribbing height for a cross-tie box crib should not exceed two times its width.

13. Prevent any forward or backward movement by positioning cribbing or step chocks upside down in a wedge-type setup in front of and behind each tire.

Commercial Trucks

Within the United States, the rules and regulations that are placed on CMVs are implemented through the DOT, which limits the amount of weight a CMV can transport, including the type of cargo being hauled with these vehicles. With literally thousands of CMVs traveling the streets and highways, the technical rescuer will in time encounter a collision involving one or multiple large commercial vehicles **Figure 11-18**. Because of the overall design, which encompasses the heavier frame structure, size, weight, and cargo, CMVs that require extrication may involve special techniques, heavy equipment, and tools to complete the job. CMV extrication is a specialized field; the NFPA outlines and requires the Level II Technician qualification, as well as the job performance requirements (JPRs) to conduct operations involving this type of incident. Proper qualified training for CMV extrication is an absolute

requirement. Having the right qualified instructor, utilizing a standardized and validated training outline with qualifying (hands-on as well as didactical) testing requirements, is the best practice model for an agency.

■ Commercial Truck Classifications

CMVs are classified in eight different weight categories, which are measured utilizing the following **Table 11-1**:

- A GVWR system
- The work duty of the engine as it relates to emissions (light duty, medium duty, and heavy duty)

The lightest commercial vehicle carries a **class 1 commercial vehicle** rating, with a GVWR ranging from 0 to 6000 lb (0 to 2722 kg); a Ford Ranger is an example of a class 1 CMV. A class 2 commercial vehicle is subdivided into two categories (class 2a and 2b). A **class 2a commercial vehicle** is considered a light-duty vehicle, with a GVWR ranging from 6001 to 8500 lb (2722 to 3856 kg); a Ford F-150 is an example of a class 2a commercial vehicle. A **class 2b commercial vehicle** is still considered by many organizations as a light-duty truck, with a GVWR of 8501 to 10,000 lb (3856 to 4536 kg); a Ford F-250 is an example of a class 2b commercial vehicle. A **class 3 commercial vehicle** has a GVWR of 10,001 to 14,000 lb (4536 to 6350 kg); an example of a class 3 commercial vehicle is the Ford F-350. A **class 4 commercial vehicle** has a GVWR of 14,001 to 16,000 lb (6351 to 7257 kg); the Ford F-450 is an example of a class 4 commercial vehicle. A **class 5 commercial vehicle** has a GVWR of 16,001 to 19,500 lb (7258 to 8845 kg); the Ford F-550 is an example of a class 5 commercial vehicle. A **class 6 commercial vehicle** has a GVWR of 19,501 to 26,000 lb (8846 to 11,793 kg); the Ford F-650 is an example of a class 6 commercial vehicle. A **class 7 commercial vehicle** has a GVWR of 26,001 to 33,000 lb (11,794 to 14,969 kg); an example of a class 7 commercial vehicle is the Ford F-750. The heaviest commercial vehicle is a **class 8 commercial vehicle** with a GVWR ranging more than 33,000 lb (more than 14,969 kg); a semi-tractor trailer is an example of a class 8 CMV. Class 7 and 8 CMVs have a GVWR that is greater than 26,000 lb (11,793 kg), requiring a minimum of a class B driver's license to operate in the United States. Other classifications that are utilized to determine weights are GAWR (gross *axle* weight rating),

Table 11-1	Weight Classifications for CMVs		
CLASS	**GVWR**	**EXAMPLE**	**WORK DUTY**
Class 1	0–6000 lb (0–2722 kg)	Ford Ranger pick-up; Toyota Tacoma	Light duty
Class 2a	6001–8500 lb (2722–3856 kg)	Ford F-150; Dodge Dakota	Light duty
Class 2b	8501–10,000 lb (3856–4536 kg)	Ford F-250; Chevrolet Silverado 2500HD or GMC Sierra 2500HD	Light duty
Class 3	10,001–14,000 lb (4536–6350 kg)	Ford F-350; Chevrolet Silverado 3500HD or GMC Sierra 3500HD	Light duty
Class 4	14,001–16,000 lb (6351–7257 kg)	Ford F-450	Medium duty
Class 5	16,001–19,500 lb (7258–8845 kg)	Ford F-550; GMC C4500 and C5500	Medium duty
Class 6	19,501–26,000 lb (8846–11,793 kg)	Ford F-650; GMC C6500	Medium duty
Class 7	26,001–33,000 lb (11,794–14,969 kg)	Ford F-750; GMC C7500	Medium/heavy duty
Class 8	above 33,000 lb (14,969 kg)	Tractor-trailer; GMC C8500 or C8500-Tandem	Heavy duty

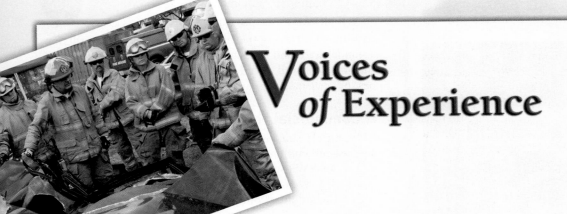

Voices of Experience

I currently serve as a Special Operations Battalion Chief for Broward Sheriff's Office Department of Fire Rescue in Broward County, Florida. Our Regional Technical Rescue and Hazardous Materials teams respond to a multitude of incidents on a regular basis.

On a calm evening in rural Broward County, a call was initiated for a semi-tractor trailer on fire in the far reaches of Alligator Alley, a remote highway that connects South Florida from the West Coast to the East, surrounded only by the Florida Everglades. Earlier in the day on this same roadway, an overturned semi box truck tied up traffic for hours, causing major delays and a road closure.

> **"Peeking from underneath the cab of the electronics semi was an ominous sight, a worst case scenario for the remote reaches of Broward County."**

The primary engine on the response was dispatched, and I was assigned as the company officer on a 2500 gallon (9,464 liters) pumper tanker that was added onto the assignment. As the first engine approached the scene, a much gloomier picture emerged as heavy fire was visible from a great distance away. Passing the overturned semi box truck from earlier in the day, the size-up was given of three semi-trucks and a passenger vehicle involved; the passenger vehicle was somewhere underneath the semi-trucks. A well composed first due officer initiated command and began to perform triage of the scene after requesting a full Technical Rescue assignment from the central part of the county.

After further investigation, we discovered that a flatbed semi, a Volkswagen beetle, and a semi-tractor trailer carrying electronics, in that order, had been stopped beyond the first incident for the road closure. Out of the darkness, a semi-tractor trailer carrying furniture struck the rear of the electronics semi at a high speed, causing catastrophic damage.

Peeking from underneath the cab of the electronics semi was an ominous sight, a worst case scenario for the remote reaches of Broward County. The Volkswagen beetle was overridden by the tractor, with only a portion of the vehicle visible. Is it possible anyone could survive this? Are there any voids? How do we get in there to assess?

Upon arrival, I met with the IC; attack lines were deployed in an attempt to keep the ensuing fire from consuming the car until it could be determined if there were any salvageable victims inside. After conferring with the first due engine in command, it was determined that there were three fatalities, two in the Volkswagen, and one in the rear semi that had been carrying furniture. The driver of the override electronics semi escaped through the windshield of his cab. After extinguishing the fire, the next task was to disassemble the entangled vehicles and extricate the unfortunate drivers.

A 120,000 lb (60 short tons) wrecker and a bobcat were utilized to lift the cab of the override semi, and remove the smaller vehicle from underneath with the assistance of our technical rescue team and on-scene companies. Fortunately, these resources were on scene nearby working on the previous accident scene otherwise the incident action plan would have been significantly delayed.

Incidents involving commercial vehicles illustrate the need for a multifaceted, coordinated plan. Knowledge of calculating weights; semi-trucks, dump-trucks, cement trucks or similar vehicles become necessary to determine how the vehicle can be moved. The standard engine/truck company cache of cribbing will not be sufficient to support the loads and cavities presented when working on commercial vehicles. In the event working underneath the load becomes necessary, determine how to safely support the load. Preplan accessing heights upwards of twelve feet to gain access to a patient trapped in a semi–tractor, and then have a game plan for getting yourself and the patient out.

Think big and small regarding hydraulics; use big tools for big metal and small tools for working in tight cavities of a day-cab. There is no one tool that does it all. Set up a cache, and bring redundancy. Think outside the box, but do it safely. Ground ladders become commonplace on these types of incidents, secure them in place when possible. Seemingly simple tasks such as disconnecting the batteries can become complex when they must be accessed via a ladder and can weigh up to 75 pounds (34 kg). Rotate personnel regularly, and be prepared for a long duration incident. On many incidents, be prepared to work for hours, not minutes, to gain access and affect patient removal. Identify the hazards and deal with them accordingly. Don't fall for the adage that "it's only diesel fuel and it's a combustible" on a hot summer roadway, where the pavement easily heats the fuel beyond its flashpoint. Be cognizant that long haul tractors can easily carry 400 plus gallons (1514 plus liters) of diesel fuel. Where is that fuel flowing when a saddle tank ruptures? Large trucks with heavy suspensions, large diesel engines and heavy structural components equal labor intensive scenes. Stay safe!

Gerard S. London
Broward Sheriff's Office Department of Fire Rescue
Broward County, Florida

GCWR (gross *combined* weight rating), and GTWR (gross *trailer* weight rating).

Commercial Truck Anatomy

Extrication involving a commercial truck requires that the technical rescuer fully understand the general anatomy for these large vehicles. Almost all commercial trucks have similar construction features that are broken down into different sections such as the cab, chassis, cargo area, and drive train. The **drive train (power train)** is a system that transfers rotational power from the engine to the wheels, which makes the vehicle move.

A **semi-truck**, or tractor, is a commercial truck capable of towing a separate trailer that has wheels only at one end (semi-trailer) **Figure 11-19 ▼**. When the semi-truck is combined with the trailer, it is known as a **semi-tractor trailer (semi-trailer)** **Figure 11-20 ▼**. A semi-truck is normally designed with three axles, one in the front for steering purposes and two tandem axles in the rear.

Figure 11-21 Rivet heads located on the outside panel of the cab or sleeper indicate the presence of heavy steel or aluminum framing directly underneath.

Figure 11-19 A semi-truck, or tractor.

Cab

The **cab** of a commercial truck is considered the enclosed space where the driver and passengers sit. The cab is usually made of a combination of steel, aluminum, and fiberglass **Figure 11-21 ▲**. There are three different types of cabs found on commercial trucks:

- **Cab-over-engine (COE)** **Figure 11-22 ▼**: Often referred to as tilt-cabs, the cab is lifted up over the engine to gain access to the engine itself. In a COE, the driver's seat is positioned over the engine and the front axle. This style

Figure 11-20 A tractor combined with a trailer is known as a semi-tractor trailer (semi-trailer).

Figure 11-22 Cab-over-engine.

Figure 11-23 Conventional cab.

Figure 11-24 Cab-beside-engine.

side or at the center directly behind the cab. Like the cab, sleepers can also be made of aluminum, steel, or a combination of both. Generally, they are designed with steel or aluminum ribs or framing with an outer shell of fiberglass or a plastic molding consisting of a material known as Melton. Between these layers are insulation and wires. Cutting into the cab and/or sleeper can require multiple tools, including but not limited to hydraulic tools, reciprocating saws, and pneumatic air chisels.

Rescue Tips

Always disconnect/disable the battery system before conducting any entry operations.

The roof sections of the cab and sleeper can contain an air-conditioning unit as well as a wind deflector, which add a considerable amount of weight to the vehicle's roof; this weight increase must be compensated for when considering a roof removal procedure for a CMV that is upright. Also, the overall height of these cabs can be greater than 15 feet (4.6 m) from ground level. The height requires that the technical rescuer build several platforms that surround the cab consisting of cross-tie box cribbing, backboards, and/or ladders, or use a flatbed towing unit positioned against the cab to gain height and stability. Working from a ground ladder placed against the cab is not the safest practice and requires multiple personnel to assist with stability.

Rescue Tips

Reciprocating saws are the most effective because of the large span of cutting that is required.

Glass

Windshield glass found in a CMV is comprised of standard laminated safety glass and is set in a rubber gasket similar to that which is found in a school bus. This type of rubber gasket

of cab was developed to maximize truck cargo space where the overall truck lengths were federally regulated and limited. This body style was also designed to aid in the mobility of the vehicle itself.

- **Conventional cab** Figure 11-23 ▲ : In this design, the driver's seat is positioned behind the engine and front axle. The front end of the vehicle extends approximately 6 to 8 feet (1.8 to 2.4 m) from the front windshield. Conventional cabs are more commonly found in the United States and Canada. To compensate for added wind resistance, a large wind/air deflector may be added to the top of the cab, which can add a lot more weight to the roof structure. The fuel tanks on these trucks also tend to be larger than on COEs.
- **Cab-beside-engine (CBE)** Figure 11-24 ▶ : In this design, the driver sits next to the engine. These trucks are mostly found in shipyards and baggage carriers within airports and seaports.

Rescue Tips

Because of the tilt cab design on a COE, for safety purposes, the cab may have to be secured to the frame of the vehicle prior to attempting any extrication procedures.

A cab may also have a **sleeper**, which is a compartment attached to the cab that allows the driver to rest while making stops during a long transport. Sleepers are most commonly found on a semi-truck. Sleepers can be added on or factory integrated to the rear of the cab with a separate access door on the

Figure 11-26 This suspension system has large steel leaf springs that may have an independent adjustment leaf spring system on each axle; this is known as an independent suspension system.

Figure 11-25 A truck chassis.

setting can be removed with a prying tool such as a large flat-head screwdriver. The glass can be removed without breaking or cutting it, although cutting the glass out with a glass saw or reciprocating saw is also a fast and viable option. Some new models of CMVs use curved windshields that are set in place and sealed using mastic-type adhesive; this setting requires that the glass be cut out for removal purposes. The side windows are comprised of tempered safety glass, which can be much thicker than the glass found in conventional vehicles.

Chassis

The chassis makes up the main structural framework of the CMV and includes the braking, steering, and suspension system Figure 11-25 ▲. The frame structure, which is the strongest part of the CMV, is constructed of heavy-gauge steel to support the weight of the vehicle's cab, suspension system, engine, and cargo. It consists of two parallel boxed, tubular, or C-shaped rails or beams that are held together with crossbeams. These crossbeams, or crossmembers, are attached to the rails by welds or bolts.

Suspension

The suspension system for a CMV is designed to protect the cargo and frame system of the vehicle from damage and wear by absorbing the impacts of the tires as they drive over uneven terrain. Suspension systems can come in the form of large steel leaf springs, which can have an independent adjustment leaf spring system per axle known as an independent suspension system Figure 11-26 ▶, and/or spiral springs in addition to air bellows

(bags), better known as an air ride system Figure 11-27 ▼. The **air ride system**, which is the same as that described in the school bus section, is designed to inflate and deflate the suspension system through onboard air pressure tanks. The suspension system connects the axles, which include the wheels, to the vehicle. When the air brake (parking brake) is engaged, an air ride suspension system will release some air but remain stable. However, in some models, it can release air and drop several inches to accommodate the resting or parked position; this can exacerbate the crushing effect on a vehicle that it has collided with (underride) and is now trapped under the trailer (as further discussed later in this chapter). Also, if a bellows bag ruptures or leaks during an incident, it can cause the vehicle to settle or shift several inches. Most air ride bellows are interconnected per axle and will release air in all the bags sharing the same axle even if just one is damaged.

There are several tools and techniques that can help the technical rescuer manage this situation. The primary tool is a

Figure 11-27 This suspension system has spiral springs in addition to air bellows (bags), better known as an air ride system.

Figure 11-28 If a bellows bag ruptures or leaks during an incident, it can cause the vehicle to settle or shift several inches. A C-type tow unit with an articulating boom applies slings to the trailer, thus lifting and stabilizing the trailer enough to conduct safe operations.

Figure 11-29 A fifth wheel.

large C-type tow unit with an articulating boom, which can apply two or more slings to the trailer, thus lifting and stabilizing the trailer enough to conduct safe operations **Figure 11-28 ▲**. Other options include placing tension stabilization struts with heavy lifting or weight management capabilities (check with the manufacturer) around the trailer, locking the struts together with a heavy cargo strap in the same fashion as stabilizing an upright school bus. This, in addition to placing cribbing in front and back of each wheel to chock and prevent any rolling, can add support to maintain and/or raise (with the appropriate strut system) the trailer's suspension system.

The other option that may apply to certain types of cargo trailers for stabilizing and maintaining the trailer's suspension system is the use of rescue-lift air bags, which can be positioned between the frame and the top of the tires. This approach requires the same amount of cribbing formations as the technique just discussed using the tension stabilization strut system. Rescue-lift air bags utilized in this manner must adhere to the safety rules discussed in Chapter 6, *Tools and Equipment*. This technique can also be used in conjunction with a tension buttress stabilization strut system.

Cargo Area

The cargo area of a commercial vehicle can be attached to the same frame as the cab, as with a box truck, or there can be a separate frame from the tractor, as with the semi-tractor trailer. The vehicle manufacturer calculates the **gross combined weight rating (GCWR)** to determine the cargo capacity limitations.

The cargo trailer of a semi-tractor trailer or semi-trailer is connected to the tractor by a large locking pin called the **kingpin**. The kingpin sits in a flat horseshoe-shaped quick-release coupling device called a **fifth wheel** or a turntable hitch mounted at the rear of the towing truck **Figure 11-29 ▶**. The kingpin and fifth wheel allow for easy hookup and release. Fifth wheels are attached to the frame rails of the tractor, and the manual release of this device is located on the driver's side of the fifth wheel itself **Figure 11-30 ▶**.

Cargo areas can contain all types of commodities and products and/or general goods (groceries, office supplies, and furniture), chemicals (liquid and solid), machinery, and other

vehicles. Semi-trailers come equipped with a stabilizing device known as **landing gear**. Landing gear is located at the front of the unit and can be lowered to stabilize the trailer when not attached to the tractor. This device is commonly operated manually by a hand crank **Figure 11-31 ▶**. In addition to the traditional cargo area on a trailer, some commercial vehicles can be equipped with a dromedary. A **dromedary (drom)** is a separate box, deck, or plate, mounted behind the cab and in front of the fifth wheel on the frame of the tractor. A dromedary

Figure 11-30 The manual release of the fifth wheel, which is on the driver's side of the fifth wheel itself.

Figure 11-31 Landing gear and landing gear crank.

Figure 11-33 Axles are designed for wheel rotation.

can be used to store products that should be kept away from the load in the cargo area or it can be used to mount a power unit such as a generator Figure 11-32 ▾ .

Rescue Tips

The wide array of cargo transported in semi-tractor trailers or box trucks can add additional concerns for the technical rescuer when conducting extrication. Not only will the technical rescuer have to be concerned with trapped victims, but cargo has the potential to wreak havoc and jeopardize safety on a scene. Cargo may include hazardous materials, livestock, or large amounts of debris.

Axles

An **axle** is a structural component or shaft that is designed for wheel rotation Figure 11-33 ▸ . Axles are categorized as either live or dead. **Live axles** are ones that transmit propulsion or cause the wheels to turn. A **dead axle** is used for load support and is commonly set in the front section of a tractor where its function is for load support and steering; hence, it is also known as a

steer axle. Another type of axle is the lift axle. The **lift axle** can be raised or lowered by an air suspension system to increase the weight-carrying capacity of the vehicle or to distribute the cargo weight more evenly across all of the axles Figure 11-34 ▾ . As mentioned previously, tractors generally have three axles. One axle is set in the front of the vehicle and has one wheel on each side; it controls the steering. The other two axles are under the rear of the tractor and have two wheels on each side; these axles can be set in tandem, with one or both used for propulsion.

Doors

Doors on CMVs are very similar to doors found on conventional vehicles. They are, however, of a heavier construction due to the increased amount of steel used. The door handle is usually located in the lower corner of the door, which makes access easier for vehicle occupants.

There may be a combination of hinges found on a CMV door such as piano-type hinges Figure 11-35 ▸ , two-piece full/solid hinges that can be hidden on the inside of the door Figure 11-36 ▸ or fully exposed on the outside, and strap hinges, which attach and extend to the outside of the door. The latching mechanism will usually consist of a Nader bolt and a latching

Figure 11-32 A dromedary deck or plate.

Figure 11-34 The lift axle.

device that grabs and locks the Nader bolt at two points rather than the standard latch that is found on a conventional vehicle. The latching mechanism and pin are normally located above the door handle about midway on the frame.

Door removal techniques apply the same philosophy as the door removal procedure in conventional vehicles, "expose and cut." Expose the latching mechanism and cut it. If using hydraulic tools, create a purchase point with a vertical spread technique in the window frame and start to spread the door down and away from the latch, creating a large enough opening to insert the hydraulic cutter and cut the latch. Gaining access through the hinge side of the door will be more complicated depending on the type of hinge you encounter. Also, remember that you will be going against the natural swing of the door and will have to gain additional access to the latch system to release it. Exterior full/solid hinges can be cut with a high-pressure air chisel, or the area surrounding them can be spread enough with a hydraulic spreader to allow cutting with a hydraulic cutter. Piano-type hinges require a lot of work because they are attached in multiple areas along a vertical line. Interior hinges must be exposed and cut with either a hydraulic cutter or a high-pressure air chisel.

Braking Systems

The braking system on a CMV utilizes compressed air to actuate the system. There are three separate types of air brakes that make up the braking system on a CMV—the service brake, emergency brake, and parking brake. The **service brake** is the normal driving brake applied by the driver to slow and/or stop the vehicle during normal driving operations. The **air brake (parking brake)** is used when the vehicle is in a fully stopped position. An air-actuated release button or switch, which is engaged by the driver, bleeds out compressed air to hold back large springs located inside sealed chambers that release and lock the brakes in place Figure 11-37 ▼ and Figure 11-38 ▶ . The

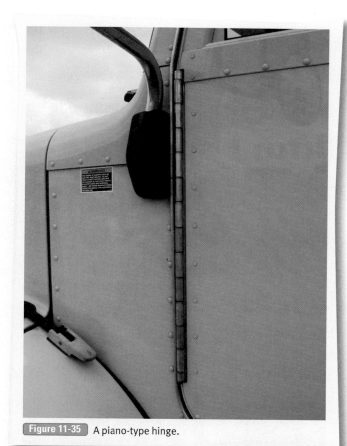

Figure 11-35 A piano-type hinge.

Figure 11-36 Two-piece full/solid hinges may be hidden on the inside of the door.

Figure 11-37 An air brake (parking brake).

Figure 11-38 The parking brake bleeds out compressed air, holding back large springs located inside the sealed chambers (shown above) that release and lock the brakes in place.

<u>emergency brakes</u> utilize a combination of the service and the parking brakes to engage when a brake failure or air line break occurs. An added safety feature to the air brake system on all trucks, tractors, and trailers is an antilock braking system (ABS), which is federally mandated by 49 CFR 393.55, *Antilock Brake Systems*, in addition to FMVSS No. 121, *Air Brake Systems*. The power cables that carry ABS power to trailers is normally color coded with a lime-green color. Other power cables may be yellow, orange, and/or black.

Air is compressed by an onboard compressor that is run by the vehicle's engine. Air is stored in tank reservoirs that are located in various areas on the tractor as well as the cargo trailer. The compressor normally stores the compressed air at approximately 120 to 125 psi (827 to 862 kPa). When the pressure drops below 100 psi (689 kPa), the compressor will restart and fill the tanks to the appropriate pressures.

The air brake system on a tractor connects the trailer by a minimum of two air lines **Figure 11-39 ▸** —the <u>service air line</u>, which is normally blue, and the <u>emergency air line (supply line)</u>, which is normally red. The emergency line controls the air brake for the trailer and is also utilized to fill the air reservoir tanks. The couplers that supply air to the braking system of the trailer are known as <u>glad hands</u> **Figure 11-40 ▸** . Some CMVs have dummy lines that are used to connect the service and emergency lines to the dummy lines utilizing glad hands to prevent dirt and/or debris from entering them when the tractor and trailer are not connected. In trailers that are designed to tow other trailers, there are shut-off valves at the rear of the towing trailer that control the service and emergency air lines and allow the closing of the air lines when not in use.

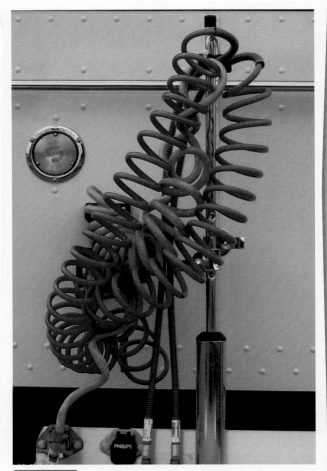

Figure 11-39 The air brake system consists of two air brake lines, usually red and blue in color, with an electrical line colored black, orange, or more commonly, lime green (as shown in the picture above).

Figure 11-40 Glad hands are couplers that supply air to the braking system of the trailer.

Rescue Tips

Battery systems for a CMV can consist of four or more 12-volt DC batteries, which are hooked together in parallel, with all positive battery terminals wired together and all negative battery terminals wired together, thus increasing the energy storage capacity. Battery trays or storage areas can be located in several areas throughout the vehicle. In semi-tractor trailers, the battery trays are generally stored underneath the step-up to make entry into the vehicle easier, but this is not a standard feature [Figure 11-41 ▾]. Battery trays can also be located behind the cab or in various other areas of the vehicle.

Figure 11-41 Battery trays are usually located underneath the step-up.

Figure 11-42 A saddle tank.

Fuels Tanks and Fuel Types

Fuel tanks for CMVs are usually constructed of aluminum but are also available in steel, ranging in sizes from 50 gallons (189 liters) to 150 gallons (568 liters). There may be a single tank or two tanks positioned on either side of the tractor frame. These tanks may be referred to as "saddle tanks" because they saddle the sides of the vehicle [Figure 11-42 ▸]. Fuel lines may lead from the top or the bottom of the tank depending on the manufacturer. These tanks are independently siphoned from the fuel pump; there is no crossover line or switch that connects each tank. The tanks also may be equipped with a shut-off valve on the tank itself or on the fuel line; the location of a shut-off valve varies among manufacturers. The fuel that is predominately utilized in the trucking industry is diesel. Bio-diesel is starting to gain some usage because it is less expensive than diesel.

Hybrid CMVs are also beginning to be manufactured, and some are very similar in design to a conventional vehicle. The **hybrid-electric commercial motor vehicle (HECMV)** utilizes either the internal combustion engine or an electric motor for propulsion. Electric power is generated through an onboard battery pack with an ultracapacitor and/or through other means such as regenerative braking.

■ Hazardous Materials

One of the greatest concerns for the technical rescuer responding to a motor vehicle accident involving a CMV is the potential for that CMV to be transporting hazardous materials, and for the cargo's or vessel's integrity to be compromised. According to the DOT Pipeline and Hazardous Materials Safety Administration, there are more than 400,000 CMVs dedicated to the transportation of hazardous materials, with 18,000,000 shipments of gasoline and 125,000 shipments of explosives conducted each year by roadway travel.

The Federal Hazardous Materials Transportation Law (49 U.S.C. § 5101 et seq.) is a statute enacted to regulate hazardous materials transportation in the United States.

Section 5103 of the Federal Hazardous Materials Transportation Law (49 U.S.C. § 5103) defines a <u>hazardous material</u> as a substance or material including an explosive; radioactive material; infectious substance; flammable or combustible liquid, solid, or gas; toxic, oxidizing, or corrosive material; and compressed gas that the Secretary of Transportation has determined is capable of posing an unreasonable risk to health, safety, and property when transported in commerce, and has designated as hazardous under Section 5103. The term hazardous material includes hazardous substances, hazardous wastes, marine pollutants, materials with an elevated temperature, materials designated as hazardous in the Hazardous Materials Table (see 49 CFR 172.101), and materials that meet the defining criteria for hazard classes and divisions in part 173 of subchapter C—Hazardous Materials Regulations (HMR).

Hazardous Materials Regulations, or HMR, refer to the regulations listed in the Code of Federal Regulations 49, parts 171 through 180, and 49 CFR, parts 100 through 185. HMR apply to the transportation of hazardous materials in interstate, intrastate, and foreign commerce by aircraft, railcar, vessel, and motor vehicle. HMR are issued by the DOT Pipeline and Hazardous Materials Safety Administration (PHMSA), which governs the transportation of hazardous materials by highway, rail, vessel, and air. The PHMSA issues rules, and develops and enforces regulations governing the safe transportation of hazardous materials.

Hazardous materials are divided into nine classifications with subdivisions, as follows:

Class 1—Explosives

- **Division 1.1**—Explosives With a Mass Explosion Hazard
- **Division 1.2**—Explosives With a Projection Hazard
- **Division 1.3**—Explosives With Predominantly a Fire Hazard
- **Division 1.4**—Explosives With No Significant Blast Hazard
- **Division 1.5**—Very Insensitive Explosives With a Mass Explosion Hazard
- **Division 1.6**—Extremely Insensitive Articles

Class 2—Gases

- **Division 2.1**—Flammable Gases
- **Division 2.2**—Nonflammable, Nontoxic Gases
- **Division 2.3**—Toxic Gases

Class 3—Flammable Liquids

Class 4—Flammable Solids, Spontaneously Combustible Materials, and Dangerous-When-Wet Materials or Water-Reactive Substances

- **Division 4.1**—Flammable Solids
- **Division 4.2**—Spontaneously Combustible Materials
- **Division 4.3**—Water-Reactive Substances or Materials Dangerous When Wet

Class 5—Oxidizing Substances and Organic Peroxides

- **Division 5.1**—Oxidizing Substances
- **Division 5.2**—Organic Peroxides

Class 6—Toxic Substances and Infectious Substances

- **Division 6.1**—Toxic Substances
- **Division 6.2**—Infectious Substances

Class 7—Radioactive Materials

Class 8—Corrosive Substances

Class 9—Miscellaneous Hazardous Materials/Products, Substances, or Organisms

At any vehicle accident involving a CMV, the technical rescuer must anticipate, plan, and be prepared for the potential of the CMV to be transporting some type of hazardous material. Roadway transportation is the most common method of hazardous material transport. There are many different vehicle/vessel types that are used to transport hazardous cargo. According to the Code of Federal Regulations, 49 CFR 171.8(2), or local jurisdictional regulations (for example,

Figure 11-43 The MC-306/DOT 406 flammable liquid tanker typically hauls flammable and combustible liquids.

Transport Canada), a **cargo tank** is considered bulk packaging that is permanently attached to or forms a part of a motor vehicle, or is not permanently attached to any motor vehicle and that, because of its size, construction, or attachment to a motor vehicle, is loaded or unloaded without being removed from the motor vehicle. One clarification on cargo tanks is that the DOT does not view tube trailers (which consist of several individual cylinders banded together and affixed to a trailer) as cargo tanks. Some of the more common CMV cargo vessels are discussed next.

One of the most utilized and reliable transportation vessels is the **MC-306/DOT 406 flammable liquid tanker** **Figure 11-43 ▲**. These cargo tanks frequently carry liquid food-grade products, gasoline, or other flammable and combustible liquids. The oval-shaped tank is pulled by a separate CMV tractor and can carry between 6000 and 10,000 gallons (between 22,712 and 37,854 liters) of product. The MC-306/DOT 406 is nonpressurized (with a working pressure between 2.65 psi and 4 psi [between 18.3 and 27.6 kPa]), is usually made of aluminum or stainless steel, and is off-loaded through valves at the bottom of the tank. These cargo tanks have several safety features, including full roll-over protection and remote emergency shut-off valves **Figure 11-44 ▼**.

Figure 11-44 The MC-306/DOT 406 cargo tanker has a remote emergency shut-off valve as a safety feature.

Figure 11-45 The MC-307/DOT 407 chemical hauler carries flammable liquids, mild corrosives, and poisons.

Figure 11-47 The MC-331 pressure cargo tanker carries materials such as ammonia, propane, Freon, and butane.

A vehicle that is similar to the MC-306/406 is the **MC-307/DOT 407 chemical hauler** **Figure 11-45**. This vessel is used to transport flammable liquids; it has a round or horseshoe-shaped tank and is capable of holding 6000 to 7000 gallons (22,712 to 37,854 liters) of liquid. The MC-307/DOT 407 also utilizes a separate CMV tractor to pull it. The MC-307/DOT 407 flammable liquid tanker typically hauls flammable and combustible liquids but can also carry mild corrosives and poisons. This type of cargo tank may be insulated (horseshoe) or noninsulated (round) and may have a higher internal working pressure than the MC-306/406, in some cases up to 35 psi (241 kPa). Cargo tanks that transport corrosives may have a rubber lining to prevent corrosion of the tank structure.

The **MC-312/DOT 412 corrosives tanker** is commonly used to carry corrosives such as concentrated sulfuric acid, phosphoric acid, and sodium hydroxide **Figure 11-46**. This cargo tank has a smaller diameter than the MC-306/DOT 406 or the MC-307/DOT 407 and is often identifiable by the presence of several heavy-duty reinforcing rings around the tank. The rings provide structural stability during transportation and in the event of a roll-over. These cargo tanks have substantial roll-over protection to reduce the potential for damage to the top-mounted valves. The MC-312/DOT 412 tanker operates at approximately 15 to 25 psi (103 to 172 kPa) and can hold up to approximately 6000 gallons (22,712 liters).

The **MC-331 pressure cargo tanker** is used to carry materials such as ammonia, propane, Freon, and butane **Figure 11-47**. The liquid volume inside the tank varies, rang-

ing from the 1000-gallon (3785-liter) delivery truck to the full-size 11,000-gallon (41,640-liter) cargo tank. The MC-331 cargo tank has rounded ends, typical of a pressurized vessel, and is commonly constructed of steel or stainless steel with a single tank compartment. The MC-331 operates at approximately 300 psi (2068 kPa), with typical internal working pressures in the vicinity of 250 psi (1724 kPa). These cargo tanks are equipped with spring-loaded relief valves that traditionally operate at 110 percent of the designated working pressure. A significant explosion hazard arises if a MC-331 cargo tank is impinged on by fire, however. Due to the nature of most materials carried in MC-331 tanks, a threat of explosion exists because of the inability of the relief valve to keep up with the rapidly building internal pressure. Responders must use great care when dealing with this type of transportation emergency.

The **MC-338 cryogenic tanker** is designed to maintain high thermal protective qualities to transport materials such as liquid nitrogen and liquid oxygen **Figure 11-48**. This low-pressure tanker relies on tank insulation to maintain the low temperatures required for the cryogens it carries. A boxlike structure containing the tank control valves is typically attached to the rear of the tanker. Special training is required to operate valves on this and any other tanker. An untrained individual who attempts to operate the valves may disrupt the normal operation of the tank, thereby compromising its ability to keep the liquefied gas cold and creating a potential explosion hazard. Cryogenic tankers have a relief valve near the valve control box. From time to time, small puffs of white vapor will be

Figure 11-46 The MC-312/DOT 412 corrosives tanker is commonly used to carry corrosives such as concentrated sulfuric acid, phosphoric acid, and sodium hydroxide.

Figure 11-48 The MC-338 cryogenic tanker maintains the low temperatures required for the cryogens it carries.

Figure 11-49 A tube trailer.

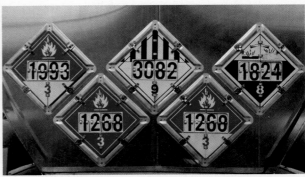

Figure 11-51 A placard is a large diamond-shaped indicator that is placed on all sides of transport vehicles that carry hazardous materials.

vented from this valve. Responders should understand that this may not be an emergency, but just a normal occurrence; the valve is designed to maintain the proper internal pressure of the vessel.

Tube trailers carry compressed gases such as hydrogen, oxygen, helium, and methane **Figure 11-49 ▲**. Essentially, they are high-volume transportation vehicles that are made up of several individual cylinders banded together and affixed to a trailer. These large-volume cylinders operate at working pressures of 3000 to 5000 psi (20,684 to 34,474 kPa). One trailer may carry several different gases in individual tubes. Typically, a valve control box is found toward the rear of the trailer with each cylinder having its own relief valve. These trailers can frequently be seen at construction sites or at facilities that use large quantities of compressed gases.

Dry bulk cargo tanks carry dry bulk goods such as powders, pellets, fertilizers, or grain **Figure 11-50 ▼**. These tanks are not pressurized but may use pressure to off load the product. Dry bulk cargo tanks are generally V-shaped with rounded sides that funnel the contents to the bottom-mounted valves.

Placards

Hazard warning placards are a type of warning system that displays the hazardous classification type on the sides of a vehicle using a diamond-shaped design. Federal law requires placards to be clearly displayed on each side and end of the vehicle.

Figure 11-50 A dry bulk cargo tank carries dry goods such as powders, pellets, fertilizers, or grain.

The **gross weight** measurement of a product is the weight of the single item package plus its contents; so, if there is one 55-gallon (208-liter) steel drum of chlorine that weighs 550 lb (249 kg), it is roughly 65 lb (29 kg) for the steel drum and 485 lb (220 kg) for the chlorine. The **aggregate weight** measurement combines all the packages to determine the total weight of the hazards. So, if you have the same 55-gallon (208-liter) drum of chlorine weighing 550 lb (249 kg) and you are also transporting 1000 lb (454 kg) of ammonia, the aggregate weight of hazards is 1550 lb (703 kg).

A vehicle transporting hazardous materials is required to post warning placards when the gross weight of a single hazard or the aggregate weight of the combined hazards totals 1001 lb (454 kg) or more **Figure 11-51 ▲**. Placards must be displayed for each product. Placards are also required for any explosives, poisonous gases, or radioactive materials, regardless of the weight. There is an exception to posting multiple placards when two or more classes of hazardous materials are loaded into the same vehicle and their aggregate weight is greater than 1001 lb (454 kg); a placard displaying the word "DANGEROUS" can be used in lieu of separate placards to identify each hazard. The only exception to this is when 2205 lb (1000 kg) or more of a hazardous material consisting of a single hazard class is loaded on the vehicle; this requires a placard representing that hazard to be displayed.

Hazardous materials classified as ORM-D materials (other regulated materials-domestic) can consist of a consumer commodity that is federally regulated but presents a limited hazard during transport and does not require hazard warning placards.

United Nations/North American Identification Numbers

United Nations/North American Hazardous Materials Code (UN/NA) identification numbers are another way to identify the hazardous material that is being transported. UN/NA identification numbers are given by the United Nations Committee of Experts on the Transport of Dangerous Goods.

When large quantities of a single hazardous material are transported by a vehicle or freight container in nonbulk packages, it must be marked on each side and each end with the UN/

NA identification number specified for that hazardous material. The UN/NA number must be displayed on an orange label or on the placard itself. The following provisions and limitations must also apply:

- Each package is marked with the same proper shipping name and identification number.
- The aggregate gross weight of the hazardous material is 8,820 lb (4,000 kg) or more.
- All of the hazardous material is loaded at one loading facility.
- The transport vehicle or freight container contains no other material, hazardous or otherwise.

Shipping Papers

Labels and placards may be helpful in identifying hazardous materials, but other sources of information are also available. Shipping papers are required by federal law to be carried on the transport vehicle when any hazardous materials are being transported. Shipping papers fully describe the content of the product(s) being transported. Several items must appear on all shipping papers, including:

- Identification of the shipper
- Identification of the receiver
- Hazard class or division
- UN/NA identification number
- Packing group number
- The total product quantity and weight with the exception of empty packages
- A 24-hour emergency contact number
- A certification statement certifying that the materials listed are the materials present
- The signature of the shipper and/or the shipper's agent verifying that all regulation requirements are met

Site Operations: Commercial Trucks

As stated earlier in this chapter, the greatest concern for the technical rescuer responding to a vehicle accident involving a CMV is hazardous materials. The cargo that the semi-tractor trailer or single frame box truck is transporting must be immediately identified before any physical approach or operation can commence. This one action, identifying a hazard, can mean the difference between a successful operation and a potential catastrophe. Complete the following steps during the scene size-up involving a CMV:

- Survey the area and identify hazardous transport cargo through placard/UN number recognition.
- Dispatch additional resources—hazardous materials teams, a TRT unit, a heavy tow unit (type C with an articulating boom), and other necessary resources.
- Locate the driver of the CMV if he or she is outside of the wreckage and is not injured, and confirm shipping cargo and location of shipping papers.
- Deploy at a minimum two 1¾-inch (44-mm) hose lines for protection and consider an appropriate foam application for any significant fuel leaks.
- When it is deemed safe to enter, conduct the inner and outer surveys to clear all hazards within the operational area.

- Create hazard zones: hot, warm, cold.
- Try to locate the shipping papers to confirm cargo before opening the cargo vessel/trailer.
- Stabilize all vehicles involved in the collision.
- Locate and disconnect the 12-volt DC battery system of the CMV and any other vehicle that may be involved.
- Disentangle and extricate the victim(s).

Victim Access: Commercial Trucks

Conventional vehicles often underride semi-trailers, but there are safe and efficient techniques for lifting a semi-trailer off of the conventional vehicle . A type C tow truck with an articulating boom can be used to lift the semi-trailer from the conventional vehicle utilizing two or more slings that are attached to the undercarriage of the trailer. This approach gives the tow operator full control of the trailer. The tow operator controls the boom's cable system with a remote controller box that he or she can operate at a safe distance in case anything should break and/or give way **Figure 11-53**. A second, smaller tow truck should be positioned to pull the vehicle out from

Figure 11-52 Conventional vehicles often underride semi-trailers.

Figure 11-53 The tow operator operates the boom's cable system with a remote controller box that he or she can operate at a safe distance.

under the trailer to commence extrication operations. The larger type C tow unit can also perform both functions by adding a snatch block. A snatch block consists of a single or multiple pulley system utilized to change cable direction (or limit line tension) toward the rear with a separate cable system, thus allowing removal of the vehicle simultaneously as the semi-trailer is raised. The larger type C tow unit must be positioned at a specific angle for this to work, which may not be possible depending on how the wreckage presents. The technical rescuer must realize the importance of dispatching a tow unit immediately without delay.

Most states have their own licensing standards and requirements that must be met to operate a towing unit. The Towing and Recovery Association of America (TRAA) is a nonprofit organization for the towing industry that offers three levels of driver certification (Level III being the highest), depending on the type or classification of vehicle being driven.

There are five types or classifications of tow units, which consist of the following:

- Class A tow truck: Equipment must have a minimum manufacturer's boom or combined boom rating of 8000 lb (4 short tons) and must be mounted on a truck chassis with a minimum manufacturer's rating of 10,000 lb (5 short tons) gross vehicle weight.
- Class B tow truck: Equipment must have a minimum manufacturer's boom or combined boom rating of 16,000 lb (8 short tons) and must be mounted on a truck chassis with a minimum manufacturer's rating of 18,000 lb (9 short tons) gross vehicle weight.

- Class C tow truck: Equipment must have a minimum manufacturer's boom or combined boom rating of 32,000 lb (16 short tons) and must be mounted on a chassis that has a minimum manufacturer's rating of 32,000 lb (16 short tons) gross vehicle weight.
- Class D tow truck: Equipment includes manufactured rollbacks and car carriers with a manufacturer's gross vehicle rating of 10,000 lb (5 short tons) and over. The rollbacks and car carriers must be mounted on a semi-tractor trailer chassis that, at a minimum, is equal to the minimum gross weight of the rollback or car carrier. Class D also includes any piece of towing equipment without a boom.
- Class E tow truck: Includes two or more tow trucks working together with a combined manufacturer's rating of a minimum of 80,000 lb (40 short tons) with access to supportive equipment, such as forklifts, banders, and air bags, for the recovery of roll-overs and wrecked, disabled, and abandoned vehicles whose cargo requires special handling. Class E refers to tow truck companies and not to tow truck equipment.

Rescue Tips

Tow units will be critical in some situations.

- Do you know how to contact this resource?
- What will be their response time?
- Have you trained with this resource?
- Do you have a backup plan?

Wrap-Up

Ready for Review

- The US DOT defines a CMV as a motor vehicle or combination of motor vehicles used in commerce to transport passengers or property if the motor vehicle has a gross vehicle weight rating (GVWR) of 26,001 lb or more (11,794 kg or more) inclusive of a towed unit(s) with a GVWR of more than 10,000 lb (4,536 kg); or a GVWR of 26,001 lb or more (11,794 kg or more); or is designed to transport 16 or more passengers, including the driver; or is of any size and is used in the transportation of hazardous materials as defined in this section.
- CMVs also include box trucks, semi-tractor trailers, concrete mixers, vehicle transporters, buses, and cranes.
- The Federal Motor Carrier Safety Administration (FMCSA) categorizes buses into carrier types or by function or purpose such as school bus, transit bus, intercity bus, and charter/tour bus.
- As an emergency responder, you may never come across a school bus accident, but it is vital to be prepared and know the makeup, structural components, and different types of school buses that are on the roadways today.
- The school bus industry designated four categories or classifications of school buses including types A, B, C, and D.
- The overall design features of the bus such as metal thickness and spacing of channel beams may vary among manufacturers, but all must meet the FMVSS for school buses.
- Many of the design features of a school bus integrate safety.
- The greatest concern for the officer in charge at a school bus extrication incident is gaining and maintaining control of the incident through proper scene management.
- Progressive agencies have preplanned and trained heavily for such an event and have preestablished MCI protocols and/or an emergency response plan in place.
- Upon your arrival at the scene of a school bus extrication, give a clear, accurate account of what is presented and conduct an inner and outer survey of the scene to formulate your action plan.
- As with most vehicles, there are basically four positions in which a school bus will present that the technical rescuer will have to stabilize: upright, on its side, on its roof, or on another vehicle.
- Within the United States, the rules and regulations that are placed on CMVs are implemented through the DOT, which limits the amount of weight a CMV can transport, including the type of cargo being hauled with these vehicles.
- CMVs are classified into eight weight categories.
- One of the greatest concerns for the technical rescuer responding to a motor vehicle accident involving a CMV is the potential for that CMV to be transporting hazardous materials, and for that cargo's or vessel's integrity to be compromised.
- Federal law requires placards to be clearly displayed on each side and end of the vehicle. United Nations/North American Hazardous Materials Code (UN/NA) identification numbers are another way to identify the hazardous material that is being transported.
- There are five types or classifications of tow units to assist in a CMV extrication.

Hot Terms

Active regeneration The second of three stages of regeneration in which fuel is injected into the system to burn and create higher temperatures of up to 1112°F (600°C).

Aggregate weight A measurement combining all packages in a CMV to determine the total weight of all the hazards.

Air brake (parking brake) A brake used on some models of CMVs that causes the suspension system to release air and drop several inches to accommodate a resting or parked position.

Air ride system A system designed to inflate and deflate the suspension through onboard air pressure tanks to protect the cargo and the frame system of the vehicle.

Axle A structural component or shaft that is designed for wheel rotation.

Bow trusses Structural steel members that run continuously from below the floor level on one side of a bus, vertically raising and bowing over to form the roof structure and then extending over and down the other side of the bus past the floor level.

Cab The enclosed space where the driver and passengers sit.

Cab-beside-engine Design in which the driver sits next to the engine. These trucks are mostly found in shipyards and baggage carriers within airports and seaports.

Cab-over-engine (COE) Often referred to as a tilt-cab, design in which the cab is lifted over the engine to gain access to the engine itself. The driver's seat is positioned over the engine and the front axle.

Cargo tank Bulk packaging that is permanently attached to or forms a part of a motor vehicle, or that is not permanently attached to any motor vehicle, and that, because of its size, construction, or attachment to a motor vehicle, is loaded or unloaded without being removed from the motor vehicle.

Charge-depleting hybrid system A plug-in hybrid electric vehicle (PHEV); one of two types of parallel hybrid systems used for types C and D school buses.

Charge-sustaining hybrid system A standard hybrid that operates on regenerative properties to charge its batteries; one of two types of parallel hybrid systems used for types C and D school buses.

Charter/tour bus A company providing transportation on a for-hire basis, usually round-trip service for a tour group or outing. The transportation can be for a specific event or can be part of a regular tour.

Class 1 commercial vehicle A vehicle with a gross vehicle weight rating ranging from 0 to 6,000 lb (0 to 2,722 kg).

Class 2a commercial vehicle A vehicle with a gross vehicle weight rating ranging from 6,001 to 8,500 lb (2,722 to 3,856 kg); a light-duty vehicle.

Class 2b commercial vehicle A vehicle with a gross vehicle weight rating ranging from 8,501 to 10,000 lb (3,856 to 4,536 kg); considered by many as a light-duty vehicle.

Class 3 commercial vehicle A vehicle with a gross vehicle weight rating ranging from 10,001 to 14,000 lb (4,536 to 6,350 kg).

Class 4 commercial vehicle A vehicle with a gross vehicle weight rating ranging from 14,001 to 16,000 lb (6,351 to 7,257 kg).

Class 5 commercial vehicle A vehicle with a gross vehicle weight rating ranging from 16,001 to 19,500 lb (7,258 to 8,845 kg).

Class 6 commercial vehicle A vehicle with a gross vehicle weight rating ranging from 19,501 to 26,000 lb (8,846 to 11,793 kg).

Class 7 commercial vehicle A vehicle with a gross vehicle weight rating ranging from 26,001 to 33,000 lb (11,794 to 14,969 kg).

Class 8 commercial vehicle A vehicle with a gross vehicle weight rating of more than 33,000 lb (more than 14,969 kg).

Commercial motor vehicle (CMV) Defined by the DOT as a motor vehicle or combination of motor vehicles used in commerce to transport passengers or property if the motor vehicle has a gross vehicle weight rating (GVWR) of 26,001 lb or more (11,794 kg or more) inclusive of a towed unit(s) with a GVWR of more than 10,000 lb (4,536 kg); or has a GVWR of 26,001 lb or more (11,794 kg or more); or is designed to transport 16 or more passengers, including the driver; or is of any size and is used in the transportation of hazardous materials.

Conventional cab Design in which the driver's seat is positioned behind the engine and front axle. The front end of the vehicle extends approximately 6 to 8 feet (1.8 to 2.4 m) from the front windshield.

Crash rail A rail designed to protect students from impact intrusions into the passenger compartment of a school bus; normally composed of 14-gauge steel extending just above the floor area between the floor and the seat rub rail, extending the entire length of the school bus.

Dead axle An axle used for load support more commonly set in the front section of a semi-truck; also functions for steering and is therefore also known as a steer axle.

Drive train (power train) A system that transfers rotational power from the engine to the wheels, which makes the vehicle move.

Dromedary (drom) A separate box, deck, or plate, mounted behind the cab and in front of the fifth wheel on the frame of a semi-truck.

Dry bulk cargo tank A tank designed to carry dry bulk goods such as powders, pellets, fertilizers, or grain. Such tanks are generally V-shaped with rounded sides that funnel toward the bottom of the tank.

Emergency air line (supply line) One of two air lines of an air brake system connecting a tractor to a trailer; the emergency air line is red.

Emergency brake A brake that utilizes a combination of both the service and parking brake to engage when a brake failure or air line break occurs.

Exhaust after-treatment device A device that replaced the standard muffler assembly, capturing and converting soot to carbon dioxide and water through the combination of a diesel particulate filter (DPF) and a diesel oxidation catalyst (DOC). This conversion process is called regeneration, and there are three stages or processes of regeneration.

Fifth wheel A turntable hitch mounted at the rear of the towing truck or semi-truck.

Glad hands Couplers that are used to suypply air to the braking system of the trailer.

Gross combined weight rating (GCWR) A rating set by the vehicle manufacturer that determines the vehicle's cargo capacity limitations.

Gross vehicle weight rating (GVWR) A rating set by the manufacturer; it shall not be less than the sum of the unloaded vehicle weight, rated cargo load, and 150 lb times the vehicle's designated seating capacity (49 CFR 567.4(g)(3)).

Gross weight The weight of the single item package plus its contents.

Hazardous material A substance or material including an explosive; radioactive material; infectious substance; flammable or combustible liquid, solid, or gas; toxic, oxidizing, or corrosive material; and compressed gas that the Secretary of Transportation has determined is capable of posing an unreasonable risk to health, safety, and property when transported in commerce, and has designated as hazardous under Section 5103.

Hybrid-electric commercial motor vehicle (HECMV) A CMV that utilizes either the internal combustion engine or an electric motor for propulsion.

Intercity bus A company providing for-hire, long-distance passenger transportation between cities over fixed routes with regular schedules.

Kingpin A large locking pin that connects the cargo trailer of a semi-tractor trailer or semi-trailer to the semi-truck or tractor.

Landing gear A stabilizing device that can be lowered to support a trailer when not attached to the tractor. This device is usually operated manually by a hand crank.

Lift axle An axle that can be raised or lowered by an air suspension system to increase the weight-carrying capacity of the vehicle, or to distribute the cargo weight more evenly across all of the axles.

Live axle An axle that transmits propulsion or causes the wheels to turn.

Manual regeneration The third of three stages of regeneration that occurs only when the parking brake is set and the engine is running.

MC-306/DOT 406 flammable liquid tanker A tanker that typically carries between 6,000 gallons and 10,000 gallons (between 22,712 and 37,854 liters) of a product such as gasoline or other flammable and combustible materials. The tank is nonpressurized.

MC-307/DOT 407 chemical hauler A tanker with a rounded or horseshoe-shaped tank capable of holding 6,000 to 7,000 gallons (22,712 to 37,854 liters) of flammable liquid, mild corrosives, and poisons. The tank has a high internal working pressure.

MC-312/DOT 412 corrosives tanker A tanker that often carries aggressive (highly reactive) acids such as concentrated sulfuric and nitric acid. It is characterized by several heavy-duty reinforcing rings around the tank and holds approximately 6,000 gallons (22,712 liters) of product.

MC-331 pressure cargo tanker A tanker that carries materials such as ammonia, propane, Freon, and butane. This type of tank is commonly constructed of steel and has rounded ends and a single open compartment inside. The liquid volume inside the tank varies, ranging from the 1,000-gallon (3,785-liter) delivery truck to the full-size 11,000-gallon (41,640-liter) cargo tank.

MC-338 cryogenic tanker A low-pressure tanker designed to maintain the low temperature required by the cryogens it carries. A boxlike structure containing the tank control valves is typically attached to the rear of the tanker.

Passive regeneration The first of three stages of regeneration that occurs automatically when the particulate matter (soot) that is caught in the diesel particulate filter is burned off naturally by the elevated temperatures of the exhaust system.

Rub rails Exterior steel attachments comprised of 16-gauge corrugated metal 4 inches (102 mm) or more in width, attached to the bow trusses, running the entire length of the school bus, and wrapping around to the rear of the vehicle.

School bus Any public or private school or district, or contracted carrier operating on behalf of the entity, providing transportation for kindergarteners through grade 12 pupils.

Semi-truck A commercial truck capable of towing a separate trailer, which has wheels only at one end (semi-trailer); may also be referred to as a tractor.

Semi-tractor trailer (semi-trailer) A semi-truck or tractor combined with a trailer; normally designed with three axles, one in the front for steering purposes and two tandem axles in the rear.

Series-operated propulsion system A system that utilizes the electric motor by itself to propel a school bus, and the combustion engine is only used to regenerate the battery pack. These systems may be found on a type A school bus.

Service air line One of two air lines of an air brake system connecting a tractor to a trailer; the emergency air line is blue.

Service brake The usual driving brake that is applied by the driver to slow and/or stop the vehicle during normal driving operations.

Sleeper A compartment attached to the cab of a truck that allows the driver to rest while making stops during a long transport.

Stringers Steel longitudinal structural members that give the bow frame truss members structural support at the roof level; they run continuously from the front of the school bus to the rear of the bus.

SWOT analysis A self-examination model that can be adjusted, adapted, and applied to any situation, incident, or project, large or small, that an organization is currently or will be involved in.

Transit bus An entity providing passenger transportation over fixed, scheduled routes, within primarily urban geographic areas.

Tube trailers A high-volume transportation device made up of several individual compressed gas cylinders banded together and affixed to a trailer. Tube trailers carry compressed gases such as hydrogen, oxygen, helium, and methane. One trailer may carry several different gases in individual tubes.

Type A school bus A conversion-type bus constructed utilizing a cutaway front section vehicle with a left side driver's door. This definition includes two subclassifications: type A-1, with a GVWR of 14,500 lb (6,577 kg) or less, and type A-2, with a GVWR greater than 14,500 lb (6,577 kg) and less than or equal to 21,500 lb (9,752 kg).

Type B school bus A school bus that is constructed utilizing a stripped chassis. The entrance door is behind the front wheels. This definition includes two subclassifications: type B-1, with a GVWR of 10,000 lb (4,536 kg) or less, and type B-2, with a GVWR greater than 10,000 lb (4,536 kg).

Type C school bus Also known as a *conventional school bus*; a school bus that is constructed utilizing a chassis with a hood and front fender assembly. The entrance door is behind the front wheels. This type of school bus has a GVWR greater than 21,500 lb (9,752 kg). Eighty-five to ninety percent of all school buses are type C or D.

Type D school bus Also known as a *"transit-style" rear or front engine school bus*; a school bus that is constructed utilizing a stripped chassis where the outer body of the bus is mounted to the bare chassis. The entrance door is ahead of the front wheels, and the face or front section of the bus is flat. Type D buses have a passenger capacity of 80 to 90 people.

Technical Rescuer *in Action*

As the officer assigned to an engine company, you are dispatched to a commercial motor vehicle accident involving a school bus and a truck. On arrival you can see that a large semi-tractor trailer has side-impacted a type C school bus. The school bus is resting on its side with occupants trapped inside.

1. What is the greatest initial concern for this incident?
   A. The number of patients.
   B. The number of personnel needed.
   C. The size of the semi-tractor trailer.
   D. Whether or not there are hazardous materials involved.

2. A type C school bus is also known as:_____
   A. a traditional bus.
   B. the largest of all school bus types.
   C. a conventional bus.
   D. a transit type bus.

3. Bow frame trusses on a bus run from floor level on one side of the bus to the: ____
   A. roof level.
   B. window sill level on the opposite side of the bus.
   C. floor level on the opposite side of the bus.
   D. rub rail on the opposite side of the bus.

4. What do exterior rivet heads on the outer panel of the school bus indicate?
   A. The panels are connected to the body.
   B. There is a structural member underneath the panel.
   C. Rivet heads are decorative only.
   D. The design is weak and the rivet head will shear off.

5. Rub rails are strategically placed with the mid rail positioned at the area of the seat cushion level and the bottom rail positioned:
   A. at the floor line.
   B. at the window line.
   C. there is no set position for this rub rail.
   D. newer buses no longer use rub rails.

6. Most wheels on a commercial motor vehicle can be flattened by cutting the air intake valve and releasing the air.
   A. True
   B. False

7. Air brake lines are colored-coded red for the emergency brakes and blue assigned for the:
   A. electrical lines.
   B. water line.
   C. parking brake.
   D. air release for the tanks.

8. An air ride system is designed to _____ through on board air pressure tanks.
   A. protect the cargo and the frame system of the vehicle.
   B. inflate and deflate the suspension
   C. inflate the passenger seats
   D. run at as little as 20 psi (138 kPa)

9. The cab of a traditional cab over engine (COE) design:
   A. can tilt forward to expose the engine.
   B. is secured to the frame and cannot tilt forward.
   C. has 4 doors.
   D. is known as a conventional design.

10. A fifth wheel located on a tractor:
    A. acts as a spare tire in case of a flat.
    B. can be dropped down in position to handle heavy loads.
    C. connects the trailer to the tractor by means of a king pin located on the trailer.
    D. is the swivel used to turn the boom of a large tow truck.

NFPA 1006 Standard

10.2.1 Plan for a commercial heavy vehicle or large machinery incident, and conduct initial and ongoing size-up, given agency guidelines, planning forms, and operations-level vehicle/machinery incident or simulation, so that a standard approach is used during training and operational scenarios, emergency situation hazards are identified, isolation methods and scene security measures are considered, fire suppression and safety measures are identified, vehicle/machinery stabilization needs are evaluated, and resource needs are identified and documented for future use. (pages 325–332)

(A) Requisite Knowledge. Operational protocols, specific planning forms, types of commercial/heavy vehicles and large machinery common to the AHJ boundaries, vehicle/machinery hazards, incident support operations and resources, vehicle/machinery anatomy, and fire suppression and safety measures. (pages 325–332)

(B) Requisite Skills. The ability to apply operational protocols, select specific planning forms based on the types of commercial/heavy vehicles and large machinery, identify and evaluate various types of commercial/heavy vehicles and large machinery within the AHJ boundaries, request support and resources, identify commercial/heavy vehicles and large machinery anatomy, and determine the required fire suppression and safety measures. (pages 325–332)

10.2.2 * Stabilize commercial/heavy vehicles and large machinery, given a vehicle and machinery tool kit and personal protective equipment, so that the vehicle or machinery is prevented from moving during the rescue operations; entry, exit, and tool placement points are not compromised; anticipated rescue activities will not compromise vehicle or machinery stability; selected stabilization points are structurally sound; stabilization equipment can be monitored; and the risk to rescuers is minimized. (pages 332–336)

(A) Requisite Knowledge. Types of stabilization devices, mechanism of heavy vehicle and machinery movement, types of stabilization points, types of stabilization surfaces, AHJ policies and procedures, and types of vehicle and machinery construction components as they apply to stabilization. (pages 332–336)

(B) Requisite Skills. The ability to apply and operate stabilization devices. (pages 332–336)

10.2.3 Determine the heavy vehicle or large machinery access and egress points, given the structural and damage characteristics and potential victim location(s), so that victim location(s) is identified; entry and exit points for victims, rescuers, and equipment are designated; flow of personnel, victim(s), and equipment is identified; existing entry points are used; time constraints are factored; selected entry and egress points do not compromise vehicle or machinery stability; chosen points can be protected; equipment and victim

stabilization are initiated; and AHJ safety and emergency procedures are enforced. (pages 325–339)

(A) Requisite Knowledge. Heavy vehicle and large machinery construction/features, entry and exit points, routes and hazards, operating systems, AHJ standard operating procedure, and emergency evacuation and safety signals. (pages 325–339)

(B) Requisite Skills. The ability to identify entry and exit points and probable victim locations, and assess and evaluate impact of heavy vehicle or large machinery stability on the victim(s). (pages 325–339)

10.2.4 Create access and egress openings for rescue from a heavy vehicle or large machinery, given vehicle and machinery tool kit, specialized tools and equipment, personal protective equipment, and an assignment, so that the movement of rescuers and equipment complements victim care and removal, an emergency escape route is provided, the technique chosen is expedient, victim and rescuer protection is afforded, and vehicle stability is maintained. (pages 325–339)

(A) Requisite Knowledge. Heavy vehicle and large machinery construction and features; electrical, mechanical, hydraulic, pneumatic systems, and alternative entry and exit equipment; points and routes of ingress and egress; techniques and hazards; agency policies and procedures; and emergency evacuation and safety signals. (pages 325–339)

(B) Requisite Skills. The ability to identify heavy vehicle and large machinery construction features, select and operate tools and equipment, apply tactics and strategy based on assignment, apply victim care and stabilization devices, perform hazard control based on techniques selected, and demonstrate safety procedures and emergency evacuation signals. (pages 325–339)

10.2.5 Disentangle victim(s), given a Level II extrication incident, a vehicle and machinery tool kit, personal protective equipment, and specialized equipment, so that undue victim injury is prevented, victim protection is provided, and stabilization is maintained. (pages 336–339)

(A) Requisite Knowledge. Tool selection and application, stabilization systems, protection methods, disentanglement points and techniques, and dynamics of disentanglement. (pages 336–339)

(B) Requisite Skills. The ability to operate disentanglement tools, initiate protective measures, identify and eliminate points of entrapment, and maintain incident stability and scene safety. (pages 336–339)

Knowledge Objectives

After studying this chapter, you will be able to:

- Describe ways to prepare for a machinery incident in your jurisdiction. (page 325)
- Describe some of the common machine hazards. (pages 325–326)

- Describe how OSHA inspection procedures vary between factories and farms. (page 326)
- Identify and describe significant hazards unique to an emergency scene involving tractors and machinery, including, but not limited to:
  - Stability (pages 329–332)
  - Fuel and other fluids (pages 329–332)
  - Electrical, hydraulic, and mechanical power (pages 329–332)
  - Environmental factors (pages 329–332)
  - Other people (pages 329–332)
- Describe a picket anchor system and its benefits and uses. (page 334)
- Explain how to isolate an injury site. (page 336)
- Describe the patient care priorities when dealing with a victim of a tractor/machinery-related injury. (pages 336–337)
- Understand the importance of concurrent rescue and EMS operations. (pages 337–339)
- Explain how to terminate an incident. (page 339)

Skills Objectives

After completing this chapter, you will be able to perform the following skills:

- Demonstrate the procedure for managing a victim whose extremity is caught in a machine. (page 327, Skill Drill 12-1)
- Demonstrate the procedure for stabilizing a tractor on its side. (pages 335–336, Skill Drill 12-2)
- Demonstrate the procedure for removing a victim from under a load. (pages 338–339, Skill Drill 12-3)

ou have been dispatched to a tractor roll-over incident in which the tractor has rolled down an embankment and pinned the operator. The caller states the operator is conscious and has been trapped for several hours down a steep embankment approximately 150 yards (137 meters [m]) off the roadway. It is a large tractor with a machine attached to the back.

1. What special resources will you summon as you are responding to this incident based on the dispatch information?

2. Are there any special patient care concerns that need to be addressed before, during, or after extrication?

Introduction

"People who work together will win, whether it be against complex football defenses, or the problems of modern society."
—Vince Lombardi

Caring for a victim of a tractor overturn or of a machinery entanglement can challenge the best-trained rescuer and incident commander (IC). Even if you live in an agricultural or industrial community, you may not be prepared for the weight of the machine that needs to be stabilized or the strength of the metal that it is made from. Providing effective patient care will be a true challenge. It is not enough to be a smart rescuer. If there are agricultural or industrial operations in your response area, specialized training is warranted.

Rescue Tips

If there are farms and other industrial facilities where tractors and/or machines are used in your response area, specialized training to deal with these emergencies is warranted.

Technical rescuers need to use response planning, or preincident planning, to determine if they have the proper capacities to stabilize and lift the equipment that is being used in their area. Construction companies, highway departments, parks and golf courses, mining operations, and the local equipment rental company are examples of entities that utilize this equipment. Technical rescue personnel should develop an understanding of what types of incidents are occurring in their community; they should also understand what types of tractors and machines are being used, in what locations, and in what capacity. Technical rescue personnel should prepare a list of the machinery dealers in their area, with contact information for each dealer. This information should include how to contact the dealers after hours or during weekends and holidays. This preplanning

information, coupled with knowledge and skills from other experiences, will help rescuers to quickly understand what will and will not work effectively to manage an agricultural or machinery rescue. This chapter will discuss agricultural machinery rescue in particular.

Machine Safety

According to the National Agricultural Safety Database (NASD), 50 percent of total farm fatalities involve tractors, and 14 percent are machine related. Tractors and machines have a great deal of power that does not discriminate between plant material and human flesh. In a farm setting, there are tremendous hazards that operators, workers, and family members are exposed to. For example, most farm functions are performed utilizing a tractor with attached machinery. Most farmers realize the dangers they, their family members, and employees face, but many feel that accidents will not happen to them. Unfortunately, the more comfortable one becomes working around hazards, the more likely they are to have an accident.

As discussed, preparation is the first step to rescue management. Rescue personnel should visit their local farming operations and other industrial plants and learn as much as possible about the equipment that is used, how it operates, and how an operator can become entrapped in the equipment. Utilizing the information in this chapter and preplanning for hazards in your area will better prepare you to manage a machinery incident.

Machine Hazards

Let's develop an understanding about machines. All machines present hazards that can cause serious injury and death to operators, bystanders, and rescue personnel. Typical machine hazards involve:

- **Pinch points:** Created when two objects move together, with at least one of them moving in a circle. Example: feed rolls. Figure 12-1 ▶

- **Wrap points:** Created when a rotating machine component is exposed. Example: power take-off shafts. **Figure 12-2 ▾**
- **Shear points:** Created when the edges of two objects move toward or next to each other closely enough to cut soft material. Example: harvester. **Figure 12-3 ▾**
- **Crush points:** Created when two objects move toward each other or when one object moves toward a stationary object. Example: hitches. **Figure 12-4 ▸**
- **Pull-in points:** Points on a machine where an individual or object may be pulled into the moving parts

Figure 12-4   Crush point.

of the machine. Commonly occurs when someone tries to remove plant material or other obstructions that have become stuck. Example: a clogged corn picker. **Figure 12-5 ▾**

One would expect that if a particular machine had a hazard point, that area would be protected with a shield or a guard. For the most part, this is the case with modern equipment. However, over time, guards are removed or damaged and not replaced; these areas become not only hazardous areas, but very dangerous ones. This is one of the big differences between machines found on the farm and machines found in a city factory. The machines found in a factory are inspected routinely by the Occupational Safety and Health Administration (OSHA) and other regulating agencies, while the majority of farms in the United States are not covered under OSHA inspections or any other inspection mandates.

Figure 12-1   Pinch point.

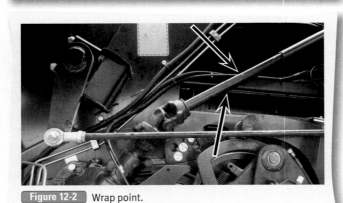
Figure 12-2   Wrap point.

Figure 12-3   Shear point.

Figure 12-5   Pull-in point.

To manage a victim whose extremity has been caught in a machine, follow the steps in **Skill Drill 12-1** ⏷ :

1. Don PPE.
2. Enter the secure work area safely.
3. Assess the scene for hazards and complete the inner and outer surveys.
4. Lay out a tarp at the edge of the secure work area for staging tools and equipment, if indicated.
5. Identify and stabilize any immediate **stored energy** that will/could be released once the power to the machine is turned off.
6. Lock out and tag out all sources of power to the machine.
7. Isolate the entanglement site by determining and securing potential sources of stored energy on each side of the entanglement site.
8. Assess patient's general condition and provide treatment per trauma protocols. Estimate length of entrapment and assess entanglement site.
9. Seek advice from machinery mechanics or co-workers on possible methods of disentanglement or disassembly.
10. Determine the priority of strategies for disentangling the patient based on the condition of the patient and availability of appropriate resources to perform the task.
11. Have tools and personnel staged to change strategies efficiently if conditions change.
12. Package the patient (and machine part if appropriate), and extricate him or her to the appropriate transport vehicle.
13. Remove and return all tools and equipment used on the machine to service. Make sure machine owners understand what actions were performed to the machine.
14. Document the incident, asking for input from all major participants in the rescue regarding hazard control, patient care, disentanglement, and extrication activities.

■ Power Sources

Like all farm emergency scenarios, machinery entanglements are so unique that an effective incident management system must be in place to coordinate all of the various team functions. It really does not matter what a particular machine does. It really does not matter if a machine is pulled behind a tractor in a corn field or sits inside a manufacturing plant in the middle of a city. What matters most is that technical rescuers recognize that machines have energy sources that will need to be managed during rescue operations. Some of these energy sources are not immediately obvious. Technical rescuers will need to anticipate and control all of these energy sources prior to and during rescue operations. Remember, to rescue means to free

from danger. If stored energy is not managed properly, it could release unexpectedly and cause serious harm to the victim as well as the rescue crew.

Machines can have one or more sources of power, and care needs to be taken to identify all of these sources. Stationary machines will typically use electric motors, gas-powered engines, or diesel-powered engines. Power within a machine can transfer through hydraulic system components, mechanical system components, and electrical system components. Electricity can travel through electric wires, switches, and motors. Take the time to study the machine to anticipate the path the energy takes through the machine, or better yet, elicit the advice from someone who knows the machine or similar machines such as a co-worker, another operator, or a machine mechanic. This assistance can be very valuable in order to manage the potential energy on each side of the victim prior to disentanglement activities. For many portable machines, the main source of power will come in the form of:

- Hydraulic power through hydraulic hoses or steel lines
- Mechanical power through the power take-off shaft
- Electrical power through electrical wiring
- Pneumatic power through air lines and tubing

Rescuers need to anticipate stored and/or potential energy that may be released during rescue efforts. This energy will need to be secured or managed before disentanglement and extrication can take place. Of particular interest is hydraulic energy that is transferred by tubing and hoses between a tractor and a machine; mechanical energy, which may be found in any of the machinery's mechanisms; and electrical energy that might be traveling through a machine.

Hydraulic Power

Hydraulics can be extremely hazardous. Many modern tractors depend on hydraulics for many functions. If a tractor stalls, power is lost to the tractor. Disastrous effects could occur because critical functions such as steering and braking can be lost. Pressure in hydraulic lines on farm equipment is normally around 1800 pounds per square inch (psi) (12,410 kilopascals [kPa]). If a hydraulic line develops a small leak, then the pressure coming through a pinhole-sized hole can exceed 7000 psi (48,263 kPa). This force could penetrate even the best rescue gloves and cause severe tissue damage. Some **hydraulic pressures**, or fluids under pressure, on farm equipment exceed 3000 psi (20,684 kPa); under such pressure, a pinhole leak could be even more devastating. Strict attention to the technical rescuer's PPE during rescue operations is essential. Full body protection must be mandated by anyone inside the hot zone, including face and eye protection, when working around hydraulic-powered equipment.

Rescuers should never use the equipment's own power to help with a rescue. Instead, hydraulic energy needs to be anticipated and controlled. Even experienced mechanics would be hard pressed to ensure which way a hydraulic lever should be pushed to achieve the desired effect. When someone's life is at stake, never take chances with an unpredictable machine.

Hydraulic pressure can be maintained within the hydraulic hoses and cylinders as long as there are no leaks in the system of hoses, fittings, and cylinders. Often this is only the case when the system is brand new. Consider the environment that farm equipment is used in. There are excessive changes in weather and severe vibrations of the machine; one can appreciate that there will be leaks in the system, which will allow the hydraulic pressure to release gradually once the engine is shut off. Because of this, it is essential that technical rescuers anticipate that anything that is held up by hydraulics will need to be secured in some mechanical way before the engine is turned off. This could be as simple as cribbing or jacking under a load, or could require chaining or strapping a wing or an arm, or placing a metal cylinder stop on an extended hydraulic cylinder Figure 12-6 ▾ .

Mechanical Power

The power take-off (PTO) shaft is a mechanical device that is used to transfer mechanical energy to and throughout a piece of machinery. For example, a PTO shaft transfers power from a tractor to a farm implement such as a spreader Figure 12-7 ▸ . A PTO shaft may also be used to transfer mechanical energy from one part of a machine to another. When facing the rear of the tractor, this shaft turns clockwise at 540 or 1000 revolutions per minute (rpm). All rescue personnel should understand the

Figure 12-6 It is essential that technical rescuers anticipate that anything that is held up by hydraulics will need to be secured in some mechanical way before the engine is turned off.

Figure 12-7 A PTO shaft transfers power from a tractor to a farm implement such as a spreader.

basic principles of the PTO system because this system is the most common system of transferring power from the tractor to a machine.

If a person becomes entangled in the PTO, it will be necessary to disassemble or cut the shaft. Unless it is a very minor entanglement, you should not attempt to unwind the victim, as this will cause further injury. Normally a minor entanglement will not occur, unless the machine was just idling when the operator came into contact with the slowly turning shaft.

Electrical Power

The first action to stabilize a machine is to secure its power source. If the power is coming from the tractor, shutting off the tractor and removing the key is the first step. Understanding how to shut off tractors is a critical part of preplanning. It is not usually a good idea to disengage the PTO control while you shut the tractor down. Doing so on some tractors could release the mechanical energy within the machine, which might cause some unexpected movement of the victim. To avoid any unexpected movement, first shut off the tractor engine so no more power is being transferred to the PTO shaft, then manually stabilize the PTO shaft with a Halligan bar or pry bar. This effort will allow you to manually control any stored energy. Once the tractor is shut off, no additional mechanical energy is being sent to the machine through the PTO shaft, and no more hydraulic fluid is being pumped through the hoses into the machine. Disconnecting the hydraulic lines from the tractor will do nothing to release hydraulic pressure within the machine.

Electrical power from the tractor can also flow to an attached machine through electrical cables or cords. Normally, securing the implement will be as simple as locating and disconnecting the electrical wire, such as pulling the plug; it is not usually necessary to disconnect the tractor's batteries. Batteries are often difficult to find and hard to access. Shutting off the electrical system to ensure that the tractor will not start and disconnecting electrical connections between the tractor and implement will be all that is needed. The receptacle for the electrical connections should be close to or on the tractor.

In any case, you will *never* use the power of the machine to remove the patient. This means you will never run the machine in reverse or use the machine's mechanical, hydraulic, or electrical power in any way to lift or move any part of the machine from the patient. There are too many unknowns and uncontrollable variables in these strategies. Rescuers need to be smarter than the machines they are working on and use their own tools, techniques, and brains to overcome the presenting forces.

Realize that when you shut off a machine, any object that is being held in an "up" position (a position subject to the forces of gravity) by hydraulic pressure should be locked in the position found prior to shutting off the engine, unless delaying shutting off the engine will have a serious effect on the patient or crew. Use cribbing, jacks, struts, straps, or chains to secure the object in as close to the position found as possible. By doing so, there will not be a sudden drop of pressure causing the object to suddenly drop with great force; however, rescuers should remain aware that a gradual drop may occur, slowly increasing pressure on the victim or quietly trapping the technical rescuers working on the scene.

Tractors

Tractors are the main workhorse on most farms. These machines are very versatile and have a variety of functions. Few farms would be in operation without at least one tractor that is used on a daily basis. Often, farmers trade in their old tractor when buying a new one. The old tractor, with few safety improvements, may be purchased by another farmer and so on. Despite the enhanced safety changes made to farm tractors over the past three decades, it is important to note that in many farming areas, the average age of tractors exceeds 30 years. Many of these tractors either have no safety mechanisms or the safety mechanisms have been broken and not fixed over the years.

One such safety mechanism may be a <u>roll-over protective structure (ROPS)</u> Figure 12-8 ▼ . According to the National Institute for Occupational Safety and Health (NIOSH), tractor overturns are the leading cause of occupational fatalities for farmers and farm workers in the United States. Since 1976, all agricultural employers have been required by OSHA to equip employee-operated tractors that were manufactured after October 25, 1976, with ROPS and safety belts. The standard is still in effect today but does not apply to family members or to family-only farms. The standard has also not been enforced on farms with fewer than 11 full-time employees in 47 of 50 states; California, Oregon, and Washington are the exceptions and cover all farms with hired workers. Because the standard does not apply in certain situations, only 8 percent of all farms in the United States are covered by the standard.

In 1985, the American Society of Agricultural Engineers (ASAE), now the American Society of Agricultural and Biological Engineers (ASABE), adopted a voluntary standard. This standard encouraged tractor manufacturers to install ROPS and seat belts on all new agricultural tractors for use in the U.S. market. The major tractor manufacturers agreed to adopt this standard, and since then, nearly all new agricultural tractors sold in the United States have been equipped with ROPS and seat belts as standard equipment, although there are no mandates in place to ensure they are used or even kept on tractors.

Although the ROPS does not prevent the tractor from rolling over, the roll bar creates a safety zone for the operator, and the seat belt holds the operator into that zone. The roll bar structure is mounted to the tractor frame and is designed to withstand the weight of the tractor for one revolution. Because the roll bar structure is an integral part of the tractor frame, it can also be used as a purchase point or lift point to assist in stabilization and/or lifting. The ROPS may come as two posts, as four posts, or as an ROPS cab or cage. You can normally find a tag mounted somewhere on the roll bar indicating that the structure is an approved ROPS Figure 12-9 ▼ . Not all cabs or roll bars with canopies are approved ROPS structures. Some are in place to provide climate comfort, typically called <u>comfort cabs</u>. These structures offer no significant support and should not be used for lifting or stability measures.

As farms get larger and more efficient, the tractors and equipment also get bigger and more powerful. While preplan-

Figure 12-8   A roll-over protective structure (ROPS).

TRACTOR MODEL	ALL 2360, 460, 2460, 510, 2510, 560, 610, & 2610 MODELS. (460-V, 560-DTE, 610-DTE, & 610-C NOT INCLUDED.)	THIS ROPS EXCEEDS THE MINIMUM REQUIREMENTS OF (BY STATIC TEST) PART 1928 OCCUPATIONAL SAFETY AND HEALTH STANDARDS FOR AGRICULTURE - SUBPART C.
R.O.P.S. PART NO.	772163	
GROSS VEH. WT.	9822 LBS.	
DATE OF MFG.		
MFG. BY: LONG MFG. N.C. INC. - TARBORO, N.C.		

Figure 12-9   You can usually find a tag mounted somewhere on the roll bar indicating that the structure is an approved ROPS structure.

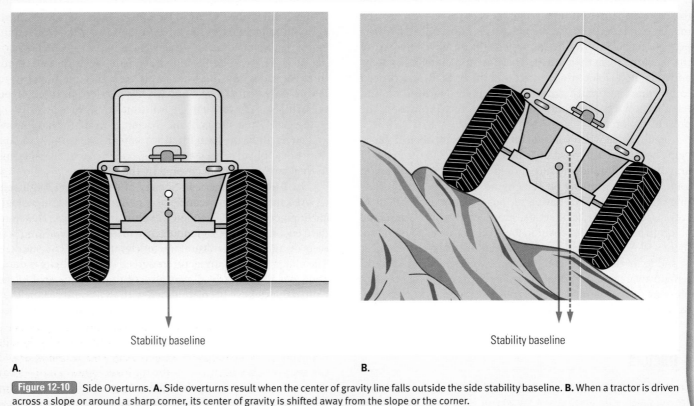

Stability baseline

A.

Stability baseline

B.

Figure 12-10 Side Overturns. **A.** Side overturns result when the center of gravity line falls outside the side stability baseline. **B.** When a tractor is driven across a slope or around a sharp corner, its center of gravity is shifted away from the slope or the corner.

ning, do not be lulled into believing that the large tractors will not upset. It is true that they will not upset as easily as smaller tractors, but operators will drive a larger more stable tractor on slopes they would never dream of driving on with a less stable tractor.

Rescue Tips

Tractors can weigh anywhere from a few hundred pounds to over 60,000 pounds (30 short tons)!

Understanding how tractors turn over can be an important first step in understanding how to initially stabilize an overturned tractor. Tractor overturns occur when the tractor's normal **center of gravity** is disrupted. When a tractor is driven across a slope, its center of gravity shifts away from the slope. When a front-end loader is raised, the tractor's center of gravity rises. When an implement is attached to the back of a tractor, the tractor's center of gravity moves backward. And when a tractor is driven around a sharp corner, its center of gravity shifts away from that corner. For a tractor to stay upright, its center of gravity must stay with the tractor's stability baseline.

Stability baselines are imaginary lines drawn between the points where the tractor tires contact the ground. Imagine a plumb line that drops straight down from the tractor's center of gravity. As long as that line falls within the tractor's stability baselines, the tractor will not turn over. As soon as that line

swings outside one of the baselines, the tractor will turn over. Side overturns result when the center of gravity line falls outside the side stability baseline (the imaginary line between the front and rear wheel on the downhill side), such as when a tractor is driven across a steep slope or when the tractor rounds a sharp corner too fast **Figure 12-10 ▲**. A rear overturn occurs when the center of gravity line falls outside the rear stability baseline (the imaginary line between the rear wheels), which happens when too heavy a load is being pulled. Obviously, the wider the baselines, the more stable the tractor is. Understanding the concept of stability baselines is important for stabilizing an overturned tractor.

Rescue Tips

In most states, there are no age restrictions for driving a tractor on a public road going from farm to field and back again. Younger drivers often lack experience in correctly handling a bad situation in a timely manner. When going too fast, which is often the case, farm equipment can become very unstable and unpredictable. Farm equipment operators should be properly trained to handle and control the equipment that they are operating.

■ Site Operations

As with all incidents, without a stable scene, rescue actions cannot take place. It is critical for rescuers to understand how to properly stabilize an agricultural extrication scene. Typical hazards at the scene of a tractor overturn are:

- Bystanders
- Instability
- Hazardous materials
- Power
- Environmental conditions

Bystanders

Consider that the bystanders who discovered the incident may have been providing care for quite some time before help arrived, and they may have made some very important decisions regarding this incident. Bystanders may include family members who discovered the incident and neighbors who assembled prior to your arrival. It would be a mistake to usher them away at this point. Allow bystanders to stay at a safe distance so they can feel that they are part of the effort that they began.

You may also be challenged with anxious family members and neighbors. For example, suppose you suspect crush syndrome, requiring the victim to be infused with fluids for 1 hour before being extricated from the machinery. You will need to educate family and neighbors on the reasons for your actions. Farmers are logical people and will surely understand your methods if you are able to explain them with confidence.

Instability

When arriving on scene to find an overturned tractor, keep in mind the central factor that caused the tractor to turn over—the center of gravity line fell outside the tractor's stability baseline. With an upset tractor, try to determine where the center of gravity exists. Determine also where the stability baselines exist. These will be the points where the tractor is resting on the ground. The first concept of managing the stability of the tractor will be to widen the stability baselines Figure 12-11 ▾ . The goals of initial stabilization are to make the scene safe for further stabilization efforts and to gain access to the patient. Do not spend so much time making the tractor stable that your rescue options are limited later and/or your patient's condition further deteriorates.

Figure 12-11   Placing cribbing can be a quick and effective means of creating a wider baseline. Placing a strut can also be effective.

Hazardous Materials

Leaking fluids such as fuel, oils, radiator fluids, battery fluids, and pesticides from attached tanks or implements will certainly be a concern with a tractor roll-over situation. One of the big differences between a tractor roll-over and a vehicle roll-over is the location of the fluids in relation to the operator. In a car, the majority of the fluids will be leaking in the front and back of the vehicle, with the operator in the middle. On a tractor, many of the fluids will be leaking from the center of the tractor, where the operator sits. A charged hose line must be deployed, if possible, in order to prepare for a combustible scene. Remember, there may be electrical energy or potentially hot surfaces that could result in ignition of fuels and other fluids. In addition, rescuers will need to assess for and control other potential hazards from pesticides, fertilizers, and other agricultural chemicals that may be carried either on the tractor or the attached equipment. Once the tractor is stabilized, it is safe to divert or dike the fluids away from the victim or rescue operations site. If possible, stop the fluids from leaking from the tractor.

Power

Once you are able to access the stabilized tractor safely, you need to secure the power source. An upset tractor most likely will not be running, but you still need to secure the potential power source. A stalled tractor with the ignition system energized is very dangerous and will need to be secured. It is important to note that turning the key to the off position may not power down the tractor as it would in a normal passenger vehicle. Understanding the tractor's fuel supply will help determine how to shut off the tractor's engine. Typical fuel systems of common farm tractors require either diesel fuel or gasoline. Occasionally you will see a propane engine.

Gasoline engines require a constant supply of electricity to energize spark plugs, so shutting off the electrical current on a gasoline-type engine will stop the engine. On diesel tractors, electrical power is only used to *start* the engine; once the diesel engine is running, electrical power is not needed to keep the engine running unless the engine has an electronic fuel pump. A diesel tractor will often have a fuel shut-off control on the console Figure 12-12 ▸ . The operation of the fuel shut-off may vary with each model tractor; some push in, some pull out, and some tractors have the fuel shut-off incorporated as part of the throttle control. On many of the newer diesel tractors, an electronic fuel pump is controlled by the key; therefore, turning the key to the off position will turn off the fuel pump and stop the engine. That is why on all tractor incidents, the first effort, once the scene is safe, should be to turn the key to the off position. Normally, the easiest way to affect the tractor's electrical circuit is by turning the key. Rescue personnel should also understand how to affect the tractor engine if they cannot get to the operator's station, or if an ignition key cannot be located.

If you cannot access the operator's console to turn off the key or operate the fuel shut-off, then you will need to find an alternate method to shut off the power. On gas-powered tractors or those engines that require electrical current to energize spark plugs, pulling the coil wire will disrupt the electrical current and shut down the engine Figure 12-13 ▸ . This action

Figure 12-12 A diesel tractor will often have a fuel shut-off control on the console.

Figure 12-14 On a diesel tractor, locating the fuel pump will reveal the fuel shut-off linkage. This linkage is the other end of the fuel shut-off control that is located on the operator's console (shown above).

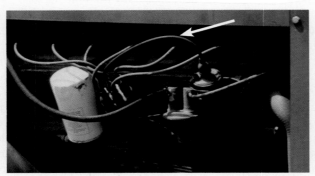

Figure 12-13 On gas-powered tractors or those engines that require electrical current to energize spark plugs, pulling the coil wire will disrupt the electrical current and shut down the engine. Caution should be exercised with this procedure if operating in rain or other wet conditions.

Figure 12-15 On vehicles powered by propane fuel, shutting off the electrical system will shut off the engine. Also, locating the propane fuel tank and turning off the fuel supply will immediately shut down these engines.

could create a spark, so discharging a carbon dioxide (CO_2) extinguisher while pulling the coil wire will absorb any fire/explosion hazard. This action could also create a shock hazard to the person pulling the coil wire.

On a diesel tractor, locating the fuel pump will reveal the fuel shut-off linkage. This linkage is the other end of the fuel shut-off control that is located on the operator's console Figure 12-14 ▶ . Shutting off the fuel supply lines will also stop the gas and the diesel engines, but this process will not yield immediate results. It may take several minutes for the gas or fuel in the fuel lines to be burned through the engine.

On vehicles powered by propane fuel, shutting off the electrical system will shut off the engine. Also, locating the propane fuel tank and turning off the fuel supply will immediately shut down these engines Figure 12-15 ▶ .

Environmental Conditions

Environmental conditions at the scene of an agricultural incident can be challenging. The discovery of a victim commonly occurs after dark. Tractor overturns and machinery incidents usually occur during bad weather when the ground is wet and slippery. These conditions can affect access and egress operations. Extra personnel will likely be needed to carry tools and equipment and to help carry out the victim once he or she has been extricated. Also, extra personnel will be needed due to the length of extrication activities that some of these incidents require.

■ Stabilization

The single most critical skill rescue personnel need to master is effective cribbing. Because of the uneven ground, large spaces, and heavy loads that will be supported, utilizing the proper technique is essential for everyone's safety. Tractor rescue operations will utilize a great deal of cribbing blocks due to the large voids that need to be shored up and the need to provide a suitable platform for lifting processes. There is no room for error with this skill. If adequate resources are available, building a box crib consisting of three blocks per layer is preferred instead of the normal box of two blocks per layer that is often used

Voices of Experience

The initial dispatch was for a male patient with his arm stuck in a chain. Upon arrival, the patient was found to have his right arm caught in a pizza dough manufacturing machine. His arm was being pinched between the sprocket and chain.

Once fire department rescue personnel arrived on scene, we made sure the electrically driven machine was locked out properly. Next, with the assistance of maintenance personnel, we were able to remove the chain guards. With these guards out of the way, we could access the chain. As we were taught in the Pennsylvania State University (PSU) agricultural rescue trainings, it was determined that there was no stored energy in the chain drive, and we were able to cut the chain. This, however, did not release the patient completely. The patient's forearm was still being pinched between the sprocket and a guide, which carried the chain to the sprocket. The sprocket, which was approximately 18 inches (457 mm) in diameter, was removed by taking the bearing block off of the frame of the machine.

As the rescue personnel worked on removing the patient's arm from the machine, EMS personnel started an IV line and administered pain medication. We were not sure what the outcome would be when the patient's arm was removed from the machine. Before removal, everyone was made aware that with this type of crushing injury there is a chance of massive blood loss when the restriction is removed. The rescuers and EMS personnel worked very well together to make sure everyone was aware of the plan to disentangle, package, and extricate the patient. Once the patient was removed, it was clear he had received a severe crushing injury from the machine. Due to the nature and severity of the injury, EMS requested air transport to a Pittsburgh trauma center.

The training and prior preplanning of the facility allowed us to be confident going into this rescue. Once on scene, we were able to see exactly what we were dealing with. The machine was fairly simple in design. With the help of on-site maintenance personnel, the lockout/tagout system was utilized. A simple flat-head screwdriver and two box-end wrenches were used to disassemble the machine with minimal damage to the equipment. Incident command was established and used properly. Stored energy was checked for and dealt with accordingly.

Without prior training on machinery rescue, the rescue would have most certainly taken longer, which may have resulted in the patient losing more blood and longer loss of circulation to his arm. Also we may not have checked for stored energy, as most rescuers are not aware of this hidden hazard.

Matt Johnston
Big Knob Fire Company
Rochester, Pennsylvania

> **The training and prior preplanning of the facility made us confident going into this rescue.**

Figure 12-16 An effective picket anchor system can be a tremendous tool in a tractor rescue effort.

for vehicle extrication. This extra block of wood per layer will provide 2¼ times the support and lifting capacity. Rough cut wood blocks are the preferable material for stabilization. The cribbing material will be exposed to wet ground and leaking oil, both of which will create a slippery surface. Treated wood or plastic will become very slippery when exposed to water and oil, creating a hazard to rescue personnel.

Picket Anchor Systems

Picket anchor systems are designed to provide a structurally significant anchor point for rigging system components when no structural or other solid points are available **Figure 12-16 ▲**. Locating an effective anchor point in the middle of a farm field can be next to impossible, especially if you are unable to drive your vehicles to the scene. Rescuers need to learn how to create a picket anchor system to effectively manage a tractor rescue operation. The picket system for farm conditions should be built with 1-inch (25-mm) rolled steel rods 4 feet (1.2 m) long that are driven into the ground at least 2 feet (0.6 m) deep (or 3 feet [0.9 m] if driven into a plowed field). The holding power of a picket anchor will depend on several factors, including the diameter and kind of material being used as a picket stake, the type of soil, the depth and angle in which the picket is driven, and the angle of the guy line that is hooked to the load in relation to the ground. Under good (loamy) soil conditions, a single picket rod can hold about 700 pounds (318 kilograms [kg]). To increase the holding capacity of your system, lash additional rods together. Adding more stakes to the system does not add to the holding capacity in direct proportions; that is, each stake added does not increase capacity by 700 pounds (318 kg). As you place picket stakes farther away from the front picket, the load to the rear picket is distributed more unevenly. Thus, the principal strength of a multiple-picket system is at the front pickets. Increase the capacity of a system by using two or more picket stakes to form the front group. This increases the bearing surface against the soil.

An effective picket anchor system can be a tremendous tool in a tractor rescue effort. Taking the time to learn how to build it correctly will be time well spent.

Wheel Spin

If any upward pressure is exerted on the tractor's wheels, as would be the case if you were lifting a tractor, you will need to secure the wheels to the frame of the tractor so they do not spin when pressure is exerted or taken off of them **Figure 12-17 ▼**. Pick two points on the tire as wide as possible to secure. All tractor wheels turn independently of the others. Consider a tractor lying on its side. The two bottom wheels touching the ground are the ones that you should be most concerned with. Unwanted turning of these wheels can be disastrous to the patient as well as the crew. You need to tie these wheels off to the frame of the tractor. It will do no good to tie off the two wheels that are in the air because the other wheels will turn if you put too much pressure on them.

Front Axle (Pivot Point)

Rescuers also should understand that the front axle of tractors has a pivot point. A **pivot point** allows the tractor to traverse uneven ground while preventing the front end from lifting off the ground **Figure 12-18 ▶**. In a rescue situation, this pivot may need to be stabilized. Securing the pivot area with cribbing will

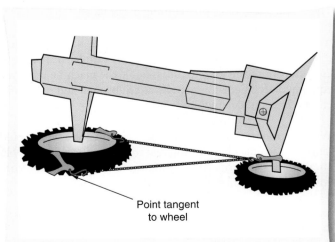

Point tangent to wheel

Figure 12-17 If any upward pressure is exerted on the tractor's wheels, you will need to secure the wheels to the frame of the tractor so they do not spin when pressure is exerted or taken off of them.

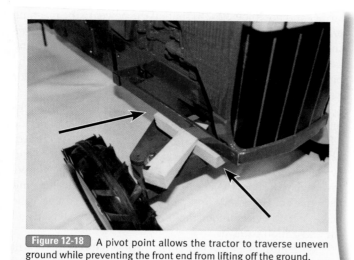

Figure 12-18 A pivot point allows the tractor to traverse uneven ground while preventing the front end from lifting off the ground.

allow the front axle to become one with the rest of the tractor, thus alleviating any unsuspected movement that could throw off the tractor's stability.

To stabilize a tractor on its side, follow the steps in **Skill Drill 12-2 ▾**:

1. Don PPE.
2. Enter the secure work area safely.

3. Assess the scene for hazards and complete the inner and outer surveys.
4. Lay out a tarp at the edge of the secure work area for staging tools and equipment, if indicated.
5. Position a safety officer at a location where he or she can view the entire stabilization operation and will be able to notice any shifting of the tractor as it is being stabilized. The officer should not touch or place a free hand on the vehicle to feel for any shifting or movement of the vehicle as the other crew members begin cribbing on either side of the vehicle; doing so may cause dangerous slipping of the tractor.
6. Determine where the tractor's stability baseline currently exists. This line is an imaginary line where the parts of the tractor are in contact with the ground.
7. Use appropriate tools and techniques to widen the baseline of the tractor. These could include cribbing under a wheel, struts placed on the frame of the tractor, cables or straps extended to anchor points, or various lifting devices (air bags, jacks, rams) strategically placed to stabilize the machine.
8. Once the tractor is stable from movement, lock out and tag out the tractor's power source(s). This will involve disabling the fuel supply and electrical system for diesel-powered tractors and the electrical system for gas-powered tractors.

NFPA 1006, (10.2.2)

Skill Drill 12-2

Stabilizing a Tractor on Its Side

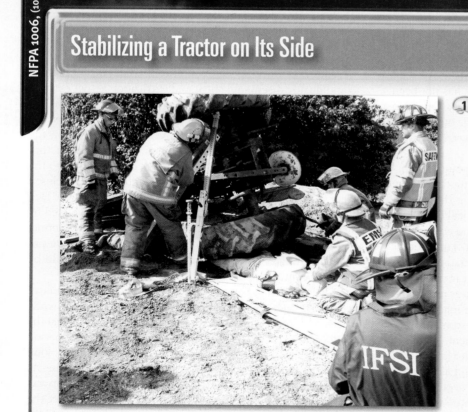

1. Don PPE. Enter the secure work area safely. Assess the scene for hazards and complete the inner and outer surveys. Lay out a tarp at the edge of the secure work area for staging tools and equipment. Position a safety officer at a location where he or she can view the entire stabilization operation and will be able to notice any shifting of the tractor. Determine where the tractor's stability baseline currently exists. Use appropriate tools and techniques to widen the baseline of the tractor. Lock out and tag out the tractor's power source(s). Disable the fuel supply and electrical system of diesel-powered tractors and the electrical system of gas-powered tractors. Use caution when removing battery cables or dismantling electrical connections on overturned tractors because of sparking hazards that could ignite fuel vapors. Consider simultaneously discharging a CO_2 fire extinguisher while performing these tasks to absorb any potential sparking. Employ appropriate diverting, damming, or diking techniques to ensure that any fluids that are leaking from the tractor do not run near the patient. Plug any leaks at the tractor source.

Care needs to be taken when removing battery cables or dismantling electrical connections on overturned tractors because of sparking hazards that could ignite fuel vapors. It would be wise to simultaneously discharge a CO_2 fire extinguisher while performing these tasks to absorb any potential sparking.

9. Employ appropriate diverting, damming, or diking techniques to ensure that any fluids that are leaking from the tractor do not run near the patient. Plug any leaks at the tractor source that you can. (**Step 1**)

■ Victim Access and Management

As discussed in previous chapters, managing the victim involves vehicle or machinery entry, victim packaging, and victim removal. One of the major differences between passenger vehicles and machines is the ability or lack of ability to remove pieces of the machine without causing damage to get to the victim. Machinery is made of strong steel and cast material that will challenge most rescue tools that are designed for vehicle extrication. Farm machines can often be dismantled more easily than they can be cut apart if the right tools are available. A great resource for advice may be another farmer or a machinery mechanic in the area. Farm machinery is made modularly, and studying it can reveal how to gain access to a trapped victim.

Once a rescuer is able to get close to the victim, he or she should observe the victim's entrapment carefully and determine how he or she is trapped. Rescue personnel need to ensure that parts of the machine on either side of the entrapped patient are secured to prevent any movement during disentanglement and extrication activities. This process is called isolating the injury site. This is the part of the machine that will receive the most focus. Once the injury site is isolated, study the machine to determine the alternative methods of disentangling the patient. At this point, the extent of the patient's injuries has not been determined. Always have alternative plans of attack. One plan may be a slow and methodical approach while the other might be aggressive, such as if the victim's condition warrants a rapid disentanglement or extrication.

While rescue personnel are developing the plan(s) for disentanglement, EMS personnel must be assessing the patient and determining an index of suspicion based on the mechanism of injury and length of entrapment. A critical difference between an entrapment with farm machinery and entrapment with any other industrial machinery is the length of time for the entrapment. In a manufacturing setting, if an operator becomes entrapped in a machine, most likely a co-worker is right there to shut off the power to that machine. Co-workers and maintenance people will then begin dismantling the machine and managing the stored energy before rescue personnel arrive. In many cases, they may actually have the patient extricated before rescue personnel arrive. Contrast that with a farmer who becomes entrapped out in a field while working alone. This patient may experience an extended delay between the time of injury and medical treatment because there is no witness to initiate the emergency response. A family member may find the farmer and not know how to shut off the power to the machine or how to dismantle it to gain access. The patient's body systems could be in terrible shape as compared to the manufacturing patient who may receive quick care and a quick release of the pressure. Crushing injuries can be a major complication for lengthy machine entrapments.

Once the patient is assessed, discussion between the technical rescuers, EMS, and the IC needs to take place to decide which approach, whether methodical or aggressive, is needed. This decision is mainly based on the patient's condition and available resources to perform the extrication. All personnel involved in the operation need to understand the priorities and objectives of the rescue and patient care concerns. Everyone needs to have a clear understanding of the operational plan and the extrication pathway.

With most machinery entanglement situations, there will be multiple strategies that can be employed for a successful rescue. Some of the more common approaches are manually unwinding the mechanism that is entrapping the victim; disassembling the parts of the machine that are entrapping the victim; and cutting, prying, or spreading the parts of the machine that are entrapping the victim. On rare occasions when the victim's condition is deteriorating rapidly, a field amputation might be the best approach. It is critical to always first stabilize the machine and the victim to be sure no further damage is done during this process. You may find that in some cases, cutting away parts of the machine will be necessary to reach the entrapped victim. If the victim is caught in a pinch or crush point, spreading may be the quickest and most effective disentanglement method. As mentioned earlier, most machines are made to be taken apart, and disassembly may in fact be quicker than cutting or spreading. Individuals with knowledge about the machinery can help you make the most effective extrication decisions.

Appropriate victim care activities need to take place prior to and during disentanglement and extrication activities. Pain control should be considered, especially for a victim who has been entrapped for longer than an hour. Crush injuries and particularly compartment syndrome should be high on the list of suspicions when assessing a machine-entrapped victim. **Compartment syndrome** is a medical term that refers to the compression of nerves, blood vessels, and muscle inside a closed space (compartment) within the body. Tissue death due to lack of oxygenation results as the blood vessels are compressed by the raised pressure within the compartment. EMS personnel should consult with medical command as to how the victim should be treated on scene and where the victim should be taken for continuing care.

One of the more critical injuries that will be seen in farm trauma is **crush injury syndrome**. This condition exists when a person becomes entrapped under a load in such a way that the weight of the load restricts the person's blood flow for a period of time (typically longer than an hour). Without adequate blood flow carrying oxygen to muscle tissue, muscle cells will die. As those cells die, they produce toxins. If those toxins build up and reach the heart, they will enter the circulatory system once the load is lifted. This can cause the heart to develop a fatal arrhythmia. For example, suppose a farmer overturns his tractor and becomes trapped underneath, with the tractor resting on his legs and pelvis. The overturn has not caused any significant internal bleeding in his abdominal cavity, his airway and breathing are not compromised, and he has no significant head or neck injury. The tractor is acting as a tourniquet to his lower extremities. If he is found immediately and extricated from under the tractor, there may not be serious complications. However, if he is not found for 3 to 4 hours or longer, his treatment needs to follow a much different course if he is going to survive. EMS personnel will need to attempt to reverse the effects of the crushing injury prior to lifting the tractor from the victim.

Refer to local protocols on crush injury, or, if protocols are not known, consult with medical command physicians prior to extrication.

Disentanglement and extrication activities need to happen quickly and methodically, and communication among everyone involved in the operation is paramount. The patient's hemodynamic condition can decline rapidly once pressure or the load is released. Once the EMS sector gives the go signal to begin disentanglement, those efforts must continue until the patient is freed and secured to the backboard or Stokes basket. Prior to these efforts, the egress path must be cleared and communicated so everyone knows their responsibility once the victim is freed. Once the victim is extricated and moved to a safe location, his or her condition needs to be reevaluated. Good communication and a verbal review prior to taking action will increase the likelihood of a smooth and orderly operation. This is the best care for a victim who has been trapped for an extended period of time.

Removing a Victim from Under a Load

Because tractor overturns are the leading cause of occupational fatalities for farmers and farm workers in the United States, it is important to know the three basic strategies to remove a patient from under a load:

1. Raise the load by lifting.
2. Stabilize the load and lower the ground (dig) to enable patient removal.
3. Combine the first two approaches, raising and digging.

In a tractor overturn scenario, the third option will probably be the best. Lifting any vehicle with a high center of gravity (like you would have with a tractor on its side) can be very unpredictable; therefore, lifting should be kept to a minimum. Rescue lift air bags are an effective tool for lifting a tractor. Their compact size allows them to be placed in tight and uneven spots, and their great lifting power can be a good match for the size of a tractor. Medium- and low-pressure rescue lift air bags and cushions also have a place in agricultural rescue operations. These large bags can take the place of scarce cribbing resources. Whatever device is used for lifting, follow these rules, which are especially important to tractor rescue:

- Know the rated capacities and limitations of lifting tools. Tools and techniques that work with vehicle extrication react differently when used for agricultural equipment.
- As with any lifting, for every inch a load is lifted, insert an inch of cribbing. There is nothing wrong with following

the lift with another lifting system. That way if one lifting system fails during a critical point in the lift, lifting can continue with the backup that is already in place.

- Know where purchase points are on a tractor. Normally this is only the frame of the tractor, inside rim of the wheels, and ROPS, if there is one and it has not been compromised by the overturn. The sheet metal or plastic material that makes up the wheel well and **cowling** of the tractor will not support the weight of the tractor and should never be used to lift or support.
- Build a very sound and level base for the lifting tool(s). Spend the time to make sure the box crib is built perfectly before committing it to a lift. In less-than-favorable conditions such as in a field, the box crib may actually push into the ground as it attempts to lift the tractor. This may occur several times until reaching solid ground. Consider using longer blocks for the cribbing base; the wider the crib base, the more stable it will be.
- If using rescue lift air bags, protect the portions of the air bag that come in contact with the tractor, in accordance with the air bag manufacturer's recommendations. If nothing else, a large piece of heavy rubber matting should be considered. Sharp metal pieces can puncture some bags, and excessive heat from an engine can melt bags.
- Realize that what is being lifted may be heavier and larger than usual. Know how your tool will react to this; be extremely cautious and ensure that all nonessential personnel are away from the scene and are paying close attention to the operation. Use extreme caution with a shifting center of gravity as the lift proceeds.

The importance of teamwork between fire and EMS interests cannot be understated in managing an agricultural incident. If these victims survive the initial trauma, their body systems have compensated for the insult, and those systems need to be managed delicately.

Rescue Tips

In a farm trauma situation, rescuers need to focus on many unfamiliar hazards as well as patient concerns that are not common. A successful rescue operation is one where the IC can effectively break down the various responsibilities and provide support to each sector of the operation. Patient care requires its own sector, and those in charge of the victim need to concentrate on assessing and treating the potential conditions based on the mechanism of injury and the amount of time that has elapsed since the onset of the trauma. Likewise, because of the unstable nature of an overturned tractor or a machinery entanglement, the management of the rescue tools and procedures should also have its own sector.

This is what is meant by teamwork. Rescue personnel may have everything in place to lift the tractor and remove the victim, but EMS personnel may need time to finish treatment before the rescue operation continues.

To remove a victim from under a load, follow the steps in **Skill Drill 12-3 ▼**:

1. Don PPE.
2. Enter the secure work area safely.
3. Assess the scene for hazards and complete the inner and outer surveys.
4. Lay out a tarp at the edge of the secure work area for staging tools and equipment, if indicated.
5. Position a safety officer at a location where he or she can view the entire stabilization operation and will be able to notice any shifting of the tractor as it is being stabilized. The officer should not touch or place a free hand on the vehicle to feel for any shifting or movement of the vehicle as the other crew members begin cribbing on either side of the vehicle; doing so may cause dangerous slipping of the tractor.
6. Perform a thorough trauma assessment of the patient's condition. This will dictate the urgency of disentangle-

ment and extrication efforts. A patient with a compromised airway and breathing issues that cannot be corrected while entrapped due to excessive loads will need immediate action. Conversely, a patient with an intact airway and no breathing issues but who has been trapped for an extended period of time should be evaluated and treated, if appropriate, for crush injury syndrome; this scenario may delay extrication times.

7. Once the patient has been assessed and is being properly treated, develop the best plan(s) of action. Appropriate methods include lifting the object from the patient, stabilizing the object and digging the patient out, and a combination of lifting and digging. The methods chosen will depend on available resources and the condition of the scene.
8. Discuss patient care and rescue strategies with the patient and his or her family. Recognize that the bystanders may have discovered this incident and performed some tasks

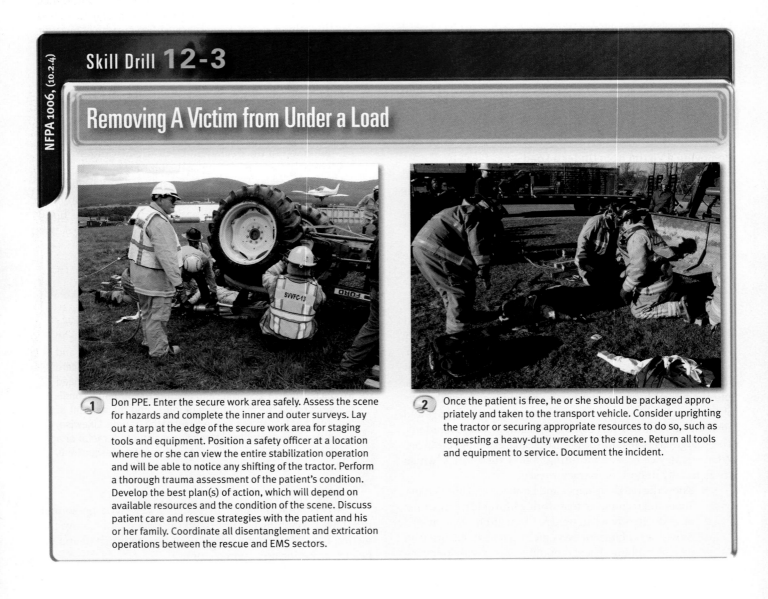

NFPA 1006, (10.2.4)

Skill Drill 12-3

Removing A Victim from Under a Load

1 Don PPE. Enter the secure work area safely. Assess the scene for hazards and complete the inner and outer surveys. Lay out a tarp at the edge of the secure work area for staging tools and equipment. Position a safety officer at a location where he or she can view the entire stabilization operation and will be able to notice any shifting of the tractor. Perform a thorough trauma assessment of the patient's condition. Develop the best plan(s) of action, which will depend on available resources and the condition of the scene. Discuss patient care and rescue strategies with the patient and his or her family. Coordinate all disentanglement and extrication operations between the rescue and EMS sectors.

2 Once the patient is free, he or she should be packaged appropriately and taken to the transport vehicle. Consider uprighting the tractor or securing appropriate resources to do so, such as requesting a heavy-duty wrecker to the scene. Return all tools and equipment to service. Document the incident.

prior to your arrival. If the bystanders are family, it is a good policy to keep them informed on the potential changes to the patient's condition.

9. Coordinate all disentanglement and extrication operations between the rescue and EMS sectors. Patient care procedures must begin prior to these efforts and must continue during them. The operation must progress efficiently and without interruption unless the EMS sector needs to halt to correct a change in the patient's condition that requires immediate attention. Rescue personnel who are operating the lifting and stabilization tools and procedures must be knowledgeable of the tools and equipment they are using and understand the limitations and capacities of them. This teamwork is essential for a successful patient outcome.

10. The IC must be ready with additional strategies that can be used should the operating plan fail for any reason. (**Step 1**)

11. Once the patient is free, he or she should be packaged appropriately and taken to the transport vehicle.

12. Once extrication activities are complete, consider uprighting the tractor or securing appropriate resources to do so, such as requesting a heavy-duty wrecker to the scene.

13. Return all tools and equipment to service. This will involve cleaning and general maintenance.

14. Document the incident according to agency protocols. (**Step 2**)

Terminating a Machinery Incident

Termination activities for an agricultural or machinery incident include making the scene as safe as possible, moving disabled equipment, or uprighting an overturned tractor. Another important step is to work with personnel or members of the farm to ensure that normal operations can be continued. This may be as simple as asking the farm personnel if they have contacted someone to fill in for the victim or care for animals while the victim is at the hospital. Remember that many industrial and agricultural operations are 24 hours a day, 7 days a week, and 365 days a year.

Return all tools used during the extrication to the station where they can be cleaned and inspected prior to being placed back in service. Most likely, the tools and equipment will be dirty and will need to be cleaned and dried before they are placed back into service.

A **postincident analysis (PIA)** should be conducted and developed by the IC or by the rescue and EMS sector commanders—or by all three parties. A PIA is a review of the incident and a way to constructively examine the positives and negatives of the incident. This review includes identifying opportunities for improvement and addressing any necessary corrective actions that may be needed to improve the organization as a whole. The analysis reports are great tools to use in an incident critique a few days after the incident. Encourage all involved with the incident to attend this critique, and offer everyone an opportunity to speak. This is not intended to be a gripe session but will allow the IC an opportunity to read the PIA report in a manner that produces meaningful discussion. Remember that these incidents are very unique and no one could be expected to have all of the procedures and techniques to make a perfect rescue. This critique is intended to address those "what if this happens again" cases. By the end of this session, you will have come full circle with the rescue steps, and you will be better prepared to handle the next call. Consideration should also be given to arrange for a **critical incident stress debriefing (CISD)**. A CISD is a structured and confidential postincident meeting designed to assist rescue personnel in dealing with psychological trauma as the result of an emergency or critical incident Figure 12-19 ▾ . Terminating an incident is discussed in more detail in Chapter 13, *Terminating the Incident*.

Figure 12-19 A critical incident stress debriefing is a structured and confidential postincident meeting designed to assist rescue personnel in dealing with psychological trauma as the result of an emergency or critical incident.

Wrap-Up

Ready for Review

- Even if you live in an agricultural or industrial community, you may not be prepared for the weight of the machine that needs to be stabilized, or the strength of the metal that it is made from.
- Technical rescue personnel should develop an understanding of what types of incidents are occurring in their community; they should also understand what types of tractors and machines are being used, at what locations, and in what capacity.
- According to the National Agricultural Safety Database (NASD), 50 percent of total farm fatalities involve tractors, and 14 percent are machine related.
- Typical machine hazards involve pinch points, wrap points, shear points, crush points, and pull-in points.
- The machines found in a factory are inspected routinely by OSHA and other regulating agencies, while the majority of farms in the United States are not covered under OSHA inspections or any other inspection mandates.
- Machines can have one or more sources of power, and care needs to be taken to identify all of them.
- Rescuers need to anticipate stored and/or potential energy that may be released during rescue efforts. This energy will need to be secured or managed before disentanglement and extrication can take place.
- For many portable machines, the main source of power will come in the form of hydraulic power, mechanical power, or electrical power.
- As with all incidents, without a stable scene, rescue actions cannot take place.
- Tractor stabilization will utilize a great deal of cribbing blocks because of the large voids that need to be shored up and the need to provide a suitable platform for lifting processes.
- Rescuers need to learn how to create a picket anchor system to effectively manage a tractor rescue operation.
- One of the major differences between passenger vehicles and machines is the ability or lack of ability to remove pieces of the machine without causing damage to get to the victim.
- Because tractor overturns are the leading cause of occupational fatalities for farmers and farm workers in the United States, it is important to know the three basic strategies to remove a patient from under a load: raise the load by lifting, stabilize the load and lower the ground (dig) to enable patient removal, or a combination of raising and digging.
- Termination activities for an agricultural or machinery incident include making the scene as safe as possible, moving disabled equipment, uprighting an overturned tractor, and working with personnel or members of the farm to ensure that normal operations can be continued.

Hot Terms

Center of gravity The point at which all of the weight of an object appears to be concentrated.

Comfort cab A tractor cab that has no roll-over protection. Its purpose is just to provide environmental protection to the operator.

Compartment syndrome A medical term that refers to the compression of nerves, blood vessels, and muscle inside a closed space (compartment) within the body. This leads to tissue death due to lack of oxygenation as the blood vessels are compressed by the raised pressure within the compartment.

Cowling A removable metal cover for an engine.

Critical incident stress debriefing (CISD) A structured and confidential postincident meeting designed to assist rescue personnel in dealing with psychological trauma as the result of an emergency or critical incident.

Crush injury syndrome A condition that results when toxins (by-products of crushed muscle) are entered into circulation after a crushing load is released from a person. This is characterized by major shock and renal failure. It can be a common injury with tractor overturns and other farm-related emergencies because of the length of time of entrapment.

Crush point A hazard created when two objects move toward each other or when one object moves toward a stationary object. Examples include hitches.

Farm implement A machine or attachment that is normally mounted to or pulled and powered by a tractor.

Hydraulic pressure Fluid under pressure that is used to power equipment.

Picket anchor system A method that provides a structurally significant anchor point using steel rods and rigging system components when no structural or other solid points are available.

Pinch point A hazard found where two rotating objects move together and at least one of them moves in a circle. Common examples include the points where drive belts contact pulley wheels, drive chains meet gear sprockets, feed rolls mesh, or gathering chains on harvesting machines draw crops into the equipment.

Pivot point A mechanism that allows the tractor to traverse uneven ground while preventing the front end of the tractor from lifting off the ground.

Postincident analysis (PIA) A review of the incident and a way to constructively examine the positive and negative outcomes of the incident; includes assessing opportunities for improvement and addressing any necessary corrective actions that may be needed to improve the organization as a whole.

Power take-off (PTO) shaft A mechanical device used to transfer mechanical energy to and throughout machinery.

Pull-in point A hazard posed by those places where an individual may be pulled into the moving parts of a machine. Commonly occurs when someone tries to remove plant material or other obstacles that have become stuck.

Roll-over protective structure (tractor or machinery) A structure that is mounted to a tractor frame that creates a safety zone for the operator in the event of a rollover.

Shear point A hazard created when the edges of two objects move toward or near each other closely enough to cut. An example may be a harvester.

Stored energy Mechanical, hydraulic, or electrical energy that can be contained within a machine after the power of the machine has been stopped.

Wrap point A hazard posed by any exposed rotating machine component. The most common examples are power take-off shafts.

Technical Rescuer *in Action*

You are in charge of a rescue operation where a 47-year-old male is being crushed between the bucket and frame of a skid steer loader. He is in severe pain and tells you that he has been trapped there for at least 2 hours before someone noticed him and called 911. You see that he is being crushed by the bucket at his pelvis and thighs. He states that he cannot feel his feet and his back and stomach hurt severely. He further states that the machine was acting up so he had gotten out to make some adjustments when the loader came down on him.

1. Your immediate rescue priority is:

A. raising the loader to free the victim.

B. ensuring the machine is cribbed and turned off.

C. providing C-spine stabilization and oxygen therapy.

D. preparing to cut the bucket from the machine.

2. What is one action that you might consider before shutting the power down to this machine?

A. Attempt to raise the hydraulic arms to free the victim.

B. Attempt to raise the hydraulics a little to offer the victim some relief.

C. Place cribbing to ensure that no more downward pressure occurs.

D. Make sure the machine is in park.

3. Based on your initial size-up of the victim's condition as described here, you immediately:

A. prepare to extricate the victim, securing him to the long backboard, and wait for the arrival of an ALS ambulance that is over 15 minutes away.

B. ask the victim and the family if they can explain how to rapidly release the bucket so you can perform a rapid extrication.

C. extricate and transport via a BLS ambulance that is on scene.

D. perform a thorough trauma assessment and consult with incoming ALS personnel for advice.

4. The greatest challenge you face in this rescue scenario is:

A. managing the hydraulic energy that is holding the bucket down.

B. managing the electrical energy that is built up in the machine.

C. managing the victim's family because you are taking so long.

D. managing the victim's airway and breathing.

5. The victim, who is in severe distress, tells you that all you need to do is raise the bucket by starting the engine and stepping on the right-hand foot lever. What should you do?

A. See if someone on the rescue crew is comfortable with this machine to perform this task.

B. See if there is a co-worker who understands how to operate the machine.

C. Inform the patient that you cannot do this action.

D. Call a farm equipment dealership to ask how to operate this machine.

6. The on-scene paramedic reports to the IC that he suspects crush injury syndrome and would like to treat the victim prior to extrication. This treatment could delay extrication by 45 to 60 minutes while treatment is taking place. What should you do?

A. Allow the paramedic to treat the victim as he requests.

B. Tell the paramedic that he needs to treat the victim in the ambulance en route to the hospital.

C. Call medical command to confirm this request.

D. Discuss this strategy with the rescue sector and the family for their opinions.

7. It is decided that methods need to be deployed to lift the bucket upward, which will raise the load from the patient. Air bags are placed under the center of the bucket to lift evenly. What other action is critical prior to beginning this lift?

A. Have backup cribbing available so that as the load is lifted, cribbing is placed.

B. Have backup lifting devices available in case the air bags fail.

C. Ensure that the victim is being properly treated prior to lifting.

D. All of these actions are equally critical.

8. You begin lifting the bucket and observe that after 2 inches of lift, the entire machine is lifting off the ground. What do you do?

A. Keep lifting, as you will soon overpower the hydraulic pressure that is holding the bucket down.

B. Stop lifting and reevaluate your actions.

C. Cut the hydraulic line to the cylinders.

D. Get a stronger lifting tool to complete the job.

9. Dismantling or breaking of a hydraulic hose can result in serious injury. Who is most vulnerable and in need of most protective measures?

A. The rescue crew

B. The victim

C. The bystanders and family members

D. The IC

10. Appropriate termination activities include:

A. making sure the machine is put back together and secured.

B. discussing the care given with the patient's family.

C. discussing with a machinery mechanic who is familiar with skid steers what the proper extrication actions are in case this call occurs again.

D. All of the above

Terminating the Incident

NFPA 1006 Standard

Chapter 5, Job Performance Requirements

5.2.7 Terminate a technical rescue operation, given an incident scenario, assigned resources, and site safety data, so that rescuer risk and site safety are managed, scene security is maintained and custody transferred to a responsible party, personnel and resources are returned to a state of readiness, record keeping and documentation occur, and postevent analysis is conducted. (pages 346–352)

(A) Requisite Knowledge. Incident command functions and resources, hazard identification and risk management strategies, logistics and resource management, personnel accountability systems, and AHJ-specific procedures or protocols related to personnel rehab. (pages 346–352)

(B) Requisite Skills. Hazard recognition, risk analysis, use of site control equipment and methods, use of data collection and management systems, and use of asset and personnel tracking systems. (pages 346–352)

Chapter 10, Vehicle and Machinery Rescue

10.1.10 Terminate a Level I vehicle/machinery incident, given personal protective equipment specific to the incident, isolation barriers, and an extrication tool kit, so that rescuers and bystanders are protected during termination operations; the party responsible for the operation, maintenance, or removal of the affected vehicle/machinery is notified of any modification or damage created during the extrication process; scene control is transferred to a responsible party; potential or existing hazards are communicated to that responsible party; and command is terminated. (pages 346–352)

Knowledge Objectives

After studying this chapter, you will be able to:

- Explain the process of terminating an incident. (pages 346–362)
- Explain the process of securing the scene. (pages 346–347)
- Explain how to secure personnel. (pages 348–350)
- Describe posttraumatic stress disorder (PTSD). (page 348)
- Discuss the tasks involved in securing equipment. (pages 347–348)
- Describe some of the negative reactions caused by stress. (pages 348–350)
- Explain common components of critical incident stress management (CISM). (page 350)
- Explain a postincident analysis review. (pages 350–357)
- Discuss the purpose of documentation and record management. (pages 351–352)

Skills Objectives

After studying this chapter, you will be able to perform the following skills:

- Terminate a vehicle extrication incident. (page 352, Skill Drill 13-1)

*A*fter returning to the station from a mentally and physically difficult and traumatic extrication incident involving a family, you notice one of your crew members acting differently, somewhat quiet and withdrawn from the rest of the crew. When you approach him and ask if everything is alright, he responds firmly that he is fine and just wants to be left alone.

1. What actions can you take?
2. Should you encourage the crew member to talk and interact with the group rather than remaining isolated?
3. Should you notify a critical incident stress management (CISM) team leader to assess the appropriate response needed?

Introduction

"It's not stress that kills us; it is our reaction to it." —Hans Selye

With the patient extricated, properly packaged, and transported to the appropriate medical facility, the victim management phase of the incident is complete, but the incident is far from over. Getting the units back in service as quickly as possible is a priority for any organization, but personnel, equipment, and the scene must be secured before units are ready to respond to the next incident. This is a basic demobilization process. This chapter will discuss the tasks that must be completed in order to terminate a vehicle extrication incident.

Terminating an Incident

Terminating an incident includes securing the scene by removing the damaged vehicle and equipment from the scene, ensuring the scene is left in a safe condition, and completing documentation and reports, incident debriefing, and stress debriefing as appropriate. After you have secured the scene and packed your equipment, it is important to return to the station and fully inventory, clean, service, and maintain all the equipment (per the manufacturer's instructions) to prepare it for the next call. Some items will inherently need repair, but most will need simple maintenance before they are placed back on the truck and considered in service.

Securing the Scene

Just because the patient has been removed from the wreckage does not mean the scene can be abandoned. Law enforcement will secure the scene by conducting their on-scene investigation. A potential crime scene should be managed by law enforcement, who will preserve and secure any evidence and close off the scene or roadway. Once a scene has been released by law enforcement, the removal of vehicles will be coordinated with a tow agency. In some cases, a vehicle may need to be turned right side up. Being proactive by standing by with a charged hose line is always a good practice during one of these procedures.

All medical waste must be accounted for and properly disposed of in approved biohazard waste containers, and any remaining body fluids such as blood, vomit, and feces must be neutralized with the appropriate solution **Figure 13-1 ▾**.

Any fluid hazards that have spilled from the vehicle, such as gasoline, motor oil, transmission fluid, or radiator fluid that was either captured with absorbent or not, will be removed by the towing agency. Towing agencies must be licensed to transport and dispose of any hazardous substances; fire rescue agencies are typically not licensed to do so. To assist the towing agency, place any vehicle parts, such as doors, roofs, and fenders, back in a heavily damaged vehicle. A word of advice is to always ask the tow agency representative if they need assistance before you

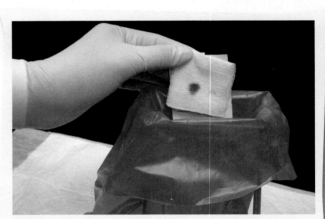

Figure 13-1 All agencies must properly secure a scene before clearing an incident. Properly disposing of used medical waste and biohazards is part of securing a scene.

start throwing items back into a vehicle. The tow agency may have procedures or a policy of their own for stowing and transporting loose objects. Being considerate of their preferences will help to maintain a good working relationship with the tow agency for future endeavors such as acquiring vehicles for use in an extrication class. It is also advisable to carry a contact list of various resources for any private and public organization or business that can offer a particular resource that can be utilized on the incident, such as public works/utilities, the department of transportation (DOT), or a heavy equipment company.

Rescue Tips

To assist the towing agency, place any vehicle parts, such as doors, roofs, and fenders, back in a heavily damaged vehicle. A word of advice is to always ask the tow agency representative if they need assistance before you start throwing items back into a vehicle. The tow agency may have procedures or a policy of their own for stowing and transporting loose objects.

The exception to these procedures will be an incident involving a fatality. Law enforcement will acquire the scene immediately to conduct a fatality investigation. There are times when a victim of a fatality may be left in the vehicle, and your agency may be called back to the scene at a later time to extricate the body from the vehicle. This is often a very traumatic experience, and it is recommended that the crew that responded initially does not return to remove the body; dispatch another unit to complete this assignment.

Securing Equipment

Accountability and maintenance of equipment after the incident has concluded are critical Figure 13-2 ▶ . Crews are normally depleted both mentally and physically after the incident, and trying to quickly gather all the equipment to get back in service can lead to problems. There must be sufficient time allotted to gather the equipment properly, with one person, normally the driver of the apparatus, in charge of inventory and overseeing all the equipment that was utilized on scene. The apparatus may need to be refueled and the equipment must be properly placed back on the apparatus in *ready to be utilized* condition for the next incident.

Decontamination is also an important factor that cannot be overlooked. Any equipment or personal protective equipment that is contaminated must be properly isolated and/or cleaned following your agency's decontamination procedures. Exposure to any biologic or chemical contaminants must be reported immediately and followed up with the proper documentation. Always refer to your departmental procedures for handling the various types of exposures.

The following checklist should be reviewed after every incident as the equipment is placed back on the apparatus:

- Ensure that all gas-powered units are topped off with gasoline.
- Ensure that the hydraulic fluid levels for all power units are full and the correct hydraulic fluid is added if needed.

Figure 13-2 Accountability and maintenance of equipment after the incident has concluded are critical.

- Conduct a quick inspection of the hydraulic hoses for any damage as they are recoiled for storage.
- Check that the couplings are in good working condition and are free of dirt or grime from the incident.
- Conduct an inspection of the hydraulic tools for any damage to the blades or arms.
- Engage the hydraulic spreader to fully close the tips and then make a quarter turn in the open position to relieve the pressure before storing it away on the apparatus. Store the hydraulic cutter with the tips just touching each other.
- For all battery-operated tools, replace batteries with charged ones.
- Examine the teeth of the reciprocating saw blade, and replace the blade with a new one if significant wear is observed.
- Examine the electrical cord of the reciprocating saw for any damaged, flattened, or opened sections. Place out of service for any damage noted.
- Examine the air chisel blade for any damage or chipping, and replace the blade if needed.
- Examine the regulator gauge for any damage.
- Lubricate pneumatic tools according to the manufacturer's recommendations.
- Change out any used air cylinders with topped-off bottles.
- Examine all wood cribbing for damage such as cracks, splitting, or recessed screws. Discard or recycle any damaged cribbing.
- Examine all strut stabilization systems for damage, and account for all securing pins and any other attachments that were included.
- Examine all cargo straps for tears or excessive exposure/absorption of any petroleum products. Place out of service for any damage noted.

- Examine chains and come along cables for any breaks in the strands or flattened sections. Place out of service for any damage noted.
- Examine the handle of the come along for any bending or deformity. Place out of service for any damage noted.
- Examine all rescue lift air bags, hoses, and regulators for any damage. Place out of service for any damage noted.

At first glance, this checklist may appear to be extreme or too time consuming, but remember that it can be completed fairly quickly as the equipment is being loaded back onto the apparatus. Any heavy maintenance such as washing, degreasing, and repairing can be completed back at the station. There is no worse scenario than arriving at another accident, pulling off the equipment, and discovering that it does not work or is missing because it was not properly accounted for at the last incident. Take the time while still on scene to properly account for all of the equipment before going back in service.

Rescue Tips

After you have secured the scene and packed your equipment, it is important to return to the station and fully inventory, clean, service, and maintain all the equipment (per the manufacturer's instructions) to prepare it for the next call.

◼ Securing Personnel

Stress is something that a rescuer experiences every day. At one time, negative reactions to stressful incidents were suppressed or hidden. Great strides in research with immediate recognition and treatment protocols have greatly reduced the negative side effects and debilitating emotional scars that can linger and impair normal everyday functions Figure 13-3 ▶ .

Stress

Stress is defined as a normal response to a stimulus, whether pleasant or unpleasant, that manifests itself in cognitive, physical, emotional, or behavioral signs. Stress is not necessarily a bad thing; it is our coping mechanism or lack thereof, that determines how stress will affect the body. There are several classifications of stress, two of which are eustress and distress. Eustress can be described as a type of stress that produces a positive response in the mind, body, and spirit, such as that experienced through physical exercise or a team sport. Eustress actually builds resistance to the negative aspects of stress. Distress is a type of stress that produces a negative response, such as that experienced through the exposure to a critical incident. The continual exposure to critical incidents will cause distress to accumulate, eventually leading to a breakdown of effectiveness and efficiency, including an erosion of concentration and self-confidence. Distress is the major contributor to most health issues. As stress builds over time, it can lead to burn-out and become a major contributor of heart failure.

Everyone reacts differently to a stressor, and a rescuer must be aware of common signs of distress. If left unresolved, distress can potentially disrupt a person's ability to properly function

Figure 13-3 Immediate recognition and treatment of stress will greatly reduce the negative side effects and debilitating emotional scars that can linger and impair normal everyday functions.

at the next emergency incident or heighten the risk for developing posttraumatic stress disorder (PTSD) or a depressive illness. PTSD is a delayed stress reaction to a prior incident. This delayed reaction is often the result of one or more unresolved issues concerning the incident.

Critical Incident Stress

Critical incident stress is a type of stress that emergency personnel are exposed to. The definition of a critical incident is an event that has the potential to create significant human distress that can overwhelm the body's normal coping mechanisms. This basically describes almost every emergency incident that a rescuer responds to. Remember that everyone reacts differently to a stressor; what may not affect you can drastically affect a coworker. Also, reactions can either occur immediately, several hours after the event, or several days later.

Rescuers who have been exposed to a traumatic or critical incident can exhibit negative stress reactions cognitively, behaviorally, emotionally, and physically. Signs and symptoms of these four categories of reactions include the following:

- **Cognitive reactions:** attention deficit, nightmares, confusion, lack of concentration, decreased ability to problem solve, constant reliving of the event through flashbacks
- **Behavioral reactions:** withdrawing from others, emotional outbursts, extreme changes in normal behavior (such as silence or hyperactivity), repeated drunkenness, negative sexual reactions, insomnia, absenteeism
- **Emotional reactions:** depression, guilt, anger, fear, anxiety, feeling of doom, grief, perceived loss of control
- **Physical reactions:** headaches, muscle twitching/tremors, dry mouth, elevated blood pressure and/or heart rate, nausea, hyperpnoea, profuse sweating, chest pains

A stress fracture left untreated will cause continual pain or, in more severe cases, a total fracture of the bone. The same holds true for a negative stress reaction to a critical incident. Proper

Voices of Experience

One early spring day we were dispatched onto the State Turnpike for a reported motor vehicle accident with entrapment. Our rescue company rolled out with an engine company for support. As we arrived on location, our size-up included a midsized four-door vehicle that ran directly into the tollbooth concrete barrier, and the force of the accident resulted in the vehicle resting on an angle, with the front of the vehicle completely pushed down into the passenger compartment and the rear of the vehicle resting on the ground at the bumper line. Inside the vehicle was a single occupant, conscious and animated about wanting to exit the vehicle, but unable to do so on his own.

Our crew got to work stabilizing the vehicle, and my partner and I began to displace the vehicle assembly to disentangle the patient from the vehicle. After the doors were removed, I found the patient's lower extremities had punctured through the firewall, resulting in double compound fractures to the lower extremities. Additional personnel assisted in opening the firewall and packaging the patient. I was located on the driver side of the vehicle, continuing to displace metal around the patient's legs. During the operation, the patient continued to offer his "technical expertise" regarding how we were performing our tasks and what he thought we could be doing better. After approximately 30 minutes, the patient was removed from the vehicle and transported to the trauma center. Our department stayed on scene while the local towing agency removed the vehicle from the tollbooth. We picked up our equipment, cleaned up the scene, returned to quarters, and secured our equipment in preparation for the next run.

The next day, the department chief called all of the responders who were at the extrication scene the previous day to report to the station immediately. Once we arrived, we were given the bad news: the patient who we extricated had full-blown active pulmonary tuberculosis, which is a very contagious disease to people who come within close proximity of those who have it. "Close proximity" can be considered living in the same house; I was within 2 feet of the patient's head for 30 minutes. I was newly married, looking forward to buying a home and starting a family; now my focus changed to health and survival. We were all sent for testing to see if any of us had contracted tuberculosis. I do not think I slept at all for 3 days. Thankfully, 72 hours later, we all received clean bills of health. We continued to go back for testing every 6 months until the potential for contamination was gone.

Our department was pretty progressive, so I thought. We were very good at most of our operations, with the obvious exceptions of postincident analysis and debriefings. After that incident, every alarm of significance was not terminated until the postincident analysis was completed *immediately* after returning to quarters. All members had an opportunity to discuss the positive and negative aspects of the incidents and to voice any concerns that may have arisen. Furthermore, the safety officer was charged with getting information on *any* potential exposure to our members, whether it was through hazardous materials exposure, bloodborne pathogens, or any other recognized hazard that was on scene. Members were briefed on what hazards were on scene, what the risks were to exposure, what symptoms may arise in the event of contamination, what to do if they became symptomatic, and where to go for medical treatment.

It is very easy for emergency responders to become complacent and "burned out" after countless alarms and incidents. Attention to detail and vigilance in safety operations on scene will continue to be the best tool in the tool cache when it comes to staying healthy on the emergency scene.

Mike Daley
Monroe Township Fire District #3
Monroe Township, New Jersey

> "After that incident, every alarm of significance was not terminated until the postincident analysis was completed immediately after returning to quarters."

and immediate treatment is needed to alleviate further potential physical and psychological problems, and it should be available to the rescuer after any exposure to a critical incident.

Critical incident stress management (CISM) is a multifaceted system of crisis intervention specifically designed to help emergency personnel who have been exposed to a traumatic event process their response to the incident in a way that validates the normal stress reactions and stabilizes the potential negative results of the individual's response. It is geared toward enhancing natural coping mechanisms and facilitating a natural resiliency and recovery from the incident. Just as we provide emergency assistance to individuals in need of help, CISM is a kind of emergency psychological first aid for responders after an exposure to a traumatic incident. It is a tremendous resource that helps personnel deal with real-time issues from traumatic events that are momentarily suppressed or openly manifested.

CISM begins with pre-incident education and strategic planning. It is designed to break down normal stress reactions to abnormal events, thus increasing the rescuer's understanding of the normal ways the mind and body react to stress. This prepares rescuers to appropriately respond when critical incidents occur. The most common components of CISM are the small group interventions consisting of defusing and debriefing.

Defusing is an informal confidential postincident meeting that allows individuals to express or release their emotions and start the process of proper coping. It should be conducted within 1 to 4 hours after the critical event. In the defusing, the individuals who were part of the response are guided to talk about the event in a safe setting in order to begin putting pieces together, which helps them as individuals and as a group, and allows them to air out some of the emotions and thoughts associated with the event. Trained members of the CISM team listen intently with the goal of assessing individual and group responses. The defusing usually lasts no more than 45 minutes and concludes with the trained members giving participants instructions specific to their needs. These instructions direct them on continuing to help themselves and each other in the hours and days ahead. The defusing sets the stage for the more formalized debriefing if one is necessary and warranted.

Critical incident stress debriefing (CISD) is a structured and confidential group discussion among those who served at a traumatic incident to address emotional, psychological, and stressful issues related to the event **Figure 13-4 ▶**. A CISD usually occurs within 12 to 72 hours of the incident. These small group interventions have proven to be the most effective and most helpful component of CISM. One of the key aspects of the small group intervention is absolute confidentiality. Some states, such as Florida, have adopted state statutes that legally protect the confidentiality of the small group CISD process. In some cases, CISD members can provide one-on-one service to anyone in immediate need. CISD is similar to triage on an emergency incident where appropriate levels of care are provided to the individuals in need. CISD is not an operational critique or a psychotherapy session. It is designed around three basic goals: (1) defuse the psychological impact of a traumatic event, (2) assist individuals in the normal recovery process, and (3) identify individuals who may need further help or require

Figure 13-4 A critical incident stress debriefing can greatly assist in helping personnel properly cope with an exposure to a traumatic incident.

professional mental health assistance. The CISD lasts from 45 to 60 minutes in most cases.

Defusings and debriefings are conducted by specifically trained, tested peers and mental health professionals. They are not psychotherapy or counseling sessions but should be considered psychological first aid and crisis intervention. They are brief and specific, and are designed to assess and assist rescuers so they may have long and healthy careers.

Broward County, Florida, is one of many counties in the nation that recognizes the importance of CISM. There is a CISM team with highly trained, seasoned peers and mental health professionals who are ready to activate 24 hours a day, 7 days a week, 365 days a year to any agency within Broward County that requests assistance. Reach out to organizations that offer CISM and ask for assistance establishing a CISM team in your region, or contact the International Critical Incident Stress Foundation for more information.

Rescue Tips

CISM is geared toward enhancing natural coping mechanisms and facilitating natural resiliency and recovery from an incident.

■ Postincident Analysis

As discussed in the previous chapter, a postincident analysis (PIA) is a review of the positive and negative aspects of an incident that identifies opportunities for improvement and addresses any necessary corrective actions that may be needed to improve the organization as a whole. There are two types of PIAs: formal and informal. A formal PIA is a well-organized event with a structured agenda where all the critical information of the incident is gathered, reviewed, and discussed with all the personnel who responded to the incident. The informal PIA can be as simple as a discussion among the crews at the scene or a discussion back at the station either after the call or on the following shift **Figure 13-5 ▶**.

Figure 13-5 An informal postincident analysis can take place anywhere.

The informal PIA allows the crew to have a heart-to-heart discussion about how the incident evolved. Talking in a no-pressure environment such as this allows everyone to determine ways to improve the response on the next call. Informal talks also build solid relationships and trust within the crew. The key is to be transparent and honest about what went right and what went wrong. Being overly critical or demeaning is destructive and has no place in these types of meetings; to keep the discussion positive and proactive, all egos and tempers must be left at the door.

The PIA is a tool that is used to build on, not break down; no one should be pointing fingers. If a deficiency is found, it needs to be addressed and corrected as a team. The PIA should be conducted after every incident, whether in a formal or informal setting; the positive results and professional development gained are astounding.

■ Documentation and Record Management

As discussed in Chapter 2, *Rescue Incident Management*, and Chapter 7, *Site Operations*, documentation or record keeping serves several important purposes, including tracking of equipment inventory, training, needs assessment, response times, and preincident planning. Adequate and accurate documentation also ensures the continuity of quality care, guarantees the proper transfer of responsibility, and fulfills the administrative needs of the department for local, state, and federal reporting requirements.

One form of documentation that is used after an incident has been terminated is an **after action report (AAR)**. An AAR is a brief summary that analyzes the overall operations and effectiveness of the agency at a particular incident, measuring its capabilities through real-time on-scene evaluations. This summary goes hand in hand with the needs assessment planning that is discussed in Chapter 2, *Rescue Incident Management*. The AAR should be a formalized document that covers topics including:

- Compliance to standard operating procedures (SOPs) and standard operating guidelines (SOGs): Were there any operational issues that were effective or not effective

relating to the current SOP/SOG for the type of incident response? Is a policy change or recommendation(s) needed?
- Medical protocols: Were there any medical interventions that are outlined in the agency's medical protocols that need to be addressed through training or by the agency's medical director?
- Staffing requirements: Was the staffing on hand at the incident sufficient to accomplish the operation in a safe and efficient manner?
- Mutual aid: Was mutual aid requested and how did the integration of personnel, equipment, and procedures work to accomplish the scene objectives?
- Equipment: Were there any equipment issues? Was the equipment on hand effective in completing all tasks? Were there any deficiencies or is there a need to upgrade the current equipment? Was there any equipment malfunction or breakage?
- Training: Are there any areas of improvement that can be and should be addressed through training? Are there areas where subject experts need to be brought in to update or train personnel?

There may be other points that can be added to this list, but understand that this basic template will give the organization a clear idea of operational needs and performance by identifying strengths and shortfalls within the agency. The authority having jurisdiction, fire chief, or chief of operations may request an AAR for significant incidents. The AAR should be revised after a PIA is conducted. There are many AAR example templates that can be downloaded from the Internet. For example, Arlington County, Virginia, has posted an AAR on their response to the September 11, 2001, terrorist attack on the Pentagon.

Rescue Tips

The after action report or documentation should include specific details about an incident such as compliance with SOPs/SOGs, medical interventions, staffing on the scene, mutual aid assistance, equipment used, and areas of training needed. The documentation helps the organization to identify strengths and shortfalls.

In addition to the AAR, all state- and national-required reporting documents such as National Fire Incident Reporting System (NFIRS) and National EMS Information System (NEMSIS) should be filled out and submitted according to the required timeline set forth by the agency or the state of jurisdiction.

Rescue Tips

Documentation or record keeping serves several important purposes, including tracking of equipment, training, needs assessment, response times, and preincident planning. It also ensures the continuity of quality care, guarantees the proper transfer of responsibility, and fulfills the administrative needs of the department for local, state, and federal reporting requirements.

To terminate an extrication incident, follow the steps in **Skill Drill 13-1**:

1. Secure the scene by returning it to its normal condition. This may involve preparing the vehicle for a towing company and removing hazardous fluids and medical waste.

2. Secure the equipment and apparatus. Account for, maintain, and decontaminate all tools and equipment before placing them back on the apparatus for the next call.

3. Secure personnel by initiating a CISM system if needed.

4. Conduct a formal or an informal PIA and complete an AAR.

Wrap-Up

■ Ready for Review

- Before units are ready to respond to the next incident, personnel, equipment, and the scene must be secured.
- A potential crime scene should be managed by law enforcement to preserve and secure any evidence.
- Accountability and maintenance of the equipment after the incident has concluded are critical. There should be one assigned person (normally the driver of the apparatus) in charge of inventorying and overseeing all the equipment that was utilized on scene.
- A basic equipment checklist can be completed fairly quickly as the equipment is being loaded back onto the apparatus. Any heavy maintenance such as washing, degreasing, and repairing can be completed back at the station.
- Making sure that you and your personnel are psychologically and emotionally sound after the incident is vital not only to being able to properly function at the next incident, but also to maintaining longevity in emergency services.
- Stress is something that a rescuer experiences every day. Great strides in research with immediate recognition and treatment protocols have greatly reduced the negative side effects and debilitating emotional scars that can linger and impair normal everyday functions.
- Critical incident stress management (CISM) is a multifaceted system of crisis intervention specifically designed to help emergency personnel who have been exposed to a traumatic event process their response to the incident in a way that validates the normal stress reactions and stabilizes the potential negative results of the individual's response.
- The most common components of CISM are the small group interventions consisting of defusing and critical incident stress debriefing (CISD).
- A postincident analysis (PIA) is a review of the positive and negative aspects of an incident that identifies opportunities for improvement and any necessary corrective actions that may be needed to improve the organization as a whole. An after action report (AAR) may be completed following this analysis.
- Documentation or record keeping aids in keeping track of equipment inventory, training, needs assessment, response times, and preincident planning.

■ Hot Terms

After action report (AAR) A brief summary that analyzes the overall operations and effectiveness of the agency at a particular incident, measuring its capabilities through real-time on-scene evaluations.

Behavioral reactions Negative reactions that may present as withdrawal from others, emotional outbursts, extreme changes in normal behavior (such as silence or hyperactivity), repeated drunkenness, negative sexual reactions, insomnia, or absenteeism.

Cognitive reactions Negative reactions that may present as attention deficit disorder, nightmares, confusion, lack of concentration, decreased ability to problem solve, or constant reliving of the event through flashbacks.

Critical incident stress management (CISM) A multifaceted system of crisis intervention specifically designed to help emergency personnel who have been exposed to a traumatic event process their response to the incident in a way that validates the normal stress reactions and stabilizes the potential negative results of the individual's response.

Defusing A less formal confidential postincident meeting that allows individuals to express or release their emotions and start the process of proper coping; it should be conducted within 1 to 4 hours after the critical event.

Distress Stress that generally produces a negative response, such as that experienced through the exposure to a critical incident.

Emotional reactions Negative reactions that may present as depression, guilt, anger, fear, anxiety, feeling of doom, grief, or perceived loss of control.

Eustress Stress that produces a positive response in the mind, body, and spirit, such as that experienced through physical exercise or a team sport. Eustress actually builds resistance to the negative aspects of stress.

Physical reactions Negative reactions that may present as headaches, muscle twitching/tremors, dry mouth, elevated blood pressure and/or heart rate, nausea, hyperpnoea, profuse sweating, or chest pains.

Posttraumatic stress disorder (PTSD) A delayed stress reaction to a prior incident. This delayed reaction is often the result of one or more unresolved issues concerning the incident.

Stress Any type of change, whether pleasant or unpleasant, that manifests itself in cognitive, physical, emotional, or behavioral signs.

Technical Rescuer *in Action*

Your crew returns from a call in which there were a number of fatalities. The crew is extremely quiet and noticeably affected by the incident.

1. _____ actually builds resistance to the negative aspects of stress.
   - **A.** Anger
   - **B.** Eustress
   - **C.** CISM
   - **D.** Distress

2. Following the CISM protocol, defusing generally lasts:
   - **A.** 1 hour up to 3 days after the incident.
   - **B.** 20 minutes maximum.
   - **C.** 1 to 4 hours.
   - **D.** up to 45 minutes.

3. A CISD usually occurs within _____ hours of the incident.
   - **A.** 2 to 4
   - **B.** 6 to 8
   - **C.** 8 to 10
   - **D.** 12 to 72

4. Securing the scene includes all of the following except:
   - **A.** equipment.
   - **B.** personnel.
   - **C.** transport.
   - **D.** the scene.

5. A delayed stress reaction to a previous incident is:
   A. distress.
   B. posttraumatic stress disorder.
   C. eustress.
   D. critical incident stress management.

6. What are the two types of postincident analyses (PIAs)?
   A. Formal and informal
   B. Official and unofficial
   C. Essential and nonessential
   D. Mandatory and volunteer

7. Cognitive reactions include:
   A. attention deficit.
   B. withdrawing from others.
   C. depression.
   D. headaches.

8. An example of an emotional reaction is:
   A. elevated blood pressure.
   B. lack of concentration.
   C. confusion.
   D. grief.

9. Due to the fatalities at this scene, _____ will acquire the scene.
   A. the morgue
   B. law enforcement
   C. an ALS crew
   D. a physician

10. These small group interventions have proven to be the most effective and most helpful component of CISM.
   A. PIAs
   B. CISDs
   C. Equipment accountability meetings
   D. Equipment maintenance meetings

NFPA 1006, *Technical Rescuer Professional Qualifications*, 2008 Edition, ProBoard Matrix Correlation

INSTRUCTIONS: In the column titled 'Cognitive/Written Test' place the number of questions from the Test Bank that are used to evaluate the applicable Job Performance Requirement (JPR), Requisite Knowledge (RK), Requisite Skills (RS), or objective. In the column titled 'Manipulative/Skills Station' identify the skill sheets that are used to evaluate the applicable JPR, RS, or objective. When the Portfolio or Project method is used to evaluate a particular JPR, RK, RS, or objective,

identify the applicable section in the appropriate column and provide the procedures to be used as outlined in the National Board on Fire Service Professional Qualifications (NBFSPQ) Operational Procedures, COA-5. Evaluation methods that are not cognitive, manipulative, portfolio, or project based should be identified in the 'Vehicle Extrication, Levels I & II: Principles and Practice' column.

Objective / JPR, RK, RS		Cognitive Written Test	Manipulative Skills Station	Portfolio	Projects	Vehicle Extrication, Levels I & II: Principles and Practice
Section	**Abbreviated Text**					
5.2.1	Identify the needed support resources					Chapter 7, pages 147–155
5.2.1(A)	RK: equipment organization and tracking methods					Chapter 7, pages 147–155
5.2.1(B)	RS: track equipment inventory					Chapter 7, pages 147–155
5.2.2	Size up a rescue incident					Chapter 7, pages 155–162
5.2.2(A)	RK: reference materials and their uses					Chapter 7, pages 155–162
5.2.2(B)	RS: use information gathering sources					Chapter 7, pages 155–162
5.2.3	Manage incident hazards					Chapter 7, pages 162–166
5.2.3(A)	RK: resource capabilities and limitations					Chapter 7, pages 162–166
5.2.3(B)	RS: resource capabilities and limitations					Chapter 7, pages 162–166
5.2.4	Manage resources					Chapter 7, pages 152–155
5.2.4(A)	RK: incident management system					Chapter 7, pages 152–155
5.2.4(B)	RS: implement an incident management system					Chapter 7, pages 152–155
5.2.5	Conduct a search					Chapter 7, pages 158–160
5.2.5(A)	RK: site-specific search environment					Chapter 7, pages 158–160
5.2.5(B)	RS: enter, maneuver in, and exit the search environment					Chapter 7, pages 158–160
5.2.6	Ground support operations for helicopter activities					Chapter 7, pages 166–168
5.2.6(A)	RK: ground support operations					Chapter 7, pages 166–168
5.2.6(B)	RS: provide ground support operations					Chapter 7, pages 166–168
5.2.7	Terminate the incident					Chapter 13, pages 346–362
5.2.7(A)	RK: incident command functions					Chapter 13, pages 346–362
5.2.7(B)	RS: hazard recognition, risk analysis					Chapter 13, pages 346–362
5.3.1	Triage victims					Chapter 9, pages 235–236
5.3.1(A)	RK: types and systems of triage					Chapter 9, pages 235–236
5.3.1(B)	RS: ability to use triage materials					Chapter 9, pages 235–236

Objective / JPR, RK, RS		Cognitive Written Test	Manipulative Skills Station	Portfolio	Projects	Vehicle Extrication, Levels I & II: Principles and Practice
Section	**Abbreviated Text**					
5.3.2	Move a victim					Chapter 9, pages 236–237
5.3.2(A)	RK: types of transport equipment and removal systems					Chapter 9, pages 236–237
5.3.2(B)	RS: ability to secure a victim					Chapter 9, pages 236–237
5.3.3	Transfer a victim to emergency medical services					Chapter 9, pages 236–237
5.3.3(A)	RK: medical protocols for victim transfer					Chapter 9, pages 236–237
5.3.3(B)	RS: report victim condition and history					Chapter 9, pages 236–237
5.4.1	Inspect and maintain hazard-specific personal protective equipment					Chapter 6, pages 102–107
5.4.1(A)	RK: function, construction, and operation of personal protective equipment					Chapter 6, pages 102–107
5.4.1(B)	RS: wear and damage indicators					Chapter 6, pages 102–107
5.4.2	Inspect and maintain rescue equipment					Chapter 6, pages 102–107
5.4.2(A)	RK: functions and operations of rescue equipment					Chapter 6, pages 102–107
5.4.2(B)	RS: identify wear and damage indicators					Chapter 6, pages 102–107
5.5.1	Tie knots, bends and hitches					Appendix D: Available at www.fire.jbpub.com
5.5.1(A)	RK: knot efficiency, proper knot utilization					Appendix D: Available at www.fire.jbpub.com
5.5.1(B)	RS: tie representative knots					Appendix D: Available at www.fire.jbpub.com
5.5.2	Single-point anchor system					Appendix D: Available at www.fire.jbpub.com
5.5.2(A)	RK: application of knots					Appendix D: Available at www.fire.jbpub.com
5.5.2(B)	RS: select rope and equipment					Appendix D: Available at www.fire.jbpub.com
5.3.3	Edge protection					Appendix D: Available at www.fire.jbpub.com
5.3.3(A)	RK: materials and devices					Appendix D: Available at www.fire.jbpub.com
5.3.3(B)	RS: select protective devices					Appendix D: Available at www.fire.jbpub.com
5.5.4	Rope mechanical advantage system					Appendix D: Available at www.fire.jbpub.com
5.5.4(A)	RK: principles of mechanical advantage					Appendix D: Available at www.fire.jbpub.com

Objective / JPR, RK, RS		Cognitive Written Test	Manipulative Skills Station	Portfolio	Projects	Vehicle Extrication, Levels I & II: Principles and Practice
Section	**Abbreviated Text**					
5.5.4(B)	RS: select rope and equipment					Appendix D: Available at www.fire.jbpub.com
5.5.5	Direct a team in the operation of a simple rope mechanical advantage system					Appendix D: Available at www.fire.jbpub.com
5.5.5(A)	RK: principles of mechanical advantage					Appendix D: Available at www.fire.jbpub.com
5.5.5(B)	RS: direct personnel effectively					Appendix D: Available at www.fire.jbpub.com
5.5.6	Direct a team					Appendix D: Available at www.fire.jbpub.com
5.5.6(A)	RK: principles of mechanical advantage,					Appendix D: Available at www.fire.jbpub.com
5.5.6(B)	RS: ability to direct personnel effectively					Appendix D: Available at www.fire.jbpub.com
5.5.7	Function as a litter tender					Appendix D: Available at www.fire.jbpub.com
5.5.7(A)	RK: task-specific selection criteria					Appendix D: Available at www.fire.jbpub.com
5.5.7(B)	RS: ability to select and use rescuer harness					Appendix D: Available at www.fire.jbpub.com
5.5.8	Construct a lowering system					Appendix D: Available at www.fire.jbpub.com
5.5.8(A)	RK: capabilities and limitations of various devices					Appendix D: Available at www.fire.jbpub.com
5.5.8(B)	RS: ability to tie knots					Appendix D: Available at www.fire.jbpub.com
5.5.9	Direct a lowering operation					Appendix D: Available at www.fire.jbpub.com
5.5.9(A)	RK: application and use of descent control devices					Appendix D: Available at www.fire.jbpub.com
5.5.9(B)	RS: ability to direct personnel					Appendix D: Available at www.fire.jbpub.com
5.5.10	Direct a lowering operation					Appendix D: Available at www.fire.jbpub.com
5.5.10(A)	RK: application and use of descent control devices					Appendix D: Available at www.fire.jbpub.com
5.5.10(B)	RS: ability to direct personnel					Appendix D: Available at www.fire.jbpub.com
5.5.11	Construct a belay system					Appendix D: Available at www.fire.jbpub.com
5.5.11(A)	RK: principles of belay systems					Appendix D: Available at www.fire.jbpub.com
5.5.11(B)	RS: ability to select a system					Appendix D: Available at www.fire.jbpub.com

Objective / JPR, RK, RS		Cognitive Written Test	Manipulative Skills Station	Portfolio	Projects	Vehicle Extrication, Levels I & II: Principles and Practice
Section	**Abbreviated Text**					
5.5.12	Operate a belay system					Appendix D: Available at www.fire.jbpub.com
5.5.12(A)	RK: application and use of belay devices					Appendix D: Available at www.fire.jbpub.com
5.5.12(B)	RS: ability to tend a belay system					Appendix D: Available at www.fire.jbpub.com
5.5.13	Belay a falling load					Appendix D: Available at www.fire.jbpub.com
5.5.13(A)	RK: application and use of belay devices					Appendix D: Available at www.fire.jbpub.com
5.5.13(B)	RS: ability to operate a belay system					Appendix D: Available at www.fire.jbpub.com
5.5.14	Conduct a system safety check					Appendix D: Available at www.fire.jbpub.com
5.5.14(A)	RK: system safety check procedures					Appendix D: Available at www.fire.jbpub.com
5.5.14(B)	RS: ability to apply and use personal protective equipment					Appendix D: Available at www.fire.jbpub.com
Level I						
10.1.1	Plan for a vehicle/machinery incident					Chapter 7, pages 148–155
10.1.1(A)	RK: operational protocols					Chapter 7, pages 148–155
10.1.1(B)	RS: ability to apply operational protocols					Chapter 7, pages 148–155
10.1.2	Establish "scene" safety zones					Chapter 7, pages 162–163
10.1.2(A)	RK: use and selection of personal protective equipment					Chapter 7, pages 162–163
10.1.2(B)	RS: ability to select and use personal protective equipment					Chapter 7, pages 162–163
10.1.3	Establish fire protection					Chapter 7, pages 163–166
10.1.3(A)	RK: types of fire and explosion hazards					Chapter 7, pages 163–166
10.1.3(B)	RS: ability to identify fire and explosion hazards					Chapter 7, pages 163–166
10.1.4	Stabilize a common passenger vehicle					Chapter 8, pages 174–190
10.1.4(A)	RK: types of stabilization devices					Chapter 8, pages 174–190
10.1.4(B)	RS: ability to apply and operate stabilization devices					Chapter 8, pages 174–190
10.1.5	Isolate potentially harmful energy sources					Chapter 8, pages 190–191
10.1.5(A)	RK: types and uses of personal protective equipment					Chapter 8, pages 190–191
10.1.5(B)	RS: ability to select and use task- and incident-specific personal protective equipment					Chapter 8, pages 190–191

Objective / JPR, RK, RS		Cognitive Written Test	Manipulative Skills Station	Portfolio	Projects	Vehicle Extrication, Levels I & II: Principles and Practice
Section	**Abbreviated Text**					
10.1.6	Determine the common passenger vehicle or small machinery access and egress points					Chapter 9, pages 198–208
10.1.6(A)	RK: common passenger vehicle or small machinery construction/features					Chapter 9, pages 198–208
10.1.6(B)	RS: ability to identify entry and exit points					Chapter 9, pages 198–208
10.1.7	Create access and egress openings					Chapter 9, pages 208–217
10.1.7(A)	RK: common passenger vehicle or small machinery construction and features					Chapter 9, pages 208–217
10.1.7(B)	RS: ability to identify common passenger vehicle or small machinery construction features					Chapter 9, pages 208–217
10.1.8	Disentangle victim(s)					Chapter 9, pages 217–232
10.1.8(A)	RK: tool selection and application					Chapter 9, pages 217–232
10.1.8(B)	RS: ability to operate disentanglement tools					Chapter 9, pages 217–232
10.1.9	Remove a packaged victim					Chapter 9, pages 236–237
10.1.9(A)	RK: patient handling techniques					Chapter 9, pages 236–237
10.1.9(B)	RS: use of immobilization, packaging, and transfer devices					Chapter 9, pages 236–237
10.1.10	Terminate a Level I vehicle/machinery incident					Chapter 13, pages 346–352
Level II						
10.2.1	Plan for a commercial heavy vehicle or large machinery incident					Chapter 11, pages 277–278, 311–315 and Chapter 12, pages 325–332
10.2.1(A)	RK: operational protocols					Chapter 11, pages 277–278, 311–315 and Chapter 12, pages 325–332
10.2.1(B)	RS: ability to apply operational protocols					Chapter 11, pages 277–278, 311–315 and Chapter 12, pages 325–332
10.2.2	Stabilize commercial/heavy vehicles and large machinery					Chapter 11, pages 278–282 and Chapter 12, pages 332–336
10.2.2(A)	RK: types of stabilization devices					Chapter 11, pages 278–282 and Chapter 12, pages 332–336
10.2.2(B)	RS: ability to apply and operate stabilization devices.					Chapter 11, pages 278–282 and Chapter 12, pages 332–336
10.2.3	Determine the heavy vehicle or large machinery access and egress points					Chapter 11, pages 282–300, 315–316 and Chapter 12, pages 325–339

Objective / JPR, RK, RS		Cognitive Written Test	Manipulative Skills Station	Portfolio	Projects	Vehicle Extrication, Levels I & II: Principles and Practice
Section	Abbreviated Text					
10.2.3(A)	RK: heavy vehicle and large machinery construction/features					Chapter 11, pages 282–300, 315–316 and Chapter 12, pages 325–339
10.2.3(B)	RS: ability to identify entry and exit points					Chapter 11, pages 282–300, 315–316 and Chapter 12, pages 325–339
10.2.4	Create access and egress openings					Chapter 11, pages 282–300, 315–316 and Chapter 12, pages 325–339
10.2.4(A)	RK: heavy vehicle and large machinery construction and features					Chapter 11, pages 282–300, 315–316 and Chapter 12, pages 325–339
10.2.4(B)	RS: ability to identify heavy vehicle and large machinery construction features					Chapter 11, pages 282–300, 315–316 and Chapter 12, pages 325–339
10.2.5	Disentangle victim(s), given a Level II extrication incident					Chapter 11, pages 282–300, 315–316 and Chapter 12, pages 336–339
10.2.5(A)	RK: tool selection and application					Chapter 11, pages 282–300, 315–316 and Chapter 12, pages 336–339
10.2.5(B)	RS: ability to operate disentanglement tools					Chapter 11, pages 282–300, 315–316 and Chapter 12, pages 336–339

NFPA 1006, *Standard for Technical Rescuer Professional Qualifications*, 2008 Edition Correlation

Chapter 4 Technical Rescuer

NFPA 1006, *Standard for Technical Rescuer Professional Qualifications*	Corresponding Textbook Chapter(s)	Corresponding Page(s)
4.1	1	Page 4
4.1.1	1	Page 4
4.1.2	1	Page 4
4.2	1	Page 4
4.3	1	Page 3
4.3.1	1	Page 3
4.3.2	1	Page 3

Chapter 5 Job Performance Requirements

NFPA 1006, *Standard for Technical Rescuer Professional Qualifications*	Corresponding Textbook Chapter(s)	Corresponding Page(s)
5.1	7	Pages 147–168
5.2	7	Pages 147–155
5.2.1	7	Pages 147–155
5.2.2	7	Pages 155–162
5.2.3	7	Pages 162–166
5.2.4	7	Pages 152–155
5.2.5	7	Pages 158–160
5.2.6	7	Pages 166–168
5.2.7	13	Pages 346–352
5.3	9	Pages 235–236
5.3.1	9	Pages 235–236
5.3.2	9	Pages 236–237
5.3.3	9	Pages 236–237
5.4	6	Pages 102–107
5.4.1	6	Pages 102–107
5.4.2	6	Pages 102–107
5.5	Appendix D	Available online at www.fire.jbpub.com
5.5.2	Appendix D	Available online at www.fire.jbpub.com
5.5.3	Appendix D	Available online at www.fire.jbpub.com
5.5.4	Appendix D	Available online at www.fire.jbpub.com
5.5.5	Appendix D	Available online at www.fire.jbpub.com
5.5.6	Appendix D	Available online at www.fire.jbpub.com
5.5.7	Appendix D	Available online at www.fire.jbpub.com
5.5.8	Appendix D	Available online at www.fire.jbpub.com
5.5.9	Appendix D	Available online at www.fire.jbpub.com
5.5.10	Appendix D	Available online at www.fire.jbpub.com
5.5.11	Appendix D	Available online at www.fire.jbpub.com
5.5.12	Appendix D	Available online at www.fire.jbpub.com
5.5.13	Appendix D	Available online at www.fire.jbpub.com
5.5.14	Appendix D	Available online at www.fire.jbpub.com

Chapter 10 Vehicle and Machinery Rescue

NFPA 1006, *Standard for Technical Rescuer Professional Qualifications*	Corresponding Textbook Chapter(s)	Corresponding Page(s)
10.1	1	Page 3
10.1.1	7	Pages 148–155
10.1.2	7	Pages 162–163
10.1.3	7	Pages 163–166
10.1.4	8	Pages 174–190
10.1.5	8	Pages 190–191
10.1.6	9	Pages 198–208
10.1.7	9	Pages 208–217
10.1.8	9	Pages 217–232
10.1.9	9	Pages 236–237
10.1.10	13	Pages 346–352
10.2	1	Page 3
10.2.1	11 and 12	Pages 277–278, 311–315; 325–332
10.2.2	11 and 12	Pages 278–282; 332–336
10.2.3	11 and 12	Pages 282–300, 315–316; 325–339
10.2.4	11 and 12	Pages 282–300, 315–316; 325–339
10.2.5	11 and 12	Pages 282–300, 315–316; 336–339

NFPA 1670, *Standard on Operations and Training for Technical Search and Rescue Incidents, 2009* Edition Correlation

Chapter 4 General Requirements

NFPA 1670, *Standard on Operations and Training for Technical Search and Rescue Incidents*	Corresponding Textbook Chapter(s)	Corresponding Page(s)
4.1	1	Page 4
4.1.1	1	Page 3
4.1.2	1	Page 3
4.1.3	1	Page 5
4.1.3.1	1	Page 5
4.1.3.2	1	Page 5
4.1.4	1	Pages 4–5
4.1.5	1	Page 5
4.1.6	1	Page 5
4.1.7	1	Pages 3–4
4.1.8	1	Pages 4–5
4.1.9	1	Page 7
4.1.10	2	Page 22
4.1.10.1	2	Page 22
4.1.10.1.1	2	Page 22
4.1.10.1.2	2	Page 22
4.1.10.2	2	Page 22
4.1.10.3	2	Page 22
4.1.10.4	2	Page 22
4.1.10.5	2	Pages 22–23
4.1.10.5.1	2	Page 22
4.1.10.5.2	2	Page 22
4.1.11	1	Page 4
4.1.12	2	Page 26
4.1.12.1	2	Page 26
4.1.12.2	2	Pages 23–25
4.1.13	2	Page 21
4.1.14	2	Page 26
4.2	2	Pages 21–22
4.2.1	2	Pages 21–22
4.2.2	2	Pages 21–22
4.2.3	2	Pages 21–22
4.2.4	2	Pages 21–22
4.2.5	2	Pages 21–22
4.2.6	2	Pages 21–22
4.2.7	2	Pages 21–22
4.2.8	2	Pages 21–22
4.3	2	Page 26
4.3.1	2	Page 26
4.3.1.1	2	Page 26
4.3.1.2	2	Page 26
4.3.2	2	Page 26
4.3.3	2	Page 26

NFPA 1670, *Standard on Operations and Training for Technical Search and Rescue Incidents*	Corresponding Textbook Chapter(s)	Corresponding Page(s)
4.3.4	2	Page 26
4.4	6	Pages 102–139
4.4.1	6	Pages 102–139
4.4.1.1	6	Pages 102–139
4.4.1.2	6	Page 119
4.4.1.3	6	Pages 102–139
4.4.2	6	Pages 102–107
4.4.2.1	6	Pages 102–107
4.4.2.2	6	Pages 102–107
4.4.2.3	6	Pages 102–107
4.4.2.4	6	Pages 106–107
4.4.2.4.1	6	Pages 106–107
4.4.2.4.2	6	Page 115
4.4.2.4.3	6	Page 115
4.4.2.4.4	6	Pages 106–107
4.5	2	Page 26
4.5.1	2	Page 26
4.5.1.1	2	Page 26
4.5.1.2	2	Page 26
4.5.1.3	2	Page 26
4.5.1.4	2	Page 26
4.5.1.5	2	Page 26
4.5.2	2	Pages 17–18
4.5.2.1	2	Pages 17–18
4.5.2.2	2	Pages 17–18
4.5.3	2	Page 14
4.5.3.1	2	Page 14
4.5.3.2	2	Pages 23–25
4.5.3.3	2	Page 23
4.5.3.4	2	Pages 16–19
4.5.3.5	2	Pages 4–5
4.5.4	2	Page 26
4.5.5	2	Pages 21–22
4.5.5.1	2	Pages 21–22
4.5.5.2	2	Pages 21–22

Chapter 8 Vehicle Search and Rescue

NFPA 1670, *Standard on Operations and Training for Technical Search and Rescue Incidents*	Corresponding Textbook Chapter(s)	Corresponding Page(s)
8.1	1	Page 4
8.2	1	Page 4
8.2.1	1	Page 4
8.2.2	1	Page 4
8.2.3	1	Page 4
8.3	1	Page 5
8.3.1	1	Page 5
8.3.2	1	Page 5
8.3.3	1	Page 5
8.3.4	1	Page 5
8.4	1	Page 5
8.4.1	1	Page 5
8.4.2	1	Page 5

Chapter 12 Machinery Search and Rescue

NFPA 1670, *Standard on Operations and Training for Technical Search and Rescue Incidents*	Corresponding Textbook Chapter(s)	Corresponding Page(s)
12.1	1	Page 4
12.2	1	Page 4
12.2.1	1	Page 4
12.2.2	1	Page 4
12.2.3	1	Page 4
12.3	1	Page 5
12.3.1	1	Page 5
12.3.2	1	Page 5
12.3.3	1	Page 5
12.3.4	1	Page 5
12.4	1	Page 5
12.4.1	1	Page 5
12.4.2	1	Page 5

Glossary

Accelerometer A sensor that detects a crash.

Active regeneration The second of three stages of regeneration in which fuel is injected into the system to burn and create higher temperatures of up to 1112°F (600°C).

Active restraint device A device that the occupant must activate; for example, a seat belt is an active device because the occupant has to engage the seat belt mechanism into the anchor unit.

Adapter A device used to convert a battery-operated tool to a general current tool.

Advanced High Strength Steel (AHSS) Steel with a *minimum* tensile strength of 73 ksi to 116 ksi (500 MPa to 800 MPa) or greater.

After action report (AAR) A brief summary that analyzes the overall operations and effectiveness of the agency at a particular incident, measuring its capabilities through real-time on-scene evaluations.

Aggregate weight A measurement combining all packages in a CMV to determine the total weight of all the hazards.

Air bag An inflatable plastic bag that inflates automatically to cushion passengers in the event of a collision.

Air brake (parking brake) A brake used on some models of CMVs that causes the suspension system to release air and drop several inches to accommodate a resting or parked position.

Air compressor A piece of equipment used to provide power to pneumatic tools or to provide breathing air.

Air impact wrench A pneumatic tool used to remove bolts/nuts of various sizes.

Air ride system A system designed to inflate and deflate the suspension through onboard air pressure tanks to protect the cargo and the frame system of the vehicle.

Air shoring Shoring extended by the use of compressed air; shoring is used where the vertical distances are too great to use cribbing or the load must be supported horizontally.

Alloyed steel Steel composed of a mixture of various metals and elements.

Alternative powered vehicle A vehicle that utilizes fuels other than petroleum or a combination of petroleum and another fuel for power.

A-post A vertical support member located closest to the front windshield of a vehicle.

Authority having jurisdiction (AHJ) An organization, office, or individual responsible for enforcing the requirements of a code or standard, or for approving equipment, materials, an installation, or a procedure.

Automatic seat belt system A seat belt system that uses a shoulder harness that automatically slides on a steel or aluminum track system on the door window frame. When the door is closed, the shoulder harness automatically slides into place. The lap section of the harness has to be manually engaged.

Awareness level The level that represents the minimum capability of organizations that provide response to technical search and rescue incidents.

Axle A structural component or shaft that is designed for wheel rotation.

Backboard slide technique An initial access technique that is used if the vehicle doors are locked, blocked, or inoperable. May be accomplished through a rear or side window.

Ballistic glass Glass that utilizes multiple layers of tempered glass, laminate material, and polycarbonate thermoplastics, all sandwiched together to the desired thickness. The weight and thickness of the glass will increase depending on each increased level of protection, which can be as high as 3 or more inches (76 or more mm).

Behavioral reactions Negative reactions that may present as withdrawal from others, emotional outbursts, extreme changes in normal behavior (such as silence or hyperactivity), repeated drunkenness, negative sexual reactions, insomnia, or absenteeism.

Beneficial systems Auxiliary-powered equipment in motor vehicles or machines that can enhance or facilitate rescues such as electric, pneumatic, or hydraulic seat positioners, door locks, window operating mechanisms, suspension systems, tilt steering wheels, convertible tops, or other devices or systems that facilitate the movement (extension, retraction, raising, lowering, conveyor control) of equipment or machinery.

Biodiesel A safe, nontoxic, biodegradable fuel used solely for diesel engines that is processed from domestic renewable resources such as plant oils, grease, animal fats, used cooking oil, and, more recently, algae. Bio-diesel can be utilized by itself as a diesel fuel or blended with petroleum diesel at varying percentages.

BLEVE Boiling liquid/expanding vapor explosion; an explosion that occurs when pressurized liquefied materials inside a closed vessel are exposed to high heat.

Body-over-frame construction Vehicle design where the body of the vehicle is placed onto a frame skeleton and the frame acts as the foundation for the vehicle. The design consists of two large beams tied together by cross member beams.

Boiling liquid, expanding vapor explosion (BLEVE) An explosion that occurs when a tank containing a volatile liquid is heated.

Bow trusses Structural steel members that run continuously from below the floor level on one side of a bus, vertically raising and bowing over to form the roof structure and then extending over and down the other side of the bus past the floor level.

B-post A vertical support member located between the front and rear doors of a vehicle.

Branches A segment within the ICS that may be functional or geographic in nature. Branches are established when the number of divisions or groups exceeds the recommended span of control for the operations section chief or for geographic reasons.

Bumper system A feature located at the front and rear of a vehicle that helps a vehicle withstand the impact of a collision.

Cab The enclosed space where the driver and passengers sit.

Cab-beside-engine Design in which the driver sits next to the engine. These trucks are mostly found in shipyards and baggage carries within airports and seaports.

Cab-over-engine (COE) Often referred to as a tilt-cab, design in which the cab is lifted over the engine to gain access to the engine itself. The driver seat is positioned over the engine and the front axle.

Cargo tank Bulk packaging that is permanently attached to or forms a part of a motor vehicle, or that is not permanently attached to any motor vehicle, and that, because of its size, construction, or attachment to a motor vehicle, is loaded or unloaded without being removed from the motor vehicle.

Center of gravity The location where all of the weight of an object is concentrated and where the load is being forced downward by Earth's gravitational pull.

Chainsaw A gasoline-powered saw capable of cutting wood, concrete, and even light-gauge steel. Standard steel chains are used to cut wood, carbide-tipped chains can cut wood and light-gauge metal, and diamond chains are used for cutting concrete.

Chain sling A sling composed of a single chain or multiple chains in various lengths, commonly in lengths of 5 and 9 feet (1.5 and 2.7 m); it is attached to a ring, either round or oblong, and used for lifting heavy loads.

Charge-depleting hybrid system A plug-in hybrid electric vehicle (PHEV); one of two types of parallel hybrid systems used for types C and D school buses.

Charge-sustaining hybrid system A standard hybrid that operates on regenerative properties to charge its batteries; one of two types of parallel hybrid systems used for types C and D school buses.

Charter/tour bus A company providing transportation on a for-hire basis, usually round-trip service for a tour group or outing. The transportation can be for a specific event or can be part of a regular tour.

Chassis The frame, braking, steering, and suspension system of a vehicle.

Circular saw An electric- or battery-powered saw that moves in a circular motion; these saws come in a variety of sizes and are used primarily for cutting wood, although special blades are available that will cut metal or masonry.

Class 1 commercial vehicle A vehicle with a gross vehicle weight rating ranging from 0 to 6000 lb (0 to 2722 kg).

Class 2a commercial vehicle A vehicle with a gross vehicle weight rating ranging from 6001 to 8500 lb (2722 to 3856 kg); a light-duty vehicle.

Class 2b commercial vehicle A vehicle with a gross vehicle weight rating ranging from 8501 to 10,000 lb (3856 to 4536 kg); considered by many as a light-duty vehicle.

Class 3 commercial vehicle A vehicle with a gross vehicle weight rating ranging from 10,001 to 14,000 lb (4536 to 6350 kg).

Class 4 commercial vehicle A vehicle with a gross vehicle weight rating ranging from 14,001 to 16,000 lb (6351 to 7257 kg).

Class 5 commercial vehicle A vehicle with a gross vehicle weight rating ranging from 16,001 to 19,500 lb (7258 to 8845 kg).

Class 6 commercial vehicle A vehicle with a gross vehicle weight rating ranging from 19,501 to 26,000 lb (8846 to 11,793 kg).

Class 7 commercial vehicle A vehicle with a gross vehicle weight rating ranging from 26,001 to 33,000 lb (11,794 to 14,969 kg).

Class 8 commercial vehicle A vehicle with a gross vehicle weight rating of more than 33,000 lb (more than 14,969 kg).

Class B foam Foam used to extinguish flammable and combustible liquid (Class B) fires.

Cognitive reactions Negative reactions that may present as attention deficit disorder, nightmares, confusion, lack of concentration, decreased ability to problem solve, or constant reliving of the event through flashbacks.

Cold zone A safe area at an incident for those agencies involved in the operations. The incident commander, command post, EMS providers, and other support functions necessary to control the incident should be located in the cold zone.

Come along A ratchet lever winching tool that can provide up to several thousand pounds of pulling force, with the standard model for extrication being 2000 to 4000 lb (907 to 1814 kg) of pulling force.

Comfort cab A tractor cab that has no roll-over protection. Its purpose is just to provide environmental protection to the operator.

Commercial motor vehicle (CMV) Defined by the DOT as a motor vehicle or combination of motor vehicles used in commerce to transport passengers or property if the motor vehicle has a gross vehicle weight rating (GVWR) of 26,001 lb or more (11,794 kg or more) inclusive of a towed unit(s) with a GVWR of more than 10,000 lb (4,536 kg); or has a GVWR of 26,001 lb or more (11,794 kg or more); or is designed to transport 16 or more passengers, including the driver; or is of any size and is used in the transportation of hazardous materials.

Commercial vehicle A type of vehicle that may be used for transporting passengers or goods.

Common passenger vehicle A light- or medium-duty passenger and commercial vehicle commonly encountered in the jurisdiction and presenting no unusual construction, occupancy, or operational characteristics to rescuers during an extrication event.

Compartment syndrome A medical term that refers to the compression of nerves, blood vessels, and muscle inside a closed space (compartment) within the body. This leads to tissue death due to lack of oxygenation as the blood vessels are compressed by the raised pressure within the compartment.

Compressed natural gas (CNG) A fuel utilized primarily in fleet vehicles; natural gas is compressed for high-pressure storage and distribution systems.

Contact point When sections of cribbing are set on top of one another, the weight-bearing section of cribbing that crosses over the other. When using a 4- by 4-inch (102- by 102-mm) piece of timber, each contact point has an estimated weight-bearing capacity of 6000 pounds (3 short tons).

Conventional cab Design in which the driver's seat is positioned behind the engine and front axle. The front end of the vehicle extends approximately 6 to 8 feet (1.8 to 2.4 m) from the front windshield.

Conventional-type vehicle A vehicle that utilizes an internal combustion engine (ICE) for power.

Cowling A removable metal cover for an engine.

C-post A vertical support member located behind the rear doors of a vehicle.

Crash rail A rail designed to protect students from impact intrusions into the passenger compartment of a school bus; normally composed of 14-gauge steel extending just above the floor area between the floor and the seat rub rail, extending the entire length of the school bus.

Crew A group of personnel working without apparatus and led by a leader or boss.

Cribbing The most common stabilization tool that gives the user several height options when trying to rapidly stabilize a vehicle.

Critical incident stress debriefing (CISD) A structured and confidential postincident meeting designed to assist rescue personnel in dealing with psychological trauma as the result of an emergency or critical incident.

Critical incident stress management (CISM) A multifaceted system of crisis intervention specifically designed to help emergency personnel who have been exposed to a traumatic event process their response to the incident in a way that validates the normal stress reactions and stabilizes the potential negative results of the individual's response.

Cross ram technique The use of a hydraulic ram to push off of the opposite door post, B-post, floor transmission hump, or inside rocker panel to move the interior of the vehicle away from the entrapped occupant.

Crumple zones Engineered collapsible zones that are incorporated into the frame of a vehicle to absorb energy during a collision.

Crush injury syndrome A condition that results when toxins (by-products of crushed muscle) are entered into circulation after a crushing load is released from a person. This is characterized by major shock and renal failure. It can be a common injury with tractor overturns and other farm-related emergencies because of the length of time of entrapment.

Crush point A hazard created when two objects move toward each other or when one object moves toward a stationary object. Examples include hitches.

Cutting torch A tool that produces an extremely high-temperature flame capable of heating steel until it melts, burns, and oxidizes, cutting through the object. This tool is sometimes used for rescue situations such as cutting through heavy steel objects.

Dash bar A steel beam or bar that runs partway or the entire width of the dash.

Dash brackets Two brackets that are bolted or welded into the floorboard of the vehicle that are designed to lock the dash in place in order to minimize any movement resulting from an impact.

Dash lift technique A technique used to lift and release a section of the dash from the front end of the vehicle using the hydraulic spreader. It is performed by making precise relief cuts in the hood's upper rail and between the hinges of the firewall area, separating the dash section from the front end of the vehicle.

Dash roll technique The standard technique for many years for displacing the dashboard. It is one of the least technical applications to apply in the field. The process involves using hydraulic rams to push or roll the entire front end of the vehicle upward and forward, including the dashboard, steering wheel, and steering column, off of the entrapped occupant.

Dead axle An axle used for load support more commonly set in the front section of a semi-truck; also functions for steering and is therefore also known as a steer axle.

Dead-man control A control feature designed to return the control of the hydraulic tool to the neutral position automatically in the event the control is released.

Defensive apparatus placement The positioning of apparatus to block and protect the scene from the flow of traffic.

Defusing A less formal confidential postincident meeting that allows individuals to express or release their emotions and start the process of proper coping; it should be conducted within 1 to 4 hours after the critical event.

Department of Transportation's *Emergency Response Guidebook* (ERG) A preliminary action guide for responders operating near hazardous materials.

Dicing A term used to describe the small pieces of glass that are produced when tempered glass is broken.

Disentanglement The spreading, cutting, or removal of a vehicle and/or machinery away from trapped or injured victims.

Distress Stress that generally produces a negative response, such as that experienced through the exposure to a critical incident.

Divisions A segment within the ICS established to divide an incident into physical or geographic areas of operation.

Door hinge A mechanism that provides the opening and closing movements for a door. Door hinges commonly range from 8- to 15-gauge metal and can be layered in a leaf system or full body.

D-ring A generic term used to describe the window frame of the door.

Drive train (power train) A system that transfers rotational power from the engine to the wheels, which makes the vehicle move.

Dromedary (drom) A separate box, deck, or plate, mounted behind the cab and in front of the fifth wheel on the frame of a semi-truck.

Dry bulk cargo tank A tank designed to carry dry bulk goods such as powders, pellets, fertilizers, or grain. Such tanks are generally V-shaped with rounded sides that funnel toward the bottom of the tank.

Electric vehicle (EV) A vehicle that is 100 percent electric, emits no air pollutants, and is propelled by one or more electric motors, which are powered by rechargeable battery packs.

Electrical generators Generators that utilize a general current or generator to operate. Primarily used to power scene lighting and to run power tools and equipment; may be portable or fixed.

Electric-powered tools Tools that utilize a general current or generator to operate.

Electronic control unit (ECU) Also known as the air bag control unit (ACU), this is the brains of an air bag system, consisting of a small processing unit generally located in the center of the vehicle.

Emergency air line (supply line) One of two air lines of an air brake system connecting a tractor to a trailer; the emergency air line is red.

Emergency brake A brake that utilizes a combination of both the service and parking brake to engage when a brake failure or air line break occurs.

Emergency response guides Booklets prepared by vehicle manufacturers to educate and assist emergency response personnel in responding to emergencies dealing with specific types and models of vehicles such as hybrid/electric, hydrogen fuel cell, and alternative fuel systems.

Emotional reactions Negative reactions that may present as depression, guilt, anger, fear, anxiety, feeling of doom, grief, or perceived loss of control.

Energy A fundamental entity of nature that is transferred between parts of a system in the production of physical change within the system and is usually regarded as the capacity for doing work.

Enhanced protective glass (EPG) A modern glass that uses both the laminating and tempering process.

Ethanol A fuel comprised of an alcohol base that is normally processed from crops such as corn, sugar, trees, or grasses.

Eustress Stress that produces a positive response in the mind, body, and spirit, such as that experienced through physical exercise or a team sport. Eustress actually builds resistance to the negative aspects of stress.

Exhaust after-treatment device A device that replaced the standard muffler assembly, capturing and converting soot to carbon dioxide and water through the combination of a diesel particulate filter (DPF) and a diesel oxidation catalyst (DOC). This conversion process is called regeneration, and there are three stages or processes of regeneration.

Expose and cut The process of creating a wide enough opening with the hydraulic spreader to expose the locking/latching mechanism or hinges and to insert a hydraulic cutter to cut.

Extended range electric vehicle (EREV) A vehicle that utilizes a series-type propulsion system that allows the vehicle to run on all battery or all electric power until it is near depletion, which occurs in the range of 40 miles or more depending on the manufacturer.

Extrication The process of removing a trapped victim from a vehicle or machinery.

Farm implement A machine or attachment that is normally mounted to or pulled and powered by a tractor.

Federal Motor Vehicle Safety Standards (FMVSSs) Safety standards enacted to protect the public from unreasonable risk of crashes, injury, or death resulting from the design, construction, or performance of a motor vehicle.

Ferrous metals Metals that contain iron, cast iron, low- and medium-alloyed steels, and specialty steels, such as tooled steels and stainless steels.

Fifth wheel A turntable hitch mounted at the rear of the towing truck or semi-truck.

Finance/administration section ICS function responsible for the accounting and financial aspects of an incident, as well as any legal issues that may arise.

Flat form rescue-lift air bags Pneumatic-filled bladders designed to retain their flat profile in the center as they are inflated to lift an object or spread one or more objects away from each other to assist in freeing a victim.

Flexible fuel vehicle (FFV) A vehicle capable of running on gasoline alone or utilizing the E85 blend of up to 85 percent ethanol and 15 percent gasoline.

Footprint A generic term used to describe an object's balance in relation to its center of gravity, as determined by how much of the object's base touches the surface and how much of the object spans the surface.

Fuel cell An electrochemical device that utilizes a catalyst-facilitated chemical reaction of hydrogen and oxygen to create electricity, which is then used to power an electric motor or generator, with the by-products of this process being water and heat.

Full hybrid A vehicle that uses either its electric motor or its internal combustion engine, or a combination of both, to propel itself.

Gas generation system An inflation system that completely fills the air bag to the appropriate inflation ratio to protect the occupant.

Glad hands Couplers that are used to supply air to the braking system of the trailer.

Golden Period The time during which treatment of shock and traumatic injuries is most critical and the potential for survival is best accomplished through rapid medical intervention.

Grab hook A device designed to take up the slack needed to make a chain the appropriate size for the task at hand; it is utilized by inserting a link of the chain into the slot of the hook. The grab hook may also be referred to as a chain shortener.

Gravitational acceleration sensor (G-sensor) A sensor that detects a vehicle's weightlessness, such as that experienced in a freefall, when the vehicle starts to roll and come down.

Gross combined weight rating (GCWR) A rating set by the vehicle manufacturer that determines the vehicle's cargo capacity limitations.

Gross vehicle weight rating (GVWR) A rating set by the manufacturer; it shall not be less than the sum of the unloaded vehicle weight, rated cargo load, and 150 lb times the vehicle's designated seating capacity (49 CFR 567.4(g)(3)).

Gross weight The weight of the single item package plus its contents.

Group A segment within the ICS established to divide an incident into functional areas of operation.

Hand tool Any tool or equipment operating from human power.

Hazard analysis The process of identifying situations or conditions that have the potential to cause injury to people, damage to property, or damage to the environment.

Hazardous material A substance or material including an explosive; radioactive material; infectious substance; flammable or combustible liquid, solid, or gas; toxic, oxidizing, or corrosive material; and compressed gas that the Secretary of Transportation has determined is capable of posing an unreasonable risk to health, safety, and property when transported in commerce, and has been designated as hazardous under Section 5103.

Heavy vehicle A heavy-duty highway, off-road, construction, or mass transit vehicle constructed of materials presenting resistance to common extrication procedures, tactics, and resources and posing multiple concurrent hazards to rescuers from occupancy, cargo, size, construction, weight, or position.

High Strength Steel (HSS) Steel with a tensile strength between 39 ksi and 102 ksi (270 MPa and 700 MPa).

High-pressure rescue-lift air bags The most commonly used bags among rescue agencies, these bags utilize a working air pressure of approximately 100 to 145 psi (689 to 1000 kPa) to lift an object or spread one or more objects away from each other to assist in freeing a victim. The high-pressure kits come with hoses, a regulator, a master control module, and various other attachments.

Hinge A mechanism that allows movable objects such as doors to join and swing open or closed.

Horizontal movement One of five directional movements; the vehicle moves forward or rearward on its longitudinal axis or moves horizontally along its lateral axis.

Hot zone The area, also known as an action zone, accessible to entry teams and rescue teams only. It immediately surrounds the dangers of the incident, and entry into this zone is restricted to protect personnel outside the zone.

Hybrid electric vehicle (HEV) A vehicle that combines two or more power sources for propulsion, one of which is electric power.

Hybrid-electric commercial motor vehicle (HECMV) A CMV that utilizes either the internal combustion engine or an electric motor for propulsion.

Hydraulic combination tool A powered rescue tool capable of both spreading and cutting.

Hydraulic cutter A powered rescue tool consisting of at least one movable blade used to cut, shear, or sever material.

Hydraulic pressure Fluid under pressure that is used to power equipment.

Hydraulic ram A powered rescue tool with a piston or other type of extender that generates extending forces or both extending and retracting forces.

Hydraulic spreader A powered rescue tool consisting of at least one movable arm that opens to move or spread apart material, or to crush or lift material.

Hydraulic tools Tools that operate by transferring energy or force from one area to another by using an incompressible fluid such as high-density oil.

Hydrogen An odorless, colorless, flammable, nontoxic gas that combines easily with other elements.

Immediate danger to life and health (IDLH) Any condition that would do one or more of the following: pose an immediate or delayed threat to life, cause irreversible adverse health effects, or interfere with an individual's ability to escape unaided from a hazardous environment.

Impact beam A steel section located within a door frame designed to absorb the impact energy of another vehicle or object and lessen the intrusion into the passenger compartment.

Incident action plan (IAP) An oral or written plan containing general objectives reflecting the overall strategy for managing an incident. It may include the identification of operational resources and assignments. It may also include attachments that provide direction and important information for management of the incident during one or more operational periods.

Incident clock A procedure where dispatch will automatically notify the incident commander at 10-minute intervals until the incident becomes static.

Incident command system (ICS) A management structure that provides a standard approach and structure to managing operations, ensuring that operations are coordinated, safe, and effective, especially when multiple agencies are working together.

Incident commander (IC) The individual with overall responsibility for the management of all incident operations.

Inclinometer sensor A tilt sensor that detects vehicle inclination or tilt with lateral acceleration (detects how fast the vehicle's tilt is changing).

Inflator One of the most critical design features for an air bag, providing the ability to fill up the bag instantaneously in milliseconds from the onset of the collision. There are two basic inflation systems—a stored compressed gas system and a gas generation system.

Initiator A device such as a squib (a pyrotechnic device) that activates the air bag through an electrical current, which becomes instantly hot and ignites the combustible material inside the containment housing, or through ignition of a burst disc, which releases compressed gas.

Inner survey A four-point inspection of the vehicle's front, driver's side, rear, and passenger's side, including the undercarriage on all sides of the vehicle. This survey is conducted approximately 3 to 5 feet (0.9 to 1.5 m) from the vehicle and is performed by the first arriving company officer or experienced personnel.

Integrated motor assist (IMA) mild hybrid A hybrid system used by Honda that is designed to start and stop the hybrid electric vehicle's internal combustion engine; in addition, it will assist the internal combustion engine when acceleration is needed.

Intercity bus A company providing for-hire, long-distance passenger transportation between cities over fixed routes with regular schedules.

Internal combustion engine An engine designed to burn a multitude of petroleum-based fuels and alternative fuels, with gasoline and diesel fuel being the most commonly used.

Junction boxes Electrical enclosures commonly found when multiple outlets are required.

Kinetic energy The energy of motion, which is based on vehicle mass (weight) and the speed of travel (velocity).

Kingpin A large locking pin that connects the cargo trailer of a tractor-trailer or semi-trailer to the semi-truck or tractor.

Ladder crib Several 2- by 4-inch (51 by 102 mm) sections of wood attached together by a strip of webbing running along the sides.

Ladder frame Body-over-frame construction that is referred to as a ladder frame because the cross members and beams resemble a ladder.

Laminated safety glass Glass that contains a layer of clear plastic film between two layers of glass.

Landing gear A stabilizing device that can be lowered to support a trailer when not attached to the tractor. This device is usually operated manually by a hand crank.

Law of Conservation of Energy A law of physics stating that energy can neither be created nor destroyed; it can only change from one form to another.

Law of Motion A law of physics describing momentum, acceleration, and action/reaction.

Level I Technical Rescuer The level that applies to individuals who identify hazards, use equipment, and apply limited techniques specified in NFPA 1006 to perform technical rescue operations. Level I rescue skills for vehicle and machinery rescue apply to those incidents that involve *common passenger vehicles*, *simple small machinery*, and environments where rescuer intervention does not constitute a high level of risk based upon the environment or other factors.

Level II Technical Rescuer The level that applies to individuals who identify hazards, use equipment, and apply advanced techniques specified in NFPA 1006 to perform technical rescue operations. Level II rescue skills for vehicle and machinery rescues apply to those incidents where *commercial* or *heavy vehicles* are involved, complex extrication processes are applied, multiple uncommon concurrent hazards are present, or heavy machinery or more than digital entrapment of a victim is involved.

Liaison officer (LO) The incident commander's point of contact for outside agencies. This officer coordinates information and resources among cooperating and assisting agencies and establishes contacts with agencies that may be capable or available to provide support.

Lift axle An axle that can be raised or lowered by an air suspension system to increase the weight-carrying capacity of the vehicle, or to distribute the cargo weight more evenly across all of the axles.

Liquefied natural gas (LNG) A colorless, odorless, nontoxic natural gas that floats on water and is lighter than air when released as a vapor.

Liquid petroleum gas (LPG) Also known as propane, a fossil fuel produced from the processing of natural gas and also produced as part of the refining process of crude oil. Propane is the third most widely utilized fuel source behind gasoline and diesel; it is commonly used with forklifts and other similar work units.

Live axle An axle that transmits propulsion or causes the wheels to turn.

Lockout/tagout systems Methods of ensuring that systems and equipment have been shut down and that switches and valves are locked and cannot be turned on at the incident scene.

Logistics section ICS function responsible for all support requirements needed to facilitate effective and efficient incident management, including providing supplies, services, facilities, and materials during the incident.

Low-pressure rescue-lift air bags Air bags with a very high lift with a maximum working air pressure of approximately 7 psi (48 kPa); they are used to lift an object or spread one or more objects away from each other to assist in freeing a victim.

Manual regeneration The third of three stages of regeneration that occurs only when the parking brake is set and the engine is running.

Marrying (vehicles) The process of joining vehicles together to eliminate any independent movement.

Mass-casualty incident (MCI) An emergency situation that involves more than one victim that places great demand on equipment or personnel, stretching the system to its limit or beyond.

MC-306/DOT 406 flammable liquid tanker A tanker that typically carries between 6000 gallons and 10,000 gallons (between 22,712 and 37,854 liters) of a product such as gasoline or other flammable and combustible materials. The tank is nonpressurized.

MC-307/DOT 407 chemical hauler A tanker with a rounded or horseshoe-shaped tank capable of holding 6000 to 7000 gallons (22,712 to 37,854 liters) of flammable liquid, mild corrosives, and poisons. The tank has a high internal working pressure.

MC-312/DOT 412 corrosives tanker A tanker that often carries aggressive (highly reactive) acids such as concentrated sulfuric and nitric acid. It is characterized by several heavy-duty reinforcing rings around the tank and holds approximately 6000 gallons (22,712 liters) of product.

MC-331 pressure cargo tanker A tanker that carries materials such as ammonia, propane, Freon, and butane. This type of tank is commonly constructed of steel and has rounded ends and a single open compartment inside. The liquid volume inside the tank varies, ranging from the 1000-gallon (3785-liter) delivery truck to the full-size 11,000-gallon (41,640-liter) cargo tank.

MC-338 cryogenic tanker A low-pressure tanker designed to maintain the low temperature required by the cryogens it carries. A boxlike structure containing the tank control valves is typically attached to the rear of the tanker.

Mechanism of injury (MOI) The way in which traumatic injuries occur; it describes the forces (or energy transmission) acting on the body that cause injury.

Medium-pressure rescue-lift air bags Air bags that have a rugged design and utilize a working air pressure of approximately 15 psi (103 kPa) used to lift an object or spread one or more objects away from each other to assist in freeing a victim. These are not as common as the low- and high-pressure rescue-lift air bags.

Methanol An alcohol-based fuel similar to ethanol. It is also known as a wood alcohol because it is processed from natural wood sources such as trees and yard clippings. It may be utilized as a flex fuel in a ratio of 85 percent methanol to 15 percent gasoline, better known as M85.

Mild hybrid A vehicle that uses electric power in conjunction with the internal combustion engine for vehicle propulsion.

Multistage inflators Also known as hybrid inflators, cylinders that can be comprised of two separate chambers of compressed gas—one with a large amount of product and the other with a smaller amount of product.

Nader bolt Named after consumer rights advocate Ralph Nader, a bolt composed of heavy-gauge metal that is round in shape with a cap at the end of it. It is a section of the latching mechanism.

Natural gas A fossil fuel primarily composed of methane that can be used as a compressed natural gas (CNG) or liquefied natural gas (LNG).

Neighborhood electric vehicle (NEV) A vehicle that is classified as a battery-operated low-speed vehicle with a top speed of 25 mph (40 kph) and that is approved for street use on public roadways with speeds posted of no greater than 35 mph (56 kph).

New technology (NT) high-pressure rescue-lift air bags Air bags that utilize a unique lifting system where each round-shaped bag can be locked together with a threaded connec-

tor, creating one bag with multiple cells that offers a distinct height advantage over traditional flat bags.

NFPA 1006 The standard that establishes the minimum requirements/qualifications necessary for fire service and other emergency response personnel that perform technical rescue. This standard establishes two skill levels (Level I and II Technical Rescuer) that describe job performance requirements.

NFPA 1670 The standard that identifies and qualifies levels of functional capabilities for safely and effectively conducting operations at technical rescue incidents. This standard outlines three skill levels—awareness, operations, and technician.

Nonferrous metals Metals or alloys free of iron, such as aluminum, copper, nickel, lead, zinc, and tin.

"O" ring or oblong ring An attachment designed to join chains together or join a chain to a come along utilizing a hook.

Occupant classification system A system consisting of three different types of sensors: the seat position sensor, which detects the proximity of the occupant to the air bag; the seat belt sensor, which detects if the occupant's seat belt is engaged and locked in the housing unit; and the occupant weight sensor, which measures the weight of the occupant, determining whether the occupant has met a preset weight threshold limit.

Operations level The level that represents the capability of organizations to respond to technical search and rescue incidents and to identify hazards, use equipment, and apply limited techniques specified in this standard to support and participate in technical search and rescue incidents.

Operations section ICS position responsible for development, direction, and coordination of all tactical operations conducted in accordance with an IAP.

Organizational analysis A process to determine if it is possible for an organization to establish and maintain a given capability.

Outer survey A survey conducted simultaneously with the inner survey; the rescuer performing the outer survey moves in the opposite direction as the rescuer performing the inner survey. Distance from the vehicle will vary with each incident, but it is generally a distance of 25 to 50 feet (7.6 to 15.2 m) starting from the perimeter of the inner survey position outward.

Parallel drive system A system that can use either the vehicle's internal combustion engine or the electric motor to power the vehicle's transmission and provide propulsion.

Passenger vehicle All sedans, coupes, and station wagons manufactured primarily for the purpose of carrying passengers, including those passenger cars pulling recreational or other light trailers.

Passive regeneration The first of three stages of regeneration that occurs automatically when the particulate matter (soot) that is caught in the diesel particulate filter is burned off naturally by the elevated temperatures of the exhaust system.

Passive restraint device A device that the occupant does not have to activate for it to function; the system is automatically activated when power is applied to the vehicle.

Physical reactions Negative reactions that may present as headaches, muscle twitching/tremors, dry mouth, elevated blood pressure and/or heart rate, nausea, hyperpnoea, profuse sweating, or chest pains.

Picket anchor system A method that provides a structurally significant anchor point using steel rods and rigging system components when no structural or other solid points are available.

Pinch point A hazard found where two rotating objects move together and at least one of them moves in a circle. Common examples include the points where drive belts contact pulley wheels, drive chains meet gear sprockets, feed rolls mesh, or gathering chains on harvesting machines draw crops into the equipment.

Pitch movement One of five directional movements; the vehicle moves up and down about its lateral axis, causing the vehicle's front and rear portions to move left or right in relation to their original position.

Pivot point A mechanism that allows the tractor to traverse uneven ground while preventing the front end of the tractor from lifting off the ground.

Planning section ICS function responsible for the collection, evaluation, dissemination, and use of information and intelligence critical to the incident.

Plug-in hybrid electric vehicle (PHEV) A hybrid vehicle that can recharge its battery system using a plug-in cord that can run off general house current in the range of 120 volts, also known as a Level 1 charging system.

Pneumatic chisel A pneumatic tool used to cut through various types and sizes of metal.

Pneumatic cut-off tool A pneumatic tool utilizing a small carbide disc, normally 3 inches (76 mm) in diameter, which rotates at a high rpm to cut through most metals.

Pneumatic tools Tools that use air under pressure to operate.

Polycarbonate A clear plastic material that is very strong and can endure impacts without breaking.

Polymer exchange membrane Also known as a proton exchange membrane, a thin membrane used in a fuel cell system that is placed between the anode and cathode and through which positive electrons are passed.

Postincident analysis (PIA) A review of the incident and a way to constructively examine the positive and negative outcomes of the incident; includes assessing opportunities for improvement and addressing any necessary corrective actions that may be needed to improve the organization as a whole.

Posttraumatic stress disorder (PTSD) A delayed stress reaction to a prior incident. This delayed reaction is often the result of one or more unresolved issues concerning the incident.

Potential energy Stored energy or the energy of position.

Pounds per square inch (psi) A unit of measure used to describe pressure; it is the amount of force that is exerted on an area equaling 1 square inch.

Power take-off (PTO) shaft A mechanical device used to transfer mechanical energy to and throughout machinery.

Pressure release device (PRD) A safety feature built into high-pressure storage cylinders that is designed to rapidly release gas contents when exposed to high temperatures, such as during a fire.

Pretensioner seat belt system A seat belt system designed to pull back and tighten when activated by a collision. The most common pretensioner seat belt system uses a pyrotechnic propulsion device to engage a gear that pulls back on the belt.

Primary access The existing openings of doors and/or windows that provide a pathway to the trapped and/or injured victim.

Primary and secondary vehicle access points The main point of entry and alternate point of entry that provides a pathway to the trapped and/or injured victim(s).

Protective ensemble Personal protective gear, including the helmet, primary eye protection, coat, pants, coveralls, gloves, footwear, and hearing protection.

Public information officer (PIO) ICS position that interfaces with the media and provides a single point of contact for information related to an incident.

Pull-in point A hazard posed by those places where an individual may be pulled into the moving parts of a machine. Commonly occurs when someone tries to remove plant material or other obstacles that have become stuck.

Purchase point The location where access can best be gained.

Ratchet strap A mechanical tensioning device with a manual gear-ratcheting drum to put tension on an object utilizing a webbing material.

Reciprocating saw A power-driven saw in which the cutting action occurs through a back-and-forth motion (reciprocating) of the blade.

Rescue-lift air bags Inflatable devices used to lift an object or spread one or more objects away from each other to assist in freeing a victim. They come in various sizes and types, such as low-pressure bags, medium-pressure bags, high-pressure bags, high-pressure flat bags, and NT (new technology) locking bag.

Response planning (preincident planning) The process of compiling, documenting, and dispersing information that will assist the organization should an incident occur at a particular location.

Risk–benefit analysis An assessment of the risk to the rescuers versus the benefits that can be derived from their intended actions.

Rocker panel A hollow section of metal running along the outer sections of the floorboard on the driver and passenger sides.

Roll movement One of five directional movements; the vehicle rocks side to side while rotating about on its longitudinal axis and remaining horizontal in orientation.

Roll-over protection system (ROPS) (vehicles) A system designed to protect occupants in vehicle roll-over incidents by means of a deployable roll bar.

Roll-over protective structure (tractor or machinery) A structure that is mounted to a tractor frame that creates a safety zone for the operator in the event of a rollover.

Roof posts Posts designed to add vertical support to the roof structure of the vehicle. These are generally labeled with an alphanumeric type description (A-B-C), starting with the first post closest to the front windshield, which is known as the A-post. Also referred to as roof pillars.

Rotary saw A fuel-powered saw capable of cutting wood, concrete, and metal; two types of blades are used on rotary saws: a round metal blade with teeth and an abrasive disc. The application of rotary saws in vehicle extrication is limited.

Rub rails Exterior steel attachments comprised of 16-gauge corrugated metal 4 inches (102 mm) or more in width, attached to the bow trusses, running the entire length of the school bus, and wrapping around to the rear of the vehicle.

Safety officer (SO) ICS position responsible for enforcing general safety rules and developing measures for ensuring personnel safety.

Safing sensor A type of air bag sensor that has a deceleration setting lower than the crash-type sensor. The sensor prevents false deployments.

Scene safety zone/operational zone Zones that are divided into hot, warm, and cold zones. These zones are strictly enforced by a designated incident safety officer.

Scene size-up The systematic and continual evaluation of information presented in either visual or audible form.

School bus Any public or private school or district, or contracted carrier operating on behalf of the entity, providing transportation for kindergarteners through grade 12 pupils.

Seat belt pretensioning system A system designed to automatically tighten or take up slack in a seat belt when a crash is detected.

Secondary access Openings created by rescuers that provide a pathway to trapped and/or injured victims.

Self-contained breathing apparatus (SCBA) A respirator with an independent air supply that allows rescuers to enter dangerous atmospheres.

Semi-tractor trailer (semi-trailer) A semi-truck or tractor combined with a trailer; normally designed with three axles, one in the front for steering purposes and two tandem axels in the rear.

Semi-truck A commercial truck capable of towing a separate trailer, which has wheels only at one end (semi-trailer); may also be referred to as a tractor.

Series drive system A system that uses the internal combustion engine alone to run an onboard generator, which in turn can either run the electric motor that turns the vehicle's transmission (providing propulsion) or be used to charge the batteries or store power in a capacitor. The internal combustion engine does not provide direct propulsion to the vehicle.

Series-operated propulsion system A system that utilizes the electric motor by itself to propel a school bus, and the combustion engine is only used to regenerate the battery pack. These systems may be found on a type A school bus.

Service air line One of two air lines of an air brake system connecting a tractor to a trailer; the emergency air line is blue.

Service brake The usual driving brake that is applied by the driver to slow and/or stop the vehicle during normal driving operations.

Shear point A hazard created when the edges of two objects move toward or near each other closely enough to cut. An example may be a harvester.

Shims Objects that are smaller than wedges used to snug loose cribbing under a load or to fill void spaces.

Shoring A stabilization technique used where the vertical distances are too great to use cribbing or the load must be supported horizontally, such as in a trench, or diagonally, such as in a wall shore.

Side-impact air bag An air bag designed to activate immediately upon impact to protect the following areas of the occupant: the head, the chest/upper torso, and a combination of the head and the chest/upper torso. There are three types of side-impact air bags. All three types can be found in the door, seat backs, roof posts, or roof rails. These air bags may be labeled HPS (head protection system), IC (inflatable curtain), SIPS (side-impact protection system), or ROI (roll-over inflator air bag).

Side-out technique A technique used to gain access to a four-door vehicle involved in a side-impact collision.

Single resource An individual vehicle and its assigned personnel.

Sleeper A compartment attached to the cab of a truck that allows the driver to rest while making stops during a long transport.

Slide crib Two sections of 4- by 4-inch cribbing positioned parallel to each other with a third section of cribbing on top traversing the two bottom sections.

Slide hook A hook that allows chain links to pass freely through the throat of the hook to tighten around an object.

Small machinery Equipment or machinery that can be disassembled simply or that is constructed of lightweight materials, presenting simple hazards, which the rescuer(s) can control.

Smart air bag system An air bag system that will automatically adjust the pressure in the air bag by utilizing multistage inflators and basing the deployment force on a number of calculated factors, such as crash severity, occupant's weight, proximity to the air bag, seat belt usage, and seat position.

Smart key A device that uses a computerized chip that communicates through radio frequencies to unlock or lock a vehicle as well as start a vehicle remotely without the requirement of traditional keys.

Space frame A frame made up of multiple lengths and angles of tubing welded into a rigid, but light, web or truss-like structure; the vehicle's outer panels are attached independently to the frame after its completion.

Spring-loaded center punch A glass removal tool used on tempered glass that, when engaged, uses a spring-loaded plunger to fire off a steel rod with a sharpened point directly into a pinpoint area of glass, causing the glass to shatter.

Squib A pyrotechnic device.

Staging area manager ICS position responsible for ensuring that all resources in the staging area are available and ready for assignment.

Standard operating procedure (SOP) An organizational directive that establishes a standard course of action (also referred to as a standard operating guideline [SOG]).

Standard seat belt harness A seat belt system that helps distribute the energy of a collision over larger areas of the body such as the chest, pelvis, and shoulders. The three-point belt mechanism uses a retractor gear that locks in place when activated. Also known as a three-point harness system.

Standards Documents developed to provide guidance on the performance of processes, products, individuals, or organizations. Compliance is voluntary, unless formally adopted by an organization or government agency.

Start/stop mild hybrid A vehicle that is not a true hybrid system by definition. The motor/generator is not used to propel the vehicle; it is designed to turn off the vehicle's internal combustion engine when the vehicle is idle and will turn the vehicle's internal combustion engine back on when the accelerator is activated.

Step chocks Specialized cribbing assemblies made out of wood or plastic blocks in a step configuration. They are typically used to stabilize vehicles.

Stored compressed gas system An inflation system comprised of a single-stage or multistage inflation process. The igniter or squib sets off a burst or rupture disc that acts as a seal, holding back the compressed gas. When activated, the disc breaks open, releasing the gas from the chamber, which expands and instantly fills the air bag.

Stored energy Mechanical, hydraulic, or electrical energy that can be contained within a machine after the power of the machine has been stopped.

Stress Any type of change, whether pleasant or unpleasant, that manifests itself in cognitive, physical, emotional, or behavioral signs.

Strike team A group of five units of the same type working on a common task or function.

Stringers Steel longitudinal structural members that give the bow frame truss members structural support at the roof level; they run continuously from the front of the school bus to the rear of the bus.

Strut tower A structural component of the suspension system that normally has both a coil spring and shock absorber. Its main function is to resist compression.

Struts Structural supports used as a "buttress" to stabilize and reinforce an object. Struts can be made of steel, aluminum, composite, and wood.

Supplemental restraint system (SRS) A system that uses supplemental restraint devices such as air bags to enhance safety in conjunction with properly applied seat belts. Seat belt pretensioning systems are also considered part of an SRS.

Supplied air respirator/breathing apparatus (SAR/SABA) A respirator in which breathing air is supplied by air line from either a compressor or stored air (bottle) system located outside the work area.

Suppression system A device that shuts down the air bag if an occupant classification system detects a child in the air bag deployment zone or if one of the sensors detects a high risk potential by acquiring the occupant's weight, height, proximity to the air bag, seat belt usage, and seat position; the system sends this information to the electronic control unit, which will then shut off the air bag if a high risk to the occupant is determined.

Swing bar A hardened section of steel that is designed to assist the door in opening and closing. It can be located between the top and bottom hinge.

SWOT analysis A self-examination model that can be adjusted, adapted, and applied to any situation, incident, or project, large or small, that an organization is currently or will be involved in.

Task force A group of up to five single resources of any type.

Technical rescue The application of special knowledge, skills, and equipment to safely resolve unique and/or complex rescue situations.

Technical rescuer A person who is trained to perform or direct a technical rescue.

Technical specialists Advisors who have the special skills required at a rescue incident.

Technician level The level that represents the capability of organizations to respond to technical search and rescue incidents, to identify hazards, use equipment, and apply advanced techniques specified in this standard necessary to coordinate, perform, and supervise technical search and rescue incidents.

Temperature relief device (TRD) A device that rapidly releases product through a small metal tube attachment when detecting excessive amounts of heat at a preset temperature.

Tempered safety glass A type of glass that has been heated and then quickly cooled; this process gives the glass its strength and resistance to impact.

Tensile strength A measurement of the amount of force required to tear a section of steel apart.

Tension buttress stabilization A strut stabilization system that uses a strap in a ratchet or jacking device to add tension to the object being stabilized, locking the vehicle in place by using a diagonal force that lowers the vehicle's center of gravity by increasing the vehicle's entire footprint.

Tepee effect The negative effect of a dash roll technique where the dash cannot release or push forward and the entire floorboard/rocker panel area, where the relief cuts were made, tepees upward and impinges on the victim.

TPI rating A rating that indicates how many teeth per inch a blade has.

Transit bus An entity providing passenger transportation over fixed, scheduled routes, within primarily urban geographic areas.

Triage The process of establishing treatment and transportation priorities according to severity of injury and medical need.

Tube trailers A high-volume transportation device made up of several individual compressed gas cylinders banded together and affixed to a trailer. Tube trailers carry compressed gases such as hydrogen, oxygen, helium, and methane. One trailer may carry several different gases in individual tubes.

Tunneling The process of gaining entry through the rear trunk area of a vehicle, a process more commonly used for a postcrash vehicle resting on its roof.

Type A school bus A conversion-type bus constructed utilizing a cutaway front section vehicle with a left side driver's door. This definition includes two subclassifications: type A-1, with a GVWR of 14,500 lb (6577 kg) or less, and type A-2, with a GVWR greater than 14,500 lb (6577 kg) and less than or equal to 21,500 lb (9752 kg).

Type B school bus A school bus that is constructed utilizing a stripped chassis. The entrance door is behind the front wheels. This definition includes two subclassifications: type B-1, with a GVWR of 10,000 lb (4,536 kg) or less, and type B-2, with a GVWR greater than 10,000 lb (4,536 kg).

Type C school bus Also known as a *conventional school bus*; a school bus that is constructed utilizing a chassis with a hood and front fender assembly. The entrance door is behind the front wheels. This type of school bus has a GVWR greater than 21,500 lb (9752 kg). Eighty-five to ninety percent of all school buses are type C or D.

Type D school bus Also known as a *"transit-style" rear or front engine school bus*; a school bus that is constructed utilizing a stripped chassis where the outer body of the bus is mounted to the bare chassis. The entrance door is ahead of the front wheels, and the face or front section of the bus is flat. Type D buses have a passenger capacity of 80 to 90 people.

U-bolt A latching mechanism made of a light-gauge steel (as compared to the Nader bolt). It is generally easy to cut through and/or release from the latch mechanism of the door.

Ultra High Strength Steel (UHSS) Steel with a tensile strength of 102 ksi (700 MPa) or greater.

Unibody construction A vehicle design with no formal frame structure; the body and frame are one piece, which is considered to be the structural integrity of the vehicle. The vehicle body is merged with the chassis. Also known as a unitized structure.

Unified command An incident management tool and process that allows agencies with different legal, geographic, and functional responsibilities to coordinate, plan, and interact effectively to manage emergencies or events. In a unified command structure, multiple agency representatives make command decisions instead of just a single IC.

Upper rail Two side beams located in the front of the vehicle that hold the hood in place and attach the front wheel strut system to the chassis.

Vehicle identification badge A type of label that vehicle manufacturers utilize to identify the type of vehicle or the fuel that is used in the vehicle.

Vertical movement One of five directional movements; the vehicle moves up and down in relation to the ground while moving along its vertical axis.

Vertical spread A door access procedure utilizing a hydraulic spreader; the tool is placed vertically in the window frame of the door and pushes off of the roof rail and window frame to create an access point to the door's latching mechanism.

Warm zone The area located between the hot zone and cold zone at an incident. The decontamination corridor is located in this zone.

Wedges Objects used to snug loose cribbing under a load or to fill a void between the crib and the object as it is raised.

Wheel well crush technique A favorable technique utilized to gain access to door hinges from the outside.

Winch Chain or cable used for a variety of lifting, pulling, and holding operations.

Window spidering An effect caused when an object breaks laminated glass and causes spiraling rings at the area of impact, resembling a spider's web.

Work A mechanism for the transfer of energy.

Working load limit (WLL) The maximum force that may be applied to an assembly in straight tension.

Wrap point A hazard posed by any exposed rotating machine component. The most common examples are power take-off shafts.

Yaw movement One of five directional movements; the vehicle twists or turns about its vertical axis, causing the vehicle's front and rear portions to move left or right in relation to their original position.

Yield strength The amount of force or stress that a section of steel can withstand before permanent deformation occurs.

Index

Credits

You Are the Technical Rescuer Courtesy of Devon Sweet

Chapter 1
Opener Courtesy of Gregory Holness; 1-1 © Rob Vomund/Dreamstime.com; 1-2 Courtesy of Mike Jachles, Broward Sheriffs Office

Chapter 2
Opener Courtesy of Scott Dornan, ConocoPhillips Alaska; 2-1 Courtesy of Robert Reul, Margate Fire Rescue; 2-2 © Steven Townsend/Code 3 Images; 2-3 © Scott Downs/Dreamstime.com; 2-6 Courtesy of Captain David Jackson, Saginaw Township Fire Department; 2-7 © Dennis Wetherhold, Jr.; 2-9 Courtesy of NIMS/FEMA

Chapter 3
Opener © GlowImages/Alamy Images; 3-2 Courtesy of Mark Woolcock; 3-3 © Jack Dagley Photography/ShutterStock, Inc.; 3-6 © Crystalcraig/Dreamstime.com; 3-8 Courtesy of Darian Russo, North Lauderdale Fire Rescue; 3-7 © Sylvie Bouchard/ShutterStock, Inc.

Chapter 4
Opener © Jim West/Alamy Images; 4-3 © David R. Frazier Photolibrary, Inc./Alamy Images; 4-6 Courtesy of Culver Company; 4-9 Courtesy of the U.S. Department of Transportation; 4-11 © Prisma Bildagentur AG/Alamy Images; 4-12 © Photos 12/Alamy Images; 4-13 © GIPhotoStock Z/Alamy Images; 4-14 © Universal Images Group Limited/Alamy Images; 4-16 © GIPhotoStock Z/Alamy Images; 4-17 © Porsche AG/AP Photos; 4-18 © Transtock, Inc./Alamy Images; 4-23 © Green Stock Media/Alamy Images; 4-24 © Jim West/Alamy Images

Chapter 5
Opener © DBURKE/Alamy Images; 5-1 © dpa/Landov; 5-2 © Insurance Institute for Highway Safety/AP Photos; 5-11 Courtesy of Volvo Cars of North America, LLC

Chapter 6
6-01C Courtesy of Pacific Helmets (NZ) Ltd.; 6-03A Courtesy of Captain David Jackson, Saginaw Township Fire Department; 6-06A © niavuli/ShutterStock, Inc.; 6-06B © Draw/ShutterStock, Inc.; 6-06C © Smolny1/Dreamstime.com; 6-10 © 2003 Berta A. Daniels; 6-16 Courtesy of SpringTools; 6-22 Courtesy of Eric J. Rickenbach; 6-25 Copyrighted image courtesy of CMC Rescue, Inc.; 6-30 Courtesy of Lift-All Company, Inc.; 6-31 Courtesy of BelAire Compressors; 6-36 Courtesy of Rescue 42, Inc.; 6-38 Courtesy of Savatech Corp.; 6-39 Courtesy of Savatech Corp.; 6-41 Courtesy of Savatech Corp.; 6-42 Courtesy of RESQTEC, Inc.; 6-43A Courtesy of American Honda Motor Co., Inc.; 6-43B Courtesy of the Berwyn Heights Volunteer Fire Department & Rescue Squad, Berwyn Heights, Maryland; 6-44 Courtesy of Akron Brass Company; 6-50 Courtesy of Akron Brass Company; 6-51 © Glen E. Ellman; 6-58 Courtesy of TNT Rescue Systems, Inc.; 6-62 © TerryM/ShutterStock, Inc.; 6-64A © Steve Redick; 6-64B Courtesy of Robert Rhea; 6-64C © Steve Redick; 6-65A Courtesy of Robert Rhea; 6-65B © Elnur/ShutterStock, Inc.; 6-67 Courtesy of CON-SPACE Communications; 6-68 Courtesy of HotStick USA, Inc.; 6-69A Courtesy of Junkin Safety; 6-69B Courtesy of Ferno Washington, Inc.; 6-69C Courtesy of Junkin Safety; 6-70A Courtesy of Ferno Washington, Inc.

Chapter 7
Opener Courtesy of Mike Jachles, Broward Sheriffs Office; 7-4 © Dan Myers; 7-5 Courtesy of Captain David Jackson, Saginaw Township Fire Department; 7-6 Courtesy of the U.S. Department of Transportation; 7-7 © Ron Hilton/Dreamstime.com; 7-8 © Jim Parkin/ShutterStock, Inc.; 7-9 Courtesy of Mike Jachles, Broward Sheriffs Office; 7-10 © San Jose Mercury News/MCT/Landov; 7-13 © Murray Wilson/Fotolia; 7-14 Courtesy of Robert Rhea; 7-18 Courtesy of Robert Kaufmann/FEMA; 7-19 Courtesy of Rob Schnepp; 7-20A © Ralph Duenas/www.jetwashimages.com; 7-20B Courtesy of Ed Edahl/FEMA; 7-21 Courtesy of Darian Russo, North Lauderdale Fire Rescue

Chapter 8
8-7 Courtesy of Devon Sweet; 8-14 Courtesy of Robert Reul, Margate Fire Rescue; 8-16 Courtesy of Robert Reul, Margate Fire Rescue

Chapter 9
9-9 Courtesy of DUALSAW®/Infusion Brands, Inc.; 9-15 Courtesy of TNT Rescue Systems, Inc.; 9-17 © Jack Dagley Photography/ShutterStock, Inc.; 9-22 © Murray Wilson/Fotolia

Chapter 11
Opener © Jess Roberson/AP Photos; 11-2 © Photodisc; 11-5 © Sylvia Pitcher Photolibrary/Alamy Images; 11-6 © Henryk Sadura/Alamy Images; 11-7 © Hemera/Thinkstock; 11-8 © Matt/Fotolia; 11-9 © Hemera/Thinkstock; 11-10 © iStockphoto/Thinkstock; 11-15 © Jess Roberson/AP Photos; 11-18 Courtesy of George Roarty/VDEM; 11-20 © iStockphoto/Thinkstock; 11-22 © Hemera/Thinkstock; 11-33 © David R. Frazier/Photo Researchers, Inc.; 11-44 Courtesy of Glen Rudner

Chapter 12
Opener Courtesy of Eric J. Rickenbach; 12-1 Courtesy of Eric J. Rickenbach; 12-3 Courtesy of Eric J. Rickenbach; 12-5 Courtesy of Eric J. Rickenbach; 12-6 Courtesy of Eric J. Rickenbach; 12-7 Courtesy of Eric J. Rickenbach; 12-8 Courtesy of Eric J. Rickenbach; 12-15 Courtesy of Travis Martin; Skill Drill 3-2 © Steven Townsend/Code 3 Images

Chapter 13
Opener Courtesy of Mike Jachles, Broward Sheriffs Office; 13-1 Courtesy of Jim Gathany/CDC

Appendix
1A © Jupiterimages/Getty Images/Thinkstock; 1B © iStockphoto/Thinkstock; 4 Courtesy of Skedco, Inc.; 6A Courtesy of SMC - Seattle Manufacturing Corporation; 6B Courtesy of SMC - Seattle Manufacturing Corporation; 7 Courtesy of SMC - Seattle Manufacturing Corporation; 8 Courtesy of Gibbs Products, Inc.; 9 Courtesy of Donald M. Colarusso, ALLHANDSFIRE.COM